Hydrology and Floodplain Analysis

Philip B. Bedient
RICE UNIVERSITY

Wayne C. Huber
UNIVERSITY OF FLORIDA

D0209911

ADDISON-WESLEY PUBLISHING COMPANY

Reading, Massachusetts ▪ Menlo Park, California ▪ New York
Don Mills, Ontario ▪ Wokingham, England ▪ Amsterdam ▪ Bonn
Sydney ▪ Singapore ▪ Tokyo ▪ Madrid ▪ Bogotá ▪ Santiago ▪ San Juan

Library of Congress Cataloging-in-Publication Data

Bedient, Philip B., 1948–
  Hydrology and floodplain analysis.

  Bibliography: p.
  Includes index.
  1. Hydrology.   I. Huber, Wayne Charles.   I. Title.
GB661.2.B42   1988   551.48   87-14308
ISBN 0-201-12056-9

Reprinted with corrections July, 1988

The programs and applications presented in this book have been included for their instructional value. They have been tested with care but are not guaranteed for any particular purpose. The publisher does not offer any warranties or representations, nor does it accept any liabilities with respect to the programs or applications.

BCDEFGHIJ-HA-898

To Cindy, Eric, and Courtney
and to my parents for their guidance
P.B.B.

To Mom and Dad,
Cathy and Lydia
W.C.H.

Preface

The Evolution of Hydrology

Hydrology is of fundamental importance to civil and environmental engineers, geologists, and other earth scientists because of the environmental significance of floods and droughts, water supply, drainage, and water quality issues. This text was written to address the computational emphasis of modern hydrology at an undergraduate or graduate level and to provide a more balanced approach to important applications in watershed analysis, floodplain computations, flood control, urban hydrology, stormwater design, and computer modeling.

In the past fifteen years, hydrology has matured as a science and engineering discipline. Following the development of traditional hydrologic theory before 1960, a tremendous increase in the number and complexity of hydrologic studies and research resulted largely from two sources in the late 1960s: the urban expansion patterns in the United States and the environmental movement, which emphasized surface water quality control. With the expansion of gaging networks for rainfall, streamflow, and water quality and the development of computer techniques for data analysis and modeling of watersheds, hydrology entered a new era in the 1970s. In the early part of the decade, a large number of sophisticated computer models were created by government agencies and university groups to address flood prediction, flood control, water resources engineering, hydraulic design, water pollution control, water quality management, and water supply issues. During the 1980s, several of the most comprehensive and best-documented computer models have become routine operating procedures for detailed hydrologic analysis of watersheds.

Increasing use and sophistication of personal computers in the early 1980s have further revolutionized the daily practice of hydrology.

Hundreds of small programs have been written to ease the computational burden on the hydrologist or engineer, and many of the earlier mainframe codes have been converted to run on powerful personal computers. The usefulness of computer models has been greatly expanded for the classroom and for the practicing engineer. This text was written to better represent the new computational emphasis in hydrology.

Evolution of the Text

The idea for the book grew from the first author's experiences in teaching Hydrology and Watershed Analysis since 1976 at Rice University. The standard texts in the field were written in the early 1970s and do not properly address the computational emphasis of hydrology as practiced today. The current text was written over a five-year period and provides a modern approach that includes detailed coverage of numerical methods and computer models routinely used in hydrology.

The author found that students were enthusiastic and learned more hydrology when they applied simple computer models to actual watershed problems. With the current availability of personal computers, such activities are becoming the rule for the training of engineers and hydrologists. This text provides a complete listing of ten simple programs in hydrology and hydraulics, which are available in FORTRAN 77 on diskette for a PC or Macintosh computer. Three (Chapters 5, 6, and 7) are devoted to the detailed theory and application of three of the most widely used computer models in hydrology, which are applied to actual watersheds and flood events. This is a particular improvement over earlier texts. The application of spreadsheet software to hydrologic computation is also illustrated (Chapters 3 and 6).

Organization of the Text

The text is divided into two main sections. The first section, consisting of the first four chapters, covers traditional topics such as precipitation, evaporation, infiltration, rainfall-runoff analysis, flood frequency, and flood routing methods. These topics provide the student with a comprehensive view of the overall hydrologic cycle in nature as well as the problem of engineering design within certain flood limits. Numerous worked examples are used to highlight theory, problem definition, solution methods, and computational approaches.

Chapter 3, Frequency Analysis, presents the standard distributions used to analyze flood-related data. Normal, lognormal, gamma, and extreme-value examples are included and compared to one another for a single set of flow data. Emphasis is placed on why specific distributions are selected and the goodness of fit to measured data. Chapter 4, Flood Routing, includes eight examples of river and reservoir routing and associated numerical methods. Kinematic wave routing equations are derived in detail, and analytical solutions are provided along with a computer program listed in Appendix E. A complete discussion is also included on more advanced hydraulic routing methods and available models.

The second section, Chapters 5 – 8, is designed to apply hydrologic theory and modern computer techniques to several areas of engineering hydrology — watershed analysis, floodplain delineation, and urban stormwater. The most current methods and computer models for flood prediction, floodplain computation, and urban storm design are emphasized in enough detail for practical use, and detailed examples are given.

Chapter 5, Hydrologic Simulation Models and Watershed Analysis, presents modern methods for flood hydrograph prediction at various points in a watershed. A comprehensive methodology organizes data input and models required in flood studies. While several well-documented and widely used hydrologic models are briefly described, the major thrust of the chapter is the use and application of the Hydrologic Engineering Center's HEC-1 model to a large urbanizing watershed. Unit hydrograph and kinematic wave methods are used and effects of urban development, channel diversion, and reservoir routing are directly considered in the detailed example. The student should gain an appreciation for the usefulness and complexity of hydrologic models for watershed analysis and for the difficulties involved in selecting input data values for calibrating large hydrologic systems.

Chapter 6, Urban Hydrology, presents standard methods and reviews available computer models for pipe and open channel storm drainage systems. Recent developments in drainage design and detention storage are included. The Storm Water Management Model (SWMM) is highlighted as the most comprehensive urban runoff model. Realistic examples present complexities that arise in urban watersheds as compared to undeveloped areas. Emphasis is placed on statistical analysis of data and the theory underlying the IDF curves used in urban drainage design. Comparisons are also made to the standard rational method design.

Chapter 7, Floodplain Hydraulics, first reviews concepts from open channel flow required to understand water surface profile computations. Next, the Hydrologic Engineering Center's HEC-2 model is described in detail because of its widespread use by practicing engineers. Water surface profiles are required in most floodplain and channel design projects, and profiles are extremely sensitive to peak flows, cross-sectional geometry, bridge locations, and roughness coefficients. Engineering judgments must often be made in selecting coefficients that greatly influence the final computed profile.

The prediction of flood peaks for a given historical or design rainfall, followed by the conversion of the peaks to water surface elevations in the floodplain, is one of the most significant problems in hydrologic analysis. Chapters 5, 6, and 7 together present a unique combination of methods often used by practicing engineers and hydrologists but never before combined in a comprehensive textbook on the subject. Seventy realistic examples have been carefully developed over many years of experience and are included along with three major case studies.

Chapter 8 presents ground water concepts, including flow in porous media, aquifer properties, well mechanics, and computer applications. This chapter addresses the importance of the subsurface component in civil and environmental engineering in terms of both water supply and contaminant transport issues. Governing equations of flow are derived and applied to a number of ground water problems, including both steady-state and transient analyses. A section on numerical methods introduces solution techniques in ground water, and examples from well mechanics using simple computer programs are described.

The text should provide the engineering or science student with all the pertinent theory to understand hydrology and floodplain analysis. The practicing engineer should find the book a modern reference for computer simulation, urban storm water design, and floodplain analysis.

Acknowledgments

The book developed over a number of years from course notes in a class in Hydrology and Watershed Analysis. During the many years of our interaction with colleagues and students, the book evolved into its present form with emphasis on modern computational methods. The authors are indebted to many individuals and research groups for their significant contributions to the overall organization and content of the text.

We are particularly indebted to the following people for their careful review of the early manuscript and for numerous suggestions and comments: Robert Johanson, University of the Pacific; Richard H. McCuen, University of Maryland; Ed Schroeder, University of California, Davis; Stuart G. Walesh, Dean, Valparaiso University; and Ralph A. Wurbs, Texas A&M University. Tom Robbins, editor at Addison-Wesley Publishing Company, was instrumental and supportive in bringing the book out of the first-draft stage to a final version. His overall management combined with Barbara Flanagan's attention to technical detail was vital to the production of a serious engineering text. The authors also thank all the professionals at Addison-Wesley for their efforts on our behalf.

Special thanks are due to Pamela Ross at Rice University for her devoted assistance and technical skill in reviewing text, figures, examples, problems, and computer programs again and again. She solved many of the homework problems contained in the instructor's manual and, simply put, the text would never have been completed without her daily efforts. Finally, we would like to thank Marilyn Kelsey-Bettschen and Diana Freeman for typing the first draft of six chapters.

November 1987

P.B.B.

W.C.H.

Contents

2

RAINFALL-RUNOFF ANALYSIS 67

5

HYDROLOGIC SIMULATION MODELS AND WATERSHED ANALYSIS 281

6

URBAN HYDROLOGY   335

8

GROUND WATER HYDROLOGY 485

1

Hydrologic Principles

1.1
INTRODUCTION TO HYDROLOGY

Hydrology is a multidisciplinary subject that deals with the occurrence, circulation, and distribution of the waters of the Earth. The domain of hydrology embraces the physical, chemical, and biological reactions of water in natural and man-made environments. Because of the complex nature of the hydrologic cycle and its relation to weather patterns, soil types, topography, and other geologic factors, the boundaries between hydrology and other earth sciences such as meteorology, geology, ecology, and oceanography are not distinct.

The **hydrologic cycle** is a continuous process in which water is evaporated from the oceans, moves inland as moist air masses, and produces precipitation if the correct conditions exist. The precipitation that falls on the land surface is dispersed via several pathways (Fig. 1.1). A portion of the **precipitation** P, or rainfall, is retained in the soil near where it falls and returns to the atmosphere by **evaporation** E, the conversion of water to water vapor, and **transpiration** T, the loss of water vapor through plant tissue. The combined loss, called **evapotranspiration** ET, is a maximum value if the water supply in the soil is adequate at all times (see Section 1.4). Another portion becomes overland flow or direct runoff, which feeds local streams and rivers. Finally, some water enters the soil system as

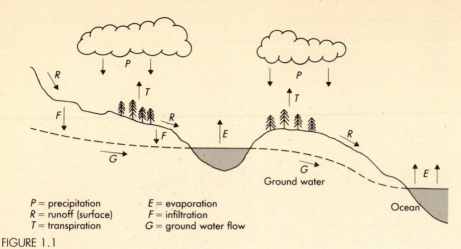

P = precipitation　　　　E = evaporation
R = runoff (surface)　　　F = infiltration
T = transpiration　　　　G = ground water flow

FIGURE 1.1

The hydrologic cycle.

infiltration F and may reenter channels later as interflow or may percolate to the deeper ground water system. Surface and ground water move toward lower elevations and may eventually discharge into the ocean. However, large quantities of surface water and portions of ground water may return to the atmosphere by evaporation and evapotranspiration.

Early History

Biswas (1972), in a concise treatment of the history of hydrology, describes the early water management practices of the Sumerians and Egyptians in the Middle East and the Chinese along the banks of the Hwang-Ho. Archeological evidence exists for hydraulic structures that were built for irrigation and other water control activities. A dam was built across the Nile about 4000 B.C., and later a canal for fresh water was constructed between Cairo and Suez.

　　　The Greek philosophers were the first serious students of hydrology, with Aristotle proposing the conversion of moist air into water deep inside mountains as the source of springs and streams. Homer suggested the idea of an underground sea as the source of all surface waters. Streamflow measurement techniques were first attempted in the water systems of Rome (97 A.D.) based on the cross-sectional area of flow. It remained for Leonardo da Vinci to discover the proper relationship between area, velocity, and flow rate during the Italian Renaissance.

　　　The first recorded measurements of rainfall and surface flow were made in the seventeenth century by Perrault, who compared measured rainfall to the estimated flow of the Seine River to show that the two were related. Perrault's findings were published in 1694. Halley, the English astronomer (1656–1742), used a small pan to estimate evaporation from the Mediterranean Sea and concluded that it was enough to account for tributary flows. Mariotte gaged the velocity of flow in the Seine River. These early beginnings of the science of hydrology provided the foundation for numerous advances in the eighteenth century, including Bernoulli's theorem, the Pitot tube, and the Chezy formula, which form the basis for hydraulics and fluid measurement.

During the nineteenth century, significant advances in ground water hydrology occurred. Darcy's law of flow in porous media, the Dupuit-Thiem well formula, and the Hagen-Poiseuille capillary flow equation were developed. In surface water hydrology, many flow formulas and measuring instruments were developed that allowed for the beginning of systematic stream gaging. Humphreys and Abbot (1861) reported on discharge measurement in the United States on the Mississippi River, and the U.S. Geological Survey set up the first systematic program of flow measurement in the United States on the Mississippi in 1888. Manning's formula was introduced in 1889 and the Price current meter was invented in 1885. In 1867, discharge measurements were organized on the Rhine River at Basel. During this period, the U.S. government founded a number of hydrologic agencies including the U.S. Army Corps of Engineers (1802), the Geological Survey (1879), the Weather Bureau (1891), and the Mississippi River Commission (1893).

The period from 1900 to 1930 was called the Period of Empiricism by Chow (1964) because of the large number of empirical formulas that were developed, many of which were found later to be unsatisfactory. Government agencies increased their effort in hydrologic research, and a number of technical societies were organized for the advancement of the science of hydrology. For example, the Bureau of Reclamation (1902), the Forest Service (1906), the U.S. Army Engineers Waterways Experiment Station (1928), and the Los Angeles County Flood Control District (1915) were organized during this period. The International Association of Scientific Hydrology (1922) and the Hydrology Section of the American Geophysical Union (AGU) were begun before 1930.

Modern History

The period from 1930 to 1950, called the Period of Rationalization (Chow, 1964), produced a significant step forward in the field of hydrology as government agencies began to develop their own programs of hydrologic research. Sherman's (1932) unit hydrograph (discussed in Chapter 2), Horton's (1933) infiltration theory (Chapter 1), and Theis's (1935) nonequilibrium equation in well hydraulics (Chapter 8) advanced the state of the art significantly. Gumbel (1941) proposed the use of extreme-value distributions for frequency analysis of hydrologic data, thus forming the basis for modern statistical hydrology (Chapter 3). In this period, the U.S. Army Corps of Engineers, the U.S. Weather Bureau (now the National Weather Service), the U.S. Department of Agriculture, and the U.S. Geological Survey (USGS) made significant contributions in hydrologic theory and the development of a national network of gages for precipitation, evaporation, and streamflow.

The National Weather Service is still largely responsible for rainfall measurements, reporting of severe storms, and other related hydrologic investigations. The U.S. Army Corps of Engineers and the U.S. Department of Agriculture (Soil Conservation Service) made significant contributions to the field of hydrology relating to flood control, reservoir development, irrigation, and soil conservation during this period. More recently, the U.S. Geological Survey has taken significant strides to set up a national network of stream gages and rainfall gages for both quantity and quality data. The USGS water supply publications and special investigations have done much to advance the field of hydrology by presenting the analysis of complex hydrologic data to develop relationships and explain hydrologic processes.

The government agencies in the United States performed vital research and provided funding for private and university research in the hydrologic area. The large dam, reservoir, and flood control projects since the 1930s are a direct result of advances in the fields of fluid mechanics, hydrologic systems, statistical hydrology, evaporation analysis, flood routing, and operations research.

Major advances in understanding the evaporation process occurred in the 1940s and 1950s related to the irrigation needs of the agricultural sector in the United States. More recently, the tremendous increase of urbanization in the United States and Europe led to better methods for predicting peak flows from floods and predicting variations in storage in water supply reservoirs. The determination and delineation of floodplain boundaries have become a major function of hydrologists as required by the Federal Emergency Management Agency (FEMA) and local flood control or drainage districts.

Computer Advances

The introduction of the digital computer into hydrology during the 1960s and 1970s allowed complex water problems to be simulated as complete systems for the first time. Large computer models can now be used to match historical data and help answer difficult control questions. The first comprehensive hydrologic model was developed by a group at Stanford University (Crawford and Linsley, 1966) and is called the Stanford Watershed Model. This model can simulate all of the major processes in the hydrologic cycle including precipitation, evaporation, evapotranspiration, infiltration, surface runoff, ground water flow, and streamflow. These terms are all defined later in Chapter 1.

Another model that has significantly altered the course of modern hydrology is HEC-1 developed by the U.S. Army Corps of Engineers Hydrologic Engineering Center (1973). This model simulates floods from rainfall data using simple loss functions and unit hydrographs (Chapter 5). A companion model, HEC-2, also developed by the U.S. Army Corps of Engineers (1976), performs water surface profile computations for a given stream geometry and peak flow rates, which can be calculated in HEC-1 (Chapter 7). The Storm Water Management Model (SWMM) was developed for the U.S. Environmental Protection Agency (Huber et al., 1981; Roesner et al., 1981) and is the most comprehensive model available for handling urban runoff in storm sewer systems (Chapter 6). The ILLUDAS model (Terstriep and Stall, 1974) was developed from a British Road Research Laboratory model and uses a simple rainfall-runoff procedure for storm drainage design and simulation (Chapter 6). Each of these models will be described in more detail in later chapters.

The recent models have come into relatively common use by investigators and engineering hydrologists and represent some of the most powerful computer tools in modern hydrology. The development of these tools over the past 20 years has helped direct the collection of the hydrologic data to calibrate or "match" the models against observation. In the process, the understanding of the hydrologic system has been greatly advanced.

Hydrologic computer models developed in the 1960s and 1970s have been applied to areas previously unstudied or only empirically defined. For example, urban stormwater, floodplain and watershed hydrology, drainage design, reservoir design and operation, flood frequency analysis, and large-river basin management have all benefited from the application of computer models.

Hydrologic simulation models applied to watershed analysis are described in detail in Chapter 5. Single-event models such as HEC-1 or SWMM are used to simulate (or calculate) the resulting storm hydrograph (discharge vs. time) from a well-defined watershed area for a given pattern of rainfall intensity. Hydrologic losses such as infiltration, evaporation, evapotranspiration, and detention storage can be directly calculated for the given watershed. Continuous models such as the Stanford Watershed Model and SWMM can account for soil moisture storage, evapotranspiration, and antecedent rainfall over long time periods. Statistical models can be used to generate a time series of rainfall or streamflow data, which can then be analyzed with flood frequency methods. Deterministic models such as HEC-1 do not consider the random nature of input rainfall directly but can be used with statistically derived storms, called design rainfalls, such as the 100-year, 24-hour event. The HEC-2 model can then be used to evaluate water surface profiles expected to occur on the average once every 100 years. This defined level of inundation is called the 100-year floodplain and is covered in detail in Chapter 7.

The advantages of computer models in hydrology include the insight gained in gathering and organizing input data and the understanding realized in attempting to "calibrate" calculated flows or water levels to observed data from an actual watershed. The exercise often guides the collection of additional data or improves the application of a specific model.

Limitations of simulation models include the danger of believing that a model will yield accurate results in all cases. An overreliance on very powerful computer models in the 1970s has led to a more skeptical approach to hydrologic modeling in the 1980s, with a return to model applications that do not exceed the availability of accurate input data. However, simulation models in hydrology, if applied correctly, still provide the most logical approach to understanding complex water resources systems, and a new era in the science of hydrology has begun. New design and operating policies have been advanced and implemented that could not have been realized or tested without the aid of sophisticated computer models.

1.2

HYDROLOGIC CYCLE

The basic components of the hydrologic cycle include precipitation, evaporation, evapotranspiration, infiltration, overland flow, streamflow, and ground water flow. The movement of water through various phases of the hydrologic cycle is erratic in time and space, giving rise to extremes of flood or drought. The magnitude and the frequency of occurrence of these extremes are of great interest to the engineering hydrologist from a design and operation standpoint. Probability theory is used extensively to analyze extremes of flow or rainfall.

The hydrologic cycle is very complex, but under certain well-defined conditions, the response of a watershed to rainfall, infiltration, and evaporation can be calculated if simple assumptions can be made. For example, if rainfall rate over a watershed is less than the rate of infiltration and if there is ample storage in soil moisture, then direct runoff from the surface and resulting streamflow will be zero. If, on the other hand, antecedent rainfall has filled soil storage and if the rainfall rate is much greater than infiltration and evaporation rates, then the volume of surface runoff will be equal to the volume of rainfall. In most cases, unfortunately, the conditions fall somewhere between these limitations and we must carefully measure or calculate more than one component of the cycle to predict watershed response.

The hydrologist must be able to calculate or estimate various components of the hydrologic cycle to properly design water resource projects. Large hydraulic projects must be designed to protect against damages from extreme floods or droughts and are usually operated with such extremes in mind. Typical issues of concern to an engineering hydrologist include the following:

1. Flood flows expected at a spillway or highway culvert
2. Reservoir capacity required to ensure adequate water for irrigation or municipal water supply during droughts
3. Effects of reservoirs, levees, and other control works on flood flows in a stream
4. Effects of urban development on future capacity of a drainage system and associated flood flows
5. Delineation of probable flood levels to improve protection of man-made projects from flooding or to promote better zoning

For any hydrologic system, a water budget can be developed to account for various flow pathways and storage components. The simplest system is an impervious inclined plane, confined on all four sides with a single outlet. A small urban parking lot follows such a model. The hydrologic continuity equation for any system is

$$I - Q = \frac{dS}{dt}, \tag{1.1}$$

where

I = inflow in vol/time

Q = outflow in vol/time

dS/dt = change in storage in vol/time

As rainfall accumulates on the surface, the surface detention increases and eventually becomes outflow from the system. Neglecting evaporation for the period of input, all input rainfall eventually becomes outflow, as shown in Fig. 1.2, but delayed somewhat in time. The difference between accumulated inflow and outflow at any time represents the change in storage. Thus, the shaded area represents detention storage volume that is eventually released from storage.

The same concept can be applied to small basins or large watersheds, with the added difficulty that all loss terms in the hydrologic budget may not be known. For a given time period, a conceptual mathematical model of the overall budget for Fig. 1.1 would become, in units of depth (in. or cm) over the basin,

$$P - R - G - E - T = \Delta S, \tag{1.2}$$

where

P = precipitation,

R = surface runoff,

(a)   Inflow (*I*)   (b)

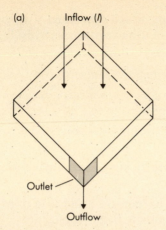

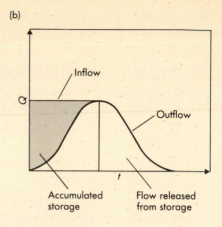

FIGURE 1.2

A simple hydrologic system (parking lot).

G = ground water flow,
E = evaporation,
T = transpiration,
ΔS = change in storage.

A **runoff coefficient** can be defined as the ratio R/P. Note that infiltration I is a loss from the surface system and a gain to the ground water and thus cancels out of the overall budget. Also, the units of inches (or cm) represent a volume of water when multiplied by the surface area of the watershed.

EXAMPLE 1.1

WATER BALANCE/UNIT CONVERSION

For a given month, a 300-acre lake has 15 cfs of inflow, 13 cfs of outflow, and a total storage increase of 16 ac-ft. A USGS gage next to the lake recorded a total of 1.3 in. precipitation for the lake for the month. Assuming that infiltration is insignificant for the lake, determine the evaporation loss, in inches, over the lake.

SOLUTION

Solving the water balance equation for inflow and outflow in a lake for evaporation gives

$$E = I - O + P - \Delta S,$$

evaporation   inflow   outflow   precipitation   change in storage

$$I = \frac{(15 \text{ ft}^3/\text{s})(\text{acre}/43{,}560 \text{ ft}^2)(12 \text{ in./ft})(3600 \text{ s/hr})(24 \text{ hr/day})(30 \text{ day/month})(1 \text{ month})}{300 \text{ acre}}$$

$$= 35.70 \text{ in.,}$$

$$O = \frac{(13 \text{ ft}^3/\text{s})(\text{acre}/43{,}560 \text{ ft}^2)(12 \text{ in./ft})(3600 \text{ s/hr})(24 \text{ hr/day})(30 \text{ day/month})(1 \text{ month})}{300 \text{ acre}}$$

$$= 30.94 \text{ in.,}$$

$$P = 1.3 \text{ in.,}$$

$$\Delta S = \frac{(16 \text{ ac-ft})(12 \text{ in./ft})}{300 \text{ acre}} = 0.64 \text{ in.,}$$

$$E = (35.70) - (30.94) + (1.3) - (0.64) \text{ in.,}$$

$$\boxed{E = 5.42 \text{ in.}}$$

This example illustrates the unit conversion problems typically encountered by the hydrologist. It is frequently necessary to convert data from one unit to another, even from metric to U.S. customary units, to solve a problem.

EXAMPLE 1.2

WATER BALANCE IN A WATERSHED

In a given year, a watershed with an area of 2500 km² received 130 cm of precipitation. The average rate of flow measured in a river draining the watershed was 30 m³/s. Estimate the amount of water lost due to the combined effects of evaporation, transpiration, and infiltration to ground water. How much runoff reached the river for the year (in cm)? What is the runoff coefficient?

SOLUTION

The water balance equation (Eq. 1.2) can be arranged to produce

$$ET + G = P - R - \Delta S.$$

Assuming that the water levels are the same for $t = 0$ and $t = 1$ year, then $\Delta S = 0$ and

$$ET + G = 130 \text{ cm}$$

$$- \frac{(30 \text{ m}^3/\text{s})(86{,}400 \text{ s/day})(365 \text{ day/yr})(100 \text{ cm/m})}{(2500 \text{ km}^2)(1000 \text{ m/km})^2}$$

$$= 130 \text{ cm} - 37.9 \text{ cm}$$

$$\boxed{ET + G = 92.1 \text{ cm.}}$$

Thus, 92.1 cm were lost to ET and G, leaving

$$\boxed{37.9 \text{ cm}}$$

as runoff to the river.

The runoff coefficient, defined as runoff divided by precipitation, is

$$R/P = 37.9 \text{ cm}/130 \text{ cm}$$

$$\boxed{R/P = 0.29.}$$

1.3

PRECIPITATION

Precipitation is the primary input quantity to the hydrologic cycle, whether in the form of rain, snow, or hail, and is generally derived from atmospheric moisture. The moisture must undergo lifting and resultant cooling, condensation, and growth of droplets before precipitation can occur. Precipitation is often classified according to conditions that generate vertical air motion: (1) **convective,** due to intense heating of air at the ground, which leads to expansion and vertical rise of the air; (2) **cyclonic,** associated with the movement of large air mass systems, as in the case of warm or cold fronts; and (3) **orographic,** due to mechanical lifting of moist air masses over mountain ranges as in the Pacific Northwest.

Atmospheric Moisture

Atmospheric moisture is a necessary source for precipitation and is derived from evaporation and transpiration. Precipitation over the United States is largely due to oceanic evaporation and subsequent transport over the continent by the atmospheric circulation system.

Common measures of atmospheric moisture, or humidity, include vapor pressure, specific humidity, mixing ratio, relative humidity, and dew point temperature. Under moist conditions, water vapor can be assumed to obey the ideal gas law, which allows derivation of simple relations between pressure, density, and temperature.

The partial pressure is the pressure that would be exerted on the surface of a container by a particular gas in a mixture. The partial pressure exerted by water vapor is called **vapor pressure** and can be derived from Dalton's law and the ideal gas law as

$$e = \frac{\rho_w RT}{0.622}, \qquad (1.3)$$

where

> e = vapor pressure in (mb),
>
> ρ_w = vapor density or absolute humidity (g/m³),
>
> R = dry air gas constant = 2.87×10^3 mb cm³/g°K,
>
> T = absolute temperature (°K).

The factor 0.622 arises from the ratio of the molecular weight of water (18) to that of air (29). Near the earth's surface the water vapor pressure is 1–2% of the total atmospheric pressure, where average atmospheric pressure is 1013.2 mb (1.013 × 10⁶ dyne/cm²). **Saturation vapor pressure** e_s is the partial pressure of water vapor when the air is completely saturated (no further evaporation occurs) and is a function of temperature.

Relative humidity (H) is approximately the ratio of water vapor pressure to that which would prevail under saturated conditions at the same temperature. It can also be stated as $H = 100 \, e/e_s$. Thus, 50% relative humidity means that the atmosphere contains 50% of the maximum moisture it could hold under saturated conditions at that temperature.

Specific humidity is the mass of water vapor contained in a unit mass of moist air (g/g) and is equal to ρ_w/ρ_m, where ρ_w is the vapor density and ρ_m is the density of moist air. Using Dalton's law and assuming that the atmosphere is composed of only air and water vapor, we have

$$\rho_m = \frac{(P-e)+0.622e}{RT} = \frac{P}{RT}(1-0.378e/P). \qquad (1.4)$$

Equation 1.4 shows that moist air is actually lighter than dry air for the same pressure and temperature. Thus,

$$q = \frac{\rho_w}{\rho_m} = \frac{0.622e}{P-0.378e}, \qquad (1.5)$$

where

> q = specific humidity (g/g),
>
> e = vapor pressure (mb),
>
> P = total atmospheric pressure (mb),
>
> ρ_m = density of mixture of dry air and moist air (g/cm³).

Finally, the **dew point temperature** T_d is the value at which an air mass just becomes saturated ($e = e_s$) when cooled at constant pressure and moisture content. An approximate relationship for saturation vapor pressure over water e_s as a function of dew point T_d is

$$e_s = 2.7489 \times 10^8 \exp\left(-\frac{4278.6}{T_d+242.79}\right), \qquad (1.6)$$

where e_s is in mb and T_d is in °C. The relationship is accurate to within 0.5% of tabulated values (List, 1966) over a range of temperatures from 0 to 40°C.

The preceding measures of atmospheric moisture are routinely used in weather analysis to predict probability of precipitation occurring over a particular region. More detail on their application and use can be found in standard textbooks (Hess, 1959; Eagleson, 1970; Wallace and Hobbs, 1977).

Phase Changes

In order for vapor to condense to water to begin the formation of precipitation, a quantity of heat known as latent heat must be removed from the moist air. The **latent heat** of condensation L_c is equal to the latent heat of evaporation L_e, the amount of heat required to convert water to vapor at the same temperature. With T measured in °C,

$$L_e = -L_c = 597.3 - 0.57(T - 0°C), \tag{1.7}$$

where L_e is in cal/g. The latent heat of melting and freezing are also related:

$$L_m = -L_f = 79.7,$$

where L_m is also in cal/g. Meteorologists use the moisture relationships and the latent heat concepts to obtain pressure-temperature relationships for cooling of rising moist air in the atmosphere. The rate of temperature change with elevation in the atmosphere is called the **ambient atmospheric lapse rate.** As moist, unsaturated air rises, the relative humidity increases, and at some elevation saturation is reached and relative humidity becomes 100%. Further cooling of the air results in condensation of the moisture at the defined lifting condensation level. Latent heat of condensation is released, warming the air and lowering the atmospheric lapse rate. This latent heat exchange is the major energy source that fuels tropical cyclones and hurricanes.

It has been observed that a relation does not necessarily exist between the amount of water vapor and the resulting precipitation over a region. Thus, condensation can occur in cloud formations without the production of precipitation at the ground surface. Other climatic processes must be considered to fully understand the mechanism of precipitation.

Causes and Mechanisms of Formation

Condensation of water vapor into cloud droplets occurs as a result of cooling of the air to a temperature below the saturation point for water vapor. This is most commonly achieved through vertical lifting to levels where pressure and temperature are lower. One-half the mass of the atmosphere lies within 18,000 ft of the surface and contains most of the clouds and moisture.

Condensation can be caused by (1) dynamic or adiabatic cooling (no heat loss to surroundings) as described earlier, (2) mixing of air masses having different temperatures, (3) contact cooling, and (4) cooling by radiation. Dynamic cooling is by far the most important producer of appreciable precipitation. Dew, frost, and fog are minor producers of precipitation caused by contact and radiational cooling.

Condensation nuclei must be present for the formation of cloud droplets. Such nuclei come from many sources, such as ocean salt, dust from clay soils, industrial combustion products, and volcanoes, and they range in size from $0.1\,\mu$ to $10\,\mu$. Cloud droplets originally average 0.01 mm in diameter, and it is only when they exceed 0.5 mm that significant precipitation occurs. It may take hours for a small raindrop (1 mm) to grow on a condensation nucleus. As vapor-laden air rises, it cools as it expands, and as saturation occurs, water vapor begins to condense on the most active nuclei. The principal mechanism for the supply of water to the growing droplet in early stages is diffusion of water vapor molecules down the vapor pressure gradient toward the droplet surface. As the droplets increase in mass, they begin to move relative to the overall cloud. However, other processes must support the growth of droplets of sufficient size (0.5 – 3.0 mm) to overcome air resistance and to fall as precipitation. These include the coalescence process and the ice crystal process.

The **coalescence process** is considered dominant in summer shower precipitation. As water droplets fall, smaller ones are overtaken by larger ones, and droplet size is increased through collision. This can produce significant precipitation, especially in warm cumulus clouds in tropical regions.

The **ice crystal process** attracts condensation on freezing nuclei because of lower vapor pressures. The ice crystals grow in size through contact with other particles, and collisions cause snowflakes to form. Snowflakes may change into rain droplets after entering air in which the temperature is above freezing. Snowfall and snowmelt are presented in Section 2.8.

Condensation nuclei can be artificially supplied to clouds to induce precipitation under certain conditions. Dry ice and silver iodide have been used as artificial nuclei. Research is continuing in this phase of weather control, and many legal and technical problems remain to be resolved regarding artificial inducement of precipitation.

As stated earlier, the process of condensation responsible for most rainfall is dynamic or adiabatic cooling. Precipitation can be classified according to conditions that generate vertical air motion: convective, cyclonic, and orographic, depicted schematically in Fig. 1.3. Orographic effects primarily relate to mountain ranges, which cause air to be lifted vertically, and precipitation or snowfall is commonly encountered on

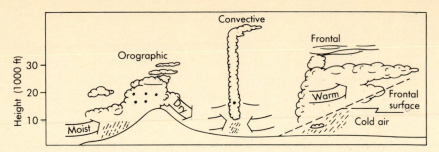

FIGURE 1.3
Precipitation mechanisms.

western slopes. Eastern slopes tend to be drier and warmer. The most intense rainfalls of concern from a local flooding standpoint are thunderstorms caused by severe convective storms. Thunderstorms begin as rising clouds and can reach heights of 25,000–40,000 ft with vertical wind speeds of 35–70 mph in updrafts. From a regional standpoint, intense cyclonic hurricanes and associated low-pressure cells can generate enormous rainfall over extended periods of time. For example, the 43 in. of rain that fell in 24 hr in Alvin, Texas, during July 1979 set a new U.S. continental record. This event was associated with hurricane Claudette in the Gulf of Mexico. Further details on mechanisms of precipitation can be found in Eagleson (1970), Gray (1970), and Wallace and Hobbs (1977).

Point Precipitation

Considerable precipitation data are available from the National Weather Service, the U.S. Army Corps of Engineers, and the U.S. Geological Survey. Interpretation of national networks of rainfall data shows the extreme variability in space and time, as can be seen in Fig. 1.4. The main source of moisture for precipitation (P) is evaporation from the oceans; thus P tends to be heavier near the coastlines, with distortion due to orographic effects, that is, effects of changes in elevation over mountain ranges. In general, amount and frequency of P is greater on the windward side of mountain barriers (the west side in the United States) and less on the lee side, also shown in Fig. 1.4. Spreen (1947) studied the effect of topography on precipitation and found reasonable correlation to four parameters in western Colorado: elevation, slope, orientation, and exposure for western Colorado.

Time variation of precipitation occurs seasonally or within a single storm, and distributions vary with storm type, intensity, duration, and time of year. Prevailing winds and relative temperature of land and bordering ocean have an effect. Maximum recorded rainfall in six major U.S. cities is shown in Table 1.1. Totals for a 24-hour period, for example,

TABLE 1.1
Maximum Recorded Rainfall in Six Major Cities, in in. (mm)

Station	Minutes			DURATION	Hours	
	5	15	30	1	6	24
New York, N.Y.	1.02 (26) 7/5/1973	1.63 (41) 7/10/1905	2.34 (59) 8/12/1926	2.97 (75) 8/26/1947	4.44 (113) 10/1/1913	9.55 (243) 10/8/1903
St. Louis, Mo.	0.88 (22) 3/5/1897	1.39 (35) 8/8/1923	2.56 (65) 8/8/1923	3.47 (88) 7/23/1933	5.82 (148) 7/9/1942	8.78 (223) 8/15/1946
New Orleans, La.	1.00 (25) 2/5/1955	1.90 (48) 4/25/1953	3.18 (81) 4/25/1953	4.71 (120) 4/25/1953	8.62 (219) 9/6/1926	14.01 (356) 4/15/1927
Denver, Colo.	0.91 (23) 7/14/1912	1.57 (40) 7/25/1965	1.99 (51) 7/25/1965	2.20 (56) 8/23/1921	2.91 (74) 8/23/1921	6.53 (166) 5/21/1876
San Francisco, Calif.	0.33 (8) 11/25/1926	0.65 (17) 11/4/1918	0.83 (21) 3/4/1912	1.07 (27) 3/4/1912	2.34 (59) 10/11/1962	4.67 (119) 1/29/1881
Houston, Tex.	– –	– –	– –	4.00 (102) 7/25/1979	15.67 (397) 7/25/1979	43.00 (1092) 7/25/1979

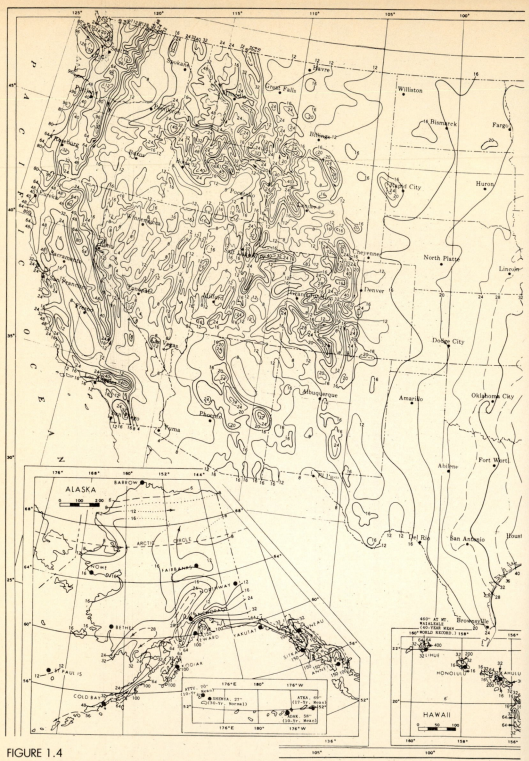

FIGURE 1.4

Distribution of average annual precipitation.

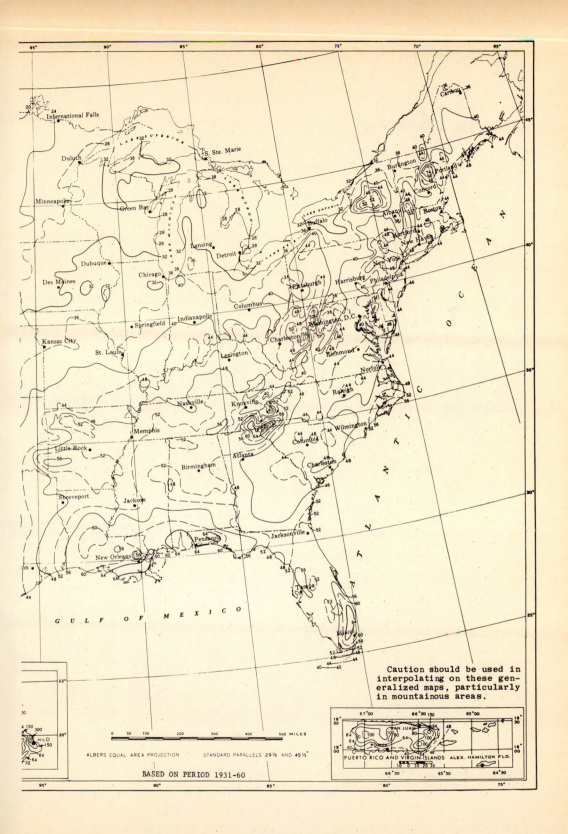

Caution should be used in interpolating on these generalized maps, particularly in mountainous areas.

ALBERS EQUAL AREA PROJECTION   STANDARD PARALLELS 29½ AND 45½°

BASED ON PERIOD 1931-60

PUERTO RICO AND VIRGIN ISLANDS   ALEX. HAMILTON FLD.

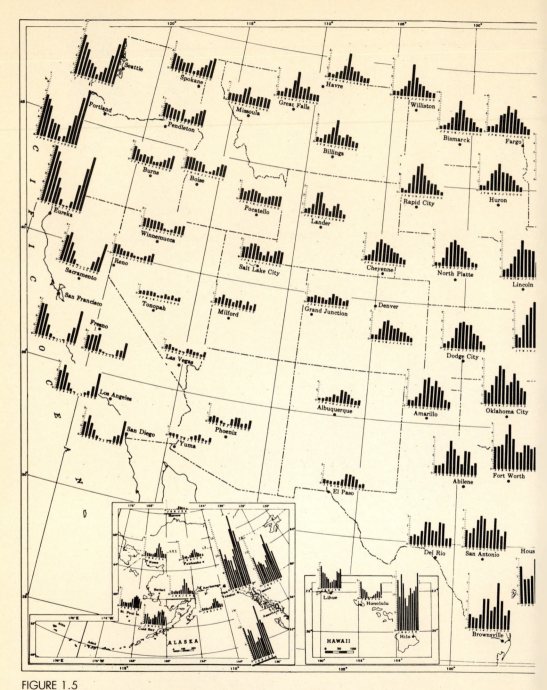

FIGURE 1.5

**Normal monthly distribution of precipitation in the United States (in.)
(1 in. = 25.4 mm). (U.S. Environmental Data Service.)**

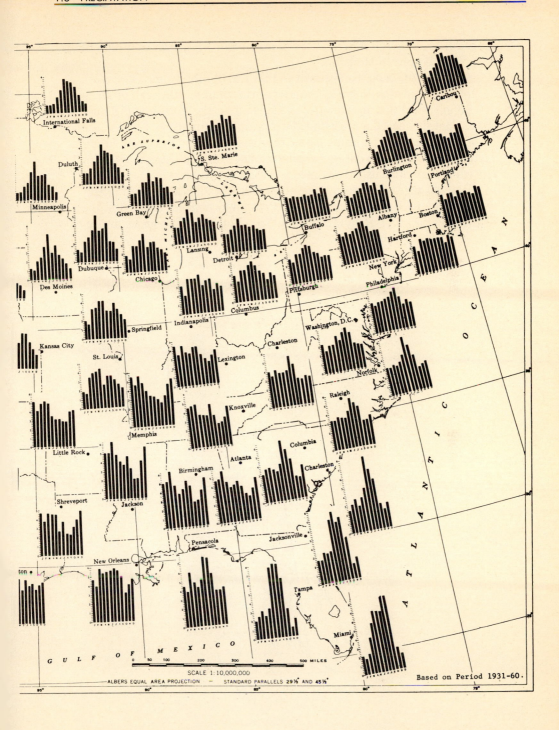

SCALE 1:10,000,000
— ALBERS EQUAL AREA PROJECTION — STANDARD PARALLELS 29½° AND 45½°

Based on Period 1931-60.

ranged from a low of 4.67 in. (119 mm) in San Francisco to a high of 43 in. (1092 mm) in Houston.

Seasonal differences are shown in Fig. 1.5, but the values shown are deceptive in that high-intensity thunderstorms or hurricanes can produce 15–30 in. of rainfall in a matter of days along the Gulf and Atlantic coasts. Hourly or even more detailed variations of rainfall (see Fig. 1.6) are often important for planning water resource projects, especially urban drainage systems. Such data are available only from sophisticated rainfall recording networks, usually located in larger urban areas.

Rainfall gage networks are maintained by the National Weather Service and the U.S. Geological Survey. Precipitation gages may be of the recording (Fig. 1.7) or nonrecording type, but recording gages are re-

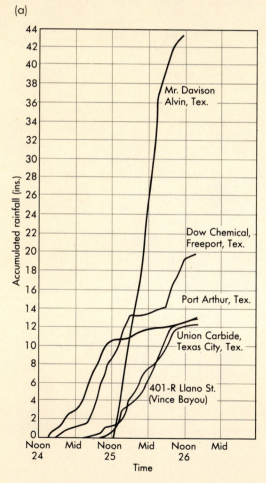

FIGURE 1.6

Storm event near Houston, Tex. (a) Accumulated rainfall, July 24–26, 1979. (b) Total rainfall contours, July 24–26, 1979.

(b)

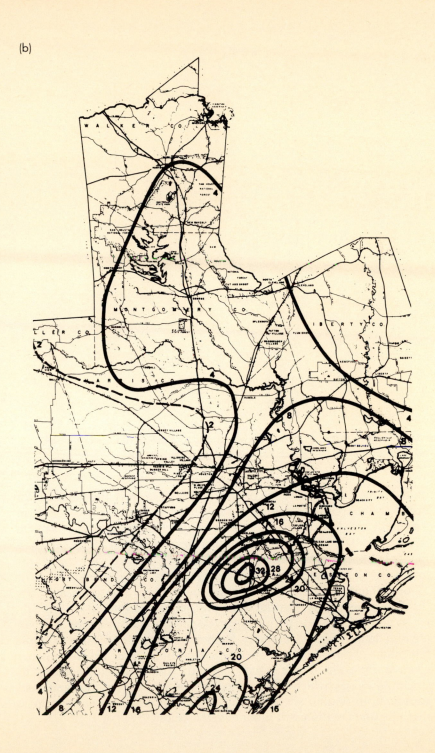

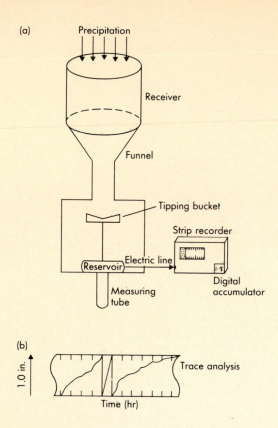

FIGURE 1.7

Recording tipping bucket gage. Trace returns to zero after each inch of rainfall. The slope of the trace registers intensity (in./hr).

quired if the time distribution of rainfall is desired, as is often the case for urban drainage or flood control works. The recording gage operates from a small tipping bucket that records on a strip chart, paper punch, or data logger every 0.01 in. of rainfall. The data are displayed in a form shown in Figure 1.7(b) as a **cumulative mass curve** and can be readily interpreted for total volume and intensity variations. Observers usually report daily or 12-hr amounts of rainfall (in in. or cm) for nonrecording gages, and thus this type of reporting provides little information on intensity.

A network of five to ten gages per 100 mi^2 is usually required in urban areas to define rainfall variability. Such networks are expensive to maintain, and equipment failures often negate portions of the network for any given storm event. These rainfall gages are usually located adjacent to streamflow gages, which are discussed in Section 1.6. Temporal and areal variations in rainfall are important in determining overall hydrologic response and are discussed in more detail below.

Point rainfall can be plotted as accumulated total rainfall or as rainfall intensity at a particular gage. The first plot is referred to as a cumulative mass curve (Fig. 1.6a), which can be analyzed for a variety of storms to determine the frequency and character of rainfall at a given site. A **hyetograph** is a plot of rainfall intensity (in./hr) with time and is depicted in Example 1.3. Hyetographs are often used as input to hydrologic computer models for a watershed.

EXAMPLE 1.3

HYETOGRAPHS AND CUMULATIVE PRECIPITATION

The chart given in Fig. E1.3(a) recorded precipitation from two U.S. Geological Survey recording gages for the storm of August 30 – September 3, 1981, for the period between 2:45 A.M. and 2:00 P.M. on August 31. For the data given, develop rainfall hyetographs and mass curves. Find the maximum-intensity rainfall for each gage in in./hr.

SOLUTION

To plot the hyetograph for each gage, we subtract the measurement for each time period from that of the previous time period. For example, for gage 4800, 3.70 was subtracted from 3.73 to give 0.03 in./15 min. We converted this to in./hr:

$$\frac{(3.73 - 3.70) \text{ in.}}{0.25 \text{ hr}} = 0.12 \text{ in./hr.}$$

Because the data were given as a cumulative reading, the mass curves are simply a plot of the data as given (see Fig. E1.3b).

The maximum intensity for gage 4800 occurred between 1000 and 1015 on August 31 and was

$$\frac{(6.98 - 6.23) \text{ in.}}{0.25 \text{ hr}} = \boxed{3.0 \text{ in/hr.}}$$

For gage 303R,

$$\frac{(6.33 - 5.55) \text{ in.}}{0.25 \text{ hr}} = \boxed{3.2 \text{ in./hr}}$$

was the maximum intensity, occurring between 0630 and 0645. These maximum intensities appear as the tallest bars on the hyetographs and as the regions of greatest slope on the cumulative precipitation curves. This illustrates the fact that the mass curve is the integral of the hyetograph as, in probability theory, the cumulative distribution function is the integral of the probability density function.

STA. NO. 08074800		STORM RAINFALL AND RUNOFF RECORD		1981 WATER YEAR	
KEEGANS BAYOU AT ROARK ROAD NEAR HOUSTON, TEX. STORM OF AUG. 30 TO SEP. 3, 1981			ACCUM. WEIGHTED PRECIP. IN.	DISCHARGE IN CFS	ACCUM. RUNOFF IN.
	G A G E N U M B E R				
DATE & TIME	4800	303R			
AUG. 31					
0245	3.70	3.23	3.42	1290.0	0.6362
0300	3.73	3.25	3.44	1310.0	0.6804
0315	3.80	3.30	3.50	1300.0	0.7241
0330	3.99	3.63	3.77	1300.0	0.7679
0345	4.00	3.78	3.87	1310.0	0.8121
0400	4.04	3.83	3.91	1320.0	0.8565
0415	4.06	3.85	3.93	1300.0	0.9003
0430	4.08	3.85	3.94	1280.0	0.9434
0445	4.23	4.05	4.12	1270.0	0.9862
0500	4.28	4.11	4.18	1260.0	1.0287
0515	4.53	4.15	4.30	1280.0	1.0718
0530	4.63	4.29	4.43	1300.0	1.1156
0545	4.76	4.33	4.50	1310.0	1.1597
0600	4.90	4.70	4.78	1330.0	1.2045
0615	4.98	5.10	5.05	1410.0	1.2520
0630	5.08	5.55	5.36	1500.0	1.3025
0645	5.20	6.33	5.88	1600.0	1.3564
0700	5.26	6.63	6.08	1710.0	1.4140
0715	5.28	6.71	6.14	1740.0	1.4727
0730	5.30	6.77	6.18	1780.0	1.5326
0745	5.38	6.85	6.26	1770.0	1.5922
0800	5.40	6.91	6.31	1770.0	1.6519
0815	5.44	6.95	6.35	1740.0	1.7105
0830	5.53	7.01	6.42	1710.0	1.7681
0845	5.58	7.07	6.47	1690.0	1.8250
0900	5.60	7.13	6.52	1680.0	1.8816
0915	5.64	7.18	6.56	1640.0	1.9369
0930	5.70	7.33	6.68	1610.0	1.9911
0945	6.08	7.53	6.95	1650.0	2.0467
1000	6.23	8.16	7.39	1690.0	2.1036
1015	6.98	8.63	7.97	1890.0	2.1673
1030	7.00	8.84	8.10	2100.0	2.2380
1045	7.28	9.00	8.31	2130.0	2.3098
1100	7.40	9.40	8.60	2170.0	2.3829
1115	7.43	9.45	8.64	2190.0	2.4566
1130	7.68	9.50	8.77	2220.0	2.5314
1145	7.86	9.52	8.86	2240.0	2.6069
1200	7.90	9.53	8.88	2270.0	2.6833
1215	7.93	9.54	8.90	2220.0	2.7581
1230	7.96	9.55	8.91	2170.0	2.8312
1245	7.96	9.56	8.92	2120.0	2.9026
1300	7.96	9.57	8.93	2070.0	2.9724
1315	7.96	9.58	8.93	2010.0	3.0401
1330	7.96	9.59	8.94	1950.0	3.1058
1345	7.96	9.60	8.94	1860.0	3.1684
1400	7.96	9.61	8.95	1780.0	3.3783

FIGURE E1.3(a)

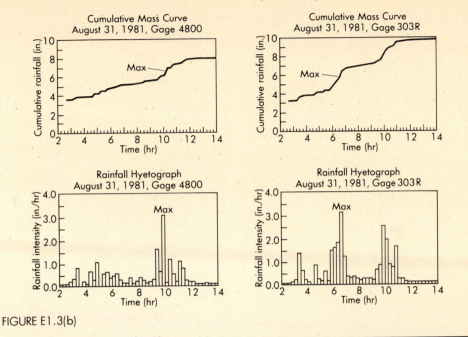

FIGURE E1.3(b)

Statistical methods (Chapter 3) can be applied to a long time series of rainfall data. For example, various duration rainfalls ranging from 5 min to 1440 min can be analyzed to develop an estimate of 10-yr probability events. These data are fitted with a line to form one of the curves on the **intensity-duration-frequency** (IDF) **curves** in Fig. 1.8. Other IDF probability lines are derived in a similar fashion for the 2-, 5-, 25-, 50-, and 100-yr design rainfalls. It can be seen that the intensity of rainfall tends to decrease with increasing duration of rainfall for each of the IDF curves. Instead of analyzing historical rainfall time series, the IDF curves can be used to derive other design rainfall events, such as the 24-hr, 25-yr storm, which can be used as input to a hydrologic model for drainage design or flood control (Chapter 6).

It is sometimes necessary to estimate point rainfall at a given location from recorded values at surrounding sites. The National Weather Service (1972) has developed a method for this based on a weighted average of surrounding values. The weights are reciprocals of the sum of squares of distances away from the point of interest. Thus,

$$D^2 = x^2 + y^2, \tag{1.8}$$

$$W = 1/D^2 = \text{Weight}, \tag{1.9}$$

$$\text{Rainfall estimate} = \sum_i P_i W_i / \sum_i W_i. \tag{1.10}$$

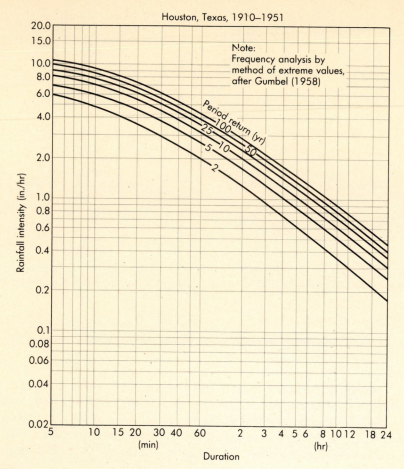

FIGURE 1.8

Expected recurrence interval for any storm.

Areal Precipitation

The average depth of precipitation over a specific area is often required to predict watershed hydrologic response or to develop frequency design storms. Three basic methods exist to derive areally averaged values: arithmetic mean, the Thiessen polygon method, and the isohyetal method (Fig. 1.9).

The simplest method is an **arithmetic mean** of point rainfalls from available gages (Fig. 1.9a). This method is satisfactory if the gages are uniformly distributed and individual variations are not far from the mean rainfall.

The **Thiessen polygon method** (Fig. 1.9b) allows for areally weighting of rainfall from each gage. Connecting lines are drawn between stations located on a map. Perpendicular bisectors are drawn to form polygons around each gage, and the ratio of the area of each polygon A_i within the watershed boundary to the total area A_T is used to weight each station's rainfall. The method is unique for each gage network and does not allow for orographic effects (those due to elevation changes) but is probably the most widely used of the three available methods.

The **isohyetal method** (Fig. 1.9c) involves drawing contours of equal precipitation (isohyets) and is the most accurate method. However, an extensive gage network is required to draw isohyets accurately. The rainfall calculation is based on finding the average rainfall between each pair of contours, multiplying by the area between them, totaling these products, and dividing by the total area. The isohyetal method can include orographic effects and storm morphology and can represent an accurate map of the rainfall pattern.

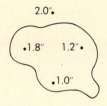

(a) Arithmetic mean:

$$\frac{1.8 + 1.2 + 1.0}{3} = 1.33 \text{ in.}$$

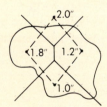

(b) Thiessen polygon method:

P_i (in.)	A_i (mi²)	A_i/A_T	$(P_i)(A_i/A_T)$ (in.)
2.0	1.5	0.064	0.13
1.8	7.2	0.305	0.55
1.2	5.1	0.216	0.26
1.0	9.8	0.415	0.42
$\Sigma =$	23.6	1.000	1.35 in.

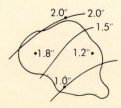

(c) Isohyetal method:

Isohyet (in.)	A (mi²)	P_{av} (in.)	V (in. $-$mi²)
>2	0.1	2.0	0.20
2	8.9	1.8	16.02
1.5	9.4	1.3	12.22
1.0	3.2	0.8	2.56
<1.0			
$\Sigma =$	21.6		31.0

Average $P = \dfrac{31.0 \text{ in.-mi}^2}{21.6 \text{ mi}^2} = 1.44 \text{ in.}$

FIGURE 1.9

Rainfall averaging methods.

EXAMPLE 1.4

RAINFALL AVERAGING METHODS

A watershed covering 34 mi² has a system of seven rainfall gages as shown
on the map in Fig. E1.4(a). Using the total storm rainfall depths given in
the accompanying table, determine the average rainfall over the watershed using (a) arithmetic averaging and (b) the Thiessen polygon method.

GAGE	RAINFALL (in.)
A	5.13
B	6.74
C	9.00
D	6.01
E	5.56
F	4.98
G	4.55

SOLUTION

a) For the arithmetic averaging method, only the gages that are
within the watershed are used, in this example the gages B and D.
Thus the arithmetic average is

$$(6.74 \text{ in.} + 6.01 \text{ in.})/2 = \boxed{6.38 \text{ in.}}$$

b) The first step in the Thiessen polygon method is to connect all the
rain gages by straight lines so that no two lines form an angle
greater than 90°. The result is a system of triangles as shown by the
dashed lines in Fig. E1.4(b). Next, we construct perpendicular
bisectors of the dashed lines. The bisectors should intersect within

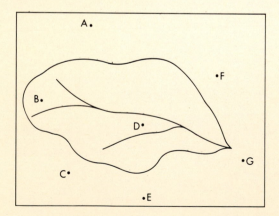

FIGURE E1.4(a)

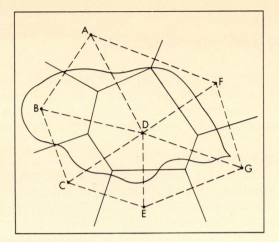

FIGURE E1.4(b)

the triangular areas. The resulting polygons around each rainfall gage are known as the Thiessen polygons.

The area of each polygon within the watershed boundary is measured by planimeter or by counting squares on graph paper, and each area is divided by the total area and multiplied by the depth of rain measured at its corresponding gage. The sum of percent area times rainfall for all the gages gives the average rainfall over the watershed. These computations are shown in the following table.

GAGE	P_i (in.)	A_i (mi²)	A_i/A_T	$(P_i)(A_i/A_T)$ (in.)
A	5.13	1.01	0.030	0.15
B	6.74	3.91	0.115	0.78
C	9.00	3.42	0.101	0.91
D	6.01	13.01	0.383	2.30
E	5.56	3.75	0.110	0.61
F	4.98	5.23	0.154	0.77
G	4.55	3.64	0.107	0.49
		34.00	1.000	6.01

Thus, the Thiessen polygon method gives an average rainfall over the basin of

6.01 in.

1.4

EVAPORATION AND TRANSPIRATION

Evaporation is the process by which water in its liquid or solid state is transformed into water vapor, which mixes with the atmosphere. Evapotranspiration is considered separately as the combined loss of water vapor from the surface of plants (transpiration) and the evaporation of moisture from soil.

Knowledge of evaporation processes is important in predicting water losses to evaporation from a lake or reservoir. Approximately 70% of the mean annual rainfall in the United States is returned to the atmosphere as evaporation or transpiration. However, variations in evaporation across the continent can be very large. Figure 1.10 shows the mean annual evaporation from shallow lakes and reservoirs in the United States. When compared with Fig. 1.4, it can be seen that annual evaporation can exceed mean rainfall, especially in arid regions of the Southwest.

For the case of evaporation from a lake surface, water loss is a function of solar radiation, temperature of the water and air, difference in vapor pressure between water and the overlying air, and wind speed across the lake. As evaporation proceeds in a closed-container system at a constant temperature, pressure in the air space increases because of an increase in partial pressure of water vapor. Evaporation continues until vapor pressure of the overlying air equals the surface vapor pressure; at this point the air space is said to be saturated at that temperature, and further evaporation ceases. This state of equilibrium would not be reached if the container were open to the atmosphere, in which case all water would eventually evaporate. Thermal energy is required to increase the free energy of water molecules to allow escape across the liquid-gas interface. The amount of heat required to convert water to vapor at constant temperature is called the latent heat of evaporation, as given in Eq. (1.7).

As vaporization continues over a flat free-water surface, an accumulation of vapor molecules causes an increase in the vapor pressure e in the air just above the water surface, until eventually condensation begins. The air is saturated when the rate of condensation equals the rate of vaporization and e equals the saturation vapor pressure. However, various convective transport processes operate to transport the vapor (by wind-driven currents) and prevent equilibrium from occurring.

Evaporation is usually of concern only in large-scale water resources planning and water supply studies. During typical storm periods, with intensities of 0.5 in./hr, evaporation is on the order of 0.01 in./hr and can thus be neglected for flood flow studies and urban drainage design applications. Evaporation has been extensively studied in the United States since 1950, beginning with the comprehensive Lake Hefner evaporation research project by Marciano and Harbeck (1954). Three primary methods are used to estimate evaporation from a lake surface: the water budget method, the mass transfer method, and the energy budget method. These are discussed in more detail in the following sections, and Brutsaert (1982) presents a modern, detailed review of evaporation mechanisms.

Water Budget Method for Determining Evaporation

The water budget method for lake evaporation is based on the hydrologic continuity equation. Assuming that change in storage ΔS, surface inflow I, surface outflow O, subsurface seepage to ground water flow GW, and precipitation P can be measured, evaporation E can be computed as

$$E = -\Delta S + I + P - O - GW. \tag{1.11}$$

The approach is simple in theory, but evaluating seepage terms can make the method quite difficult to implement. The obvious problems with the method result from errors in measuring precipitation, inflow, outflow, change in storage, and subsurface seepage. Good estimates using the method were obtained for Lake Hefner, Oklahoma, with 5–10% error. It should be emphasized that Lake Hefner was selected from more than one hundred lakes and reservoirs as one of three or four that best met water budget requirements.

Mass Transfer Method

Mass transfer techniques are based primarily on the concept of turbulent transfer of water vapor from a water surface to the atmosphere. Numerous empirical formulas have been derived to express evaporation rate as a function of vapor pressure differences and wind speed above a lake or reservoir. Many such equations can be written in the form of Dalton's law:

$$E = (e_o - e_a)(a + bu), \tag{1.12}$$

where

> e_o = vapor pressure at the water surface,
>
> e_a = vapor pressure at some fixed level above the water surface,
>
> u = wind speed,
>
> a, b = empirical constants.

An obstacle to comparing different evaporation formulas is the variability in measurement heights for v and e_a. Reducing all formulas to the same measurement level of 2 m (6.5 ft) for wind speed and vapor pressure and taking into account the 30% difference between reservoir and pan

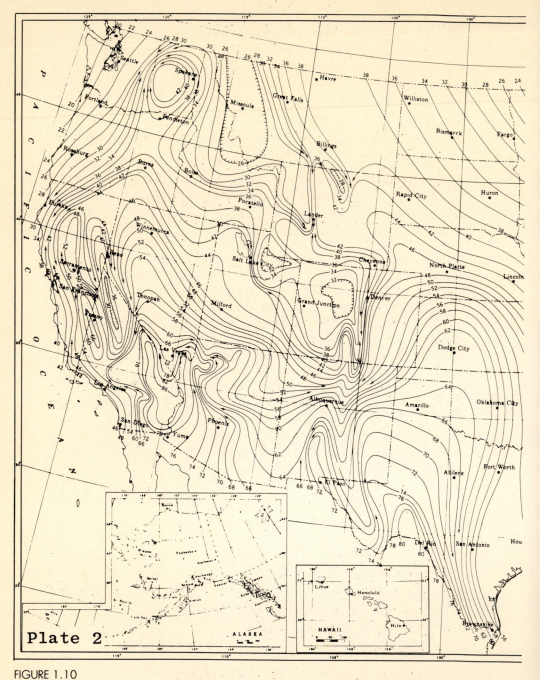

FIGURE 1.10

Average annual lake evaporation (in.). (U.S. National Weather Service.)

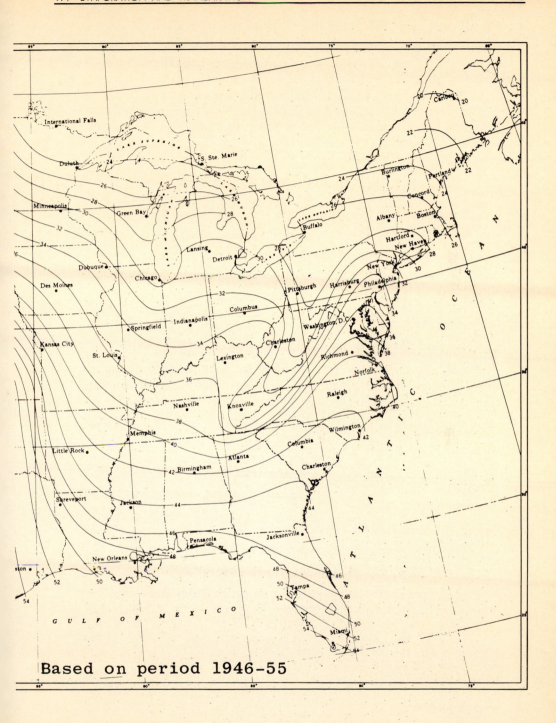

Based on period 1946-55

evaporation, the scatter between the common formulas is considerably reduced. The formula with the best data base is the Lake Hefner formula given by Marciano and Harbeck (1954), which also performed well at Lake Mead (Harbeck, 1958). Several formulas use a nonzero value of the constant as in Eq. (1.12), due probably to small local air movements with velocities sufficient to remove excess vapor from above a pan surface. Harbeck and Meyers (1970) present the formula

$$E = Nu_2(e_o - e_2),\qquad(1.13)$$

where

E = evaporation (cm/day),

N = 0.012 for Lake Hefner, 0.0118 for Lake Mead,

e_o = vapor pressure at water surface (mb),

e_2 = vapor pressure 2 m above water surface (mb),

u_2 = wind speed 2 m above the water surface (m/s).

Energy Budget Method

The most accurate and complex method for determining evaporation uses the energy budget of a lake (Fig. 1.11). The overall energy budget for a lake can be written in langley/day, where 1 langley = 1 cal/cm^2:

$$Q_N - Q_h - Q_e = Q_\theta - Q_v,\qquad(1.14)$$

where

Q_N = net radiation absorbed by the water body,

Q_h = sensible heat transfer (conduction and convection to the atmosphere),

Q_e = energy used for evaporation,

Q_θ = increase in energy stored in the water body,

Q_v = advected energy of inflow and outflow.

Letting L_e represent the latent heat of vaporization (cal/g) and R the ratio of heat loss by conduction to heat loss by evaporation, Eq. 1.14 becomes

$$E = \frac{Q_N + Q_v - Q_\theta}{\rho L_e(1 + R)},\qquad(1.15)$$

where E is the evaporation (cm/day) and ρ is the density of water (g/cm^3). The Bowen ratio (R) is used as a measure of sensible heat transfer and can be computed by

$$R = 0.66 \left(\frac{T_s - T_a}{e_s - e_a}\right)\left(\frac{P}{1000}\right) = \gamma \frac{T_s - T_a}{e_s - e_a},\qquad(1.16)$$

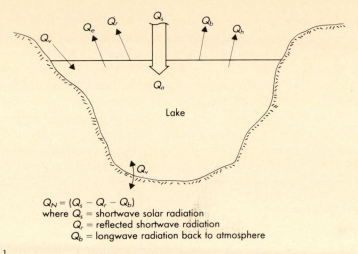

$Q_N = (Q_s - Q_r - Q_b)$
where Q_s = shortwave solar radiation
       Q_r = reflected shortwave radiation
       Q_b = longwave radiation back to atmosphere

FIGURE 1.11

Lake energy budget.

where

P = atmospheric pressure (mb),

T_a = air temperature (°C),

T_s = water surface temperature (°C),

e_a = vapor pressure of the air (mb),

e_s = saturation vapor pressure at surface water temperature (mb),

γ = the psychometric constant $0.66P/1000$.

Application of the energy budget method requires measurements of total incoming or net radiation. The Bowen ratio was conceived because sensible heat transfer cannot be readily computed. The method was applied to Lake Hefner and Lake Mead and was used to evaluate empirical coefficients for the mass transfer method and to interpret pan evaporation data at Lake Hefner. The energy budget method is theoretically the most accurate, but it requires the collection of large amounts of detailed atmospheric data. The radiation energy budget is described in more detail in Appendix D.1. To get around this problem, other methods, such as pan evaporation methods, have been developed to estimate shallow-lake evaporation.

Pan Evaporation

Evaporation can be measured from a standard weather bureau class A pan, an open galvanized iron tank 4 ft in diameter and 10 in. deep mounted 12 in. above the ground. To estimate evaporation, the pan is filled to a depth of 8 in. and must be refilled when the depth has fallen to

7 in. The water surface level is measured daily, and evaporation is computed as the difference between observed levels, adjusted for any precipitation measured in a standard rain gage. Alternatively, water is added each day to bring the level up to a fixed point. Pan evaporation rates are higher than actual lake evaporation and must be adjusted to account for radiation and heat exchange effects. The adjustment factor is called the **pan coefficient,** which ranges from 0.64 to 0.81 and averages 0.70 for the United States. However, the pan coefficient varies with exposure and climatic conditions and should be used only for rough estimates of lake evaporation. Pan evaporation data are available at a number of stations in the United States (Farnsworth and Thompson, 1982).

Combined Methods

Penman (1948) first used the best features of the mass transport and energy budget methods to derive a water surface evaporation relation that is fairly easy to compute. The Penman equation is

$$E = \frac{\Delta}{\Delta + \gamma} Q_N + \frac{\gamma}{\Delta + \gamma} E_a, \tag{1.17}$$

where

$\Delta = (e_s - e_{sz})/(T_s - T_z)$ is the slope of the e_s vs. T curve, which is tabulated as γ/Δ vs. T_z in Brutsaert (1982),

Q_N = net radiation absorbed (same units as E_a),

e_s = saturation vapor pressure at surface T_s (mb),

e_{sz} = saturation vapor pressure at elevation z (mb) (where $T = T_z$),

γ = constant in Bowen ratio (Eq. 1.16),

E_a = value of E in Eq. 1.13.

This approach provides a reasonably reliable means of estimating lake evaporation for daily or monthly periods assuming net advection is not appreciable, which should apply for large bodies of water. Example 1.5 shows the use of Fig. 1.12, which is based on the Penman equation (Kohler et al., 1955).

EXAMPLE 1.5

DETERMINING EVAPORATION USING THE PENMAN EQUATION

For a given shallow lake, the mean daily air temperature T_a is 70°F. The solar radiation is measured as 650 langley/day (1 langley = 1 cal/cm^2). The mean daily dew point temperature is 50°F, and the wind movement at six in. above the pan rim is 40 mi/day. Using the nomograph for shallow lake evaporation shown in Fig. 1.12, determine the daily lake evaporation in inches.

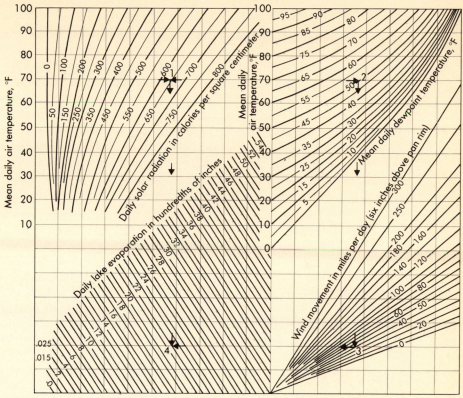

FIGURE 1.12

**Shallow-lake evaporation as a function of solar radiation, air tempera-
ture, dew point, and wind movement. (Reproduced with permission
from Kohler et al., 1955.)**

SOLUTION

Point 1 on Fig. 1.12 shows the point where $T_a = 70°F$ and the solar
radiation is 650 langley/day. Point 2 is drawn at the intersection of the
curve of $T_s = 50°F$ and $T_a = 70°F$. A vertical line is drawn from point 2
to the curve representing wind movement equal to 40 mi/day. This is
point 3. Point 4 is found at the intersection of a horizontal line projected
from point 3 and a vertical line projected from point 1. The daily lake
evaporation is then read from the graph as

$$E = 0.21 \text{ in.}$$

Evapotranspiration

For a water budget over the whole watershed, one is usually concerned with the total evaporation from all free-water surfaces, plus **transpiration,** the loss of vapor through small openings (stomata) in plant tissue (e.g., leaves). For most plants, transpiration occurs only during daylight hours during photosynthesis, which can lead to diurnal variations in the shallow ground water table in heavily vegetated areas. The combined evaporation and transpiration loss is called **evapotranspiration** (ET) and is a maximum if the water supply to both the plant and soil surface is unlimited. The maximum possible loss is limited by meteorologic conditions and is called **potential ET** (Thornthwaite, 1948); potential ET is approximately equal to the evaporation from a large, free-water surface such as a lake. Thus methods discussed in the previous section for evaporation can be used to predict potential ET.

Large surface area and high temperature of leaves can easily create transpiration rates that equal or even exceed potential ET. Exceedance is possible in instances of an "oasis" of well-watered vegetation, such as a crop or golf course, located in a larger area in which actual ET is less than the potential. Actual transpiration and, hence, actual ET are usually limited by moisture supply to the plants. Obviously the effect of limited moisture will depend on plant characteristics such as root depth and the ability of the soil to transport water to the roots. These effects have led to various empirical factors that may be applied to potential ET computed

TABLE 1.2
Pan Coefficients for Evapotranspiration Estimates

TYPE OF COVER	PAN COEFFICIENT	REFERENCE
St. Augustine grass	0.77	Weaver and Stephens
Bell peppers	0.85–1.04	(1965)
Grass and clover	0.80	Brutsaert (1982, p. 253)
Oak-pine flatwoods (East Texas)	1.20	Englund (1977)
Well-watered grass turf		Shih et al. (1983)
Light wind, high relative humidity	0.85	
Strong wind, low relative humidity	0.35	
Everglades agricultural areas	0.65	
(75% sugar cane, 25% truck crops and pasture)		
Irrigated grass pasture (Central California)	0.76	Hargreaves and Samani (1982)

using one of the evaporation methods discussed previously. For example, Penman (1950) determined that ET from a vegetated land surface in Great Britain was 60–80% of potential ET computed using his method; Shih et al. (1983) used a value of 70% for crops in southern Florida. On the other hand, Priestly and Taylor (1972) multiplied Penman equation evaporation by 1.26 for well-water crops to compute ET. Other empirical formulas that incorporate radiation, air temperature, and precipitation are given by Thornthwaite (1948), Turc (1954), and Hamon (1961), and in agricultural practice the Blaney-Criddle method is widely used (Blaney and Criddle, 1950; Criddle, 1958).

Once again, pan evaporation can be used to estimate ET if coefficients are known for the specific vegetation. There is a wide range of pan coefficient values, as indicated in Table 1.2. Caution should be used with any pan coefficient method for prediction of ET since the method is weak

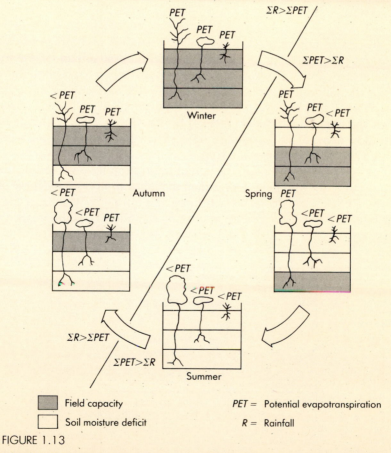

FIGURE 1.13

Idealized annual soil moisture cycle for three vegetational types.

to begin with (Brutsaert, 1982) and coefficients will experience a strong seasonal variation that follows vegetational growth patterns.

The complex annual relationship among potential ET, rainfall, and soil moisture is illustrated in Fig. 1.13 for plants with three different root depths. The **field capacity** of the soil is the moisture content above which water will be drained by gravity. The **wilting point** of the soil is the moisture content below which plants cannot extract further water. As the soil moisture content is reduced below field capacity, actual ET becomes less than potential ET. If the soil water content is reduced below the wilting point, the plant may die. This complicated process depends on the plant, soil type, meteorology, and season. A full discussion is beyond the scope of this text; further details may be found in Brutsaert (1982), Hillel (1982), and Salisbury and Ross (1969).

1.5

INFILTRATION

The process of infiltration has been widely studied and represents an important mechanism for movement of water into the soil under gravity and capillarity forces. Horton (1933) showed that when the rainfall rate i exceeds the infiltration rate f, water infiltrates the surface soils at a rate that generally decreases with time. For any given soil, a limiting curve defines the maximum possible rates of infiltration vs. time. The rate of infiltration depends in a complex way on rainfall intensity, soil type, surface condition, and vegetal cover.

For excess rate of rainfall, the actual infiltration rate will follow the limiting curve shown in Fig. 1.14, called the **infiltration capacity** curve of the soil. The capacity decreases with time and ultimately reaches a constant rate. The decline is caused mainly by the filling of soil pores with

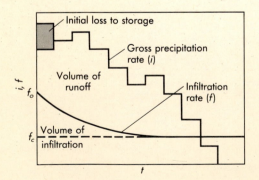

FIGURE 1.14

Horton's infiltration concept.

water, thus reducing capillary suction. For example, it has been shown through controlled tests that the decline is more rapid and the final constant rate is lower for clay soils than for sandy soils.

Equations for Infiltration Rate

The hydrologic concept of infiltration capacity is empirically based on observations at the ground surface. Horton (1940) suggested the following form of the infiltration equation, where rainfall intensity $i > f$ at all times:

$$f = f_c + (f_o - f_c)e^{-kt}, \tag{1.18}$$

where

> f = infiltration capacity (in./hr),
> f_o = initial infiltration capacity (in./hr),
> f_c = final capacity (in./hr),
> k = empirical constant (hr^{-1}).

Rubin and co-workers (1963, 1964) showed that Horton's observed curves can be theoretically predicted given the rainfall intensity, the initial soil moisture conditions, and a set of unsaturated characteristic curves for the soil. They showed that the final infiltration rate is numerically equivalent to the saturated hydraulic conductivity of the soil. Furthermore, Rubin showed that ponding at the surface will occur if rainfall duration is greater than the time required for soil to become saturated at the surface. Horton's equation is depicted graphically in Fig. 1.14, and Example 1.6 illustrates its use.

EXAMPLE 1.6

HORTON'S INFILTRATION EQUATION

The initial infiltration capacity f_o of a watershed is estimated as 4.5 in./hr, and the time constant is taken to be 0.35 hr^{-1}. The equilibrium capacity f_c is 0.4 in./hr. Use Horton's equation to find (a) the values of f at $t = 10$ min, 30 min, 1 hr, 2 hr, and 6 hr, and (b) the total volume of infiltration over the 6-hr period.

SOLUTION

Horton's equation (Eq. 1.18) is

$$f = f_c + (f_o - f_c)e^{-kt}.$$

Substituting the values for f_o, f_c, and k gives

$$f = 0.4 \text{ in./hr} + 4.1(e^{-0.35t}) \text{ in./hr}.$$

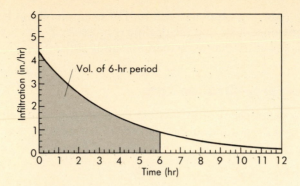

FIGURE E1.6

Solving for each value of t gives the following table.

t (hr)	f (in./hr)
1/6	4.27
1/2	3.84
1	3.29
2	2.44
6	0.90

The volume (in inches over the watershed) can be found by plotting the curve given by the table of values and then finding the area under the curve bounded by $t = 0$ and $t = 6$ hr. The plot is shown in Fig. E1.6. The curve is given by the equation

$$f = 0.4 \text{ in./hr} + 4.1(e^{-0.35t}) \text{ in./hr.}$$

Integrating over the interval $t = 0$ to $t = 6$ hr gives:

$$\begin{aligned}
\text{Vol} &= \int f \, dt \\
&= \int (0.4 + 4.1e^{-0.35t}) \, dt \\
&= [0.4t + (4.1/-0.35)e^{-0.35t}]_0^6 \\
&= [2.4 - (4.1/0.35)e^{-2.1}] - [0 - (4.1/0.35)e^0]
\end{aligned}$$

$$\boxed{\text{Vol} = 12.68 \text{ in. over the watershed.}}$$

Other equations have been developed by investigators utilizing analytical solutions to the unsaturated flow equation from soil physics (see the next section). Philip (1957) in a classic set of papers developed an

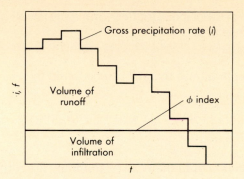

FIGURE 1.15

φ index method.

equation of the form

$$f = (0.5)At^{-1/2} + B, \tag{1.19}$$

$$F = At^{1/2} + Bt, \tag{1.20}$$

where

$f =$ infiltration capacity (in./hr),

$F =$ cumulative infiltration volume (in.),

$A, B =$ constants related to soil type and water movement.

Without detailed measurements of actual loss rates and because urban watersheds have high imperviousness, empirical approaches can sometimes give satisfactory answers. Infiltration has been observed to account for a variable percentage of gross rainfall that falls on a watershed. Most urban drainage and flood control studies rely on either the Horton equation or simpler methods to predict average losses during storm passage. The **φ index** is the simplest method and is calculated by finding the loss difference between gross precipitation and observed surface runoff measured as a hydrograph. The φ index method assumes that the loss is uniformly distributed across the rainfall pattern (Fig. 1.15). The use of φ index methods for infiltration is illustrated in Example 1.7.

EXAMPLE 1.7

φ INDEX METHOD FOR INFILTRATION

A watershed covering 600 acres had a storm in which rain fell at a rate of 0.5 in./hr for the first 2 hr, 0.3 in./hr for the next 5 hr, and 0.4 in./hr for the following hour. Determine (a) the total depth of rain that fell during the storm in inches over the watershed; (b) the volume of rain that fell, in acre-feet; and (c) the ϕ index for the storm if the runoff coefficient is 0.80. Neglect the effect of ET.

SOLUTION

a) The depth of rain that fell can be found by

$$d = \Sigma i_i t_i$$
$$= (0.5 \text{ in./hr})(2 \text{ hr}) + (0.3 \text{ in./hr})(5 \text{ hr}) + (0.4 \text{ in./hr})(1 \text{ hr})$$

$\boxed{d = 2.90 \text{ in.}}$

b) The volume of rainfall can be found by using the relation

$$V = (d)(A)$$
$$= (2.90 \text{ in.})(600 \text{ ac})(1 \text{ ft}/12 \text{ in.})$$

$\boxed{V = 145 \text{ ac-ft.}}$

c) The ϕ index is found by dividing the infiltration volume by the area times the duration. Neglecting the effects of ET, infiltration is equal to the rainfall minus the runoff. Since the runoff coefficient is 0.80, the runoff is equal to 0.8 times the rainfall, giving infiltration as 0.2 times the rainfall. Thus,

$$\phi = 0.2 \text{ Rainfall}/(\text{Area})(\text{Duration})$$
$$= (0.2)(145 \text{ ac-ft})/(600 \text{ ac})(8 \text{ hr})(1 \text{ ft}/12 \text{ in.})$$

$\boxed{\phi = 0.073 \text{ in./hr.}}$

Theoretical Infiltration Methods*

Theoretical approaches include the solution of the governing equation of continuity and Darcy's law (Chapter 8) in an unsaturated porous media. The governing equation, presented in more detail in Appendix D.2, takes the form

$$\frac{\partial \theta}{\partial t} = -\frac{\partial}{\partial z}\left[K(\theta)\frac{\partial \psi(\theta)}{\partial z}\right] - \frac{\partial K(\theta)}{\partial z}, \tag{1.21}$$

where

θ = volumetric moisture content (0/0),

z = distance below the surface (cm),

$\psi(\theta)$ = capillary suction (pressure) (cm of water),

$K(\theta)$ = unsaturated hydraulic conductivity (cm/s).

* This section may be omitted without loss of continuity in text material.

Hydraulic conductivity $K(\theta)$ relates velocity and hydraulic gradient in Darcy's law. **Moisture content** θ is defined as the ratio of the volume of water to the total volume of a unit of porous media. For saturated ground water flow, θ equals the **porosity** of the sample n, defined as the ratio of volume of voids to total volume of sample; for unsaturated flow above a water table, $\theta < n$. The **water table** defines the boundary between the unsaturated and saturated zones and is defined by the surface on which the fluid pressure P is exactly atmospheric, or $P = 0$. Hence, the total hydraulic head $\phi = \psi + z$, where $\psi = P/\rho g$, the pressure head.

The value of ψ is greater than zero in the saturated zone below the water table and equals zero at the water table. It follows that ψ is less than zero in the unsaturated zone, reflecting the fact that water is held in soil pores under surface-tension forces. Soil physicists refer to $\psi < 0$ as the tension head or capillary suction head, which can be measured by an instrument called a tensiometer.

To further complicate the analysis of unsaturated flow, the moisture content θ and the hydraulic conductivity K are functions of the capillary suction ψ. Also, it has been observed experimentally that the θ-ψ relationships differ significantly for different types of soil. Figure 1.16 summarizes unsaturated zone parameters and relationships.

Equation (1.21) is called Richards (1931) equation and is a very difficult partial differential equation. Both numerical and analytical solutions exist for certain special cases. The most difficult part of the procedure is determining the characteristic curves for a soil, which relate unsaturated hydraulic conductivity K and moisture content θ to capillary suction ψ. The characteristic curves reduce to the fundamental hydraulic parameters K and n in the saturated zone and remain as functional relationships in the unsaturated zone.

Philip (1957) solved Eq. (1.21) analytically for the condition of excess water at the surface and given characteristic curves. His coefficients can be predicted in advance from soil properties and do not have to be fitted to field data. However, the more difficult case where the rainfall rate is less than the infiltration capacity cannot be handled by Philip's equation.

One of the most interesting and useful approaches to solving the governing equation was originally advanced by Green and Ampt (1911). In this method, water is assumed to move into dry soil as a sharp wetting front. At the location of the front, the average capillary suction head $\psi = S_{av}$, is used to represent the characteristic curve. The moisture content profile at the moment of surface saturation is shown in Fig. 1.17(a). The area above the moisture profile is the amount of infiltration up to surface saturation F_s and is represented by the shaded area of depth L_s in Fig. 1.17(a). Thus, $F_s = (\theta_s - \theta_i)L_s = M_d L_s$, where θ_i is the initial moisture content, θ_s is the saturated moisture content, and $M_d = \theta_s - \theta_i$.

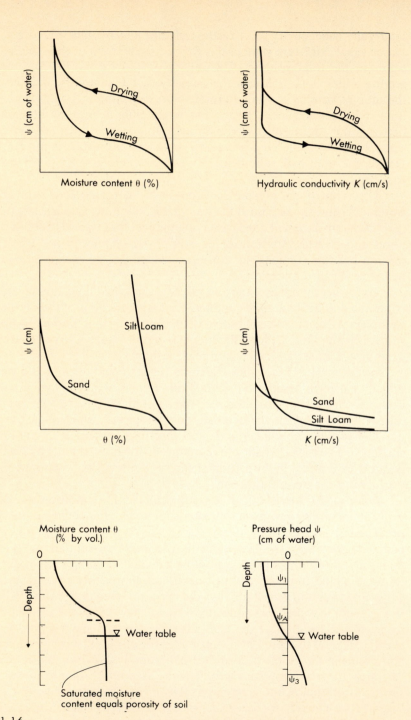

FIGURE 1.16

Typical θ-ψ relationships in the unsaturated zone.

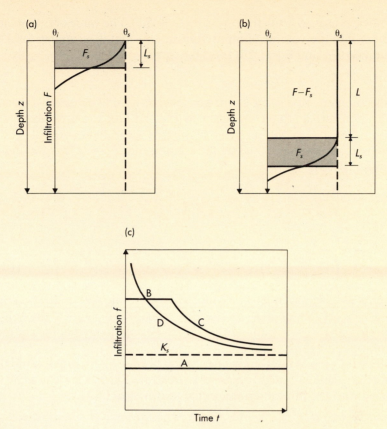

FIGURE 1.17

Moisture and infiltration relations. (a) Moisture profile at moment of surface saturation. (b) Moisture profile at later time. (c) Infiltration behavior under rainfall. (Adapted from Mein and Larson, 1973.)

Darcy's law (Chapter 8) is then used with the unsaturated value for K as

$$q = -K\frac{\partial h}{\partial z} = -K(\theta)\left(\frac{\phi_2 - \phi_1}{z_2 - z_1}\right), \tag{1.22}$$

where

q = velocity (cm/s),

z = depth below surface (cm),

ϕ = potential or head (cm),

$K(\theta)$ = capillary conductivity (cm/s),

S_{av} = average capillary suction (cm),

$L_s = (z_2 - z_1)$ (cm),

and if we assume $\phi_1 = 0$, then $\phi_2 = -(L_s + S_{av})$. The value of L_s can then be related to the volume of infiltration storage that is occurring for K at saturation K_s, and Eq. (1.22) takes the form

$$I = K_s(S_{av} + L_s)/L_s \tag{1.23}$$

at the moment of surface saturation, where rainfall intensity I equals infiltration rate q. Referring again to Fig. 1.17(a), the area above the moisture profile at the moment of surface saturation is equal to the volume of infiltration F_s. Using the relation that $F_s = M_d L_s$, we obtain from Eq. (1.23):

$$F_s = \frac{S_{av} M_d}{(I/K_s) - 1}, \tag{1.24}$$

or

$$I = K_s\left(1 + \frac{M_d S_{av}}{F_s}\right), \tag{1.25}$$

which is the original Green and Ampt equation. Note that infiltration rate I must be greater than K_s to use this equation; if $M_d = 0$ the soil is saturated, and $F_s = 0$. Equation (1.25) can be used to predict the amount of infiltration before runoff.

For the case of infiltration after runoff begins, Fig. 1.17(b) shows the moisture profile as it might appear at a later time. Darcy's law can again be applied with the infiltration rate equal to the capacity f:

$$f = \frac{K_s(S_{av} + L_s + L)}{L_s + L}. \tag{1.26}$$

If $L_s = F_s/M_d$, then $L_s + L = F/M_d$, where F is the cumulative infiltration at any time and F_s is the value at the moment of surface saturation. We obtain

$$f = K_s\left(1 + \frac{S_{av} M_d}{F}\right), \tag{1.27}$$

which is again identical to the Green and Ampt equation.

Three possible conditions exist for application of Eq. (1.27) with rainfall intensity I, saturated conductivity K_s, and infiltration capacity f. In case A, $I < K_s$, and all rainfall infiltrates with no runoff generated (curve A in Fig. 1.17c). In case B, $K_s < I \le f$, all rainfall infiltrates, and the soil moisture level increases near the surface (curve B in Fig. 1.17c). In case C, $K_s < f \le I$, and runoff is being generated as shown by curves C and D in Fig. 1.17(c). In case C, the infiltration rate is at capacity and is decreasing (see Example 1.8).

The advantage of the Green and Ampt equation is that it can be related to measurable properties and can be applied to the case of excess water supply at the surface. Finally, since the conductivity for θ_i is usually very small, we can integrate over the range of moisture content $\theta_i - \theta_s$ to obtain an estimate of S_{av},

$$S_{av} = \int_0^1 S \, dK_r, \tag{1.28}$$

where K_r is relative conductivity, $K(\theta)/K$; this area under the S-K_r curve between $K_r = 0$ and $K_r = 1$ provides an estimate for S_{av}, the average capillary suction.

The method requires the use of suction vs. moisture content and suction vs. relative conductivity for specific soils of interest. Mein and Larson (1973) found excellent agreement between this model and numerical solutions of Richards equation and experimental soil data.

EXAMPLE 1.8

GREEN AND AMPT INFILTRATION EQUATION

For the following soil properties, develop a plot of infiltration rate f vs. infiltration volume F using the Green and Ampt equation:

$K_s = 1.97$ in./hr,

$\theta_s = 0.518$,

$\theta_i = 0.318$,

$S_{av} = 9.37$ in.,

$I = 7.88$ in./hr.

SOLUTION

Noting that $M_d = \theta_s - \theta_i$, we can solve Eq. (1.24) to obtain the volume of water that will infiltrate before surface saturation is reached:

$$F_s = \frac{S_{av} M_d}{(I/K_s - 1)}$$

$$= \frac{(9.37 \text{ in.})(0.518 - 0.318)}{[(7.88 \text{ in./hr})/(1.97 \text{ in./hr})] - 1}$$

$$\boxed{F_s = 0.625 \text{ in.}}$$

Until 0.625 in. has infiltrated, the rate of infiltration is equal to the rainfall rate. After that point (surface saturation) the rate of infiltration is given by the equation (Eq. 1.27)

$$f = K_s(1 + S_{av} M_d/F).$$

Solving this equation for various values of F gives the graph shown in Fig. E1.8.

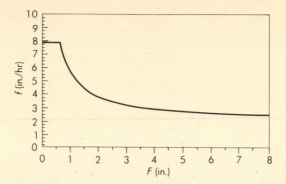

FIGURE E1.8

Measurement of Infiltration

An infiltrometer is a tube or other device designed to isolate a section of soil. The cylinder is flooded and f is measured by the drop in water level. Sprinkling techniques are used to simulate heavy and uniform rainfall, and infiltration is computed as a difference between water applied and measured surface runoff. Carefully controlled infiltration experiments also require the use of tensiometers to measure capillary suction, electrical resistance to measure soil moisture content, and multiple-depth wells to measure the response in the water table and below. Thus, the actual measurement of infiltration is complex and expensive and is usually limited to experimental basins.

In large heterogeneous watersheds, infiltration varies greatly and is usually evaluated from considerations of the overall water balance, with the difference between rainfall and streamflow approximating the loss to infiltration. If extended time periods are involved, then evapotranspiration must also be estimated as a loss from the watershed. In most practical cases, the ϕ index method or a form of Horton's equation is used to calculate rates of infiltration.

1.6

STREAMFLOW

When rainfall strikes the land surface, it may initially distribute to fill depression storage, infiltrate to fill soil moisture and ground water, or travel as interflow to a receiving stream. Depression storage capacity is usually satisfied early in storm passage, followed by soil moisture capacity (see Fig. 1.18). Eventually, overland flow and surface runoff commence.

The classic concept of streamflow generation by **overland flow** is due to Horton (1933), who proposed that overland flow is common and

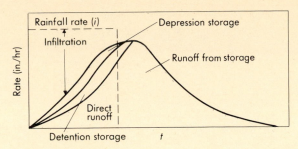

FIGURE 1.18
Surface flow distribution.

areally widespread. Later investigators incorporated the great heterogene-ity that exists across natural watersheds and developed the partial area contribution concept (Betson, 1964). This concept recognized that only certain portions of a watershed regularly contribute overland flow to streams and that no more than about 10% of a watershed contributes overland flow. In urban environments with large areas of impervious cover, the overland flow percentage may be larger.

A second important concept of surface runoff generation is subsur-face stormflow moving through a shallow soil horizon without reaching the zone of saturation, called **interflow.** Freeze (1972) concluded that subsurface stormflow could be a significant component only on convex hillslopes that feed deeply incised channels, and then only for very highly permeable soils. On concave slopes, saturated valley wetlands created by rising water tables become large and contribute overland flow, which usually exceeds subsurface stormflow.

Overland flow quickly moves downgradient toward the nearest small rivulet or channel, which then flows into larger streams and channels as open channel flow. Once in a stream, velocities and depths can be mea-sured at a particular cross-section through time, which allows the calcula-tion of the **hydrograph,** or discharge vs. time curve. After rainfall ceases, storage areas drain out, completing the storm cycle.

The channel may contain a certain amount of **base flow** coming from ground water and soil contributions even in the absence of rainfall. Dis-charge from rainfall excess, after losses have been subtracted, makes up the direct runoff hydrograph. The total hydrograph is direct runoff plus base flow. The duration of rainfall determines the portion of watershed area contributing to the peak, and the intensity of rainfall determines the resulting peak flow rate. If rainfall maintains a constant intensity for a very long time, maximum storage is achieved and a state of equilibrium discharge is reached, as shown in Fig. 1.19. The condition of equilibrium discharge is seldom attained in nature because of the variability of actual rainfall over a basin.

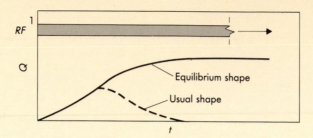

FIGURE 1.19

Equilibrium discharge hydrograph.

Components of the Hydrograph

The shape of the hydrograph is produced by components from (1) surface runoff, (2) interflow, (3) ground water or base flow, and (4) channel precipitation. Surface runoff occurs only after rainfall intensity exceeds infiltration capacity ($i > f$). The actual shape and timing of the hydrograph is determined largely by the size, shape, slope, and storage in the basin and by the intensity and duration of input rainfall. These factors are discussed in more detail in Chapter 2 along with the effects of land use changes and urbanization.

The hydrograph integrates variations in rainfall and watershed physiography into a unique picture of hydrologic response for a given set of conditions. Direct runoff from most storms derives from a relatively small source area closest to stream channels, but to accurately describe the location of such areas is beyond the scope of present hydrologic capability. Comparisons of measured hydrographs to measured rainfalls yield the most reliable approaches to runoff predictions, as presented in Chapter 2.

Streamflow Measurement

Streamflow is generally measured by observing stage, or elevation above a specified datum, in a channel and then relating stage to discharge via a rating curve. A staff gage is a scale set so that a portion is immersed in water and can be read manually during storm passage. A wire-weight gage is another type of manual gage that is lowered from a bridge structure to the water surface.

Most automatic recording gages, such as those used by the U.S. Geological Survey for routine streamflow monitoring, use a float type device to measure stage (or a gas bubbler to measure pressure) (Fig. 1.20). These recording gages are usually installed in protective shelters so that continuous strip charts, paper punch tapes, or solid-state data loggers can be used

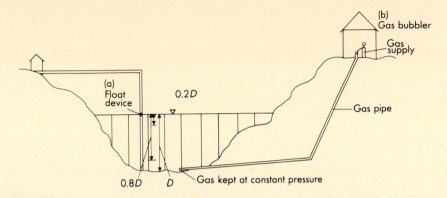

FIGURE 1.20

Typical streamflow gages.

to record stage data. In some urban basins, such data are often telemetered directly into computers for flood evacuation purposes. The more modern gages record the pressure required to maintain a small flow of gas from a tube submerged in the channel.

Selection of a site for gaging must include consideration of access, channel controls where Q and depth are related, and seasonal changes in vegetation. It is best to avoid locations that are affected by backwater from a dam, an intersecting stream, or tidal action. Often it is necessary to construct a low concrete weir with a V-notch to maintain a stable low water rating curve. Most USGS sites have been carefully selected to incorporate these factors, and the USGS guidelines should be consulted before attempting to develop any new sites.

Once a station has been established, a **rating curve** must be developed between stage and discharge by actually measuring flow in the channel, usually with a current meter. The current meter is suspended from a bridge or held by a rod in shallow water. The recommended procedure for determining mean velocity is to take measures at $0.2D$ and $0.8D$, where D is stream depth, and average the two values (see Fig. 1.20). For shallow water, a single reading at $0.6D$ is acceptable. The selection of these velocity measurement depths accounts for the logarithmic shape of the velocity distribution.

The total discharge is found by dividing the channel into several sections, as shown in Fig. 1.20. The average velocity of each section is multiplied by its associated area (width $\times$ depth of section), and these are summed across the channel to yield total discharge Q corresponding to a particular stage z, obtaining one point, as graphed in Fig. 1.21.

The development of a rating curve requires a plot of several stage and discharge readings for a particular station (Fig. 1.21). If the channel control is reasonably permanent and the slope of the channel fairly constant, a simple rating curve is probably acceptable. However, it will be shown in

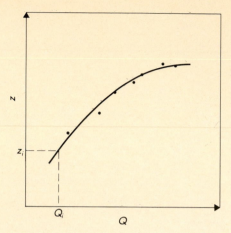

FIGURE 1.21

Rating curve.

Chapter 4 that actual rating curves can be looped because of storage and hydraulic effects in the channel, and the single-valued rating curve is at best an approximation to the actual relationship. The Chezy formula (1769) or Manning's formula (1889) can be used to describe flow of water in an open channel. They relate velocity or flow rate to cross-sectional area, depth, channel slope, and a roughness coefficient (Chapters 4 and 7).

1.7

SUBSURFACE FLOW

Subsurface flow from ground water or springs usually accounts for flow in a channel during periods of little or no rainfall. During a storm event, this base flow component of the hydrograph increases due to increased infiltration to the saturated zone. However, it is very difficult to quantify this increase in base flow. Several methods are available to separate base flow from the remaining direct runoff hydrograph. These methods are quite arbitrary and are presented in more detail in Chapter 2.

Base flow contributions are usually minimal for heavily urbanized areas in which channels have been improved or lined with concrete. Also, for extreme events, direct runoff will completely dominate the peak of the hydrograph, and errors in base flow separation are relatively unimportant. For the case of moderate storms over relatively pervious basins, base flow contributions may be significant and should be carefully analyzed from a representative group of storm hydrographs. Development of unit hydrographs assumes that base flow can be separated from the direct runoff hydrograph.

SUMMARY

The components of the hydrologic cycle are shown in Fig. 1.22; they include precipitation, evaporation, evapotranspiration, infiltration, interflow, overland flow, streamflow, and ground water flow, with all outflows eventually reaching the ocean or atmosphere. Man-made alterations for water resources development, water supply, flood control, irrigation, drainage, navigation, recreation, and water quality enhancement usually change the natural hydrologic cycle. It is important for the hydrologist to understand the various storage and transport pathways in the cycle so that effects of changes can be predicted.

Chapter 1 presents a concise treatment of mechanisms of precipitation, evaporation, evapotranspiration, and infiltration, including governing equations and examples of hydrologic water balances. Streamflow is introduced in Chapter 1 but is covered in more detail in Chapter 3, "Frequency Analysis," Chapter 4, "Flood Routing," and Chapter 7, "Floodplain Hydraulics."

Precipitation is the most important variable in the cycle since its variation in intensity and duration largely determines the outflow re-

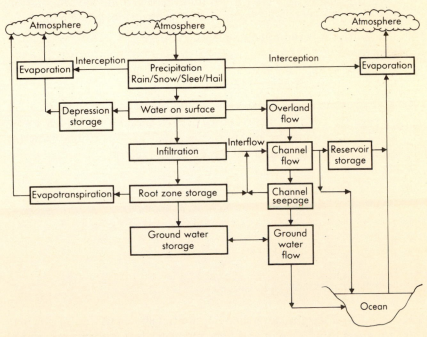

FIGURE 1.22

Components of the hydrologic cycle.

sponse from a watershed. With the complexity of the cycle, computer methods have recently been developed for watershed analysis in both undeveloped and urban areas. The remainder of the text includes presentations of "Hydrologic Simulation Models and Watershed Analysis" (Chapter 5) and "Urban Hydrology" (Chapter 6). Ground water hydrology and well mechanics are addressed in Chapter 8.

PROBLEMS

1.1. A lake with a surface area of 525 acres was monitored over a period of time. During a one-month period the inflow was 30 cfs, the outflow was 27 cfs, and a 1.5-in. seepage loss was measured. During the same month, the total precipitation was 4.25 in. Evaporation loss was estimated as 6.0 in. Estimate the storage change for this lake during the month.

1.2. Clear Lake has a surface area of 708,000 m². For a given month the lake has an inflow of 1.5 m³/s and an outflow of 1.25 m³/s. A storage change of + 708,000.0 m³ was recorded. If a precipitation gage recorded a total of 22.5 cm for this month, determine the evaporation loss (in cm) for the lake. Assume that seepage loss is negligible.

1.3. Figure P1.3 lists rainfall data recorded by the USGS for the storm of September 1, 1979, at the gage on the Hummingbird Street Ditch. Using these data, develop a rainfall hyetograph. What time period had the highest-intensity rainfall?

1.4. A small urban watershed has four rainfall gages as located in Fig. P1.4 on page 58. Total rainfall recorded at each gage during a storm event is listed. Compute the mean areal rainfall for this storm using (a) arithmetic averaging and (b) the Thiessen method.

GAGE	RAINFALL (in.)
A	3.26
B	2.92
C	3.01
D	3.05

1.5. Mud Creek has the watershed boundaries shown in Fig. P1.5. There are seven rain gages in and near the watershed, and the amount of rainfall at each one during a storm is given in the accompanying table. Using 10-square-to-the-inch graph paper and a scale of 1 in. = 1000 m, determine the mean rainfall of the given storm using the

STA. NO. 08074910	STORM RAINFALL AND RUNOFF RECORD		1979 WATER YEAR	
HUMMINGBIRD STREET DITCH AT HOUSTON, TEXAS STORM OF SEP. 1, 1979		ACCUM. WEIGHTED PRECIP. IN.	DISCHARGE IN CFS	ACCUM. RUNOFF IN.
	GAGE NUMBER			
DATE & TIME	4910			
SEPT. 1				
0000	0.0	0.0	0.0	0.0
1610	0.0	0.0	0.0	0.0
1615	0.10	0.10	0.0	0.0
1620	0.40	0.40	0.5	0.0002
1625	0.60	0.60	5.0	0.0022
1630	1.10	1.10	22.0	0.0111
1635	1.40	1.40	60.0	0.0353
1640	1.60	1.60	90.0	0.0716
1645	1.80	1.80	102.0	0.1128
1650	1.90	1.90	111.0	0.1576
1655	2.00	2.00	119.0	0.2056
1700	2.20	2.20	124.0	0.2556
1705	2.30	2.30	130.0	0.3081
1710	2.40	2.40	134.0	0.3622
1715	2.50	2.50	137.0	0.4175
1720	2.50	2.50	138.0	0.4731
1725	2.60	2.60	137.0	0.5284
1730	2.60	2.60	135.0	0.6101
1740	2.60	2.60	128.0	0.7651
1800	2.60	2.60	111.0	0.9891
1830	2.60	2.60	79.0	1.1804
1900	2.60	2.60	46.0	1.2917
1930	2.60	2.60	24.0	1.3498
2000	2.60	2.60	11.0	1.3765
2030	2.60	2.60	5.9	1.3908
2100	2.60	2.60	3.5	1.3992
2130	2.60	2.60	2.0	1.4041
2200	2.60	2.60	1.2	1.4070
2230	2.60	2.60	0.7	1.4087
2300	2.60	2.60	0.4	1.4101
2400	2.60	2.60	0.2	1.4106

FIGURE P1.3

Thiessen method. Compare the result to the mean rainfall found using arithmetic averaging.

GAGE NUMBER	RAINFALL (cm)
1	25.15
2	21.34
3	26.80
4	24.38
5	20.70
6	25.10
7	23.75

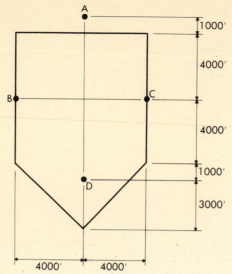

FIGURE P1.4

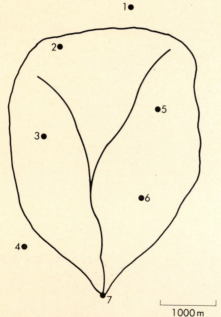

FIGURE P1.5

1.6. a) Using the ideal gas law ($P = \rho RT$), find the density of dry air at 25°C with a pressure of 950 mb. The gas constant R has the value 2.786×10^6 cm^2/s^2 °K for dry air, when pressure is in mb.

   b) Find the density of moist air at the same pressure and temperature if the relative humidity is 65%.

1.7. The following questions refer to Fig. 1.4.
   a) Which state has the point with the highest average annual precipitation?
   b) Compare the average annual rainfall of the lower West Coast with that of the upper West Coast.
   c) What is the average annual precipitation for your city?

1.8. The mass curve in Fig. P1.8 represents the weighted precipitation reported for Brays Bayou at Gessner Drive for November 29 – 30, 1981.
   a) What time period had the highest-intensity rainfall?
   b) What is the intensity for the time period of part (a)?
   c) What was the total precipitation from hours 0 to 24 (on November 29)?
   d) What was the total precipitation from hours 24 to 48 (on November 30)?

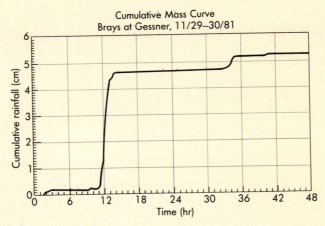

FIGURE P1.8

1.9. Using Fig. 1.12, find the daily evaporation from a small lake with the following characteristics:

Mean daily temperature	$= 25.6°C,$
Daily solar radiation	$= 550 \text{ cal/cm}^2,$
Mean daily dew point	$= 4.4°C,$
Wind movement (6 in. above pan)	$= 5.5 \text{ ft/s}.$

1.10. A class A pan is maintained near a small lake to determine daily evaporation. The level in the pan is observed every day and water is added if the level falls near 7 in. Determine the daily lake evaporation if the pan coefficient is 0.70.

DAY	RAINFALL (in.)	WATER LEVEL (in.)
1	0	8.00
2	0.23	7.92
3	0.56	7.87
4	0.05	7.85
5	0.01	7.76
6	0	7.58
7	0.02	7.43
8	0.01	7.32
9	0	7.25
10	0	7.19
11	0	7.08*
12	0.01	7.91
13	0	7.86
14	0.02	7.80

* Refilled at this point.

1.11. Given an initial rate of infiltration equal to 1.25 in./hr and an equilibrium capacity of 0.25 in./hr, use Horton's equation (Eq. 1.18) to find the infiltration capacity at times $t = 10$ min, 15 min, 30 min, 1 hr, 2 hr, 4 hr, and 6 hr. Assume a time constant $k = 0.25$ hr^{-1}.

1.12. A sketch of the infiltration curve obtained using Horton's equation is shown in Fig. P1.12. Prove that $k = (f_o - f_c)/F'$ if F' is the area between the curve and the f_c line.

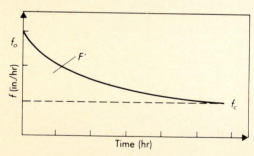

FIGURE P1.12

1.13. The incremental rainfall data in the table were recorded at a recording rainfall gage on a small urban watershed.
a) Plot the rainfall hyetograph.
b) Determine the total storm rainfall volume.

c) If the net rainfall for the storm was 30.5 cm, find the ϕ index (cm/hr).

TIME (hr)	RAINFALL (cm)
0	0
0.5	5.0
1.0	5.5
1.5	10.0
2.0	12.0
2.5	10.0
3.0	4.5
3.5	3.0
4.0	2.5
4.5	0

1.14. Determine the ϕ index for the November 29, 1981 storm described in Problem 1.8. The cumulative runoff recorded for that day is 1.37 cm.

1.15. a) How many calories of heat would it take to evaporate 8 liters of water at 15°C?

b) The amount of heat required to raise a unit mass of any substance by 1° is defined as the specific heat of that substance. How many grams of ice would be melted by the amount of heat found in part (a) if the ice were at −9°C? (Specific heat of ice = 0.5 cal/g°K.)

1.16. The following questions refer to Fig. 1.8.

a) What is the return period of a storm that recorded 3.1 in./hr for 2 hr in Houston, Texas?

b) What amount of rain (in.) would have to fall in a 6-hr period to be considered a 100-yr storm in Houston?

c) What is the return period of a storm that lasts 1 hr and records 3.5 in. of rainfall?

Problems 1.17−1.19 refer to the Hummingbird Street Ditch data used in Problem 1.3.

1.17. a) Plot the cumulative mass curves for rainfall and runoff on the same graph.

b) Using the mass curves, determine the ϕ index for this storm.

1.18. What is the runoff coefficient for the September 1, 1979, storm?

1.19. Plot the hydrograph for this storm (discharge vs. time). Label the following:

a) peak flow Q_p,

b) time to peak t_p (distance from center of mass of rainfall to peak flow),

c) time of rise T_R (distance from start of discharge to peak flow),

d) time base T_B (distance from start of discharge to end of discharge).

REFERENCES

Betson, R. P., 1964, "What Is Watershed Runoff?" *J. Geophys. Res.,* vol. 69, pp. 1541–1551.

Biswas, A. K., 1972, *History of Hydrology,* North Holland Publishing Co., Amsterdam.

Blaney, H. F., and W. D. Criddle, 1950, "Determining Water Requirements in Irrigated Areas from Climatological and Irrigation Data," U.S. Dept. Agriculture, Soil Conservation Service, Tech. Pap. 96, Washington, D.C.

Brutsaert, W. H., 1982, *Evaporation into the Atmosphere,* D. Reidel Publishing Co., Boston.

Chézy, 1769, referred to in Rouse, H., and S. Ince, 1957, *History of Hydraulics,* Iowa Institute of Hydraulic Research, State University of Iowa.

Chow, V. T. (editor), 1964, *Handbook of Applied Hydrology,* McGraw-Hill Book Company, New York.

Crawford, N. H., and R. K. Linsley, Jr., 1966, *Digital Simulation in Hydrology: Stanford Watershed Model IV,* Tech. Rept. No. 39, Department of Civil Engineering, Stanford University, Stanford, California.

Criddle, W. D., 1958, "Methods of Computing Consumptive Use of Water," *J. Irrigation and Drainage Division, Proc. ASCE,* vol. 84, no. IR1, January, p. 27.

Eagleson, P. S., 1970, *Dynamic Hydrology,* McGraw-Hill Book Company, New York.

Englund, C. B., 1977, "Modeling Soil Water Hydrology Under a Post Oak-Shortleaf Pine Stand in East Texas," *Water Resources Res.,* vol. 13, no. 3, June, pp. 683–686.

Farnsworth, R. K., and E. S. Thompson, 1982, "Mean Monthly, Seasonal, and Annual Pan Evaporation for the United States," NOAA Tech. Rept. NWS34, Washington, D.C.

Freeze, R. A., 1972, "Role of Subsurface Flow in Generating Surface Runoff, 2: Upstream Source Areas," *Water Resources Res.,* vol. 8, no. 5, pp. 1272–1283.

Gray, D. M., 1970, *Principles of Hydrology,* Water Information Center, New York, New York.

Green, W. H., and G. A. Ampt, 1911, "Studies of Soil Physics, 1: The Flow of Air and Water Through Soils," *J. of Agriculture Science,* vol. 4, no. 1, pp. 1–24.

Gumbel, E. J., 1958, *Statistics of Extremes,* Columbia University Press, New York.

Hamon, W. R., 1961, "Estimating Potential Evapotranspiration," *J. Hydraulics Division, Proc. ASCE,* vol. 87, no. HY3, May, pp. 107–120.

Harbeck, G. E., 1958, *Water-Loss Investigations: Lake Mead Studies,* U.S. Geol. Surv. Prof. Pap. 298, pp. 29–37.

Harbeck, G. E., and J. S. Meyers, 1970, "Present Day Evaporation Measurement Techniques," *Proc. ASCE, J. Hyd. Div.,* vol. 96, no. HY7, pp. 1381–1389.

Hargreaves, G. H., and Z. A. Samani, 1982, "Estimating Potential Evapotranspiration," *J. Irrigation and Drainage Division, Proc. ASCE,* vol. 108, no. IR3, September, pp. 225–230.

Hess, S. L., 1959, *Introduction to Theoretical Meteorology,* Holt, Rinehart and Winston, New York.

Hillel, D., 1982, *Introduction to Soil Physics,* Academic Press, New York.

Horton, R. E., 1933, "The Role of Infiltration in the Hydrologic Cycle," *Trans. Am. Geophys. Union,* vol. 14, pp. 446–460.

Horton, R. E., 1940, "An Approach Towards a Physical Interpretation of Infiltration Capacity," *Proc. Soil Sci. Soc. Am.,* vol. 5, pp. 399–417.

Huber, W. C., J. P. Heaney, S. J. Nix, R. E. Dickinson, and D. J. Polmann, 1981, *Storm Water Management Model User's Manual, Version III,* EPA-600/2-84-109a (NTIS PB84-198423), Environmental Protection Agency, Athens, GA, November.

Humphreys, A. A., and H. L. Abbot, 1861, "Report upon the Physics and Hydraulics of the Mississippi River," *Profess. Papers Corps Topograph. Engrs.,* 1861, Philadelphia, 2nd ed., 1876, Washington, D.C.

Hydrologic Engineering Center, 1973, *HEC-1 Flood Hydrograph Package: User's Manual* and *Programmer's Manual,* updated 1981, United States Army Corps of Engineers, Davis, California.

Hydrologic Engineering Center, 1976, *HEC-2 Water Surface Profiles: User's Manual,* updated 1982, United States Army Corps of Engineers, Davis, California.

Kohler, M. A., T. J. Nordenson, and W. E. Fox, 1955, "Evaporation from Pans and Lakes," U.S. Weather Bureau Res. Paper 38, Washington, D.C.

List, R. J., 1966, *Smithsonian Meteorological Tables,* 6th ed., Smithsonian Institution, Washington, D.C.

Manning, R., 1889, "On the Flow of Water in Open Channels and Pipes," *Trans. Inst. Civil Engrs., Ireland,* vol. 20, pp. 161–207, 1891, Dublin, Ireland; *Supplement,* vol. 24, pp. 179–207, 1895. (The formula was first presented on Dec. 4, 1889, at a meeting of the Institution.)

Marciano, J. J., and G. E. Harbeck, 1954, "Mass Transfer Studies," *Water Loss Investigations: Lake Hefner,* USGS Prof. Paper 269, Washington, D.C.

Mein, R. G., and C. L. Larson, 1973, "Modeling Infiltration During a Steady Rain," *Water Resources Res.,* vol. 9, no. 2, pp. 384–394.

National Weather Service, 1972, *National Weather Service River Forecast System Forecast Procedures,* NOAA Tech. Memo. NWS HYDRO 14, Silver Spring, Maryland.

Penman, H. L., 1948, "Natural Evaporation from Open Water, Bare Soil, and Grass," *Proc. R. Soc. London,* ser. A, vol. 27, pp. 779–787.

Penman, H. L., 1950, "Evaporation over the British Isles," *Quart Jl. Roy. Met. Soc.,* vol. 76, no. 330, pp. 372–383.

Philip, J. R., 1957, "The Theory of Infultration: I. The Infiltration Equation and Its Solution," *Soil Sci.,* vol. 83, pp. 345–357.

Priestly, C. H. B., and R. J. Taylor, 1972, "On the Assessment of Surface Heat Flux and Evaporation Using Large Scale Parameters," *Mon. Weather Rev.,* vol. 100, pp. 81–92.

Richards, L. A., 1931, "Capillary Conduction of Liquids Through Porous Mediums," *Physics,* vol. 1, pp. 318–333.

Roesner, L. A., R. P. Shubinski, and J. A. Aldrich, 1981, *Storm Water Management Model User's Manual, Version III: Addendum I, EXTRAN,* EPA-600/2-84-109b (NTIS PB84-198431), Environmental Protection Agency, Cincinnati, November.

Rubin, J., and R. Steinhardt, 1963, "Soil Water Relations During Rain Infiltration, I: Theory," *Soil Sci. Soc. Amer. Proc.,* vol. 27, pp. 246–251.

Rubin, J., R. Steinhardt, and P. Reiniger, 1964, "Soil Water Relations During Rain Infiltration, II: Moisture Content Profiles During Rains of Low Intensities," *Soil Sci. Soc. Amer. Proc.,* vol. 28, pp. 1–5.

Salisbury, F. B., and C. Ross, 1969, *Plant Physiology,* Wadsworth Publishing Co., Belmont, California.

Sherman, L. K., 1932, "Streamflow from Rainfall by the Unit-Graph Method," *Eng. News-Rec.,* vol. 108, pp. 501–505.

Shih, S. F., L. H. Allen, Jr., L. C. Hammond, J. W. Jones, J. S. Rogers, and A. G. Smajstrla, "Basinwide Water Requirement Estimation in Southern Florida," *Trans. American Soc. Agricultural Engineers,* vol. 26, no. 3, pp. 760–766.

Spreen, W. C., 1947, "Determination of the Effect of Topography upon Precipitation," *Trans. Am. Geophys. Union,* vol. 28, no. 2, pp. 285–290.

Terstriep, M. L., and J. B. Stall, 1974, *The Illinois Urban Drainage Area Simulator, ILLUDAS,* Bull. 85, Illinois State Water Survey, Urbana, Illinois.

Theis, C. V., 1935, "The Relation Between the Lowering of the Piezometric Surface and the Rate and Duration of a Well Using Ground-Water Recharge," *Trans. Am. Geophys. Union,* vol. 16, pp. 519–524.

Thornthwaite, C. W., 1948, "An Approach Toward a Rational Classification of Climate," *Geograph. Rev.,* vol. 38, pp. 55–94.

Turc, L., 1954, *Calcul du Bilan de L'eau Evaluation en Fonction des Précipitations et des Températures,* IASH Rome Symp. 111 Pub No. 38, pp. 188–202.

Wallace, J. M., and P. V. Hobbs, 1977, *Atmosphere Science: An Introductory Survey,* Academic Press, New York.

Weaver, H. A., and J. C. Stephens, 1963, "Relation of Evaporation of Potential Evapotranspiration," *Trans. American Soc. Agricultural Engineers,* vol. 6, pp. 55–56.

2

Rainfall-Runoff Analysis

2.1
RAINFALL-RUNOFF RELATIONSHIPS

When rainfall or snowmelt exceeds the infiltration rate at the surface, excess water begins to accumulate as surface storage in small depressions governed by surface topography. Eventually, overland flow or sheet flow may occur in portions of a watershed, and the flow quickly concentrates into small rivulets or channels, which then flow into larger streams. As mentioned in Section 1.6, subsurface streamflow and base flow can also contribute to the overall discharge hydrograph from a rainfall event.

Hydrologists are concerned with the amount of runoff generated in a watershed for a given rainfall pattern, and attempts have been made to statistically analyze historical rainfall, evaporation, and streamflow data to develop predictive relationships. Factors such as antecedent rainfall, soil moisture, variable infiltration, and variable runoff response with season make the development of such relationships difficult.

A number of investigators have attempted to develop rainfall-runoff relationships that could apply to any region or watershed under any set of conditions. However, these methods must be used with extreme caution because of all of the variable factors that affect the calculation of runoff from a known volume of rainfall. Kohler and Linsley (1951) presented a coaxial relationship involving rainfall duration and the **antecedent pre-cipitation index** (API), which accounts for soil moisture depletion during periods of no rainfall. The Soil Conservation Service (1964) presented a useful set of rainfall-runoff curves that also include land cover, soil type, and initial losses (abstraction) in determining direct runoff (Section 2.4).

The U.S. Geological Survey is responsible for extensive hydrologic gaging networks, and data gathered on an hourly or daily basis can be plotted for a given watershed to relate rainfall to direct runoff for a given year. Annual rainfall-runoff relationships remove seasonal effects and other storage effects so that the relationship of rainfall minus losses can often be approximated by a linear regression line with runoff.

Simple rainfall-runoff relationships should be used in water resources planning studies only where rough water balances are required. A detailed knowledge of the magnitude and time distribution of both rainfall and runoff or streamflow is required for most flood control or floodplain studies, especially in urban watersheds.

One of the simplest rainfall-runoff formulas is called the **rational method,** which allows for the prediction of peak flow Q_p (cfs) from

$$Q_p = CIA,$$

where

    C = runoff coefficient, variable with land use,

    I = intensity of rainfall of chosen frequency for a duration equal to time of concentration t_c (in./hr),

    t_c = time for rainfall at the most remote portion of the basin to travel to outlet (min, hr)

    A = area of watershed (acres)

The rational method is usually attributed to Kuichling (1889) and Lloyd-Davies (1906), but Mulvaney (1851) clearly outlined the procedure in a paper in Ireland. The concept behind the method lies in the assumption that a steady, uniform rainfall rate will produce maximum runoff when all parts of a watershed are contributing to outflow, a condition that is met after the time of concentration t_c has elapsed. For example, Mulvaney points out that a basin with $t_c = 3$ hr would yield a greater runoff with 1 in. falling in 3 hr than with 2 in. falling in 24 hr.

The method has been extended to include nonuniform rainfall and irregular areas through the use of **time-area methods,** which involve a time-area curve indicating the distribution of travel times from different parts of the basin (see Section 2.2). The rational method was the forerun-ner of the concept of the storm hydrograph, and the remainder of this chapter is devoted to the development of hydrograph theory and the application of hydrograph methods for the analysis of complex rainfalls on large watersheds.

2.2

HYDROGRAPH ANALYSIS

Surface Runoff Phenomena

A **hydrograph** is a continuous plot of instantaneous discharge vs. time. It results from a combination of physiographic and meteorologic conditions in a watershed and represents the integrated effects of climate, hydrologic losses, surface runoff, interflow, and ground water flow. Climatic factors that influence the hydrograph shape and volume of runoff include (1) rainfall intensity and pattern, (2) areal distribution of rainfall over the basin, and (3) duration of the storm event. Physiographic factors of importance include (1) size and shape of the drainage area, (2) nature of the stream network, (3) slope of the land and the main channel, and (4) storage detention in the watershed (Sherman, 1932).

During a given rainfall, hydrologic losses such as infiltration, depression storage, and detention storage must first be satisfied before surface runoff begins. As the depth of surface detention increases, overland flow may occur in portions of a basin. Water eventually moves into small rivulets, small channels, and finally the main stream of a watershed. Some of the water that infiltrates the soil may move laterally through upper soil zones until it enters a stream channel. This portion of runoff is called **interflow** or **subsurface stormflow.** Some precipitation may percolate to the water table, usually many feet below the ground surface, and contribute to a stream as base flow if the water table intersects the stream channel. Figure 2.1 illustrates the distribution of a prolonged uniform rainfall.

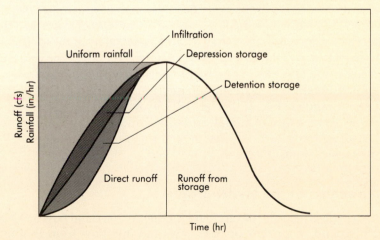

FIGURE 2.1

Distribution of uniform rainfall.

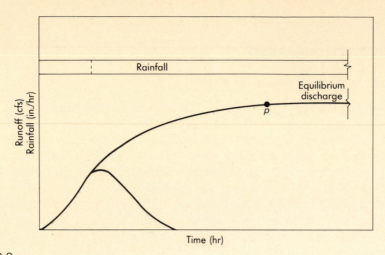

FIGURE 2.2

Equilibrium hydrograph.

The hydrograph consists in most cases of a rising limb, crest segment, and falling portion or recession. If rainfall continues at a constant intensity, then an **equilibrium discharge** can be reached so that inflow and outflow are equal (Fig. 2.2). The point P indicates the time at which the entire discharge area contributes to the flow t_c. The condition of equilibrium discharge is seldom observed in nature, except for very small basins, because of natural variations in rainfall intensity and duration.

Base flow in a natural channel is due to contributions from shallow ground water and is another hydrograph component. In large natural watersheds, base flow may be a significant fraction of streamflow, while it can often be neglected in small, urbanized streams where overland flow predominates. Base flow is separated and subtracted from the total storm hydrograph to derive the **direct runoff** (DRO) **hydrograph.** The volume of water under the DRO hydrograph must correspond to rainfall excess or net rainfall, obtained by deducting all infiltration and storage losses from total storm rainfall (Fig. 2.3). The DRO hydrograph represents the response of the watershed to rainfall excess, with the shape and timing of the DRO hydrograph related to duration and intensity of rainfall as well as physiographic (storage) factors.

Components of the Hydrograph

A hydrograph is made up of various components: overland flow or surface runoff, ground water flow or base flow, and interflow from infiltrated water that is temporarily stored in the soil and laterally enters the stream via upper soil layers. The relative contribution of each component to the

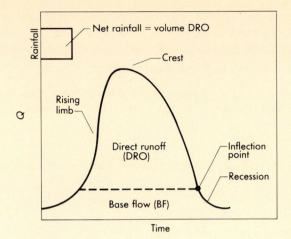

FIGURE 2.3
Hydrograph relations.

hydrograph is dependent on rainfall rate i relative to the infiltration rate f as well as on the level of **soil moisture storage** S_D vs. **field capacity** F, which is defined as the amount of water held in soil after excess gravitational water has drained (Horton, 1935).

No overland runoff occurs for the case $i < f$, and interflow and ground water flow are zero if $F < S_D$ since soil moisture storage still exists unless ground water is coming from long-term storage. The case where both $i > f$ and $F > S_D$ is typical of a large storm event in which direct runoff, interflow, and ground water flow all contribute. Channel precipitation is also a component, although it is usually a very small fraction of total flow.

Surface runoff is an important factor whenever it rains substantially. High-intensity rainfall or urbanization simply magnifies the peak and shortens the time to peak. Interflow may be a large factor in storms of moderate intensity over basins with relatively thin soil covers overlying rock or hardpan. If ground water flows toward the stream during periods of heavy rainfall, the stream is called **effluent.** If flow is from the stream to the ground water system, as in the case of drought conditions, the stream is called **influent.**

In practice, it is customary to consider the total flow to be divided into only two parts: direct runoff (DRO) and base flow (BF), as shown in Fig. 2.3. DRO may include a substantial portion of interflow, whereas base flow is considered to be mostly from ground water flow.

The hydrograph in Fig. 2.3 consists of a rising limb, crest segment, and recession. The slope of the rising limb is largely determined by the storm intensity, and the point of inflection on the recession generally marks the time at which surface inflow ceases and water is thereafter withdrawn from basin storage.

Recession and Base Flow Separation

Several techniques exist to separate DRO from base flow based on the analysis of ground water recession curves. In some cases, the recession curve can be described by an exponential depletion equation of the general form

$$q_t = q_o e^{-kt} \tag{2.1}$$

where

q_o = specified initial discharge,

q_t = discharge at a later time t,

k = recession constant.

Equations of this form are used often in engineering to describe first-order depletion. Equation (2.1) will plot as a straight line on semilogarithmic paper, and the difference between this curve and the total hydrograph plotted on the same paper represents the DRO. In practice, the base flow recession is normally extended back under the inflection point of the hydrograph and then connected to the point at which surface runoff begins (ABD in Fig. 2.4). Several other methods also exist for base flow separation, including a simple horizontal line from point A to point D in Fig. 2.4. These methods all suffer the disadvantage of being quite arbitrary and somewhat inaccurate. Base flow separation is more an art than a science at this time and, in many cases of practical interest such as urban drainage, base flow is often neglected because it represents such a small fraction of total flow. Base flow is usually more important in natural

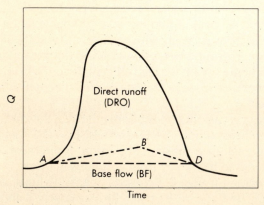

FIGURE 2.4

Base flow separation.

streams and large rivers because of the contribution along the banks from a water table.

Whatever method is selected for separation, the hydrologist should be consistent so that hydrographs can be compared from storm to storm and basin to basin.

Net Storm Rainfall and the Hydrograph

Surface runoff phenomena have been discussed, and Fig. 2.1 presented the distribution of gross rainfall into components of infiltration, depression storage, detention storage, and direct runoff. We can write the continuity equation for this process as

$$\text{Gross rainfall} = \text{Depression storage} + \text{Evaporation} + \text{Infiltration} + \text{Surface runoff,}$$

where detention storage is included as eventual surface runoff. It is often important to determine the time distribution for net storm rainfall, or rainfall excess, which equals direct surface runoff plus detention storage draining out over a longer period of time. Generally, the methods employed to determine rainfall excess include the Horton and Green-Ampt infiltration methods, with initial loss for depression storage, or the infiltration ϕ index method, which is a constant loss rate during the period of storm rainfall. These methods were described in detail in Section 1.5. The method used in the HEC-1 model (see Chapter 5) is an exponential decay function that depends on the rainfall intensity and antecedent losses. In practice, the loss rate coefficients are difficult to estimate, and the simpler ϕ index method is used most often probably because of lack of data on infiltration distribution in time (Fig. 2.5). Note that the ϕ index tends to

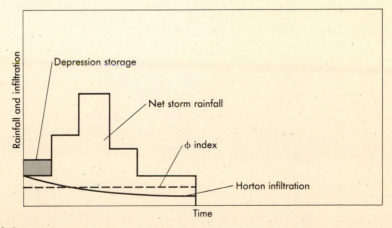

FIGURE 2.5

Infiltration loss curves.

underestimate losses at the beginning of the storm and overestimates losses at the end. Example 2.1 depicts the calculation of net storm rainfall and hydrograph volume.

EXAMPLE 2.1

NET STORM RAINFALL

Rain falls as shown in the rainfall hyetograph (intensity vs. time) in Fig. E2.1(a). The ϕ index for the storm is 0.5 in./hr. Plot the net rainfall on the hydrograph (flow vs. time) given in Fig. E2.1(a). Determine the total volume of runoff over the watershed.

SOLUTION

First we draw the net rainfall hyetograph shown in Fig. E2.1(b). Then this is added to the hydrograph in the upper left corner (Fig. E2.1c). Note that although the net rainfall from time $t = 4$ to $t = 5$ is 0 in./hr, the duration of the rainfall is still considered to be 5 hr.

The volume of runoff is equal to the area under the hydrograph. To determine the volume of runoff, we can use $\Sigma \overline{Q}/dt$. This estimates the

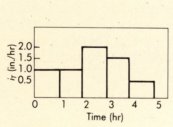

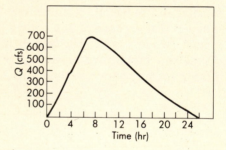

FIGURE E2.1(a)

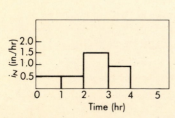

FIGURE E2.1(b)

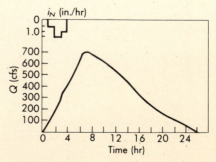

FIGURE E2.1(c)

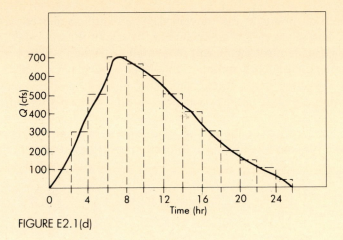

FIGURE E2.1(d)

volume as the bar graph shown in Fig. E2.1(d). Calculations are tabulated in the following table.

TIME (hr)	$\bar{Q}$ (cfs)	VOL (cfs-hr)
0–2	100	200
2–4	300	600
4–6	500	1000
6–8	700	1400
8–10	650	1300
10–12	600	1200
12–14	500	1000
14–16	400	800
16–18	300	600
18–20	200	400
20–22	150	300
22–24	100	200
24–26	50	100

$$\Sigma \bar{Q} \, dt = \boxed{9100 \text{ cfs-hr}}$$
$$= \boxed{9100 \text{ ac-in.}}$$

Other methods that could be used with Fig. E2.1(d) include use of a grid overlay, use of a planimeter with the area under the curve equal to the volume of runoff, and mathematical methods of integration such as Simpson's rule.

Once net rainfall has been determined for a watershed, it then becomes a central problem of engineering hydrology to convert net rainfall into direct surface runoff. The resulting hydrograph is basically built up from contributions of overland flow and channel flow arriving at different times from all points in the watershed. The relative times of travel of overland and channel flow are related to the size of the watershed; overland flow time is more significant in a small watershed whereas time of travel in the channel predominates in a large watershed.

An interesting way to understand how rainfall excess is converted into a hydrograph is to use the concept of the **time-area histogram.** This method assumes that the outflow hydrograph results from pure translation of direct runoff to the outlet, ignoring any storage effects in the watershed. If a rainfall of uniform intensity is distributed over the water-

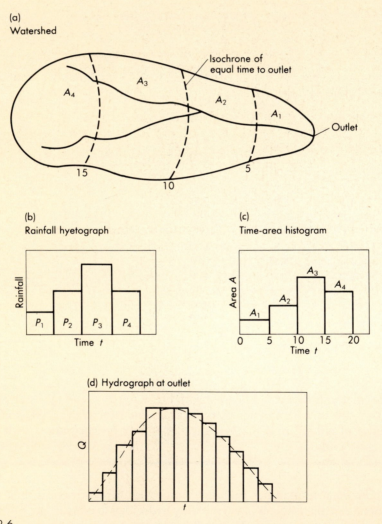

FIGURE 2.6

Time-area method.

shed area, water first flows from areas immediately adjacent to the outlet, and the percentage of total area contributing increases progressively in time. For example, in Fig. 2.6 the surface runoff from area A_1 reaches the outlet first, followed by contributions from A_2, A_3, and A_4, in that order. One can deduce

$$Q_n = R_i A_1 + R_{i-1} A_2 + \cdots + R_1 A_j, \qquad (2.2)$$

where

Q_n = hydrograph ordinate at time n (cfs),

R_i = excess rainfall ordinate at time i (ft/s),

A_j = time-area histogram ordinate at time j (ft^2).

(Note that the number of hyetograph ordinates need not be equal to the number of histogram ordinates.) Runoffs from storm period R_1 on A_3, R_2 on A_2, and R_3 on A_1 arrive at the outlet simultaneously to produce Q_3. The total hydrograph is developed by evaluating Q_1, Q_2, Q_3, $\ldots$, Q_n. The time-area concept provides useful insight into the surface runoff phenomena, but its application is limited because of the difficulty of constructing isochronal lines and because the hydrograph must be further adjusted or routed to represent storage effects in the watershed (see Example 2.2).

A more general concept in actual practice is the theory of the **unit hydrograph,** still recognized as one of the most important contributions to hydrology related to surface runoff prediction. This theory, combined with infiltration methods and flood routing in stream channels and reservoirs, is sufficient to handle input rainfall variability and storage effects in small and large watersheds. It should be noted that the time-area method is a special case of the unit hydrograph approach.

EXAMPLE 2.2

TIME-AREA HISTOGRAM

	A	B	C	D
Area (ac)	1	2	3	1
Time to gage G (hr)	1	2	3	4

A watershed is divided into sections as given in Fig. E2.2(a). Runoff from each section will contribute to flow at gaging station G as shown in the accompanying table. Consider a rainfall intensity of 0.5 in./hr falling uniformly for 5 hr. Assume no losses, and derive a bar hydrograph such as Fig. E2.2(b) for the response of the storm at gage G.

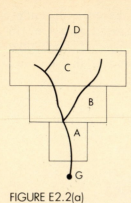

FIGURE E2.2(a)

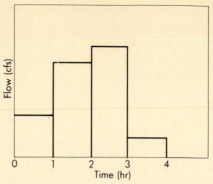

FIGURE E2.2(b)

SOLUTION

The time-area histogram method uses Eq. (2.2):

$$Q_n = R_i A_1 + R_{i-1} A_2 + \cdots + R_1 A_j.$$

For $n = 5$, $i = 5$, and $j = 4$,

$$Q_5 = R_5 A_1 + R_4 A_2 + R_3 A_3 + R_2 A_4$$
$$= (0.5 \text{ in./hr})(1 \text{ ac}) + (0.5 \text{ in./hr})(2 \text{ ac}) + (0.5 \text{ in./hr})(3 \text{ ac})$$
$$+ (0.5 \text{ in./hr})(1 \text{ ac})$$

$$Q_5 = 3.5 \text{ ac-in./hr}.$$

Noting that 1 ac-in./hr $\approx$ 1 cfs, we have

$$Q_5 = 3.5 \text{ cfs}.$$

The resulting hydrograph is given in Fig. E2.2(c) and the accompanying table.

t (hr)	Q (cfs)
0	0
1	0.5
2	1.5
3	3.0
4	3.5
5	3.5
6	3.0
7	2.0
8	0.5
9	0

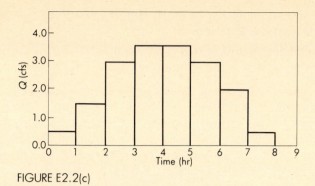

FIGURE E2.2(c)

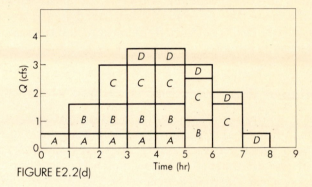

FIGURE E2.2(d)

The contribution of each subarea can be better visualized by indicating the portion of flow resulting from each area, as done in Fig. E2.2(d).

2.3

UNIT HYDROGRAPH THEORY

Sherman (1932) originally advanced the theory of the unit hydrograph (UH), defined as "basin outflow resulting from one inch (one centimeter) of direct runoff generated uniformly over the drainage area at a uniform rainfall rate during a specified period of rainfall duration." Several assumptions inherent in the unit hydrograph tend to limit its application (Johnstone and Cross, 1949) for a given watershed:

1. Rainfall excesses of equal duration are assumed to produce hydrographs with equivalent time bases regardless of the intensity of the rain.

2. Direct runoff ordinates for a storm of given duration are assumed directly proportional to rainfall excess volumes. Thus, twice the rainfall produces a doubling of hydrograph ordinates.

3. The time distribution of direct runoff is assumed independent of antecedent precipitation.

4. Rainfall distribution is assumed to be the same for all storms of equal duration, spatially and temporally.

The classic statement of unit hydrograph theory can be summarized briefly: the hydrologic system is linear and time-invariant (Dooge, 1973). The property of proportionality and the principle of superposition both apply to the unit hydrograph but were not seriously questioned until the mid-1950s. While the assumptions of linearity and time invariance are not strictly correct for a watershed, we adopt them as long as they are useful. Nonlinear examples exist in open channel flow, laboratory runoff models, and field demonstrations. However, the linear assumptions are still useful because they are relatively simple and are the best-developed methods, and the results obtained from linear methods are acceptable for most engineering purposes.

Derivation of Unit Hydrographs: Gaged Watersheds

A typical storm hydrograph and rainfall **hyetograph** for a drainage basin are shown in Fig. 2.7(a). The storm hydrograph is a plot of discharge vs. time. Timing parameters such as rainfall excess **duration** D and **time to peak** t_p are also illustrated. The hydrograph is characterized by a **rising limb, crest segment,** and **recession curve** (see Fig. 2.3). The timing aspects of the hydrograph are characterized by the following parameters:

1. **Lag time** (L or t_p): time from the center of mass of rainfall excess to the peak of the hydrograph

2. **Time of rise** (T_R): time from the start of rainfall excess to the peak of the hydrograph

3. **Time of concentration** (t_c): time of equilibrium of the watershed, where outflow is equal to net inflow, also the time for a wave to propagate from the most distant point in the watershed to the outlet.

4. **Time base** (T_b): total duration of the direct runoff (DRO) hydrograph.

The following general rules should be observed in developing unit hydrographs from gaged watersheds:

1. Storms should be selected with a simple structure with relatively uniform spatial and temporal distributions.

2. Watershed sizes should generally fall between 1000 ac and 1000 mi^2.

3. Direct runoff should range from 0.5 to 2.0 in.

4. Duration of rainfall excess D should be approximately 25–30% of lag time t_p.

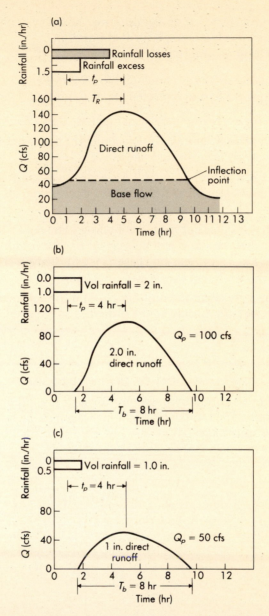

FIGURE 2.7

Unit hydrograph determination. (a) Total storm hydrograph. (b) Hydrograph minus base flow, rainfall minus losses. (c) Hydrograph adjusted to be a 2-hr unit hydrograph.

5. A number of storms of similar duration should be analyzed to obtain an average unit hydrograph for that duration.

6. Step 5 should be repeated for several different durations.

The following are the essential steps for developing a unit hydrograph from a single storm hydrograph (see Fig. 2.7 and Example 2.3):

1. Analyze the hydrograph and separate base flow (Section 2.2).

2. Measure the total volume of DRO under the hydrograph and convert this to inches (cm) over the watershed.

3. Convert total rainfall to rainfall excess through infiltration methods and evaluate duration D for the DRO hydrograph and the unit hydrograph.

4. Divide the ordinates of the DRO hydrograph (Fig. 2.7b) by the volume in inches (cm) and plot these results as the unit hydrograph for the basin (Fig. 2.7c). The time base T_b is assumed constant for storms of equal duration and thus it will not change.

5. Check the volume of the unit hydrograph to make sure it is 1.0 in. (1.0 cm), and graphically adjust ordinates as required.

EXAMPLE 2.3

DETERMINATION OF UNIT HYDROGRAPH

Convert the direct runoff hydrograph shown in Fig. E2.3(a) into a 2-hr unit hydrograph. The rainfall hyetograph is given in the figure and the ϕ index for the storm was 0.5 in./hr. The base flow in the channel was 100 cfs (constant). What are t_p and T_B for the storm?

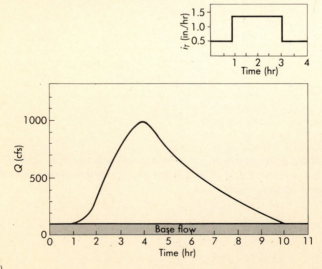

FIGURE E2.3(a)

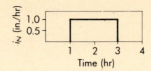

FIGURE E2.3(b)

SOLUTION

First we find the net rainfall. After subtracting the ϕ index, we plot the rainfall excess hyetograph shown in Fig. E2.3(b). This represents 2.0 in. of rainfall or 1 in./hr for 2 hr. Then we subtract base flow from all the flow values. Finally, the hydrograph must be converted to 1 in. of direct runoff over the watershed, or 0.5 in./hr of rain for 2 hr. To do so, we take each ordinate minus its base flow and multiply it by 1/2. This entire procedure is tabulated as follows.

TIME (hr)	Q (cfs)	$Q - $ BF (cfs)	2-hr UH, Q
0	0	0	0
1	100	0	0
2	300	200	100
3	700	600	300
4	1000	900	450
5	800	700	350
6	600	500	250
7	400	300	150
8	300	200	100
9	200	100	50
10	100	0	0
11	0	0	0

The 2-hr UH graphs as shown in Fig. E2.3(c). T_B, the time base of the storm, is

$$10 - 1 = \boxed{9 \text{ hr}}$$

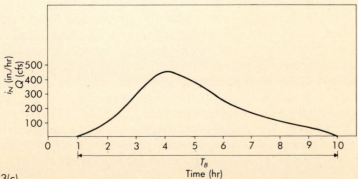

FIGURE E2.3(c)

and t_p, the time to peak, is measured from the center of mass of rainfall to the peak:

$$4 - 2 = \boxed{2 \text{ hr.}}$$

We emphasize that because of assumptions of linearity inherent in the unit hydrograph development, care must be used in applying unit hydrographs under conditions that tend to violate linearity. Large variations in duration and intensity of rainfall are reflected in the shape of the resulting hydrograph. Short-duration unit hydrographs are often added and lagged together to generate long-duration (24-hr) storm hydrographs. If intensity variations are large over the long-duration storm, assumptions of linearity may be violated.

Amount of runoff and areal distribution of runoff can cause variations in the shape of the hydrograph. For large basins, it is usually best to develop unit hydrographs for subareas and then add and lag them together to generate an overall hydrograph. Peaks of unit hydrographs derived from very small events are often lower than those derived from larger storms because of differences in interflow and channel flow times. Assumptions of linearity can usually be verified by comparing hydrographs from storms of various magnitudes. If nonlinearity does exist, then derived unit hydrographs should be used only for generating events of similar magnitude, and caution should be exercised in using unit hydrographs to extrapolate extreme events.

Despite these limitations, when unit hydrographs have been used in conjunction with modern flood routing methods, resulting flood prediction for small and large basins has become quite accurate over the past decade. Detailed applications are presented in Chapter 5.

S-Curve Method

A unit hydrograph for a particular watershed is defined for a specific duration of precipitation excess. The linear property of the unit hydrograph can be used to generate a unit hydrograph of larger or smaller duration. For example, given a 1-hr unit hydrograph for a particular watershed, a unit hydrograph resulting from a 2-hr unit storm can be developed by adding two 1-hr unit hydrographs, the second one lagged by 1 hr, adding ordinates, and dividing the result by 2. In this way the 1 in. of rainfall in 1 hr has been distributed uniformly over 2 hr in deriving the 2-hr unit hydrograph (Fig. 2.8). This lagging procedure is restricted to integer multiples of the original duration.

The **S-curve method** allows construction of a unit hydrograph of any duration. Assume that a unit hydrograph of duration D is known and that we wish to generate a unit hydrograph for the same watershed with dura-

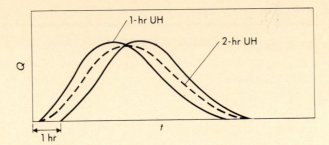

FIGURE 2.8

Unit hydrograph lagging.

tion D'. The first step is to generate the S-curve hydrograph by adding a series of unit hydrographs of duration D, each lagged by D (Fig. 2.9, curve a). This corresponds to the runoff hydrograph resulting from a continuous rainfall excess intensity of $1/D$ in./hr, where D is in hr.

By shifting the S-curve in time by D' hr and subtracting ordinates between the two S-curves, the resulting hydrograph (Fig. 2.9, curve b) must be due to rainfall of $1/D$ in./hr that occurs for D' hr. Thus, to convert curve b to a unit hydrograph, we must multiply all the hydrograph ordinates by D/D', resulting in the unit hydrograph of duration D'. A tabular procedure is presented in Example 2.4 for a specific case of generating a 3-hr unit hydrograph from a 2-hr unit hydrograph. A computer program for generating S-curves is contained in Appendix E.

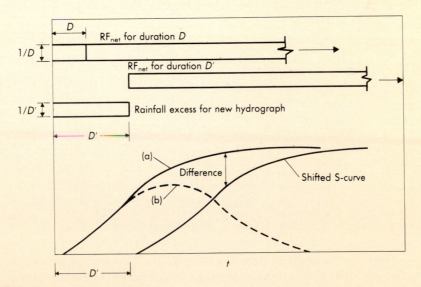

FIGURE 2.9

S-curve hydrograph method.

EXAMPLE 2.4

S-CURVE METHOD

Convert the following tabulated 2-hr unit hydrograph to a 3-hr unit hydrograph using the S-curve method.

TIME (hr)	2-HR UH ORDINATE (cfs)
0	0
1	75
2	250
3	300
4	275
5	200
6	100
7	75
8	50
9	25
10	0

SOLUTION

First, the 2-hour unit hydrograph must be used to create the S-curve. This is done by lagging the 2-hr curve by 2 hr and adding this to the original curve. The S-curve represents an infinite number of these additions. However, as the following tabulation shows, it is generally necessary to repeat the process only until a relatively constant value is reached. (Oscillation sometimes occurs.)

TIME (hr)	2-HR UH	2-HR LAGGED UH'S				SUM
0	0					0
1	75					75
2	250	0				250
3	300	75				375
4	275	250	0			525
5	200	300	75			575
6	100	275	250	0		625
7	75	200	300	75		650
8	50	100	275	250	0	675
9	25	75	200	300	75	675
10	0	50	100	275	250	0	675
11		25	75	200	300	75	675

Once the S-curve has been constructed, the 3-hr unit graph can be derived from it. This is done by lagging the S-curve by 3 hr and subtracting the lagged S-curve from the original S-curve. These values must then be multiplied by the ratio of the duration D of the original unit hydrograph to the duration D' of the desired unit hydrograph, or D/D'. This step gives the ordinates of the desired unit hydrograph. Here,

$D/D' = 2/3$.

This procedure is done graphically in Figs. E2.4(a), (b), and (c).

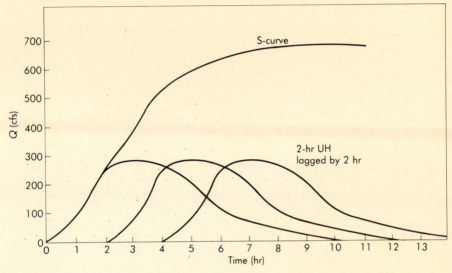

FIGURE E2.4(a)

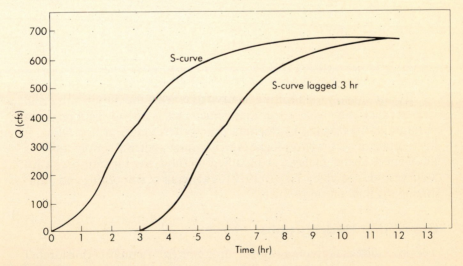

FIGURE E2.4(b)

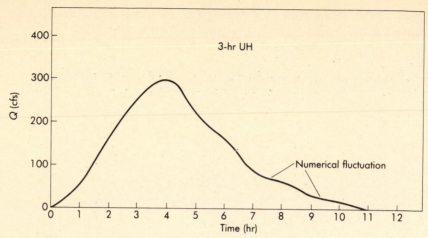

FIGURE E2.4(c)

TIME (hr)	S-CURVE ORDINATE	S-CURVE LAGGED 3 HR	DIFFERENCE	3-HR UH ORDINATE
0	0		0	0
1	75		75	50.0
2	250		250	166.7
3	375	0	375	250.0
4	525	75	450	300.0
5	575	250	325	216.7
6	625	375	250	166.7
7	650	525	125	83.3
8	675	575	100	66.7
9	675	625	50	33.3
10	675	650	25	16.7
11	675	675	0	0

It is not uncommon for the S-curve to reach a maximum value and then oscillate somewhat about that value. This is due to slight errors in duration of the original hydrograph or to other nonlinearities. Often, if fluctuations arise, the hydrologist exercises judgment and simply smooths the S-curve in the analysis. A modified method, the forward-backward S-curve, is discussed by Tauxe (1978) as a means of smoothing and minimizing the fluctuation problem.

Use of Unit Hydrographs

The procedure of deriving a storm hydrograph from a multiperiod rainfall excess is illustrated in Example 2.5 and represents a more general case of the time-area method (Eq. 2.2). Unit hydrograph ordinates are multiplied by rainfall excess and added and lagged in sequence to produce the resulting storm hydrograph. Each unit hydrograph is added at the time corresponding to the rainfall spike that produced a response. Base flow could be added to produce a more realistic storm hydrograph if typical base flow values are available for the watershed under study. Care should be taken in this calculation to ensure that the time increments of rainfall excess correspond to the duration of the unit hydrograph. The governing equation for the storm hydrograph in discrete form is called the **convolution equation:**

$$Q_n = \sum_{i=1}^{n} P_i U_{n-i+1}, \quad \text{or}$$
$$Q_n = P_n U_1 + P_{n-1} U_2 + P_{n-2} U_3 + \cdots + P_1 U_n, \tag{2.3}$$

where Q_n is the storm hydrograph ordinate, P_i is rainfall excess, and U_j ($j = n - i + 1$) is the unit hydrograph ordinate. Periods of no rainfall can also be included as shown in Example 2.5.

EXAMPLE 2.5

STORM HYDROGRAPH FROM THE UNIT HYDROGRAPH

Given the rainfall excess hyetograph and the 1 hr unit hydrograph in Fig. E2.5(a), derive the storm hydrograph for the watershed for the event. Assume no losses to infiltration or evapotranspiration.

SOLUTION

Equation (2.3) states:

$$Q_n = P_n U_1 + P_{n-1} U_2 + P_{n-2} U_3 + \cdots + P_1 U_n.$$

Here,

$$P_n = [0.5, 1.0, 1.5, 0.0, 0.5] \text{ in.}$$

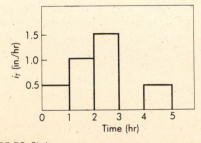 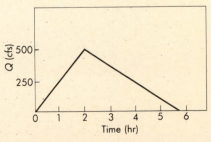

FIGURE E2.5(a)

from the rainfall hyetograph. Using intervals of 1 hr, we have

$U_n = [0, 250, 500, 375, 250, 125, 0]$ cfs

from the unit hydrograph. By Eq. (2.3),

$Q_0 = (0.5)(0) = 0,$

$Q_1 = (1.0)(0) + (0.5)(250) = 125$ cfs,

$Q_2 = (1.5)(0) + (1.0)(250) + (0.5)(500) = 500$ cfs, . . .

The value derived from these calculations can be tabulated as follows:

TIME (n hr)	$P_1 U_n$	$P_2 U_n$	$P_3 U_n$	$P_4 U_n$	$P_5 U_n$	Q_n
0	0					0
1	125	0				125
2	250	250	0			500
3	187.5	500	375	0		1062.5
4	125	375	750	0	0	1250
5	62.5	250	562.5	0	125	1000
6	0	125	375	0	250	750
7		0	187.5	0	187.5	375
8			0	0	125	125
9				0	62.5	62.5
10					0	0

Q_n is equal to the sum across the row for time n. The resulting storm hydrograph, Q_n, can be plotted as in Fig. E2.5(b).

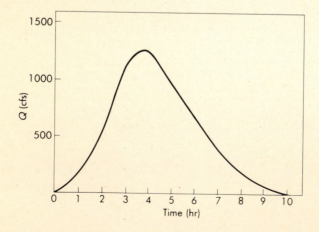

FIGURE E2.5(b)

The reverse procedure allows us to determine a unit hydrograph from a direct storm hydrograph produced by a multiperiod rainfall excess. For a four-period rainfall (Fig. 2.10),

$$Q_1 = P_1 U_1,$$
$$Q_2 = P_2 U_1 + P_1 U_2,$$
$$Q_3 = P_3 U_1 + P_2 U_2 + P_1 U_3,$$
$$Q_4 = P_4 U_1 + P_3 U_2 + P_2 U_3 + P_1 U_4,$$
$$Q_5 = P_4 U_2 + P_3 U_3 + P_2 U_4 + P_1 U_5,$$
$$Q_6 = P_4 U_3 + P_3 U_4 + P_2 U_5 + P_1 U_6,$$
$$Q_7 = P_4 U_4 + P_3 U_5 + P_2 U_6 + P_1 U_7,$$
$$Q_8 = P_4 U_5 + P_3 U_6 + P_2 U_7,$$
$$Q_9 = P_4 U_6 + P_3 U_7,$$
$$Q_{10} = P_4 U_7. \tag{2.4}$$

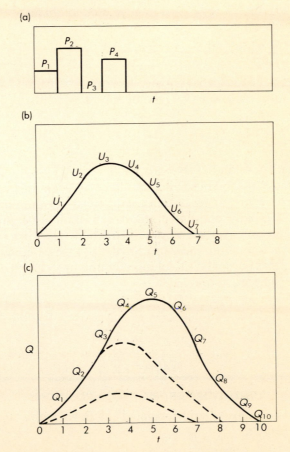

FIGURE 2.10
Unit hydrograph from multiperiod storm.

The relationship between the number of storm hydrograph periods n, the unit hydrograph ordinates j, and the rainfall excess periods i is

$$n = j + i - 1. \tag{2.5}$$

The general equation (Eq. 2.4) can be expressed in a more compressed form as

$$[Q] = [P][U], \tag{2.6}$$

where $[P]$ is a matrix that can be solved for $[U]$. Such a solution requires the inverse matrix $[P]^{-1}$, which must be a square matrix with a nonzero determinant. The matrix $[P]$ does not meet this condition; but by using the transpose matrix $[P^T]$, a square symmetrical matrix $[P^T P]$ can be generated. The matrix $[P^T P]$ can then be used to obtain

$$[P^T P][U] = [P^T][Q]. \tag{2.7}$$

This equation generates simultaneous normal equations for a least-squares solution. Equation (2.7) can be solved explicitly for $[U]$ using

$$[U] = [P^T P]^{-1}[P^T][Q] \tag{2.8}$$

The procedure developed by Newton and Vineyard (1967) solves Eq. 2.6 using the Gauss method of solving simultaneous equations. As with any numerical procedure, under certain conditions the error in the values of $[U]$ can grow rapidly, and unreal values can result, especially for later ordinates of the unit hydrograph. Several methods have been proposed to overcome this limitation, including Snyder's (1955) method. Example 2.6 demonstrates the analytical solution for unit hydrograph decomposition from a complex storm.

EXAMPLE 2.6

UNIT HYDROGRAPH DECOMPOSITION

For the total storm hydrograph tabulated as follows, with a constant base flow (BF) of 10 cfs,

a) find the 1-hr unit hydrograph (UH),

b) determine the size of the drainage basin.

TIME (hr)	NET RAINFALL (in.)	Q (cfs)
0	1.0	10
1	2.0	20
2	0.0	130
3	1.0	410
4		570
5		510
6		460
7		260
8		110
9		60
10		10

SOLUTION

a) First, we separate base flow from the total storm hydrograph by simply subtracting 10 cfs from Q since the base flow is constant.

TIME (hr)	$Q -$ BF (cfs)
0	0
1	10
2	120
3	400
4	560
5	500
6	450
7	250
8	100
9	50
10	0

Noting that $i = 4$, $n = 11$, and $n = j + i - 1$, where n indicates the number of storm hydrograph ordinates, i the number of rainfall excess periods, and j the number of unit hydrograph ordinates, we have

$$11 = j + 4 - 1, \quad \text{or}$$
$$j = 8.$$

Expanding Eq. (2.3) yields the following set of equations:

$$Q_1 = P_1 U_1,$$
$$Q_2 = P_2 U_1 + P_1 U_2,$$
$$Q_3 = P_3 U_1 + P_2 U_2 + P_1 U_3,$$
$$Q_4 = P_4 U_1 + P_3 U_2 + P_2 U_3 + P_1 U_4,$$
$$Q_5 = P_4 U_2 + P_3 U_3 + P_2 U_4 + P_1 U_5,$$

$$Q_6 = P_4U_3 + P_3U_4 + P_2U_5 + P_1U_6,$$
$$Q_7 = P_4U_4 + P_3U_5 + P_2U_6 + P_1U_7,$$
$$Q_8 = P_4U_5 + P_3U_6 + P_2U_7 + P_1U_8,$$
$$Q_9 = P_4U_6 + P_3U_7 + P_2U_8,$$
$$Q_{10} = P_4U_7 + P_3U_8,$$
$$Q_{11} = P_4U_8.$$

Solving each equation in order gives the following values for the 1-hr unit hydrograph ordinates. Note that it is necessary to solve only the first eight equations. The remaining equations may be used to check the solutions. The 1-hr UH becomes

j	TIME (hr)	UH (cfs)
1	0	0
2	1	10
3	2	100
4	3	200
5	4	150
6	5	100
7	6	50
8	7	0

b) The area under the unit hydrograph of Fig. E2.6 can be found by calculating the integral

$$\int_0^7 u(t)\,dt$$

or approximated by the sum

$$\sum_{i=1}^{7} u_i\Delta t,$$

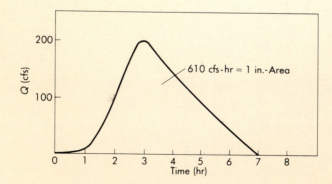

FIGURE E2.6

where $\Delta t = 1$ hr. Using the values found in part (a), we have

$$\sum_{i=1}^{7} u_i \Delta t = 610 \text{ cfs-hr.}$$

By definition of the unit hydrograph, the value of the area under the unit hydrograph represents 1 in. of direct runoff over the watershed. Thus, the area of the basin if

$$610 \text{ cfs-hr} = (1 \text{ in.})(\text{Area } A),$$

$$A = \left(\frac{610 \text{ cfs-hr}}{1 \text{ in.}}\right)\left(\frac{12 \text{ in.}}{\text{ft}}\right)\left(\frac{3600 \text{ s}}{\text{hr}}\right)\left(\frac{1 \text{ ac}}{43,560 \text{ ft}^2}\right),$$

$$\boxed{A = 605 \text{ ac.}}$$

2.4

SYNTHETIC UNIT HYDROGRAPH DEVELOPMENT

While the procedures presented in Section 2.3 for deriving unit hydrographs from gaged watersheds have been applied with success, streamflow and rainfall data are available for relatively few watersheds. Hence, methods for deriving unit hydrographs for ungaged basins have evolved based on theoretical or empirical formulas relating hydrograph peak flow and timing to basin characteristics. These are usually referred to as **synthetic unit hydrographs** and offer the hydrologist or engineer a multitude of methods for developing a unit hydrograph for a particular basin. However, the formulas all have certain limiting assumptions and should be applied to new areas with extreme caution. Some calibration to adjacent watersheds where streamflow gages exist should be attempted if possible.

Synthetic unit hydrographs developed along two main lines; one assumed that each watershed had a unique unit hydrograph, and the second assumed that all unit hydrographs could be represented by a single family of curves or a single equation.

The first line of development was based on the rational method modified to include the time-area curve (Section 2.2) for a particular watershed. Clark (1945) assumed that watershed response would be given by routing the time-area curve through an element of linear storage (see Chapter 4), which tends to attenuate and time-lag the hydrograph. Each unit hydrograph would be unique for a watershed, and this method thus represented a significant improvement over the time-area method.

The second approach to the unit hydrograph (UH) development assumed mathematical representations for the shape of the UH. The most popular one was advanced by the Soil Conservation Service (1964), which represented a dimensionless UH of discharge vs. time by a gamma function (discussed later in this section). Since volume is fixed, only one parameter is required to determine the entire UH, either the t_p or the peak flow rate.

In the late 1950s, both approaches began to converge after the important contributions of O'Kelly (1955) and Nash (1958, 1959). O'Kelly's work was based on replacing the time-area curve by a UH in the shape of an isosceles triangle. Nash later suggested that the two-parameter gamma distribution gave the general shape of an **instantaneous unit hydrograph** as the output from a cascade of linear reservoirs. (A linear reservoir is a reservoir with outflow that is linearly proportional to the storage volume.) Gray (1962) based his UH method on the same gamma distribution.

Empirical expressions were necessary to transform the methods of Clark (1945) into usable UH techniques for actual basins. Johnstone and Cross (1949) proposed one of the first relationships for t_c, the time of concentration in hours,

$$t_c = 5.0(L/\sqrt{S})^{0.5}, \tag{2.9}$$

and for **storage delay time** K, in hours,

$$K = 1.5 + 90(A/LR)$$

where L is the length of the main stream in mi, A is the area in mi^2, S is the slope in ft/mi, and R is an overland slope factor.

Investigators have realized that a number of parameters are important in determining the shape and timing of the UH for a watershed. The lag time to peak t_p, the time of rise t_R, and the time of concentration t_c are often used (Section 2.3). The time base T_b is included to define the duration of direct runoff. These timing parameters must be related to watershed characteristics in developing a synthetic unit hydrograph.

The discharge parameter most often used is the peak discharge Q_p. A **routing parameter** K is sometimes included when the hydrograph has been routed through a linear reservoir with storage delay time K. The K value also characterizes the recession portion of the UH when it can be represented as a declining exponential, i.e., $K = k^{-1}$ of Eq. (2.1). Watershed parameters of most concern include area A, main channel length L, length to watershed centroid L_c, and slope of main channel S.

Assumptions and Modifying Factors

Most synthetic unit hydrograph methods assume that the unit hydrograph of a basin represents the integrated effect of size, slope, shape, and storage characteristics. As long as these factors are constant between two basins and do not vary with time, the unit response will be identical for the two basins. For two basins of the same size, if the slope of one is greater or if the basin shape of one is more concentrated (if the length/width ratio is lower), the shape of the hydrograph will shift, as shown by the shift from a

to b in Fig. 2.11. Storage characteristics of the basin relate to slope, soil type, topography, channel resistance, and shape. If upstream reservoir storage is combined with downstream channelization, the unit hydrograph may shift to curve b due to a more rapid rise time from concrete channels and less infiltration loss from urban development. Finally, if an elongated basin were fully developed with no upstream reservoir storage, one might expect a shift to curve c, where time of rise is much shorter and peak flow is much greater than in the original unit hydrograph a. Such a case is shown in Fig. 2.12 for an actual developing watershed, Brays Bayou in Houston, Texas. Earlier synthetic hydrograph methods generally do not consider urbanization effects, but more modern empirical formulas usually account for urban channels, percent impervious (paved) area, or percent storm-sewered area.

Most of the methods for synthetic unit hydrographs relate lag time t_p or time of rise t_R of the hydrograph to measures of length and shape of the main channel in the basin. Some methods also relate timing to the inverse of the slope of the main channel or land. Thus, the longer the basin and the

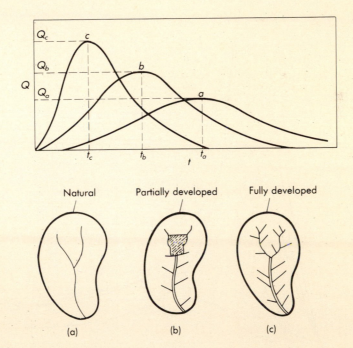

FIGURE 2.11

Modifying factors on unit hydrographs. (a) Natural watershed development, represented by curve a in the top part of the figure. (b) Partial development, represented by curve b. (c) Fully developed watershed, represented by curve c.

smaller the slope, the longer the time of rise of the hydrograph, as expected.

A second relation is usually presented between peak flow Q_p and area of basin and between Q_p and the inverse of the t_p or t_R of the hydrograph, thus indicating that larger areas produce higher Q_p. From continuity, the higher the peak flow, the smaller t_p must be to keep the volume of the unit hydrograph as 1.0 in. (1.0 cm) of direct runoff. This is easily seen by referring to Fig. 2.11, where curves a, b, and c are all unit hydrographs.

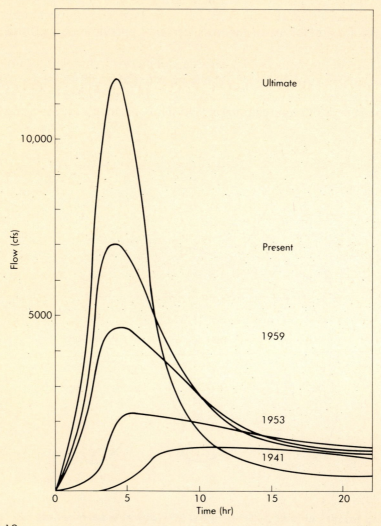

FIGURE 2.12

Brays Bayou unit hydrograph changes as a function of land use changes.

The following sections describe a few of the more popular synthetic unit hydrograph methods.

Snyder's Method

Snyder (1938) was the first to develop a synthetic unit hydrograph based on a study of watersheds in the Appalachian Highlands. In basins ranging from 10 to 10,000 mi^2, Snyder's relations are

$$t_p = C_t(LL_c)^{0.3}, \tag{2.10}$$

where

t_p = basin lag (hr),

L = length of the main stream from the outlet to the divide (mi),

L_c = length along the main stream to a point nearest the watershed centroid (mi),

C_t = coefficient usually ranging from 1.8 to 2.2 (C_t has been found to vary from 0.4 in mountainous areas to 8.0 along the Gulf of Mexico),

$$Q_p = 640C_pA/t_p, \tag{2.11}$$

where

Q_p = peak discharge of the unit hydrograph (cfs),

A = drainage area (mi^2),

C_p = storage coefficient ranging from 0.4 to 0.8 where larger values of C_p are associated with smaller values of C_t,

$$T_b = 3 + t_p/8, \tag{2.12}$$

where T_b is the time base of the hydrograph, in days. For small watersheds, Eq. (2.12) should be replaced by multiplying t_p by a value of from 3 to 5 as a better estimate of T_b. Equations (2.10), (2.11), and (2.12) define points for a unit hydrograph produced by an excess rainfall of duration $D = t_p/5.5$. For other rainfall excess durations D', an adjusted formula for t_p becomes

$$t_p' = t_p + 0.25 (D' - D), \tag{2.13}$$

where t_p' is the adjusted lag time (hr) for duration D' (hr). Once the three quantities t_p, Q_p, and T_b are known, the unit hydrograph can be sketched so that the area under the curve represents 1.0 in. of direct runoff from the watershed. Taylor and Schwartz (1952) performed a similar study on watersheds ranging in size from 20 to 1600 mi^2 in the Atlantic states.

Caution should be used in applying Snyder's method to a new area without first deriving coefficients for gaged streams in the general vicinity of the problem basin. The coefficients C_t and C_p have been found to vary considerably from one region to another. Example 2.7 illustrates Snyder's method.

EXAMPLE 2.7

SNYDER'S METHOD

Use Snyder's method to develop a unit hydrograph for the area of 100 mi² described below. Sketch the approximate shape. What duration rainfall does this correspond to?

$$C_t = 1.8, \qquad L = 18 \text{ mi},$$
$$C_p = 0.6, \qquad L_c = 10 \text{ mi}$$

SOLUTION

The unit hydrograph is sketched in Fig. E2.7. By Eq. (2.10),

$$t_p = C_t(LL_c)^{0.3}$$
$$t_p = 1.8(18 \cdot 10)^{0.3} \text{ hr}$$
$$\boxed{t_p = 8.6 \text{ hr.}}$$

By Eq. (2.11),

$$Q_p = 640(C_p)(A)/t_p$$
$$= 640(0.6)(100)/8.6$$
$$\boxed{Q_p = 4465 \text{ cfs.}}$$

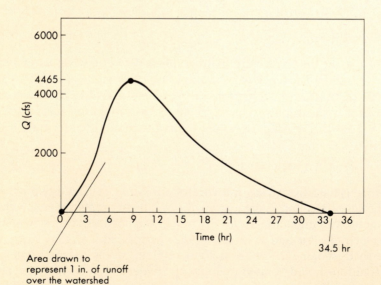

FIGURE E2.7

Since this is a small watershed,

$$T_b \approx 4t_p = 4(8.6) \text{ hr}$$

$$\boxed{T_b = 34.4 \text{ hr.}}$$

And the duration of rainfall

$$D = t_p/5.5$$
$$= 8.6/5.5 \text{ hr}$$

$$\boxed{D = 1.6 \text{ hr.}}$$

SCS Method

The method developed by the Soil Conservation Service (SCS, 1957) is based on a **dimensionless hydrograph,** developed from a large number of unit hydrographs ranging in size and geographic location. The hydrograph is represented as a simple triangle (Fig. 2.13), with rainfall duration D (hr), time of rise T_R (hr), time of fall B(hr), and peak flow Q_p (cfs). The volume of direct runoff is

$$\text{Vol} = \frac{Q_p T_R}{2} + \frac{Q_p B}{2}, \qquad \text{or}$$

$$Q_p = \frac{2 \text{ Vol}}{T_R + B} \tag{2.14}$$

From a review of a large number of hydrographs, it was found that

$$B = 1.67 \, T_R. \tag{2.15}$$

Therefore, Eq. (2.14) becomes, for 1.0 in. of rainfall excess,

$$\begin{aligned}
Q_p &= \frac{0.75 \text{ Vol}}{T_R} \\
&= \frac{(0.75)(640)A(1.008)}{T_R} \\
&= \frac{484A}{T_R}, \tag{2.16}
\end{aligned}$$

where

$A = $ area of basin (sq mi),

$T_R = $ time of rise (hr).

Capece et al. (1984) found that a factor as low as 10–50 holds for flat, high-water-table watersheds rather than the value 484 presented here. Thus, care must be used when applying this method.

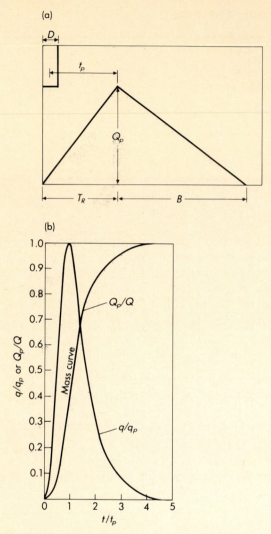

FIGURE 2.13

(a) SCS triangular unit hydrograph. (b) SCS dimensionless unit hydrograph (SCS, 1964).

From Fig. 2.13 it can be shown that

$$T_R = D/2 + t_p, \qquad (2.17)$$

where

D = rainfall duration (hr),

t_p = lag time from centroid of rainfall to Q_p (hr).

Lag time t_p is estimated from any one of several empirical equations used by the SCS, such as

$$t_p = \frac{\ell^{0.8}(S+1)^{0.7}}{1900y^{0.5}},$$

(2.18)

where

t_p = lag time (hr),
ℓ = length to divide (ft),
y = average watershed slope (%),
S = 1000/CN − 10,
CN = curve number for various soil/land use (see Table 2.1).

The SCS dimensionless unit hydrograph can be used to develop a curved hydrograph, using the same t_p and Q_p as the triangular hydrograph in Fig. 2.13.

Soil Conservation Service (1964) runoff estimates assume a relationship between accumulated total storm rainfall P, runoff Q, and infiltration plus initial abstraction $(F + I_a)$. It is assumed that

$$F/S = Q/P_e,$$

(2.19)

where F is infiltration occurring after runoff begins, S is **potential abstraction,** Q is direct runoff in in., and P_e is effective storm runoff $(P - I_a)$. With $F = (P_e - Q)$ and $P_e = (P - I_a) = (P - 0.2S)$ based on data from small watersheds,

$$Q = \frac{(P - 0.2S)^2}{P + 0.8S}.$$

(2.20)

The SCS method uses the **runoff curve number** CN, which is related to storage by

$$CN = 1000/(S + 10),$$

(2.21)

where potential abstraction S (in.) becomes

$$S = (1000/CN) - 10.$$

(2.22)

Figure 2.14 presents the SCS solution in graphical form for a range of CNs and rainfall amounts. Runoff curve numbers for selected land uses are presented in Table 2.1. The hydrologic soil group varies from A for sandy, well-drained soils to D for clayey, poorly drained soils. The SCS report *Urban Hydrology for Small Watersheds* (1986) provides a simple graphical and tabular procedure for determining peak flows for urban areas. Example 2.8 illustrates the SCS UH method.

TABLE 2.1

Runoff Curve Numbers for Selected Agricultural, Suburban, and Urban Land Use (Antecedent moisture condition II; $I_a = 0.2S$)

LAND USE DESCRIPTION	HYDROLOGIC SOIL GROUP			
	A	B	C	D
Cultivated land[1]				
Without conservation treatment	72	81	88	91
With conservation treatment	62	71	78	81
Pasture or range land				
Poor condition	68	79	86	89
Good condition	39	61	74	80
Meadow				
Good condition	30	58	71	78
Wood or forest land				
Thin stand, poor cover, no mulch	45	66	77	83
Good cover[2]	25	55	70	77
Open spaces, lawns, parks, golf courses, cemeteries, etc.				
Good condition: grass cover on 75% or more of the area	39	61	74	80
Fair condition: grass cover on 50–75% of the area	49	69	79	84
Commercial and business areas (85% impervious)	89	92	94	95
Industrial districts (72% impervious)	81	88	91	93
Residential[3]				
Average lot size Average % Impervious[4]				
1/8 ac or less 65	77	85	90	92
1/4 ac 38	61	75	83	87
1/3 ac 30	57	72	81	86
1/2 ac 25	54	70	80	85
1 ac 20	51	68	79	84
Paved parking lots, roofs, driveways, etc.[5]	98	98	98	98
Streets and roads				
Paved with curbs and storm sewers[5]	98	98	98	98
Gravel	76	85	89	91
Dirt	72	82	87	89

1. For a more detailed description of agricultural land use curve numbers, refer to *National Engineering Handbook,* Section 4, "Hydrology," Chapter 9, Aug. 1972.

2. Good cover is protected from grazing and litter and brush cover soil.

3. Curve numbers are computed assuming that the runoff from the house and driveway is directed toward the street with a minimum of roof water directed to lawns where additional infiltration could occur.

4. The remaining pervious areas (lawn) are considered to be in good pasture condition for these curve numbers.

5. In some warmer climates of the country a curve number of 95 may be used.

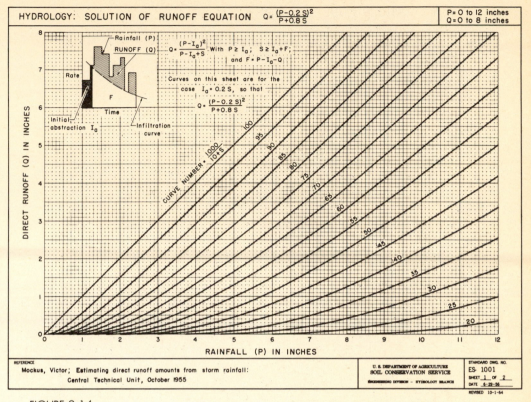

FIGURE 2.14

Graphical solution of rainfall-runoff equation.

EXAMPLE 2.8

SCS UNIT HYDROGRAPH

For the watershed of Example 2.7, develop a unit hydrograph using the SCS method. The watershed consists of meadows in good condition with soil group D. The average slope in the watershed is 100 ft/mi. Assume the same duration of rainfall as found in Example 2.7. Sketch the resulting hydrograph.

SOLUTION

Equation (2.18) gives the following relationship for t_p:

$$t_p = \frac{\ell^{0.8}(S+1)^{0.7}}{1900y^{1/2}}.$$

From Table 2.1, the SCS curve number is found to be 78. Therefore,

$$S = 1000/CN - 10$$
$$= 1000/78 - 10,$$

$$\boxed{S = 2.82 \text{ in.}}$$

From Example 2.7, $L = 18$ mi, so

$$\ell = (18 \text{ mi})(5280 \text{ ft/mi}) = \boxed{95,040 \text{ ft.}}$$

The slope is 100 ft/mi, so

$$y = (100 \text{ ft/mi})(1 \text{ mi}/5280 \text{ ft})(100\%)$$
$$= 1.9\%,$$

and

$$t_p = \left[\frac{(95,040)^{0.8}(2.82 + 1)^{0.7}}{1900 \sqrt{1.9}}\right]$$

$$= \boxed{9.4 \text{ hr.}}$$

From Eq. (2.17) and with $D = 1.6$ hr from Example 2.7,

$$T_R = D/2 + t_p$$
$$= (1.6/2) + 9.4 \text{ hr}$$
$$= \boxed{10.2 \text{ hr,}}$$

and Eq. (2.16) gives:

$$Q_p = \frac{484A}{T_R}$$
$$= \frac{484(100)}{10.2} \text{ cfs,}$$

$$\boxed{Q_p = 4,745 \text{ cfs.}}$$

To complete the graph, it is also necessary to know the time of fall B. The volume is known to be 1 in. of direct runoff over the watershed, so

$$\text{Vol} = (100 \text{ mi}^2) \left(\frac{5280 \text{ ft}}{\text{mi}}\right)^2 \left(\frac{\text{ac}}{43,560 \text{ ft}^2}\right)(1 \text{ in.}) = 64,000 \text{ ac-in.}$$

From Eq. (2.14),

$$\text{Vol} = \frac{Q_p T_R}{2} + \frac{Q_p B}{2} = 64,000 \text{ ac-in.} = 64,000 \text{ cfs-hr,}$$

$$64,000 \text{ cfs-hr} = \frac{(4745 \text{ cfs} \times 10.2 \text{ hr})}{2} + \frac{(4745 \text{ cfs})(B \text{ hr})}{2},$$

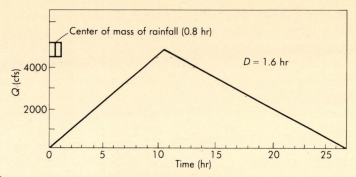

FIGURE E2.8

so

$$\boxed{B = 16.8 \text{ hr.}}$$

The triangular unit hydrograph is shown in Fig. E2.8 and should be compared with Fig. E2.7.

Espey Methods

Many regional unit hydrograph methods have been developed based on statistical analysis of gaged watersheds in a particular area of the United States. These methods generally relate time to peak and peak flow statistically to watershed characteristics such as length, slope, area, and percent imperviousness. Espey and Winslow (1968) developed the following best-fit equations for urban and rural watersheds in Texas for the 30-min unit hydrograph. The equations have been applied with success in the Houston metropolitan area (Smith & Bedient, 1980).

$$t_p = (16.4 \, \phi L^{0.316} S^{-0.0488} I^{-0.49}) - 15, \qquad (2.23)$$
$$Q_p = 3.54 \times 10^4 \, (t_p + 15)^{-1.1} \, A, \qquad (2.24)$$

where

t_p = time to peak from centroid of rainfall (min),
Q_p = peak flow (cfs)
ϕ = dimensionless conveyance factor (0.8 – 1.3),
L = length of main channel (ft),
S = dimensionless slope of main channel,
I = impervious area (%),
A = drainage area (mi^2).

The value of ϕ depends on the type of channel and ranges from 0.8 for concrete lined channels to 1.3 for natural channels. Care should be taken in applying these equations to watersheds with slopes, soils, or land use different from the ones used to derive the equations. Example 2.9 illustrates the use of the Espey concept for a small watershed of given land use.

Espey and Altman (1977) extended the original method and developed a nomograph method for 10-min unit hydrographs for small urban watersheds. They evaluated 41 small watersheds (18 from Texas) ranging in size from 0.014 to 15 mi^2. Size of drainage area A, main channel length L, main channel slope S, extent of impervious cover I, and a dimensionless channel conveyance factor ϕ were selected as the most important parameters describing the physiographic characteristics of each watershed. The empirical method can be applied to a small watershed where adequate relationships have not been previously developed. Table 2.2 presents the final empirical equations for the 10-min unit hydrographs along with width and time base factors for shaping. Nomographs are available to allow a simple procedure for graphical determination.

TABLE 2.2

10-min Unit Hydrograph Equations for Espey-Altman Method

EQUATIONS	TOTAL EXPLAINED VARIATION	EQUATION NUMBER
$T_R = 3.1 L^{0.23} S^{-0.25} I^{-0.18} \phi^{1.57}$	0.802	(2.25)
$Q = 31.62 \times 10^3 A^{0.96} T_R^{-1.07}$	0.936	(2.26)
$T_B = 125.89 \times 10^3 A Q^{-0.95}$	0.844	(2.27)
$W_{50} = 16.22 \times 10^3 A^{0.93} Q^{-0.92}$	0.943	(2.28)
$W_{75} = 3.24 \times 10^3 A^{0.79} Q^{-0.78}$	0.834	(2.29)

$L =$ total distance (ft) along the main channel from the point being considered to the upstream watershed boundary.

$S =$ main channel slope (ft/ft) as defined by $H/(0.8L)$, where L is the main channel length as described immediately above and H is the difference in elevation between two points A and B. A is a point on the channel bottom at a distance $0.2L$ downstream from the upstream watershed boundary. B is a point on the channel bottom at the downstream point being considered.

$I =$ impervious area within the watershed (%).

$\phi =$ dimensionless watershed conveyance factor as described in the text.

$A =$ watershed drainage area (mi^2).

$T_R =$ time of rise of the unit hydrograph (min).

$Q =$ peak flow of the unit hydrograph (cfs).

$T_B =$ time base of the unit hydrograph (min).

$W_{50} =$ width of the hydrograph at 50% of Q (min).

$W_{75} =$ width of the unit hydrograph at 75% of Q (min).

<hr>

EXAMPLE 2.9

ESPEY-WINSLOW METHOD

Use the Espey-Winslow method to develop a 30-min triangular unit hydrograph for a watershed with the following characteristics:

$\phi = 1.2$,

$L = 20,000$ ft,

$S = 0.0019$ ft/ft,

$I = 25\%$,

$A = 4.7$ mi^2.

SOLUTION

From Eqs. (2.23) and (2.24),

$$t_p = (16.4 \, \phi L^{0.316} S^{-0.0488} I^{-0.49}) - 15 \text{ min},$$
$$Q_p = 3.54 \times 10^4 (t_p + 15)^{-1.1} A \text{ cfs},$$

so

$$t_p = [(16.4)(1.2)(20,000)^{0.316}(0.0019)^{-0.0488}(25)^{-0.49}] - 15$$
$$= 111 \text{ min} = 1.9 \text{ hr}$$

and

$$Q_p = 3.54 \times 10^4 (111 + 15)^{-1.1}(4.7)$$
$$= \boxed{814 \text{ cfs.}}$$

Assuming a triangular hydrograph,

$$\text{Vol} = \frac{1}{2} Q_p T_B = (1 \text{ in.})(A), \qquad A \text{ in ac}$$

$$T_B = 2A/Q_p$$
$$= \frac{(2 \text{ in.})(4.7 \text{ mi}^2)(640 \text{ ac/mi}^2)}{814 \text{ cfs}}$$

$$= \boxed{7.4 \text{ hr.}}$$

The 30-min UH will have 2 in./hr falling for 1/2 hr:

$$T_R = D/2 + t_p$$
$$= 0.25 + 1.9 \text{ hr}$$
$$= 2.15 \text{ hr.}$$

The 30-min UH is shown in Fig. E2.9.

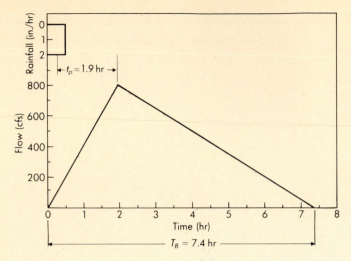

FIGURE E2.9

2.5

APPLICATIONS OF UNIT HYDROGRAPHS

Once a unit hydrograph of given duration based on known storms or synthetic methods has been developed for a particular basin under a given set of physiographic conditions, it can be further utilized for the following hydrologic calculations (Fig. 2.15):

1. Design storm hydrographs for selected recurrence interval storms (10-yr, 25-yr, 100-yr) can be developed through convolution adding and lagging procedures (Fig. 2.15a).

2. Effects of land use changes, channel modifications, storage additions, and other variables can be tested to determine changes in the unit hydrograph (Fig. 2.11).

3. Hydrographs for watersheds consisting of several subbasins can be simulated by adding, lagging, and routing the flows produced by unit hydrographs for each subbasin (Chapter 4). Effects of various rainfall patterns and land use distributions can be tested on overall hydrologic response of the large watershed (Fig. 2.15b). See Example 2.10.

4. Storage routing methods (Chapter 4) can be used to translate inflow unit hydrographs through a reservoir or detention basin of particular size to attenuate peak flow or lag time to peak of the hydrograph.

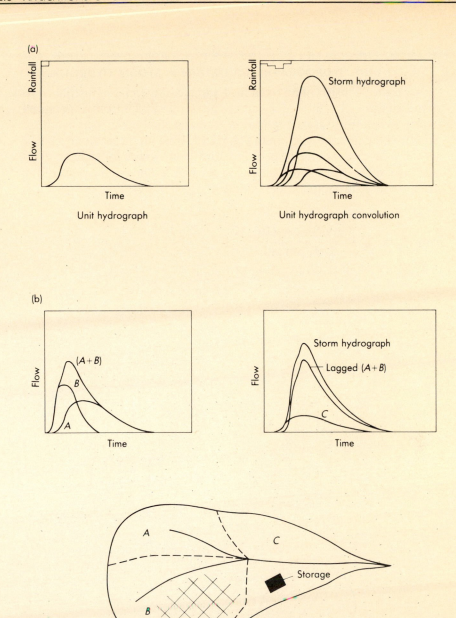

FIGURE 2.15

Unit hydrograph applications. (a) Development of design storm hydrograph. (b) Development of watershed hydrograph.

EXAMPLE 2.10

USE OF UNIT HYDROGRAPHS TO OBTAIN A STORM HYDROGRAPH

A watershed consists of an open field and a commercial industrial complex, each covering 50% of the area. The input rainfall from a measured
storm event, the average infiltration loss, and the known 1-hr synthetic
unit hydrographs are as given in Fig. E2.10. Find (a) the area of each half
of the watershed; (b) the peak of the storm hydrograph for subarea A; (c)
the peak of the storm hydrograph for subarea B; and (d) the peak of the
total storm hydrograph at the outfall.

SOLUTION

a) $A_A = A_B = \dfrac{\text{area under UH}}{1 \text{ in. rainfall}}$

$A_B = A_A = \left(\dfrac{\frac{1}{2}(125)(4) \text{ cfs-hr}}{1 \text{ in.}}\right)\left(\dfrac{1 \text{ ac-in.}}{\text{cfs-hr}}\right)$

$= \boxed{250 \text{ ac.}}$

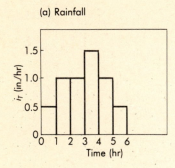

(a) Rainfall   (b) Infiltration

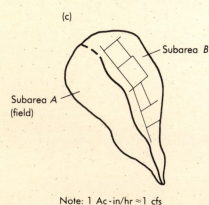

(c)

Subarea B

Subarea A
(field)

Note: 1 Ac-in/hr ≈ 1 cfs

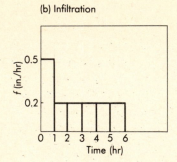

UH for B

UH for A

FIGURE E2.10

b) To find Q_p for A, we must find the total storm hydrograph. First we calculate P_{net} (rainfall excess). The volume of rainfall for each time period is found by substituting values from the table into the following equation:

$$P_{net} = (i_T - f)(\Delta t).$$

TIME (hr)	i_T (in./hr)	f (in./hr)	Δt (hr)	P_{net} (in.)
1	0.5	0.5	1	0.0
2	1.0	0.2	1	0.8
3	1.0	0.2	1	0.8
4	1.5	0.2	1	1.3
5	1.0	0.2	1	0.8
6	0.5	0.2	1	0.3

The peak of the unit hydrograph for A occurs at 2.5 hr, so 30-min time intervals should be used. Otherwise, Q_p will be underestimated. Using the tabular method of Example 2.5, we have the following:

TIME (hr)	U_n (cfs)	P_1U_n	P_2U_n	P_3U_n	P_4U_n	P_5U_n	P_6U_n	Q (cfs)
0	0	0						0
0.5	20	0						0
1.0	40	0	0					0
1.5	60	0	16					16
2.0	80	0	32	0				32
2.5	100	0	48	16				64
3.0	80	0	64	32	0			96
3.5	60	0	80	48	26			154
4.0	40	0	64	64	52	0		180
4.5	20	0	48	80	78	16		222
5.0	0	0	32	64	104	48	0	248
5.5			16	48	130	64	6	264 Q_p
6.0			0	32	104	80	12	228
6.5				16	78	64	18	176
7.0				0	52	48	24	124
7.5					26	16	30	72
8.0								
8.5								

Q_p for area $A = 264$ cfs.

c) For subarea B, $P_{net} = i_T \Delta t$ since there are no losses (100% impervious). Increments of 1 hr will suffice since the peak of the unit hydrograph occurs at $t = 2$ hr.

TIME (hr)	U_n (cfs)	P_1U_n	P_2U_n	P_3U_n	P_4U_n	P_5U_n	P_6U_n	Q (cfs)
0	0	0						0
1	62.5	31.25	0					31.25
2	125	62.5	62.5	0				125
3	62.5	31.25	125	62.5	0			218.75
4	0	0	62.5	125	93.75	0		281.25
5			0	62.5	187.50	62.5	0	312.5 Q_p
6				0	93.75	125	31.25	250
7					0	62.5	62.5	125
8								

Q_p for area $B = 312.5$ cfs.

d) The peak outflow for the total storm hydrograph will occur at the same time as the peak for either subarea A or subarea B since the total storm hydrograph is obtained by adding the two subarea storm hydrographs. Subarea B has a peak at time $t = 5$ hr. The total storm hydrograph ordinate at this time is $(312.5 \text{ cfs} + 248 \text{ cfs}) = 560.5$ cfs. The storm hydrograph for subarea A peaks at time $t = 5.5$ hr. The total storm hydrograph ordinate for this point is $(281.25 \text{ cfs} + 264 \text{ cfs}) = 545.25$ cfs (where 281.25 cfs is interpolated). Therefore, the peak outflow from the watershed for this storm event is

560.5 cfs.

2.6
CONCEPTUAL MODELS

Instantaneous Unit Hydrographs (IUH)

A useful mathematical extension of unit hydrographs of finite duration can be developed if the duration of rainfall excess D approaches zero while the quantity (unit depth) remains constant. The runoff produced by this instantaneous rainfall is called the **instantaneous unit hydrograph.** The IUH is a response function for a particular watershed to a unit impulse of rainfall excess.

The IUH is assumed to be a unique function for a watershed, independent of time or antecedent conditions. The output function $Q(t)$, or total storm discharge, is produced by summing the outputs due to all instantaneous inputs $i(t)$. If the input is a succession of inputs of volume $i(\tau)d\tau$, then each adds its contribution $i(\tau)u(t - \tau) \, d\tau$ to the rate of output Q at time t. Stated mathematically, the runoff rate at any fixed time t is

$$Q(t) = \int_0^t i(\tau)u(t-\tau)\, d\tau, \qquad (2.30)$$

where $i(\tau)$ is the rainfall excess at time τ (Fig. 2.16) and $u(t-\tau)$ can be viewed as a weighting function given to rainfall intensities that occurred at time $(t-\tau)$ before. The integral, known as the convolution integral, gives the output runoff as a continuous function of time t. Equation (2.3) is the discrete form of Eq. (2.30).

The real usefulness of the IUH as a mathematical concept can be seen in relation to the S-curve (Section 2.3). Consider an S-curve formed by a continuous rainfall of unit intensity, which can be an infinite series of time units each separated by τ. By summing all the individual unit hydrographs of duration τ, the S-curve ordinates become

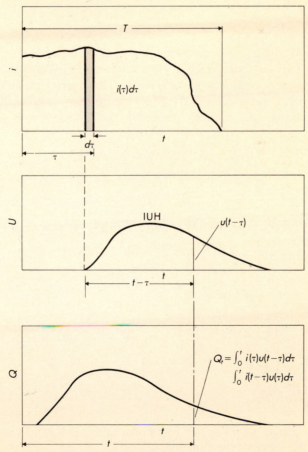

FIGURE 2.16
Use of the IUH to generate a hydrograph.

$$S(t) = \tau[u(t) + u(t - \tau) + u(t - 2\tau) + \cdots] \qquad (2.31)$$

since τ also equals the depth for a unit intensity. Then, as $\tau \to 0$ in the limit for the IUH,

$$S(t) = \int_0^t u(t)dt. \qquad (2.32)$$

Therefore, the S-curve is the integral of the IUH, and the IUH is the first derivative of the S-curve. The slope of the S-curve, dS/dt, is proportional to the ordinate of the IUH.

The IUH has been widely used, although it suffers the same disadvantage as the UH in that rainfall-runoff responses may be nonlinear and dependent on antecedent conditions. Nonlinear models for the unit hydrograph have been developed in the general literature and are reviewed by Chow (1964) and Raudkivi (1979).

Linear Models

Nash (1958) advanced the conceptual model of a watershed as a cascade of n reservoirs in series, each with a linear storage-discharge relation $S = KQ$, where K is the average delay time. This routing method through n reservoirs is a special case of the Muskingum method (Chapter 4). The continuity equation is combined with the storage relation to yield

$$I - Q = \frac{dS}{dt} = K\frac{dS}{dt}, \qquad \text{or} \qquad (2.33)$$

$$Q + K\frac{dQ}{dt} = I. \qquad (2.34)$$

Multiplication by $e^{t/k}$ allows the differential equation to be solved:

$$\frac{d}{dt}(Qe^{t/K}) = \frac{I}{K}e^{t/K}, \qquad \text{or} \qquad (2.35)$$

$$Qe^{t/K} = \frac{1}{K}\int (Ie^{t/K})\,dt + C_1. \qquad (2.36)$$

For an instantaneous inflow of unit volume to the reservoir,

$$Q = e^{-t/K}\left[\frac{1}{K}\int e^{t/K}\,\delta(0)dt + C_1\right]$$

$$= \frac{1}{K}(e^{-t/K}), \qquad t > 0, \qquad (2.37)$$

where the integral is a Laplace transform of the δ-function that simply picks out the value of the function at $t = 0$:

$$L(\delta(0)) = \int \delta(0)e^{-pt}dt = e^{po} = 1.$$

Thus, the response of a linear reservoir to the pulse input is a sudden jump at the moment of inflow followed by an exponential decline. If this reservoir is allowed to flow into a second reservoir, where Q_1 is inflow and Q_2 is outflow, then

$$Q_1 - Q_2 = K \frac{dQ_2}{dt} \tag{2.38}$$

whose solution is

$$Q_2 = \frac{1}{K} \left(\frac{t}{K} \right) e^{-t/K}. \tag{2.39}$$

It can be shown that for the nth reservoir,

$$Q_n = \frac{1}{K\Gamma(n)} \left(\frac{t}{K} \right)^{n-1} e^{-t/K}, \tag{2.40}$$

where $\Gamma(n)$ is the gamma function and Eq. (2.40) is the probability density function of the gamma distribution, as discussed in Chapter 3. Gray (1962) also based his unit hydrograph method on the two-parameter gamma distribution. Nash (1958) suggested that this model represents the general equation of the IUH.

The S-curve from the Nash model can be derived from Eq. (2.32) as

$$S(t) = \int_0^t u(t)dt = \frac{1}{K} \int_0^t \left(\frac{t}{K} \right)^{n-1} \frac{e^{-t/K}}{(n-1)!} \, dt, \tag{2.41}$$

which is a gamma distribution $\Gamma(t/K, n-1)$. The IUH can be interpreted as the frequency distribution of arrival times of water particles at the basin outlet, given an instantaneous unit rainfall spread uniformly over the basin. The expected value or average time is the time lag between centroids of the IUH and $i(\tau)$, the input rainfall hyetograph:

$$E(t) = \int_0^\infty u(t)t \, dt = nK \int_0^\infty \left(\frac{t}{K} \right)^n \frac{e^{-t/K}}{n!} \, d(t/K), \qquad \text{or}$$

$$E(t) = nK. \tag{2.42}$$

$E(t)$ also represents the first moment of the IUH about $t = 0$.

The variance Var(t) can also be derived as

$$\text{Var}(t) = E(t^2) - [E(t)]^2 = \int_0^\infty u(t)t^2 \, dt - (nK)^2$$

$$= K^2 n(n+1) - n^2 K^2 = K^2 n. \tag{2.43}$$

Thus, the values of n and K in the Nash (1959) model can be evaluated by the method of moments according to Chapter 3:

$$n = \frac{(\mu'_{1Q} - \mu'_{1i})^2}{\mu_{2Q} - \mu_{2i}} \tag{2.44}$$

and

$$K = \frac{\mu_{2Q} - \mu_{2i}}{\mu'_{1Q} - \mu'_{1i}},$$ (2.45)

where μ'_{1Q} and μ'_{1i} are first moments of output and input, respectively, and μ_{2Q} and μ_{2i} are second central moments of output and input, respectively. Nash analyzed catchments in the United Kingdom and found that the following relations gave good results:

$$\mu'_1 = 27.6(A/S)^{0.3}$$

and

$$\frac{\mu_2}{(\mu'_1)^2} = \frac{0.41}{L^{0.1}},$$ (2.46)

where A is area (mi^2), S is slope (parts per thousand), L is main channel length (mi), and $\mu'_1 = nK$ and $\mu_2 = nK^2$. The first relation had a multiple correlation (R) of 0.90, while the second relation had $R = 0.45$.

In 1959, Dooge attempted to produce a general conceptual model of the unit hydrograph. Since the unit hydrograph existed for a linear system, he argued that only linear storage elements and linear distortionless channels should be components of the model. Both translation effects and storage or attenuation effects exist in an actual watershed. Dooge assumed that since the model is linear, the translation effects were lumped together as linear channels and the storage effects were lumped together as linear reservoirs (Nash model). Because of linear superposition, channels and reservoirs could be interchanged or reordered without affecting the response of the system. The shape of a hydrograph routed through a linear channel remains unchanged. For inflow $I = f(t)$, the outflow is $Q = f(t - \tau)$. Thus, the impulse response to a linear reservoir routed through a linear channel yields

$$Q = (1/K) \exp[-(t - \tau)/K], \qquad t > \tau.$$ (2.47)

Further simplifications of the Dooge linear model were made by Diskin (1964), who modeled the watershed response with two cascades of equal linear reservoirs in parallel. Kulandaiswamy (Chow, 1964) produced a nonlinear response function using a nonlinear storage expression, based on an extension of the Muskingum model. He assumed

$$S = \sum_n a_n(Q, I) \frac{d^n Q}{dt^n} + \sum_m b_m(Q, I) \frac{d^m I}{dt^m}.$$ (2.48)

The nonlinear approach may be a more realistic description of watershed behavior, but it is not used much in practice because of the difficulty in obtaining the required coefficients in Eq. (2.48).

2.7
KINEMATIC WAVE METHODS FOR OVERLAND FLOW

Henderson and Wooding (1964) and Wooding (1965) developed a theory for the overland and the stream hydrograph. All flows are assumed to obey the equations of continuity and momentum. (A complete derivation of the concept of **kinematic waves** is presented in Chapter 4.) The continuity and momentum equations for overland kinematic waves reduce to the following two equations, respectively:

$$\frac{\partial y}{\partial t} + \frac{\partial q}{\partial x} = i - f = i_e \tag{2.49}$$

and

$$q = \alpha y^m = \frac{1.49}{N} \sqrt{S_o}\, y^{5/3}, \tag{2.50}$$

where

$y = y(x, t) =$ depth of overland flow (ft),

$q = q(x, t) =$ rate of overland flow/unit width (ft^2/s),

$i - f = i_e =$ net rainfall rate (ft/s),

$\alpha =$ conveyance factor $= (1.49/N)\sqrt{S_o}$ when obtained from Manning's equation,

$m = 5/3$ when obtained from Manning's equation,

$N =$ effective roughness coefficient,

$S_o =$ average overland flow slope,

$y_o =$ mean depth of overland flow.

Equation (2.50) is a form of Manning's equation applied to an overland flow plane, using Manning's equation with N values for overland flow (see Table 4.2). Solutions basically couple the continuity condition with a uniform flow equation (such as Manning's) for the momentum equation. Solution methods to Eqs. (2.49) and (2.50) using the method of characteristics are described in detail by Lighthill and Whitham (1955), Eagleson (1970), Overton and Meadows (1976), and Raudkivi (1979). Most practical applications of kinematic wave methods require the use of numerical methods because of nonuniform rainfall and basin characteristics (Chapter 4).

Some insights into the shape of kinematic waves can be gained by referring to Fig. 2.17 from Wooding (1965), which shows the basin discharge due to a steady uniform rainfall with steady infiltration. T' is the ratio of the duration of rainfall excess T to the time of concentration t_c. For $T' = 2$, the flow increases until $t = t_c$ and the duration of rainfall

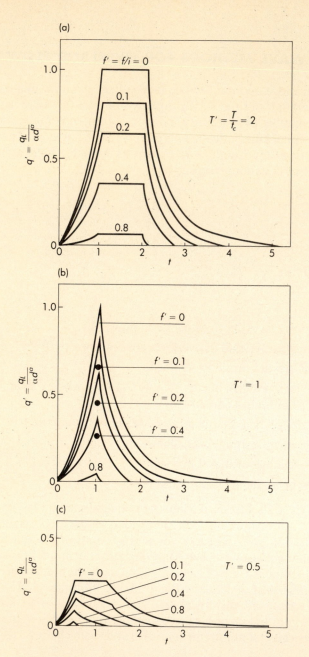

FIGURE 2.17

Effect of steady infiltration $f' = f/i$ on the catchment discharge $q' = q_L/(\alpha d^a)$ due to a steady rainfall of given intensity. (From Wooding, 1965.)

equals the time of concentration, or maximum travel time in the basin. Outflow is constant until $t = 2t_c$, or until the cessation of rainfall excess. Outflow is reduced below the peak condition when $T' = 0.5$. For all rainfall durations, the effect of infiltration is to reduce the peak outflow approximately proportional to the ratio f/i, where f is infiltration rate and i is rainfall intensity. Detailed equations describing the hydrographs in Fig. 2.17 can be found in Eagleson (1970), Overton and Meadows (1976), and Raudkivi (1979).

Kinematic wave routing can be used to derive overland flow hydrographs, which can be added to produce collector hydrographs and, eventually, can be routed as channel or stream hydrographs. In recent years, the kinematic wave method has been added to several available computer models, such as the HEC-1 Flood Hydrograph Package from the U.S. Army Corps of Engineers (Hydrologic Engineering Center, 1981). The concept uses a number of interconnected elements, including overland flow planes, collector channels or pipes, and main channels, to describe an overall watershed. Explicit numerical methods (Section 4.6) are employed to solve Eqs. (2.49) and (2.50) for each element, and overland flow becomes input to the collector system, which eventually forms the lateral input hydrograph to a main channel. Chapter 4 presents kinematic wave flood routing in detail, and Chapter 5 presents detailed case studies for an actual watershed.

2.8
SNOWFALL AND SNOWMELT

The distribution of mean annual snowfall in the United States is shown in Fig. 2.18, but errors may be large in mountainous regions because of lack of measurements in remote areas. There is a gradual increase in snowfall with latitude and elevation, and up to 400 in./yr (1000 cm/yr) occurs in the Sierra Nevada and Cascade Range of the western United States.

Snow measurements are obtained using standard and recording rain gages, snow stakes, and snow boards, which measure accumulation over a period of time. Snow surveys at a number of spaced points along a snow course are necessary to account for drifting and blowing snow. The Soil Conservation Service coordinates many snow surveys in the western United States.

Snow can be a dominant source of streamflow, especially in the western United States, but significant percentages of runoff in the Northeast and Midwest also originate as snow. Because the storage and melting

FIGURE 2.18

Mean annual snowfall in the United States (in.). (From U.S. Environmental Data Service.) ➡

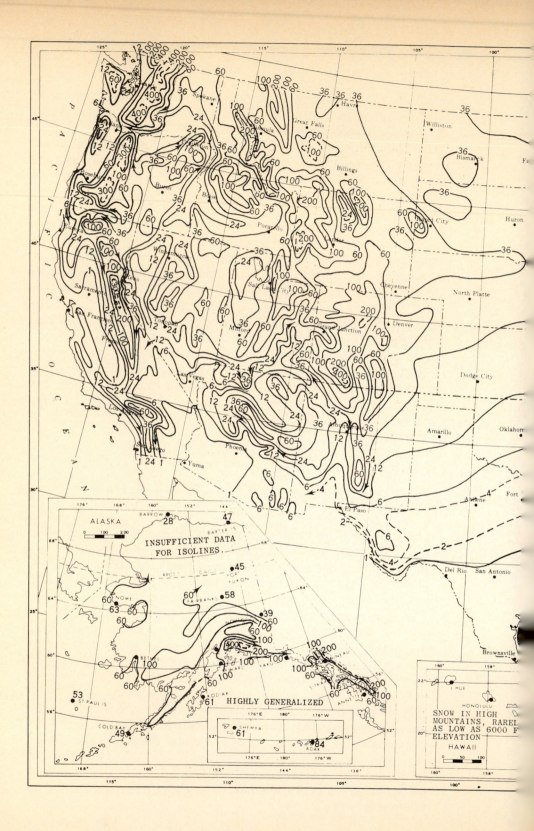

INSUFFICIENT DATA
FOR ISOLINES.

HIGHLY GENERALIZED

SNOW IN HIGH
MOUNTAINS, RAREL
AS LOW AS 6000 F
ELEVATION
HAWAII

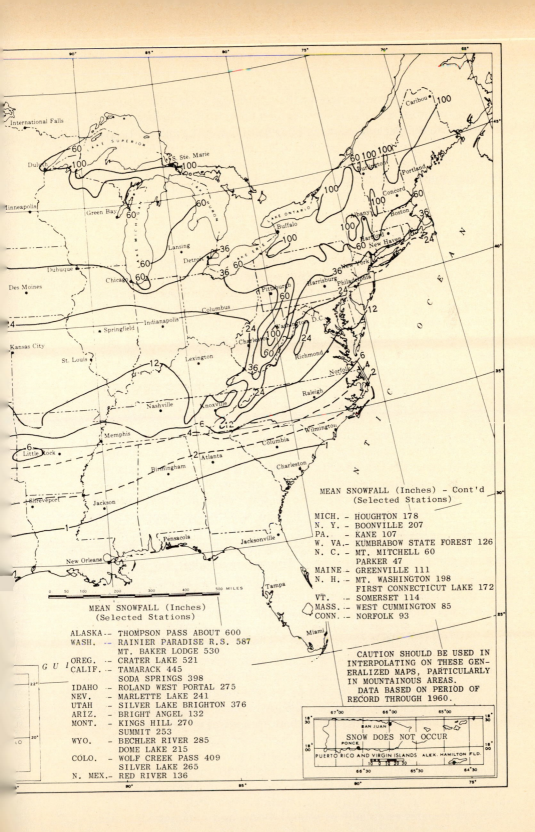

MEAN SNOWFALL (Inches) - Cont'd
(Selected Stations)

MICH. - HOUGHTON 178
N. Y. - BOONVILLE 207
PA. - KANE 107
W. VA. - KUMBRABOW STATE FOREST 126
N. C. - MT. MITCHELL 60
         PARKER 47
MAINE - GREENVILLE 111
N. H. - MT. WASHINGTON 198
        FIRST CONNECTICUT LAKE 172
VT. - SOMERSET 114
MASS. - WEST CUMMINGTON 85
CONN. - NORFOLK 93

CAUTION SHOULD BE USED IN
INTERPOLATING ON THESE GEN-
ERALIZED MAPS, PARTICULARLY
IN MOUNTAINOUS AREAS.
    DATA BASED ON PERIOD OF
RECORD THROUGH 1960.

MEAN SNOWFALL (Inches)
(Selected Stations)

ALASKA -- THOMPSON PASS ABOUT 600
WASH. -- RAINIER PARADISE R.S. 587
         MT. BAKER LODGE 530
OREG. -- CRATER LAKE 521
CALIF. -- TAMARACK 445
          SODA SPRINGS 398
IDAHO - ROLAND WEST PORTAL 275
NEV. - MARLETTE LAKE 241
UTAH - SILVER LAKE BRIGHTON 376
ARIZ. - BRIGHT ANGEL 132
MONT. - KINGS HILL 270
        SUMMIT 253
WYO. - BECHLER RIVER 285
       DOME LAKE 215
COLO. - WOLF CREEK PASS 409
        SILVER LAKE 265
N. MEX. - RED RIVER 136

SNOW DOES NOT OCCUR

PUERTO RICO AND VIRGIN ISLANDS   ALEX. HAMILTON FLD.

of snow play an important role in the hydrologic cycle of some areas, hydrologists must be able to reliably predict the contribution of snowmelt to overall runoff. Because snowmelt begins in the spring and the derived runoff is out of phase with periods of greatest water demand, control schemes such as storage reservoirs have been implemented in some areas for water supply. Under certain conditions, snowmelt can also contribute to flooding problems, especially in mountainous areas.

Physics of Snowmelt

The energy exchanged between the snowpack, the atmosphere, and the earth is the controlling factor in rates of snowmelt. Other geographic, topographic, and surface cover factors are also important. As snow accumulates on the ground over a period of time, its density increases with the settling and compaction of the snowpack. Newly fallen snow has a density about 10% that of liquid water, but the density can increase to 50% with aging. Hence a rule of thumb for new snow is that the depth of water equivalent is approximately 1/10 the depth of the snowpack. *Ripe snow* holds all the liquid water it can against the action of gravity. Hence, any heat added to ripe snow will produce runoff. Heat added to a snowpack at a temperature below freezing will act to ripen it.

Snowmelt and evaporation are both thermodynamic processes and can be studied with an energy-balance approach. The energy for snowmelt is derived from (1) net solar radiation, (2) net longwave radiation exchange, (3) conduction and convection transfer of sensible heat to or from the overlying air, (4) condensation of water vapor from the overlying air, (5) conduction from the underlying soil, and (6) heat supply by incident rainfall. The heat exchange at the snow-air interface dominates the snowmelt process, and heat exchange with the soil is secondarily important. Figure 2.19 shows the typical daily energy budget during active snowmelt.

For each gram of water melted in a snowpack at $0°C$ ($32°F$), 80 cal/g of latent heat must be supplied. Thus, if the density of water is 1 g/cm, 80 cal will melt 80 cm³ (water equivalent) of snow. Additionally, 80 cal/cm³ = 80 cal/cm²-cm = 80 ly/cm, where ly denotes a langley, or 1 cal/cm². Hence, 80 ly are required to melt 1 cm of snow, or 2.54 × 80 = 203.2 ly/in. of snow (water equivalent). Meltwater may refreeze with cooling, or it may drain from the snowpack, contributing to soil moisture, streamflow, or ground water recharge. Snowmelt computations are made difficult by the variations in solar radiation received by the surface and variations in the **albedo,** the reflection coefficient. Although only 5 – 10% of incoming shortwave radiation is reflected by a water surface, up to 83% is reflected by a clean, dry snow surface. As the snow ages, its albedo can drop to less than 50% because of structural changes, and shortwave radiation penetrates to varying depths, depending on snow density.

Snow radiates essentially as a black body, and outgoing longwave

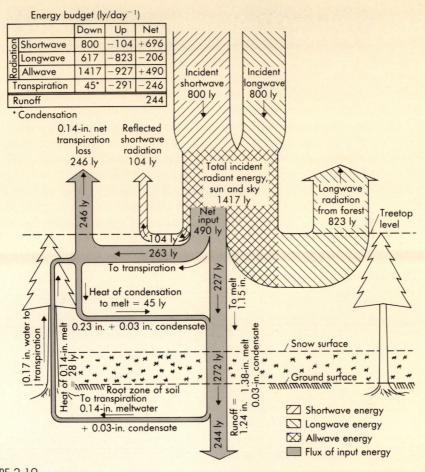

Energy budget (ly/day⁻¹)

		Down	Up	Net
Radiation	Shortwave	800	−104	+696
	Longwave	617	−823	−206
	Allwave	1417	−927	+490
Transpiration		45*	−291	−246
Runoff				244

* Condensation

FIGURE 2.19

Typical daily energy budget during active snowmelt. (From U.S. Army Corps of Engineers, 1956.)

radiation at 0°C is equivalent to 8.4 cm (3.3 in.) of melt over a 1-day period. Because of reradiation from the atmosphere back to the Earth, longwave radiation loss is equivalent to only about 2 cm (0.8 in.) of melt per day under clear skies and near-freezing temperatures.

Heat exchange between a snowpack and the atmosphere is affected by conduction, convection, condensation, and evaporation (U.S. Army Corps of Engineers, 1956; Anderson, 1976). Wind velocity and air temperature are primary factors affecting the density of new snow. The rate of transfer of sensible heat (convection) is proportional to the temperature difference between the air T_a and the snow T_s, and the wind velocity v. Condensation melt is proportional to the vapor pressure difference between the snow surface and the atmosphere and to the wind velocity. The transfer of heat from the underlying soil to the snowpack is small on a daily basis but may accumulate to several cm of melt during an entire season. This may be enough to keep the soil saturated and to produce a rapid response of runoff when melting occurs. Raindrop temperature is reduced as it enters a snowpack and an equivalent amount of heat is imparted to the snow based on the surface wet-bulb temperature. For example, the heat available in 10 mm of rain at 10°C will melt about 1.2 mm of water from the snowpack.

In many basins of very high relief, the lower limit of snow cover (the snow line) is dynamic and moves up and down the slope. As melting proceeds, snow cover recedes more rapidly on southerly and barren slopes, with the rate of snowmelt decreasing with elevation. Given the temperature at an index station, an **area-elevation curve** for the basin, and the average snow line elevation, one can compute the area subject to melting based on the rate of change of temperature with elevation, about 1°C per 100 m (5°F per 1000 ft) increase in elevation.

The Snowpack Energy Budget

The snowpack energy budget is written in terms of the energy flux to and from the snowpack, with units of energy per unit area per unit time. The most convenient units are cal/cm²-day, or ly/day. In terms of the energy budget components mentioned earlier,

$$\Delta H = Q_N + Q_g + Q_c + Q_e + Q_p, \tag{2.51}$$

where

ΔH = change in heat storage in snowpack,

Q_N = net (incoming minus reflected minus outgoing) shortwave and longwave radiation,

Q_g = conduction of heat to snowpack from underlying ground,

Q_c = convective transport of sensible heat from air to snowpack (negative if from snowpack to air),

Q_e = release of latent heat of vaporization by condensation of water vapor onto snowpack,

Q_p = advection of heat to snowpack by rain.

If the change in storage is positive, it will act either to ripen the snow or to produce melt. Details for all terms cannot be included here. However, computation of net shortwave and longwave radiation is discussed in detail in Appendix D.1. Convective heat transport and condensation melt are both diffusive processes; the heat flux is proportional to the gradient of temperature for the former and to the gradient of vapor pressure for the latter. Diffusivities or transfer coefficients are functions of wind velocity and surface roughness; Eagleson (1970) summarizes equations developed from data gathered from California's Sierra Nevada by the U.S. Army

Corps of Engineers (1956). Heat transfer to or from the ground H_g is usually neglected unless the snowpack and ground surface temperatures are both known. Heat advected by rain into the snowpack is simply proportional to the difference between the rain temperature (usually assumed equal to the wet-bulb temperature) and the temperature of the snow.

Assuming the snow is ripe, the amount of melt from a positive value of H is

$$M = \Delta H/80 \text{ cm/day} \quad \text{or}$$
$$M = \Delta H/203.2 \text{ in./day.} \tag{2.52}$$

Energy budget methods represent the best techniques for predicting snowmelt when all components can be estimated. The National Weather Service (Anderson, 1973, 1976) has developed such a model for runoff forecasting. Unfortunately, some energy budget components are usually missing, leading to the use of simpler but less accurate degree-day methods based solely on temperature data.

Degree-Day or Temperature-Index Melt Equations

Degree-day or temperature-index equations are empirical or can result from a linearization of the energy budget equation (Huber et al., 1981). Daily snowmelt is assumed proportional to the difference between average daily temperature and a base temperature:

$$M = D_f(T_a - T_B), \tag{2.53}$$

where

$M =$ daily melt (in./day; cm/day),

$T_a =$ average daily air temperature (°F; °C),

$T_B =$ base melt temperature (°F; °C),

$D_f =$ degree-day factor, (in./day-°F; cm/day-°C).

Empirical degree-day factors range from 0.05 to 0.15 in./day-°F (0.2–0.7 cm/day-°C). Equation (2.53) is basically a regression equation; U.S. Army Corps of Engineers (1956) results are shown in Fig. 2.20. Base melt temperatures can be below freezing since Eq. (2.53) represents a linearization about reference conditions well above freezing; i.e., the equations shown in Fig. 2.20 should not be used for $T_a < 34°F$.

Degree-day equations may be made more sophisticated by accounting for rainfall, wind, radiation, and other factors in the coefficients; Viessman et al. (1977) summarize equations prepared in this manner by the U.S. Army Corps of Engineers (1956). However, in most cases, if those kinds of ancillary data are available, a better plan would be to proceed with the full energy budget method.

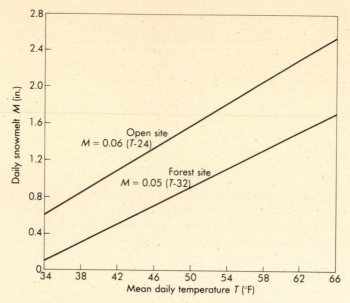

Note: Equations are applicable only to range of temperatures shown on the diagram.

FIGURE 2.20

Empirical degree-day equations. (From U.S. Army Corps of Engineers, 1956.)

Computation of Runoff from Snowmelt

Synthesis of runoff hydrographs associated with snowmelt can be short-term, for a few days, or long-term, for a complete melt season. Short-term forecasting is used in preparing operational plans for reservoirs or other flood control works and for calculating design floods. To forecast a few days in advance, only the present condition of a snowpack and stream-flow are required. For long-term forecasting over a whole season, initial conditions are required as well as reliable predictions of meteorologic parameters. Typical data used in the computation of runoff generated from snowmelt are shown in Fig. 2.21.

Snowmelt occurs only from those portions of a watershed covered by snow, and elevation is significant since rates of snowmelt generally de-crease with elevation due to a general reduction in temperature. Elevation effects can be considered by dividing a basin into a series of elevation zones where the snow depth, losses, and snowmelt are assumed uniform in each. In most practical cases, snowmelt is estimated by the index methods presented in the previous section. The budget method itself is used in design flood studies. Once inflow water has been calculated,

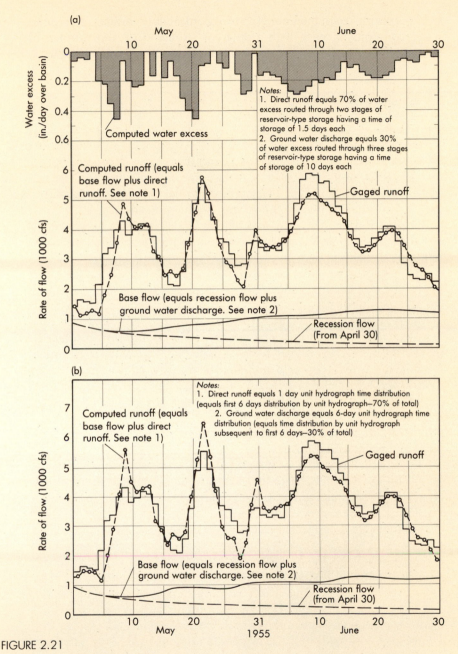

FIGURE 2.21

**Hydrometeorological data and computation of water generated.
(From U.S. Army Corps of Engineers, 1956.)**

storage routing techniques are used to accommodate surface and ground water components, each with a different storage time and number of routing units. Storage routing methods are treated in detail in Chapter 4.

SUMMARY

Chapter 2 has presented the concept of rainfall-runoff analysis, a central problem of engineering hydrology. Gross rainfall must be adjusted for losses to infiltration, evaporation, and depression storage to obtain rainfall excess, which equals direct surface runoff. The concept of the unit hydrograph allows for the conversion of rainfall excess into a basin hydrograph through a linear adding and lagging procedure called hydrograph convolution. The unit hydrograph is defined as 1 in. (1 cm) of direct runoff generated uniformly over a basin for a specified period of rainfall.

Unit hydrographs can be derived from actual storm data for gaged basins. S-curve methods allow unit hydrographs of one duration to be converted to another duration for a given watershed, and decomposition methods allow unit hydrographs to be derived from multiperiod storms. Synthetic unit hydrographs represent theoretical or empirical approaches that can be applied to ungaged basins. Snyder's (1938) method was one of the first and is still used in a modified form today. Most synthetic unit hydrograph formulas relate parameters such as time to peak or lag time (t_p) to measures of channel length, slope, watershed size, and watershed shape. A second relation usually correlates peak flow to basin area and t_p. Unit hydrographs are greatly affected by urban or agricultural development and channelization.

Conceptual models were developed in the late 1950s in an effort to establish a theoretical basis for the unit hydrograph. Linear models based on a cascade of serial reservoirs were used to represent the response of a watershed. Kinematic wave methods were developed in the 1960s for overland flow hydrographs and channel hydrographs and were based on solutions of continuity and Manning's equation for a given geometry. Kinematic wave methods are becoming more useful with the current popularity of numerical methods and computer models such as HEC-1 by the U.S. Army Corps of Engineers Hydrologic Engineering Center (1981).

The last major section of Chapter 2 presents snowfall and snowmelt theory as it relates to the rainfall-runoff process. The physics of snowmelt is briefly covered, and empirical degree-day equations are shown that can be used to estimate melt rates in in./day or cm/day as a function of air temperature and a melt coefficient. Snowmelt can be a significant factor in flood flows and water supply estimates.

PROBLEMS

2.1. a) Explain the concept of the time-area method.

b) What physical factors affect the shape and timing of the unit hydrograph?

c) Rework Example 2.2 for a rainfall intensity of 1/3 in./hr falling uniformly for 3 hr. Develop the bar hydrograph.

2.2. Determine the storm hydrograph resulting from the rainfall pattern in Fig. P2.2(a) using the unit hydrograph given in Fig. P2.2(b).

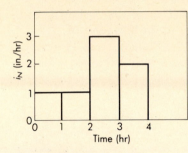

FIGURE P2.2(a)   FIGURE P2.2(b)

2.3. a) Given a triangular 2-hr unit hydrograph with

$$T_B = 12 \text{ hr},$$
$$T_R = 4 \text{ hr},$$
$$Q_p = 2 \text{ cfs},$$

where

T_B = time base of the unit hydrograph,
T_R = time of rise,
Q_p = peak flow,

graph an S-curve for this basin.

b) Develop the 1-hr unit hydrograph from a tabular computation of the S-curve using $\Delta t = 1$ hr. (It may be necessary to smooth the hydrograph.)

2.4. A watershed has the characteristics given below. Find the peak discharge Q_p, the basin lag time t_p, and the time base of the unit hydrograph T_B using Snyder's method.

$$A = 150 \text{ mi}^2,$$
$$C_t = 1.70,$$
$$L = 27 \text{ mi},$$

$L_{ca} = 15$ mi,

$C_p = 0.7$

2.5. A sketch of the Buffalo Creek Watershed is shown in Fig. P2.5. Areas A and B are identical in size, shape, slope, and channel length. Unit hydrographs (1 hr) are given for the existing conditions and full urbanization for both areas.

   a) Assuming existing conditions for both areas, evaluate the peak outflow at point 1 if 2.5 in./hr of rain falls for 2 hr. Assume a total infiltration loss of 1 in.

   b) Assume that area B has reached full urbanization and area A has remained at existing conditions. Determine the outflow hydrograph at point 1 if a net rainfall of 2 in./hr falls for 1 hr.

   c) Sketch the outflow hydrograph for the Buffalo Creek Watershed under complete urbanization (A and B both urbanized) for the rainfall given in part (b).

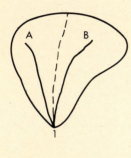
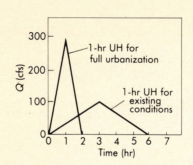

FIGURE P2.5

2.6. Given the storm hydrograph and rainfall pattern in Fig. P2.6 and the accompanying table, construct a 4-hr unit hydrograph. What is the area of the basin? (*Note:* The rainfall duration of the actual storm is

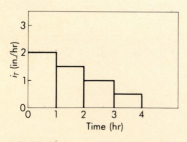
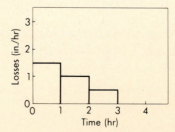

FIGURE P2.6

4 hr. Therefore, the storm hydrograph has the shape of a unit hydrograph. See Example 2.3.)

TIME (hr)	0	1	2	3	4	5	6	7	8	9	10	11
Q (cfs)	0	50	150	350	500	400	300	200	150	100	50	0

2.7. The storm hydrograph and rainfall in Fig. P2.7 and the accompanying table were recorded for a particular storm event. Assuming no infiltration losses, determine (a) the 1-hr unit hydrograph, (b) the S-curve, (c) the 30-min unit hydrograph.

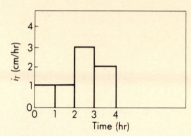

FIGURE P2.7

TIME (hr)	0	1	2	3	4	5	6	7	8	9
Q (m³/s)	0	7.5	22.5	48.8	78.8	75.0	48.8	26.3	7.5	0

2.8. The USGS recorded a storm event on June 15, 1976. The data for this storm are plotted in Fig. P2.8 on page 134 with linear approximations for the rainfall and runoff cumulative mass curves between the points indicated by black dots. The drainage area is 88.4 mi².
   a) Determine the duration and intensity of the approximated uniform rainfall.
   b) What is the time to peak for this storm?
   c) Find the ϕ index for this storm using the approximated rainfall and runoff.

2.9. Storm data (Fig. P2.9) were recorded for the Hummingbird Street Ditch on September 1, 1979 by the USGS. Linear approximations for the rainfall and runoff cumulative mass curves are shown by the black dots in the figure. The drainage area of this watershed is 0.32 mi².
   a) Determine the duration and intensity of the approximated rainfall.
   b) What is the time to peak for this storm?
   c) Find the ϕ index for this storm using the approximated rainfall and runoff.
   d) Develop a unit hydrograph for this watershed using the duration of part (a).

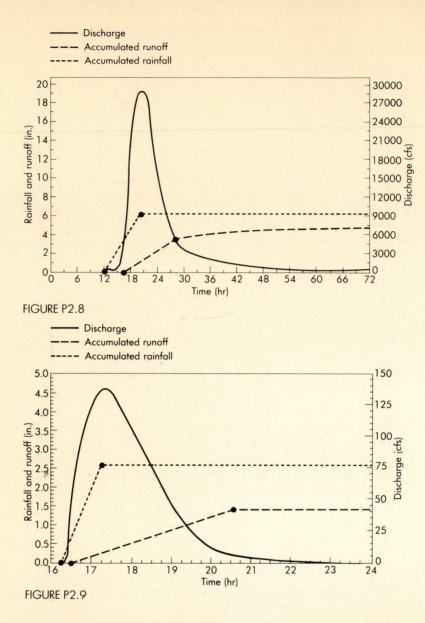

FIGURE P2.8

FIGURE P2.9

2.10. A watershed has the following characteristics:

$A = 2600$ ac,

$L = 4$ mi,

$S = 53$ ft/mi,

$I = 40\%$,

and the channel is lined with concrete.

a) Using the Espey-Winslow equations, determine the 30-min unit hydrograph (t_p and Q_p).

b) Using the Espey-Altman equations or nomographs, determine the 10-min unit hydrograph (T_R, T_B, Q_p, W_{50}, W_{75}). Sketch the hydrograph.

2.11. The watershed of Problem 2.10 is a residential area with 1/4-ac lots. The soil is categorized as soil group B. Determine the unit hydrograph for this area for a storm duration of 1 hr using SCS methods. Assume that the average watershed slope is the same as the channel slope.

2.12. Using the relations obtained from Eq. (2.3), develop a storm hydrograph for the rainfall pattern given in Fig. P2.12 using the 30-min unit hydrograph given in the accompanying table.

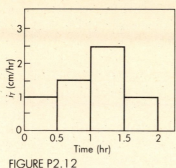
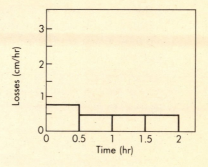

FIGURE P2.12

TIME (hr)	0	0.5	1.0	1.5	2.0	2.5	3.0	3.5
U (m³/s)	0	30	80	50	30	15	5	0

2.13. Using the FORTRAN program listed in Appendix E, develop S-curves for the following unit hydrographs.

a) $D = 30$ min

TIME (hr)	0	0.5	1.0	1.5	2.0	2.5	3.0	3.5
U (m³/s)	0	30	80	50	30	15	5	0

b) $D = 3$ hr

TIME (hr)	0	3	6	9	12	15	18	21	24	27	30	33
U (cfs)	0	175	200	150	100	75	50	40	20	10	5	0

c) $D = 6$ hr

TIME (hr)	0	6	12	18	24	30	36	42	48
U (cfs)	0	80	260	240	110	75	50	10	0

2.14. Use the S-curve program listed in Appendix E to develop
   a) the 1-hr unit hydrograph from the 3-hr unit hydrograph of Problem 2.13(b),
   b) the 3-hr unit hydrograph from the 6-hr unit hydrograph of Problem 2.13(c).

2.15. Using the add-lag program listed in Appendix E, repeat Problem 2.2 and compare the results.

2.16. Repeat Problem 2.12 with the add-lag program listed in Appendix E. Compare the program results with the hand calculations.

REFERENCES

Anderson, E. A., 1973, "National Weather Service River Forecast System: Snow Accumulation and Ablation Model," *NOAA Tech. Memo. NWS HYDRO-17,*U.S. Dept. of Commerce, Washington, D.C.

Anderson, E. A., 1976, *A Point Energy and Mass Balance of a Snow Cover,* NOAA Tech. Rept. NWS 19, U.S. Dept of Commerce, Washington, D.C.

Capece, J. C., K. L. Campbell, and L. B. Baldwin, 1984, "Estimating Runoff Peak Rates and Volumes from Flat, High-Water-Table Watersheds," Paper no. 84-2020, ASAE, St. Joseph, Missouri.

Chow, V. T. (editor), 1964, *Handbook of Applied Hydrology,* McGraw-Hill Book Company, New York.

Clark, C. O., 1945, "Storage and the Unit Hydrograph," *ASCE Trans.,* vol. 110, pp. 1419–1446.

Diskin, M. H., 1964, *A Basic Study of the Linearity of the Rainfall Runoff Process in Watersheds,* Ph.D. Thesis, University of Illinois, Urbana.

Dooge, J. C. I., 1959, "A General Theory of the Unit Hydrograph," *J. Geophys. Res.,* vol. 64, no. 2, pp. 241–256.

Dooge, J. C. I., 1973, *Linear Theory of Hydrologic Systems,* Agr. Res. Ser. Tech. Bull. No. 1468, United States Department of Agriculture, Washington, D.C.

Eagleson, P. S., 1962, "Unit Hydrographs for Sewered Areas," *Proc. ASCE, J. Hyd. Div.,* vol. 88, no. HY2, pp. 1–25.

Eagleson, P. S., 1970, *Dynamic Hydrology,* McGraw-Hill Book Company, New York.

Espey, W. H., Jr., D. G. Altman, and C. B. Graves, 1977, *Nomographs for Ten-Minute Unit Hydrographs for Small Urban Water-*

sheds, ASCE Urban Water Resources Research Program, Tech. Memo 32 (NTIS PB-282158), ASCE, New York.

Espey, W. H., Jr., and D. E. Winslow, 1968, *The Effects of Urbanization on Unit Hydrographs for Small Watersheds, Houston, Texas,* TRACOR for the Office of Water Resources Research, U.S. Department of the Interior, Austin, Texas.

Gray, D. M., 1962, *Derivation of Hydrographs for Small Watersheds from Measurable Physical Characteristics,* Iowa State University, Agr. and Home Econ. Expt. Sta. Res. Bull. 506, pp. 514–570.

Henderson, F. M., and F. A. Wooding, 1964, "Overland Flow and Groundwater Flow from Steady Rainfall of Finite Duration," *J. Geophys. Res.,* vol. 69, no. 8, pp. 1531–1539.

Horton, R. E., 1935, *Surface Runoff Phenomena,* Edwards Brothers, Inc., Ann Arbor, Michigan.

Huber, W. C., J. P. Heaney, S. J. Nix, R. E. Dickinson, and D. J. Polmann, 1981, *Storm Water Management Model User's Manual, Version III,* EPA-600/2-84-109a (NTIS PB84-198423), EPA, Athens, Georgia.

Hydrologic Engineering Center, 1981, *HEC-1 Flood Hydrograph Package: User's Manual* and *Programmer's Manual,* updated 1985, U.S. Army Corps of Engineers, Davis, California.

Johnstone, D., and W. P. Cross, 1949, *Elements of Applied Hydrology,* Ronald Press Company, New York.

Kohler, M. A., and R. K. Linsley, Jr., 1951, *Predicting the Runoff from Storm Rainfall,* U.S. Weather Bureau, Research Paper 34.

Kuichling, E., 1889, "The Relation Between the Rainfall and the Discharge of Sewers in Populous Districts," *ASCE Trans.,* vol. 20, pp. 1–56.

Lighthill, M. J., and G. B. Whitham, 1955, "On Kinematic Waves, Part I: Flood Movement in Long Rivers," *Roy. Soc. (London) Proc.,* ser. A, vol. 229, no. 1178, pp. 281–316.

Lloyd-Davies, D. E., 1906, "The Elimination of Storm Water from Sewerage Systems," *Inst. Civ. Eng. Proc.,* vol. 164, pp. 41–67.

Mulvaney, T. J., 1851, "On the Use of Self-Registering Rain and Flood Gauges," *Inst. Civ. Eng. (Ireland) Trans.,* vol. 4, no. 2, pp. 1–8.

Nash, J. E., 1958, "The Form of the Instantaneous Unit Hydrograph," *General Assembly of Toronto, Internatl. Assoc. Sci. Hydrol. (Gentbrugge) Pub. 42,* Compt. Rend. 3, pp. 114–118.

Nash, J. E., 1959, "Systematic Determination of Unit Hydrograph Parameters," *J. Geophys. Res.,* vol. 64, no. 1, pp. 111–115.

Newton, D. W., and J. W. Vineyard, 1967, "Computer-Determined Unit Hydrograph from Floods," *Proc. ASCE, J. Hyd. Div.,* vol. 93, no. HY5, pp. 219–236.

O'Kelly, J. J., 1955, "The Employment of Unit-Hydrographs to Determine the Flows of Irish Arterial Drainage Channels," *Inst.Civ. Eng. (Ireland) Proc.,* vol. 4, no. 3, pp. 365–412.

Overton, D. E., and M. E. Meadows, 1976, *Stormwater Modeling,* Academic Press, New York.

Raudkivi, A. J., 1979, *Hydrology,* Pergamon Press, Elmsford, New York.

Sherman, L. K., 1932, "Streamflow from Rainfall by the Unit-Graph Method," *Eng. News-Rec.,* vol. 108, pp. 501–505.

Smith, D. P., Jr., and P. B. Bedient, 1980, "Detention Storage for Urban Flood Control," *J. Water Res. Plan., Man. Div., ASCE,* vol. 106, no. WR2, pp. 413–425.

Snyder, F. F., 1938, "Synthetic Unit Graphs," *Trans. AGU,* vol. 19, pp. 447–454.

Snyder, W. M., 1955, "Hydrograph Analysis by the Method of Least Squares," *ASCE, J. Hyd. Div.,* vol. 81, pp. 1–25.

Soil Conservation Service, 1957, *Use of Storm and Watershed Characteristics in Synthetic Hydrograph Analysis and Application,* U.S. Department of Agriculture, Washington, D.C.

Soil Conservation Service, 1964, *SCS National Engineering Handbook, Section 4: Hydrology,* U.S. Department of Agriculture, Washington, D.C.

Soil Conservation Service, 1986, *Urban Hydrology for Small Watersheds,* 2nd ed., Tech. Release No. 55 (NTIS PB87-101580), U.S. Department of Agriculture, Washington, D.C.

Tauxe, G. W., 1978, "S-Hydrographs and Change of Unit Hydrograph Duration," *ASCE Tech. Notes, J. Hyd. Div.,* vol. 104, no. HY3, pp. 439–444.

Taylor, A. B., and H. E. Schwartz, 1952, "Unit Hydrograph Lag and Peak Flow Related to Basin Characteristics," *Trans. Am. Geophys. Union.,* vol. 33.

United States Army Corps of Engineers, North Pacific Division, 1956, *Snow Hydrology,* Portland, Oregon.

Viessman, W., J. W. Knapp, G. L. Lewis, and T. E. Harbaugh, 1977, *Introduction to Hydrology,* 2nd ed., Harper & Row, New York.

Wooding, R. A., 1965, "A Hydraulic Model for the Catchment-Stream Problem, Part I: Kinematic Wave Theory," *J. Hydrol.,* vol. 3, pp. 254–267.

3

Frequency Analysis

3.1
INTRODUCTION

Scope of the Chapter

Many processes in hydrology must be analyzed and explained in a probabilistic sense because of their inherent randomness. For instance, it is not possible to predict streamflow and rainfall on a purely deterministic basis in either the past ("hindcasting") or future (forecasting) since it is impossible to know all their causal mechanisms quantitatively. Fortunately, statistical methods are available to organize, present, and reduce observed data to a form that facilitates their interpretation and evaluation. This chapter presents methods by which the uncertainty in hydrologic data may be quantified and presented in a standard probabilistic framework.

Random Variables

A **random variable** is a parameter (e.g., streamflow, rainfall, stage) that cannot be predicted with certainty; that is, a random variable is the outcome of a random or uncertain process. Such variables may be treated statistically as either **discrete** or **continuous**. Most hydrologic data are continuous and are analyzed probabilistically using continuous frequency distributions. For example, values of flow in the hydrograph

139

sketched in Fig. 3.1(a) can equal any positive real number, to the accuracy of the flow meter; that is, the data are continuous. However, the data themselves are often presented in a discrete form because of the measurement process. Daily streamflows may be estimated to the nearest cfs. This form of data presentation is called **quantized-continuous;** that is, the continuous observations are assigned to quantiles. This is also illustrated in Fig. 3.1(a), in which flows are assigned to the nearest cfs.

Discrete random variables can take on values only from a specified domain of discrete values. For example, flipping a coin will produce either a head or tail; rolling a die produces an integer from 1 to 6. The output from a tipping bucket rain gage is a hydrologic example (Fig. 3.1b): there either is or is not a tip during any time interval. Quantized-continuous data could be treated as discrete; indeed, they are discretized whenever

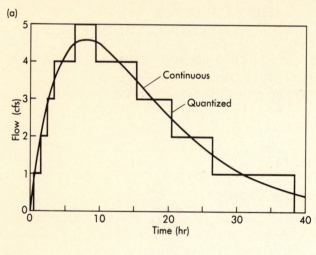

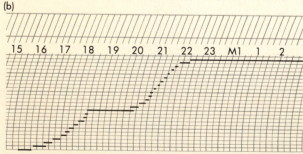

FIGURE 3.1

Continuous and discrete data. (a) Continuous and quantized data. (b) Discrete data. Strip chart output from a tipping bucket rain gage. Each vertical increment is 0.01 in. of rain.

tabulated data are processed since numerical values must be truncated (e.g., to the nearest 1 cfs for flow or to the nearest 0.01 in. for rainfall). However, for frequency analysis this is usually very inconvenient because of the enormous range of values that would have to be considered. For example, if streamflows are measured to the nearest cfs and there is a range of flows from 0 to 5000 cfs, then 5000 discrete intervals would have to be considered. Approximating the record as continuous is much easier. Although discrete frequency distributions are sometimes applied to continuous variables (e.g., storm rainfall depths), most applications of discrete distributions in hydrology are to a random variable that represents the *number* of events satisfying a certain criterion, for example, the number of floods expected to exceed a certain magnitude during a period of years.

Presentation of Data

Quantized-continuous data are often presented in the form of a *bar chart* or **histogram.** The height and general shape of the histogram are useful for characterizing the data and lend insight into the selection of a frequency distribution to be applied to the data, for example, whether or not the distribution should be symmetric or skewed. Using streamflows as an example, the range of flows is first divided into **class intervals** and the number of observations **(frequency)** corresponding to each class interval is tabulated. The width of the class intervals should be small enough so that the underlying pattern of the data may be seen but large enough so that such a pattern does not become confusing. The value used for the class interval can alter the viewer's impression of the data (Benjamin and Cornell, 1970); this value may conveniently be altered in many computer plotting programs so that the engineer can compare several different options. As an aid, Panofsky and Brier (1968) suggest

$$k = 5 \log_{10} n, \tag{3.1}$$

where k is the number of class intervals and n is the number of data values. Class intervals do not have to be of constant width if it is more convenient to group some data into a larger, aggregated interval.

If the ordinates of the histogram are divided by the total number of observations, **relative frequencies** (probabilities) of each class interval result, such that the ordinates sum to 1.0, providing an alternative method of plotting the data. Still a third way is in the form of a **cumulative frequency distribution,** which represents a summation of the histogram relative frequencies up to a given interval and is the probability that a value along the abscissa is less than or equal to the magnitude at that point. Both relative frequencies and cumulative frequencies are extensively used in hydrology and are best illustrated by an example.

EXAMPLE 3.1

FREQUENCY HISTOGRAMS

Records of 31 years of annual maximum flows for Cypress Creek, near Houston, Tex., are presented in Table 3.1. Equation (3.1) indicates that 7 or 8 class intervals would be appropriate. Eight is used here since it permits convenient boundaries of 2000 cfs. (This criterion is more important than any other rule of thumb for the number of class intervals.)

TABLE 3.1

Data and Frequency Computations for Cypress Creek, near Houston, Tex.

UNRANKED DATA		RANKED DATA		UNRANKED DATA		RANKED DATA	
Year	Flow (cfs)	Rank	Flow (cfs)	Year	Flow (cfs)	Rank	Flow (cfs)
1945	9840	1	15600	1961	6260	17	3310
1946	5170	2	10300	1962	1360	18	3210
1947	1620	3	9840	1963	1000	19	3000
1948	235	4	7760	1964	2770	20	2820
1949	15600	5	6560	1965	1400	21	2770
1950	4740	6	6260	1966	3210	22	2520
1951	427	7	5440	1967	1110	23	1900
1952	3310	8	5230	1968	5230	24	1620
1953	4400	9	5170	1969	4300	25	1400
1954	7760	10	4740	1970	2820	26	1360
1955	2520	11	4710	1971	1900	27	1110
1956	340	12	4400	1972	3980	28	1000
1957	5440	13	4300	1973	6560	29	427
1958	3000	14	3980	1974	4710	30	340
1959	3690	15	3690	1975	3460	31	235
1960	10300	16	3460				

CLASS INTERVAL (cfs)	CLASS MARK (cfs)	FREQUENCY	RELATIVE FREQUENCY	CUMULATIVE FREQUENCY
0–2000	1000	9	0.29	0.29
2000–4000	3000	9	0.29	0.58
4000–6000	5000	7	0.23	0.81
6000–8000	7000	3	0.10	0.91
8000–10,000	9000	1	0.03	0.94
10,000–12,000	11,000	1	0.03	0.97
12,000–14,000	13,000	0	0.00	0.97
14,000–16,000	15,000	1	0.03	1.00
		$\Sigma = 31$		

Frequencies, relative frequencies, and cumulative frequencies are also given in the table and are plotted in Figs. 3.2 and 3.3. For example, in Fig. 3.2, the probability that a flow lies between 2000 and 4000 cfs is 0.29. From the cumulative frequency histogram (Fig. 3.3), the probability that a flow is less than or equal to 4000 cfs is 0.58. Note that the relative frequencies sum to 1.0 as indicated in Table 3.1 and by the final ordinate shown in Fig. 3.3.

One way of discretizing these continuous flow data would be to assign a discrete random variable to each class interval. Any flow falling within a class interval would then be assigned the discrete value corresponding to that interval, usually the midpoint of the interval or the **class mark**. In this case, just the class mark would be plotted on the abscissa (Fig. 3.4). A corresponding probability statement is that the probability that a flow equals 3000 cfs is 0.29 (this approximation is made even though it is known that a continuous random variable can never exactly equal any specific numerical value).

The positive **skew** of the Cypress Creek flows is apparent (Figs. 3.2 and 3.4); that is, there is a decided "tail" to the right. This will be quantified and related to various distributions in further examples.

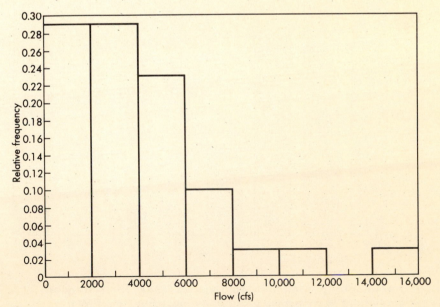

FIGURE 3.2

Relative frequency histogram for Cypress Creek, near Houston, Tex.

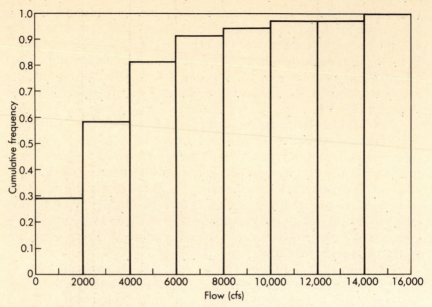

FIGURE 3.3

Cumulative frequency histogram for Cypress Creek.

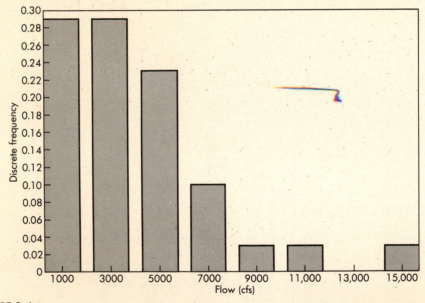

FIGURE 3.4

Discrete frequencies (probabilities) for Cypress Creek. These frequencies are equivalent to a discrete probability mass function (PMF).

3.2

PROBABILITY CONCEPTS

Consider an experiment with N possible outcomes, X_1, X_2, . . . , X_i, . . . , X_N. The outcomes are **mutually exclusive** if no two of them can occur simultaneously. They are **collectively exhaustive** if they account for all possible outcomes of the experiment. The **probability** of an event X_i may be defined as the relative number of occurrences of the event after a very large number of trials. This probability may be estimated as $P(X_i) = n_i/n$, where n_i is the number of occurrences (frequency) of event X_i in n trials. Thus, n_i/n is the relative frequency or probability of occurrence of X_i.

A **discrete probability** is simply the probability of a discrete event. If one defines $P(X_i)$ as the probability of the random event X_i, the following conditions hold on the discrete probabilities of these events when considered over the sample space of all possible outcomes:

$$0 \le P(X_i) \le 1, \tag{3.2}$$

$$\sum_{i=1}^{N} P(X_i) = 1. \tag{3.3}$$

The probability of the **union** (occurrence of either, symbolized by $\cup$) of two mutually exclusive events is the sum of the probabilities of each:

$$P(X_1 \cup X_2) = P(X_1) + P(X_2). \tag{3.4}$$

Two events X_1 and Y_1 are **independent** if the occurrence of one does not influence the occurrence of the other. The probability of the **intersection** (occurrence of both, symbolized by $\cap$) of two independent events is the product of their individual probabilities:

$$P(X_1 \cap Y_1) = P(X_1) \cdot P(Y_1). \tag{3.5}$$

For events that are neither independent or mutually exclusive,

$$P(X_1 \cup Y_1) = P(X_1) + P(Y_1) - P(X_1 \cap Y_1). \tag{3.6}$$

The conditional probability of event X_1 given that event Y_1 has occurred is

$$P(X_1|Y_1) = P(X_1 \cap Y_1)/P(Y_1). \tag{3.7}$$

If events X_1 and Y_1 are independent, then Eqs. (3.5) and (3.7) combine to give

$$P(X_1|Y_1) = P(X_1) \cdot P(Y_1)/P(Y_1) = P(X_1). \tag{3.8}$$

These concepts are often illustrated on Venn diagrams (Fig. 3.5), on which areas are proportional to probabilities, with the total area corresponding to a probability of 1.0, or 100%.

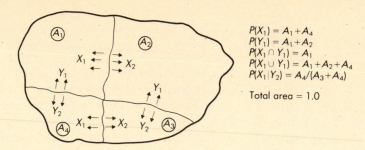

FIGURE 3.5

Venn diagram for illustration of probabilities.

EXAMPLE 3.2

CONDITIONAL PROBABILITIES

Let event Y_1 be the condition that a rainstorm occurs on a given day and event X_1 be the condition that lightning is observed on a given day.

Let probabilities of these events be

$$P(X_1) = 0.3 \quad \text{(Probability of lightning} = 30\%),$$
$$P(Y_1) = 0.1 \quad \text{(Probability of rain} = 10\%),$$
$$P(Y_1|X_1) = 0.5 \quad \text{(If there is lightning, the conditional probability of rain is 50\%).}$$

What is the probability that *both* rain and lightning occur (i.e., the probability of the intersection of X_1 and Y_1)? From Eq. (3.7),

$$P(X_1 \cap Y_1) = P(Y_1|X_1) \cdot P(X_1) = 0.15.$$

But what if both events were independent, such that $P(Y_1|X_1) = P(Y_1) = 0.1$? Then

$$P(X_1 \cap Y_1) = P(X_1) \cdot P(Y_1) = 0.03.$$

The probability of the joint occurrence of independent events will always be less than or equal to the probability of their joint occurrence if they are dependent.

3.3

RANDOM VARIABLES AND PROBABILITY DISTRIBUTIONS

Discrete and Continuous Random Variables

The behavior of a random variable may be described by its probability distribution. Every possible outcome of an experiment is assigned a numerical value according to a discrete probability mass function (PMF) or a continuous probability density function (PDF). In hydrology, discrete random variables are most commonly used to describe the *number* of occurrences that satisfy a certain criterion, e.g., the number of floods that exceed a specified value, the number of storms that occur at a given location, etc. Examples later in this chapter will be of this type. As a rule, discrete random variables are associated only with parameters that can assume only integer values. However, it is possible to round continuous variables to the nearest integer or to rescale to integers. For example, 2.18 in. of rain is 218 hundredths of an inch of rain.

Occasionally, it is convenient to treat a continuous variable in a discrete sense, as for the flows described in Fig. 3.4. Let the notation $P(x_1)$ mean the probability that the discrete random variable X takes on the value x_1. Letting X represent "discrete" streamflows, we can assign relative frequencies from Fig. 3.4:

$$P(1000) = 0.29, \qquad P(9000) \;\; = 0.03,$$
$$P(3000) = 0.29, \qquad P(11000) = 0.03,$$
$$P(5000) = 0.23, \qquad P(13000) = 0.00,$$
$$P(7000) = 0.10, \qquad P(15000) = 0.03.$$

Note that these values satisfy the probability axioms of Eqs. (3.2) and (3.3). Furthermore, for discrete probabilities,

$$P(a \le x \le b) = \sum_{a \le x_i \le b} P(x_i). \tag{3.9}$$

The cumulative distribution function (CDF) is defined as

$$F(x) = P(X \le x) = \sum_{x_i \le x} P(x_i). \tag{3.10}$$

From the values tabulated above, $F(7000) = 29 + 29 + 23 + 10 = 91\%$.

Continuous random variables are usually used to represent hydrologic phenomena such as flow, rainfall, volume, depth, and time. Values are not restricted to integers, although continuous variables might be commonly rounded to integers. For a continuous random variable, the *area* under the **probability density function** $f(x)$ represents probability. Thus (see Fig. 3.6),

$$P(x_1 \le x \le x_2) = \int_{x_1}^{x_2} f(x)\, dx, \tag{3.11}$$

and the area under the PDF equals 1.0:

$$\int_{-\infty}^{\infty} f(x)\, dx = 1.0. \tag{3.12}$$

The PDF itself is *not* a probability and has units the inverse of the units of X, e.g., cfs^{-1}. However, unlike other engineering calculations, it is

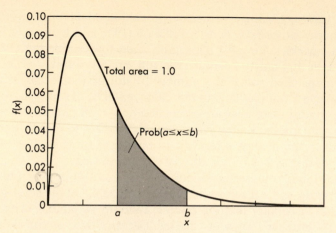

FIGURE 3.6
Continuous probability density function (PDF).

not customary to list the units of the PDF. In fact, the numerical value of the PDF is rarely needed. Rather, it is the **cumulative density function** (CDF) that is of interest since it *is* a probability. The continuous CDF is defined similarly to its discrete counterpart:

$$F(x_1) = P(-\infty \le x \le x_1) = \int_{-\infty}^{x_1} f(x)\, dx. \tag{3.13}$$

Other properties include

$$0 \le F(x) \le 1.0 \tag{3.14}$$

and

$$P(x_1 < x \le x_2) = F(x_2) - F(x_1). \tag{3.15}$$

The PDF and CDF are related by Eq. (3.13), whose inverse is

$$\frac{dF(x)}{dx} = f(x) \tag{3.16}$$

The histogram of Fig. 3.2 may be represented by a continuous PDF if the relative frequencies are divided by the class interval Δx. The area under the histogram is then 1.0. For example, if the ordinates of the relative frequency histogram of Fig. 3.2 were divided by 2000 cfs, the figure would correspond to a PDF. This illustrates that PDFs may have any nonnegative and single-valued shape; they need not look like smooth curves.

It is possible to have mixed distributions in which a discrete probability mass represents the probability that a variable takes on a specific (discrete) value, while a continuous PDF represents the rest of the range, with an area equal to 1.0 minus the discrete probability mass. For example, a mixed distribution is shown in Fig. 3.7, in which the probability is 0.15 that the flow is zero.

The choice of which continuous PDF to use to represent the data is difficult since several candidates often are able to mimic the shape of the frequency histogram (Fig. 3.2). Representative PDFs commonly used for hydrologic variables will be described subsequently.

Moments of a Distribution

The concept of moments is common from engineering mechanics. A PMF or PDF is a functional form whose moments are related to its parameters. Thus, if moments can be found, then generally so can the parameters of the distribution. The moments themselves are also indicative of the shape of the distribution.

For a discrete distribution, the Nth moment about the origin can be defined as

$$\mu_N' = \sum_{i=-\infty}^{\infty} x_i^N P(x_i) \tag{3.17}$$

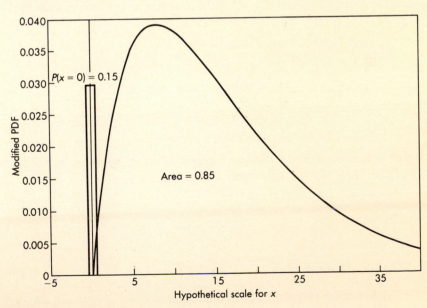

FIGURE 3.7

Mixed frequency distribution. Use discrete probability mass function (PMF) for probability of zero value and continuous probability density function (PDF) (scaled to area = 0.85) for probabilities for values greater than zero.

and for a continuous distribution as

$$\mu_N' = \int_{-\infty}^{\infty} x^N f(x)\, dx. \tag{3.18}$$

The first moment about the origin is the mean or average or expected value, denoted by $E(\ )$ for "expectation." Thus,

$$E(x) \equiv \mu = \sum_{-\infty}^{\infty} x_i\, P(x_i) \qquad \text{(for a discrete PMF)} \tag{3.19}$$

and

$$E(x) \equiv \mu = \int_{-\infty}^{\infty} x f(x)\, dx \qquad \text{(for a continuous PDF)}. \tag{3.20}$$

The mean is a measure of central tendency and is also called a *location parameter* because it indicates where along the x-axis the bulk of the distribution is located.

Often the distribution of one variable will be known and information will be desired about a related variable. For example, the distribution of flows may be known and information desired about stage, which is a function of flow. The expected value of a function $g(x)$ of a random variable x may be defined in a similar manner as for the original variable x:

$$E[g(x)] = \sum_{-\infty}^{\infty} g(x_i)\, P(x_i) \tag{3.21}$$

when x is a discrete random variable, and

$$E[g(x)] = \int_{-\infty}^{\infty} g(x)f(x)\, dx \tag{3.22}$$

when x is a continuous random variable. The expectation is a linear operator, such that if a and b are constants,

$$E(a) = a, \tag{3.23}$$
$$E(bx) = bE(x), \tag{3.24}$$

and

$$E(a + bx) = a + bE(x). \tag{3.25}$$

Higher-order moments about the origin of distributions are usually not needed. Instead, **central moments** about the mean may be defined for a discrete PMF as

$$\mu_N = \sum_{-\infty}^{\infty} (x_i - \mu)^N P(x_i) \tag{3.26}$$

and for a continuous PDF as

$$\mu_N = \int_{-\infty}^{\infty} (x - \mu)^N f(x) \, dx. \tag{3.27}$$

These moments are simply the expected value of the difference between x and the mean, raised to the Nth power. Clearly, the first central moment is zero. The second central moment is called the **variance** and is very important:

$$\text{Var}(x) \equiv \sigma^2 = \mu_2 = E[(x - \mu)^2] = \sum_{-\infty}^{\infty} (x_i - \mu)^2 P(x_i) \tag{3.28}$$

for discrete random variables and

$$\text{Var}(x) \equiv \sigma^2 = \mu_2 = E[(x - \mu)^2] = \int_{-\infty}^{\infty} (x - \mu)^2 f(x) \, dx \tag{3.29}$$

for continuous random variables. The variance is the expected value of the squared deviations about the mean and represents the scale, or spread, of the distribution. An equivalent measure is the **standard deviation** σ, which is simply the square root of the variance. From the properties of the expectation,

$$\begin{aligned}
\text{Var}(x) &= E[(x - \mu)^2] = E[x^2 - 2x\mu + \mu^2] \\
&= E[x^2] - E[2\mu x] + E[\mu^2] \\
&= E[x^2] - 2\mu E[x] + \mu^2 = E[x^2] - 2\mu^2 + \mu^2 \\
&= E[x^2] - \mu^2 \\
&= E[x^2] - [E(x)]^2. \tag{3.30}
\end{aligned}$$

The variance is *not* a linear operator. Useful relationships include

$$\text{Var}(a) = 0, \tag{3.31}$$
$$\text{Var}(bx) = b^2 \text{Var}(x), \tag{3.32}$$
$$\text{Var}(a + bx) = b^2 \text{Var}(x), \tag{3.33}$$

where a and b are constants.

Higher moments can be defined if needed, but the only one commonly used in hydrology is the **skewness**, which is the third central moment normalized by dividing by the cube of the standard deviation:

$$g \equiv \mu_3 / \sigma^3. \tag{3.34}$$

The skewness is a shape parameter and is illustrated in Fig. 3.8. If the distribution is symmetric, the skewness is zero.

It is sometimes useful to have a normalized measure of the scale of the distribution. The coefficient of variation, defined as the ratio of the standard deviation to the mean, may be used for this purpose:

$$CV = \sigma / \mu. \tag{3.35}$$

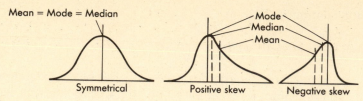

FIGURE 3.8

Effect of skewness on probability density function (PDF) and relative locations of mean, median, and mode. (From Haan, 1977, Fig. 3.3.)

An additional measure of central tendency is the **median** x_m, which is not a moment but rather the value of x for which the CDF equals 0.5:

$$F(x_m) = 0.5. \tag{3.36}$$

Another parameter of occasional interest that is not a moment is the **mode** of the distribution. This is the value of x at which the PDF (or PMF) is a maximum. The relationship between the mean, median, and mode is also illustrated in Fig. 3.8. Most (but not all) distributions of interest are unimodal. (The mixed distribution of Fig. 3.7 is bimodal.)

The moments and parameters discussed in this subsection refer to the underlying probability distributions and may be derived analytically. Functional forms for PMFs or PDFs can be substituted into the summations or integrals and the moments evaluated in terms of the parameters of the distribution. This task is not illustrated here since the relationships between moments and distribution parameters will be given for each distribution discussed. The relationships provide a simple method of obtaining estimates of distribution parameters if the moments are known. For this purpose, estimates of the moments must be made from the data.

Estimates of Moments from Data

Given numerical values for the parameters of a distribution, it is possible to generate a series $x_1, x_2, \ldots, x_n$ of random variables that belong to a given PMF or PDF. Such a series of infinite length would constitute the **population** of all random variables belonging to a given PMF or PDF with a given set of parameters. Similarly, the parameters define the moments since they are related, as discussed above (and demonstrated subsequently). Observed hydrologic data often result from a mixture of physical processes (e.g., runoff may result from a mixture of rainfall and snowmelt) and hence may incorporate a mixture of underlying probability distributions. In addition, observed data are ordinarily subject to errors of observation and will not fit any distribution perfectly. Hence, the underlying population values of the moments calculated from such data will remain unknown. However, *estimates* of their values may readily be obtained from the data, as shown below for the three moments of most importance in hydrology. If the number of independent samples of a random variable is n, an estimate of the mean is

$$\hat{\mu} = \bar{x} = \frac{1}{n} \sum_{i=1}^{n} x_i. \qquad (3.37)$$

Higher moments are subject to **bias** in their estimates. An unbiased estimate is one for which the expected value of the estimate equals the population value. It can be shown (Benjamin and Cornell, 1970) that $E(\bar{x}) = \mu$, as desired. For the variance, an unbiased estimate is

$$\hat{\sigma}^2 \equiv S^2 = \frac{1}{n-1} \sum_{i=1}^{n} (x_i - \bar{x})^2$$

$$= \frac{\sum x_i^2 - n\bar{x}^2}{n-1}, \qquad (3.38)$$

where the divisor $n - 1$ (instead of the intuitive value n) eliminates the bias. (Unless stated otherwise, all summations are from 1 to n.) Clearly, for large samples, a divisor of n would yield almost the same answer, and both divisors may be found in practice, although the unbiased estimate is usually preferred. Computationally, the second form of Eq. (3.38) is often preferable.

Since the moment estimates are functions of random variables, they themselves are random variables. The variance of the estimate of the mean may be computed to be

$$\mathrm{Var}(\bar{x}) \equiv S_{\bar{x}}^2 = \frac{S_x^2}{n}. \qquad (3.39)$$

Thus, if the variance of the mean is interpreted as a measure of the error in the estimate of the mean, it is reduced as the sample size increases. This is generally true for all moment estimates.

The skewness presents special problems since it involves a summation of the cubes of deviations from the mean and is therefore subject to larger errors in its computation. An approximately unbiased estimate is

$$C_{s1} \equiv \hat{g} = \frac{n}{(n-1)(n-2)} \cdot \frac{\sum (x_i - \bar{x})^3}{S_x^3}. \qquad (3.40)$$

with S_x^2 given by Eq. (3.38). Unfortunately, the appropriate bias correction depends on the underlying distribution (Bobee and Robitaille, 1975), but a correction to C_{s1} suitable for most hydrologic applications (Tasker and Stedinger, 1986) is

$$C_{s2} = (1 + 6/n) \cdot C_{s1} \qquad (3.41)$$

This is strictly for the use of the Pearson type 3 distribution (Section 3.5) but will also serve as an improvement over C_{s1} for use with other distributions. The skewness estimate computed using Eq. (3.41) is called the **station estimate,** meaning that the estimate incorporates data values only from the gaging station of interest.

Error and bias in the skewness estimate increase as the number of observations n decreases. As an alternative to the station estimate for skewness, the Interagency Advisory Committee on Water Data (1982) recommends using **regional skew coefficients** for flood flow analysis instead of the station estimate if there are fewer than 25 years of records at a station and using a weighted estimate for a period of record n between 25 and 100. The station skew is used for n greater than 100. Regional skew coefficients, C_{sr} (for logs of streamflows) for the United States are shown in Fig. 3.9. The weighted estimate is thus

$$C_{sw} = C_{s2} \frac{n-25}{75} + C_{sr} \left(1 - \frac{n-25}{75} \right) \tag{3.42}$$

Improvements to the regional skew coefficients shown in Fig. 3.9 can be made by performing a regional analysis of skewness data at a local (e.g., state, river basin) level.

Urbanizing watersheds pose difficult problems in the estimation of moments since the runoff tends to increase with time. Hence, the frequency distributions may be **nonstationary** (time-varying). Simulation models, described in Chapters 5 and 6, may be used to develop frequency distributions for flows in rapidly urbanizing basins at a given stage of development.

EXAMPLE 3.3

MOMENTS OF AN ANNUAL MAXIMUM SERIES

The series of 31 annual maximum flows on Cypress Creek north of Houston, Tex., is shown again in Table 3.2. Evaluate the mean and standard deviation of the original data and of the logs (base 10) of the data using Eqs. (3.37) and (3.38). Compare the various skewness estimates.

SOLUTION

The results of our evaluation are as follows.

	ORIGINAL DATA (cfs)	$\log_{10}$ DATA (log cfs)
Mean	4144	3.463
Standard deviation	3311	0.424
Skewness (Eq. 3.40)	1.659	−0.936
Skewness (Eq. 3.41)	1.981	−1.117

Considering the $\log_{10}$ values, using the regional data from Fig. 3.9 gives $C_{sr} = -0.3$ for Houston. A weighted average using Eq. 3.42 gives an

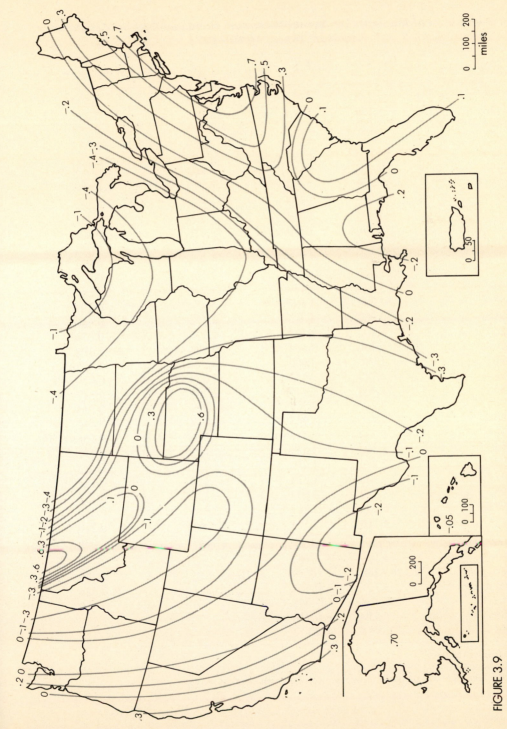

FIGURE 3.9

Generalized skew coefficients of logarithms of annual maximum streamflow. (From Interagency Advisory Committee on Water Data, 1982.)

TABLE 3.2

Computation of Moments for Annual Maximum Floods on Cypress Creek, near Houston, Texas, 1945 – 1975.

YEAR	FLOW (cfs)	log 10 FLOW	FLOW – MEAN	log(FLOW) – MEAN*
1945	9,840	3.993	5,696	0.530
1946	5,170	3.713	1,026	0.251
1947	1,620	3.210	−2,524	−0.253
1948	235	2.371	−3,909	−1.092
1949	15,600	4.193	11,456	0.730
1950	4,740	3.676	596	0.213
1951	427	2.630	−3,717	−0.832
1952	3,310	3.520	−834	0.057
1953	4,400	3.643	256	0.181
1954	7,760	3.890	3,616	0.427
1955	2,520	3.401	−1,624	−0.061
1956	340	2.531	−3,804	−0.931
1957	5,440	3.736	1,296	0.273
1958	3,000	3.477	−1,144	0.014
1959	3,690	3.567	−454	0.104
1960	10,300	4.013	6,156	0.550
1961	6,260	3.797	2,116	0.334
1962	1,360	3.134	−2,784	−0.329
1963	1,000	3.000	−3,144	−0.463
1964	2,770	3.442	−1,374	−0.020
1965	1,400	3.146	−2,744	−0.317
1966	3,210	3.507	−934	0.044
1967	1,110	3.045	−3,034	−0.417
1968	5,230	3.719	1,086	0.256
1969	4,300	3.633	156	0.171
1970	2,820	3.450	−1,324	−0.013
1971	1,900	3.279	−2,244	−0.184
1972	3,980	3.600	−164	0.137
1973	6,560	3.817	2,416	0.354
1974	4,710	3.673	566	0.210
1975	3,460	3.539	−684	0.076
Mean	4,144	3.463	0	0
Variance†	10,961,550	0.179	10,961,550	0.179
Stand. Dev.†	3,311	0.424	3,311	0.424
Sum	128,462	107.345	0	0
CS1	1.659	−0.936		
CS2	1.981	−1.117		

* Mean of the logs.
† Unbiased.
Note: This table was prepared using microcomputer spreadsheet software.

alternative estimate for the skewness of the logs of -0.35, considerably less than the corrected station value given by Eq. 3.41. Which value is more nearly correct could be determined from a study of other stations in the Houston area. For purposes of examples in this chapter, the value of -1.117 will be used (Eq. 3.41).

Fitting a Distribution to Data

An intuitive use for moment estimates is to fit probability distributions to data by equating the moment estimates obtained from the data to the functional form for the distribution. For example, the normal distribution has parameters mean μ and variance σ^2. The **method of moments** fit for the normal distribution is simply to use the estimates from the data for the mean and variance. As another example, the parameter λ of the exponential distribution is equal to the reciprocal of the mean of the distribution. Hence, the method of moments fit is $\hat{\lambda} = 1/\hat{\mu}$.

Graphical methods and the method of maximum likelihood are two alternative procedures to the method of moments for fitting distributions to data. Graphical methods will be discussed in Section 3.6. Although the method of maximum likelihood is superior to the method of moments by some statistical measures, it is generally much more computationally complex than the method of moments and is beyond the scope of this text. Maximum likelihood methods for several distributions used in hydrology are presented by Kite (1977).

The main objective of determining the parameters of a distribution is usually to evaluate its CDF. In some cases, however, the same purpose may be accomplished without calculating the actual distribution parameters. Instead, the distribution is evaluated using frequency factors K, defined as

$$K = \frac{x - \bar{x}}{S_x}. \tag{3.43}$$

The value of K is a function of the desired value for the CDF and may also be a function of the skewness. Hence, if K is known for the calculated skewness and desired CDF value, the corresponding value of x can be calculated. Frequency factors will be illustrated subsequently for several of the distributions to be discussed.

3.4

RETURN PERIOD OR RECURRENCE INTERVAL

The most common means used in hydrology to indicate the probability of an event is to assign a **return period** or **recurrence interval** to the event. The return period is defined as follows:

An annual maximum event has a return period (or recurrence interval) of T years if its magnitude is equaled or exceeded once, *on the average,* every T years. The reciprocal of T is the exceedance probability of the event, that is, the probability that the event is equaled or exceeded in any one year.

Thus, the 50 yr flood has a probability of 0.02, or 2%, of being equaled or exceeded in any single year. It is imperative to realize that the return period implies nothing about the actual time sequence of an event. The 50-yr flood does not occur like clockwork once every 50 yr. Rather, one expects that on the average, about twenty 50-yr floods will be experienced during a 1000-yr period. There could in fact be two 50-yr floods in a row (with probability $0.02 \times 0.02 = 0.0004$ for independent events).

The concept of a return period implies independent events and is usually found by analyzing the series of maximum annual floods (or rainfalls, etc.). The largest event in one year is assumed to be independent of the largest event in any other year. But it is also possible to apply such an analysis to the n largest independent events from an n-yr period, regardless of the year in which they occur. In this case, if the second largest event in one year was greater than the largest event in another year, it could be included in the frequency analysis. This series of n largest (independent) values is called the series of **annual exceedances,** as opposed to an annual maximum series. Both series are used in hydrology, with little difference at high return periods (rare events). There are likely to be more problems of ensuring independence when using annual exceedances, but for low return periods annual exceedances give a more realistic lower return period for the same magnitude than do annual maxima. The relationship between return period based on annual exceedances T_e and annual maxima T_m is (Chow, 1964)

$$T_e = \frac{1}{\ln T_m - \ln(T_m - 1)}. \tag{3.44}$$

This relationship is shown in Fig. 3.10.

EXAMPLE 3.4

ANNUAL EXCEEDANCES AND ANNUAL MAXIMA

Consider the following hypothetical sequence of river flows.

YEAR	THREE HIGHEST INDEPENDENT FLOWS (cfs)		
1	700,	300,	150
2	900,	600,	100
3	550,	400,	200
4	850,	650,	350
5	500,	350,	100

The sequences of ranked annual maxima and annual exceedances are as follows.

RANK	ANNUAL MAXIMA	ANNUAL EXCEEDANCES
1	900	900
2	850	850
3	700	700
4	550	650
5	500	600

A further option is to use all the data in the historic time series, whether or not they are independent. This is called the complete series; an example is the series of 365 daily streamflows (from which annual maxima and annual exceedances are determined). Frequency information derived from such a complete series is usually shown in a flow-duration curve (Searcy, 1959), which is a plot of magnitude vs. percent of time the magnitude is equaled or exceeded. The information from such an analysis cannot be directly related to return period since the values in the complete series are not necessarily independent.

Return periods may be assigned to minimum events (e.g., droughts) in an entirely analogous manner, with the interpretation that "equaled or

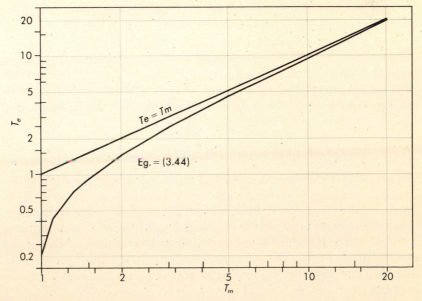

FIGURE 3.10

Relationship between return period for annual exceedances T_e and return period for annual maxima T_m (Eq. 3.44).

exceeded" means "equal to or more severe than." Thus, a 20-yr low flow is one for which the probability is 5% that the flow will be less than or equal to the given value in any one year.

Finally, return periods need not be limited to units of years. As long as the events are independent, months or even weeks can be used. The 6-month rainfall thus has a probability of 1/6 of being equaled or exceeded in any one month.

3.5
COMMON PROBABILISTIC MODELS

References and Objectives

Many discrete probability mass functions and continuous probability density functions are used in hydrology, but this text can focus on only a few of the most common. Others are covered in references on statistical hydrology such as Gumbel (1958), Chow (1964), Benjamin and Cornell (1970), Haan (1977), and Kite (1977). It is especially difficult to pare down the number of continuous distributions for inclusion since at least ten could be selected that have been applied to flood flows. However, the most common are the normal, gamma (Pearson type 3), lognormal, log-gamma (log Pearson type 3) and Gumbel (extreme value type I); only these are included here. In addition, the exponential distribution is discussed because of its simplicity and its application to interevent times.

The objective of a discrete analysis is most often to assign probabilities to the number of occurrences of an event, whereas the objective of a continuous analysis is most often to determine the probability of the magnitude of an event. For the discrete analysis, there may be interest in both the PMF and the CDF, but for continuous analysis the value of the PDF itself is rarely of interest. Rather, only the CDF for the continuous random variable need be evaluated. These distinctions will be seen as the various distributions are presented.

The Binomial Distribution

It is common to examine a sequence of independent events for which the outcome of each can be either a success or a failure; e.g., either the T-yr flood occurs or it does not. Such a sequence consists of Bernoulli trials, independent trials for which the probability of success at each trial is a constant p. The binomial distribution answers the question What is the probability of exactly x successes in n Bernoulli trials? This will be the only discrete distribution considered in this text; it is very commonly used in hydrology.

The probability that there will be x successes followed by $n - x$ failures is just the product of the probabilities of the n independent events:

$p^x(1 - p)^{n-x}$. But this represents just one possible sequence for x successes and $n - x$ failures; all possible sequences must be considered, including those in which the successes do not occur consecutively. The number of possible ways (combinations) of choosing x events out of n possible events is given by the binomial coefficient (Parzen, 1960)

$$\binom{n}{x} = \frac{n!}{x!(n - x)!}. \tag{3.45}$$

Thus, the desired probability is the product of the probability of any one sequence and the number of ways in which such a sequence can occur:

$$P(x) = \binom{n}{x} p^x(1 - p)^{n-x}, \qquad x = 0, 1, 2, 3, \ldots, n. \tag{3.46}$$

The notation $B(n, p)$ indicates the binomial distribution with parameters n and p; example PMFs are shown in Fig. 3.11. From Eqs. (3.19) and (3.28), the mean and variance of x are

$$E(x) = np, \tag{3.47}$$

$$\mathrm{Var}(x) = np(1 - p). \tag{3.48}$$

The skewness is

$$g = \frac{1 - 2p}{[np(1 - p)]^{0.5}}. \tag{3.49}$$

Clearly, the skewness is zero and the distribution is symmetric if $p = 0.5$. The cumulative distribution function is

$$F(x) = \sum_{i=0}^{x} \binom{n}{i} p^i(1 - p)^{n-i}. \tag{3.50}$$

Evaluation of the CDF can get very cumbersome for large values of n and intermediate values of x. It is tabulated by the Chemical Rubber Company (n.d.) and by the National Bureau of Standards (1950). For large values of n, the relationship between the binomial and beta distribution may be used (Abramowitz and Stegun, 1964; Benjamin and Cornell, 1970), or it may be approximated by the normal when $p \approx 0.5$.

EXAMPLE 3.5

RISK AND RELIABILITY

Over a sequence of n yr, what is the probability that the T-yr event will occur at least once? The probability of occurrence in any one year (event) is $p = 1/T$, and the number of occurrences is $B(n, p)$. The Prob(at least one occurrence in n events) is called the **risk**. Thus, the risk is the sum of the probabilities of 1 flood, 2 floods, 3 floods, $\ldots$, n floods occurring

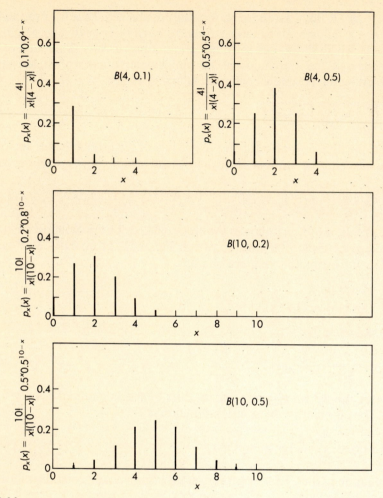

FIGURE 3.11

Binomial probability mass function (PMF). (From Benjamin and Cornell, 1970, Fig. 3.3.1.)

during the n-yr period, but this would be a very tiresome way in which to compute it. Instead,

$$\text{Risk} = 1 - P(0)$$
$$= 1 - \text{Prob(no occurrence in } n \text{ yr)}$$
$$= 1 - (1 - p)^n$$
$$= 1 - (1 - 1/T)^n. \qquad (3.51)$$

Reliability is defined as $1 - \text{risk}$. Thus

$$\text{Reliability} = (1 - p)^n$$
$$= (1 - 1/T)^n. \tag{3.52}$$

This concept of risk and reliability is very important for hydrologic design. Equation (3.51) can be used to determine the return period required for a given design life and level of risk. Values are shown in Table 3.3 that illustrate the very high return periods required for low risk for a long design life.

EXAMPLE 3.6

CRITICAL FLOOD DESIGN

Consider the 50-yr flood ($p = 0.02$).

a) What is the probability that at least one 50-yr flood will occur during the 30-yr lifetime of a flood control project? This is just the risk of failure discussed above, and the distribution of the number of failures is $B(30, 0.02)$. Thus, from Eq. (3.51),

$$\text{Risk} = 1 - (1 - 0.02)^{30}$$
$$= 1 - 0.98^{30}$$
$$= 1 - 0.545$$
$$= 0.455.$$

If this risk is too great, the engineer might design for the 100-yr event for which the risk is

$$\text{Risk} = 1 - 0.99^{30} = 0.26$$

and the reliability is 0.74. For this latter circumstance, there is a 26% chance of occurrence of the 100-yr event over the 30-yr lifetime of the project.

b) What is the probability that the 100-yr flood will not occur in 10 yr? In 100 yr?

For $n = 10$, $P(x = 0) = (1 - p)^{10} = 0.99^{10} = 0.92$,
For $n = 100$, $P(x = 0) = (1 - p)^{100} = 0.99^{100} = 0.37$.

Thus, there is a 37% chance (37% reliability) that the 100-yr flood will not occur during a sequence of 100 yr.

c) In general, what is the probability of having no floods greater than the T-yr flood during a sequence of T yr?

$$P(x = 0) = (1 - 1/T)^T$$

TABLE 3.3

Return Periods for Various Degrees of Risk and Expected Design Life (Eq. 3.51)

RISK (%)	RELIABILITY (%)	EXPECTED DESIGN LIFE, n (yr)							
		2	5	10	15	20	25	50	100
75	25	2.0	4.1	7.7	11.3	14.9	18.5	36.6	72.6
63	37	2.6	5.5	10.6	15.6	20.6	25.6	50.8	101.1
50	50	3.4	7.7	14.9	22.1	29.4	36.6	72.6	144.8
40	60	4.4	10.3	20.1	29.9	39.7	49.4	98.4	196.3
30	70	6.1	14.5	28.5	42.6	56.6	70.6	140.7	280.9
25	75	7.5	17.9	35.3	52.6	70.0	87.4	174.3	348.1
20	80	9.5	22.9	45.3	67.7	90.1	112.5	224.6	448.6
15	85	12.8	31.3	62.0	92.8	123.6	154.3	308.2	615.8
10	90	19.5	48.0	95.4	142.9	190.3	237.8	475.1	949.6
5	95	39.5	98.0	195.5	292.9	390.4	487.9	975.3	1950.1
2	98	99.5	248.0	495.5	743.0	990.5	1238.0	2475.4	4950.3
1	99	199.5	498.0	995.5	1493.0	1990.5	2488.0	4975.5	9950.4
0.5	99.5	399.5	998.0	1995.5	2993.0	3990.5	4988.0	9975.5	19950.5

As T gets large, this approaches $1/e$ (Benjamin and Cornell, 1970, p. 234). Thus for large T,

$$P(x = 0) \approx e^{-1} = 0.368 \qquad \text{(the reliability)},$$

and

$$P(x > 0) \approx 1 - e^{-1} = 0.632 \qquad \text{(the risk)}.$$

As an approximation, the T-yr flood will occur within a T-yr period with approximately a 2/3 probability.

EXAMPLE 3.7

COFFERDAM DESIGN

A cofferdam has been built to protect homes in a floodplain until a major channel project can be completed. The cofferdam was built for the 20-yr flood event. The channel project will require 3 yr to complete. Hence, the process is $B(3, 0.05)$. What are the probabilities that

a) The cofferdam will not be overtopped during the 3 yr (the reliability)?

Reliability $= (1 - 1/20)^3 = 0.95^3 = 0.86$

b) The cofferdam will be overtopped in any one year?

Prob $= 1/T = 0.05$

c) The cofferdam will be overtopped exactly once in 3 yr?

$$P(x = 1) = \binom{3}{1} p^1 (1 - p)^2 = 3 \cdot 0.05 \cdot 0.95^2 = 0.135$$

d) The cofferdam will be overtopped at least once in 3 yr (the risk)?

Risk $= 1 - $ Reliability $= 0.14$

e) The cofferdam will be overtopped only in the third year?

Prob $= (1 - p)(1 - p)p = 0.95^2 \cdot 0.05 = 0.045$

The Exponential Distribution

Consider a process of random arrivals such that the arrivals (events) are independent, the process is stationary (the parameters of the process do not change with time), and it is not possible to have more than one arrival at an instant in time. These conditions describe a Poisson process (Benjamin and Cornell, 1970), which is often representative of the arrival of storm events. If the random variable t represents the interarrival time (the time between events), it is found to be exponentially distributed with PDF

$$f(t) = \lambda e^{-\lambda t}, \qquad t \geq 0. \tag{3.53}$$

This PDF is sketched in Fig. 3.12.

The mean of the distribution is

$$E(t) = 1/\lambda \tag{3.54}$$

and the variance is

$$\text{Var}(t) = 1/\lambda^2. \tag{3.55}$$

Thus, this distribution has the interesting property that its mean equals its standard deviation, or $CV = 1.0$, where CV is the coefficient of variation. The distribution is obviously skewed to the right, with a constant skewness, $g = 2$.

The CDF may easily be evaluated analytically:

$$F(t) = \int_0^t \lambda e^{-\lambda \tau} d\tau = 1 - e^{-\lambda t}. \tag{3.57}$$

Clearly, when $t = \infty$, $F(\infty) = 1$, as it should (the area under the PDF must equal 1).

The exponential distribution may be easily manipulated analytically and as a result is sometimes used to approximate more complex skewed

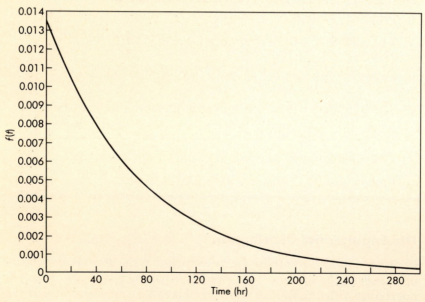

FIGURE 3.12

Exponential probability density function (PDF). Parameters for Example 3.8.

distributions such as the gamma or extreme value. It is occasionally used to fit rainfall totals or river flows, but is most often applied to interevent times. Closely related distributions are the Poisson, which is the PMF for the number of arrivals in time t, and the gamma, which is the PDF for the time between k events (Benjamin and Cornell, 1970).

EXAMPLE 3.8

STORM INTERARRIVAL TIMES

During the course of a year, about 110 independent storm events occur at Gainesville, Fla., and their average duration is 5.3 hr. Ignoring seasonal variations, in a year of 8760 hr, the average interevent time is

$$\bar{t} = \frac{8760 - 110 \cdot 5.3}{110} = 74.3 \text{ hr.}$$

The exponential distribution may thus be fit by the method of moments with $\hat{\lambda} = 1/\bar{t} = 0.0135 \text{ hr}^{-1}$.

a) What is the probability that at least 4 days = 96 hr elapse between storms?

$\text{Prob}(t \geq 96) = 1 - F(96) = e^{-0.0135 \cdot 96} = 0.27$

b) What is the probability that the separation between two storms will be exactly 12 hr?

$\text{Prob}(t = 12) = 0$ since the probability that a continuous variable exactly equals a specific value is zero.

c) What is the probability that the separation between two storms will be less than or equal to 12 hr?

$\text{Prob}(t \leq 12) = F(12) = 1 - e^{-0.0135 \cdot 12} = 1 - 0.85 = 0.15$

The Normal Distribution

The normal distribution is also known as the Gaussian distribution or normal error curve and is fundamental to the entire realm of probability and statistics. One reason is that the central limit theorem states that under very general conditions, as the number of variables in the sum becomes large, the distribution of the sum of a large number of random variables will approach the normal distribution, regardless of the underlying distribution (Benjamin and Cornell, 1970). Many physical processes can be conceptualized as the sum of individual processes. Hence, the normal distribution finds many applications in hydrology as well as in the statistical areas of hypothesis testing, confidence intervals, and quality control.

The PDF for the normal (the bell-shaped curve) is shown in Fig. 3.13 and is given by

$$f(x) = \frac{1}{\sqrt{2\pi}\,\sigma}\, e^{-(1/2)[(x-\mu)/\sigma]^2}, \qquad -\infty \le x \le \infty. \tag{3.58}$$

The parameters of the distribution are the mean μ and variance σ^2 themselves, and the skewness is zero. The method of moments fit is therefore simply to use the estimates $\bar{x}$ and S_x^2 in the distribution. Random variables having the distribution of Eq. (3.58) are denoted by $N(\mu, \sigma^2)$. Although the symmetric nature of the normal distribution usually makes it unsuitable for flood flows (or other extremes), it often describes annual totals (e.g., annual runoff in ac-ft) quite well.

The CDF is evaluated after a change of variable to

$$z \equiv (x - \mu)/\sigma, \tag{3.59}$$

where z is known as the standard normal variate and is $N(0, 1)$. The CDF is then

$$F(z) = \int_{-\infty}^{z} \frac{1}{\sqrt{2\pi}}\, e^{-u^2/2}\, du. \tag{3.60}$$

Unfortunately, the integration cannot be performed analytically, but

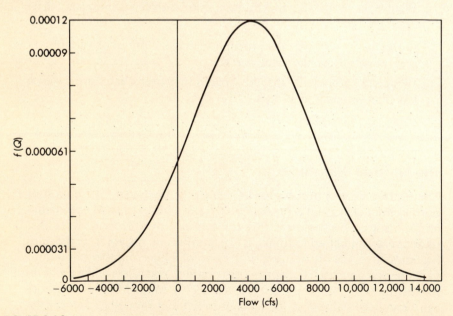

FIGURE 3.13

Normal probability density function (PDF). Parameters for Example 3.9.

tables of $F(z)$ vs. z are found in virtually every statistics text and in this book in Appendix D.3. The similarity of Eq. (3.59) to Eq. (3.43) for the frequency factor is not accidental. In fact, z corresponds exactly to the frequency factor K, discussed in Section 3.3. The magnitude of x for a given return period can be found easily by the following procedure:

1. $F(z) = 1 - 1/T$.
2. Obtain z from Appendix D.3.
3. Then $x = \bar{x} + z \cdot S_x$.

EXAMPLE 3.9

APPLICATION OF NORMAL DISTRIBUTION TO FLOOD FLOWS

The normal distribution is to be fit to the data for Cypress Creek of Table 3.2 and Example 3.3. The fit is not expected to be very good since the data are so skewed.

a) What is the 100-yr flood?

$$F(Q_{100}) = 1 - 1/T = 0.99$$

From Appendix D.3, $z = 2.326$. Thus,

$$Q_{100} = \bar{Q} + z \cdot S_Q = 4144 + 2.326 \cdot 3311 = 11{,}850 \text{ cfs.}$$

b) What is the probability that a flood will be less than or equal to 10,000 cfs?

$$z = \frac{10{,}000 - 4144}{3311} = 1.769.$$

From Appendix D.3,

$$F(1.769) = 0.9615 = \text{Prob}(Q \leq 10000).$$

The return period of a flow of 10,000 cfs is thus

$$T = \frac{1}{1 - F} = 26 \text{ yr.}$$

The Lognormal Distribution

Consider a hypothetical runoff calculation, in which runoff is equal to the *product* of functions of several random factors such as rainfall, contributing area, loss coefficient, evaporation, etc. In general, if a random variable x results from the product of a large number of other random variables (i.e., a "multiplicative mechanism"), then the distribution of the logarithm of x will approach normality since the logarithm of x will consist of the sum of the logs of the contributing factors. In hydrology, it is easy to conceive of many contributing factors to runoff that are inherently random and about which there is very little deterministic information. Hence, a multiplicative mechanism for runoff may be a very reasonable assumption. In general, a random variable has a **lognormal** distribution if the log of the random variable is distributed normally. For example, if $y = \log x$ and if y is distributed normally, then x is distributed lognormally.

The lognormal PDF is sketched in Fig. 3.14. Because it is bounded on the left by zero, has a positive skewness, and is easily used through its relationship to the normal, it is widely applied in hydrology. The skewness is a function of the coefficient of variation:

$$g = 3\ CV + CV^3. \tag{3.61}$$

The skewness of the transformed variable y is, of course, zero since y is distributed normally.

Although the PDF of x (the untransformed variable) can easily be derived (Benjamin and Cornell, 1970), it is seldom needed. Instead, the moments of $y = \log x$ (either base 10 or natural logs may be used) are found, and the normal distribution is used for y. This may be done in two ways. The method of moments requires that the moments of the untransformed data be calculated and related to the moments of the logs. When the natural log is used (i.e., $y = \ln x$), these relationships are

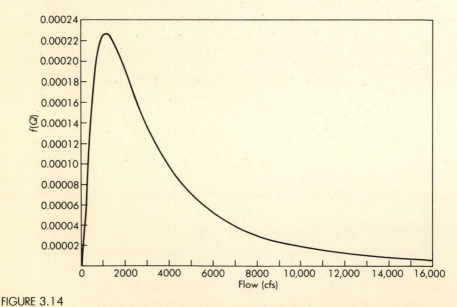

FIGURE 3.14

Lognormal probability density function (PDF). Parameters for Example 3.10.

$$CV_x^2 \equiv \frac{\sigma_x^2}{\mu_x^2} = e^{\sigma_y^2} - 1 \tag{3.62}$$

and

$$\mu_x = e^{\mu_y - \sigma_y^2/2}, \tag{3.63}$$

which can be solved for μ_y and σ_y^2. (If base 10 logarithms are used, Eqs. (3.62) and (3.63) can be used with powers of 10 instead of powers of e, or the mean and variance can be converted to natural logs; see Example 3.10, part c.) Note that the log of the mean is *not* the mean of the logs. Rather,

$$\mu_y = \log x_m, \tag{3.64}$$

where x_m is the median of x. The method of moments using Eqs. (3.62) and (3.63) will preserve the moments of x, should it be necessary (e.g., for simulation of river flows; see Fiering and Jackson, 1971).

The second method is more common and amounts to the maximum likelihood estimates for the parameters of y. In this case, logarithms of all the flows are calculated first. Then the mean and variance of the log-transformed data are used. If moments of the transformed data are calculated by both methods, their correspondence will increase as the sample size increases.

The three-parameter lognormal distribution has an additional parameter that is a nonzero lower bound on the value of x. Three moments are required to estimate the three parameters; applications to hydrology are given by Sangal and Biswas (1970).

EXAMPLE 3.10

APPLICATION OF LOGNORMAL DISTRIBUTION TO FLOOD FLOWS

Repeat Example 3.9 using the lognormal distribution. Parameters of the logs (base 10) were calculated in Example 3.3.

a) What is the 100-yr flood?
The z value was found in Example 3.9 to be 2.326. Thus

$$y_{100} = 3.463 + 2.326 \cdot 0.424 = 4.449$$

and

$$Q_{100} = 10^{y_{100}} = 28,130 \text{ cfs.}$$

This value is much higher than the estimate (11,850 cfs) from the normal distribution. But it is a better estimate because the lognormal distribution more accurately reflects the skewness of the actual flow data.

b) What is the probability that a flow will be less than or equal to 10,000 cfs?

The value of $y = \log_{10}(10000) = 4.000$.

$$z = \frac{4.000 - 3.463}{0.424} = 1.267$$

From Appendix D.3, $F(1.267) = 0.8975$. The return period of a 10,000-cfs flood is thus $1/(1 - 0.8975) = 9.8$ yr.

c) How do the moment estimates from the log-transformed data compare with those found using Eqs. (3.62) and (3.63)? These two equations must be solved for the moments of the logs. Using symbols for the estimated mean and variance and rearranging the equations gives values for natural logs:

$$S_y^2 = \ln(CV_x^2 + 1) = \ln(0.799^2 + 1) = 0.494 \tag{3.65}$$

and

$$\bar{y} = \ln \bar{x} - S_y^2/2 = \ln 4144 - 0.494/2 = 8.083 \tag{3.66}$$

(The coefficient of variation of the untransformed data is $CV_x = 3311/4144 = 0.799$.) Converting to base 10 logs by dividing the mean by 2.301 and the variance by 2.301^2 gives the following results.

	USING MOMENT RELATIONSHIPS	USING TRANSFORMED DATA
$\bar{y}$	3.513	3.463
S_y^2	0.093	0.180
S_y	0.305	0.424

Although the values are not wildly different, use of the moment relationships would certainly give different results than use of the transformed data. Again, the latter procedure is most common and preferable.

The Gamma (Pearson Type 3) Distribution

This distribution receives extensive use in hydrology simply because of its shape (and its well-known mathematical properties). The three-parameter gamma distribution is sketched in Fig. 3.15 and has the pleasing properties of being bounded on the left and of positive skewness. (The distribution can also be used with negative skewness.) Although the three parameters of the PDF are simple functions of the mean, variance, and skewness (Kite, 1977), it is more common in hydrologic applications to evaluate the CDF with frequency factors (Section 3.3), for which the PDF parameters are unnecessary. The frequency factors K are a function of the

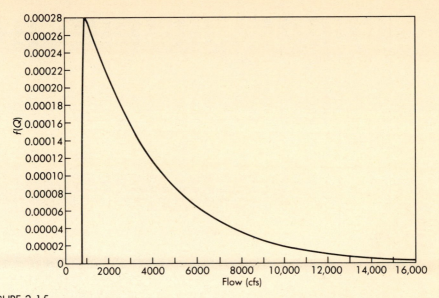

FIGURE 3.15

Three-parameter gamma probability density function (PDF). The lower limit on the abscissa is 801 cfs. Parameters for Example 3.11.

skewness and return period (or CDF) and are given in Table 3.4. Thus, to evaluate the T-yr flood, the moments of the data are computed and

$$Q_T = \overline{Q} + K(C_s, T) \cdot S_Q. \tag{3.67}$$

This procedure is satisfactory for return periods listed in Table 3.4, but interpolation between return periods is not appropriate. Instead, the magnitudes should be plotted vs. return period or vs. the CDF and graphical interpolation performed. (Graphical procedures will be discussed in Section 3.6.) This is also the procedure to use for the inverse problem of finding the probability (or return period) corresponding to a given flood magnitude. Alternatively, the three parameters of the distribution may be found and tables of the CDF used to determine either the magnitude or the probability. The gamma distribution is directly related to the chi-squared distribution, for which many tables are available (Abramowitz and Stegun, 1964; Benjamin and Cornell, 1970; Haan, 1977).

The two-parameter gamma distribution corresponds to setting the left boundary (Fig. 3.15) to zero. In this case the sample skewness need not be calculated. Instead, the skewness of the two-parameter gamma should be used:

$$g = 2CV^2 \qquad \text{or} \qquad C_s = 2 S_x^2/\overline{x}^2. \tag{3.68}$$

This skewness can then be used in Table 3.4.

TABLE 3.4

Frequency Factors K for Gamma and log Pearson Type 3 Distributions (Haan, 1977, Table 7.7)

	RECURRENCE INTERVAL IN YEARS							
	1.0101	2	5	10	25	50	100	200
	PERCENT CHANCE (≥) = $1 - F$							
SKEW COEFFICIENT C_s	99	50	20	10	4	2	1	0.5
3.0	−0.667	−0.396	0.420	1.180	2.278	3.152	4.051	4.970
2.9	−0.690	−0.390	0.440	1.195	2.277	3.134	4.013	4.904
2.8	−0.714	−0.384	0.460	1.210	2.275	3.114	3.973	4.847
2.7	−0.740	−0.376	0.479	1.224	2.272	3.093	3.932	4.783
2.6	−0.769	−0.368	0.499	1.238	2.267	3.071	3.889	4.718
2.5	−0.799	−0.360	0.518	1.250	2.262	3.048	3.845	4.652
2.4	−0.832	−0.351	0.537	1.262	2.256	3.023	3.800	4.584
2.3	−0.867	−0.341	0.555	1.274	2.248	2.997	3.753	4.515
2.2	−0.905	−0.330	0.574	1.284	2.240	2.970	3.705	4.444
2.1	−0.946	−0.319	0.592	1.294	2.230	2.942	3.656	4.372
2.0	−0.990	−0.307	0.609	1.302	2.219	2.912	3.605	4.298
1.9	−1.037	−0.294	0.627	1.310	2.207	2.881	3.553	4.223
1.8	−1.087	−0.282	0.643	1.318	2.193	2.848	3.499	4.147
1.7	−1.140	−0.268	0.660	1.324	2.179	2.815	3.444	4.069
1.6	−1.197	−0.254	0.675	1.329	2.163	2.780	3.388	3.990
1.5	−1.256	−0.240	0.690	1.333	2.146	2.743	3.330	3.910
1.4	−1.318	−0.225	0.705	1.337	2.128	2.706	3.271	3.828
1.3	−1.383	−0.210	0.719	1.339	2.108	2.666	3.211	3.745
1.2	−1.449	−0.195	0.732	1.340	2.087	2.626	3.149	3.661
1.1	−1.518	−0.180	0.745	1.341	2.066	2.585	3.087	3.575
1.0	−1.588	−0.164	0.758	1.340	2.043	2.542	3.022	3.489
.9	−1.660	−0.148	0.769	1.339	2.018	2.498	2.957	3.401
.8	−1.733	−0.132	0.780	1.336	1.993	2.453	2.891	3.312
.7	−1.806	−0.116	0.790	1.333	1.967	2.407	2.824	3.223
.6	−1.880	−0.099	0.800	1.328	1.939	2.359	2.755	3.132
.5	−1.955	−0.083	0.808	1.323	1.910	2.311	2.686	3.041
.4	−2.029	−0.066	0.816	1.317	1.880	2.261	2.615	2.949

.3	−2.104	−0.050	0.824	1.309	1.849	2.211	2.544	2.856
.2	−2.178	−0.033	0.830	1.301	1.818	2.159	2.472	2.763
.1	−2.252	−0.017	0.836	1.292	1.785	2.107	2.400	2.670
0	−2.326	0	0.842	1.282	1.751	2.054	2.326	2.576
−.1	−2.400	0.017	0.846	1.270	1.716	2.000	2.252	2.482
−.2	−2.472	0.033	0.850	1.258	1.680	1.945	2.178	2.388
−.3	−2.544	0.050	0.853	1.245	1.643	1.890	2.104	2.294
−.4	−2.615	0.066	0.855	1.231	1.606	1.834	2.029	2.201
−.5	−2.686	0.083	0.856	1.216	1.567	1.777	1.955	2.108
−.6	−2.755	0.099	0.857	1.200	1.528	1.720	1.880	2.016
−.7	−2.824	0.116	0.857	1.183	1.488	1.663	1.806	1.926
−.8	−2.891	0.132	0.856	1.166	1.448	1.606	1.733	1.837
−.9	−2.957	0.148	0.854	1.147	1.407	1.549	1.660	1.749
−1.0	−3.022	0.164	0.852	1.128	1.366	1.492	1.588	1.664
−1.1	−3.087	0.180	0.848	1.107	1.324	1.435	1.518	1.581
−1.2	−3.149	0.195	0.844	1.086	1.282	1.379	1.449	1.501
−1.3	−3.211	0.210	0.838	1.064	1.240	1.324	1.383	1.424
−1.4	−3.271	0.225	0.832	1.041	1.198	1.270	1.318	1.351
−1.5	−3.330	0.240	0.825	1.018	1.157	1.217	1.256	1.282
−1.6	−3.388	0.254	0.817	0.994	1.116	1.166	1.197	1.216
−1.7	−3.444	0.268	0.808	0.970	1.075	1.116	1.140	1.155
−1.8	−3.499	0.282	0.799	0.945	1.035	1.069	1.087	1.097
−1.9	−3.553	0.294	0.788	0.920	0.996	1.023	1.037	1.044
−2.0	−3.605	0.307	0.777	0.895	0.959	0.980	0.990	0.995
−2.1	−3.656	0.319	0.765	0.869	0.923	0.939	0.946	0.949
−2.2	−3.705	0.330	0.752	0.844	0.888	0.900	0.905	0.907
−2.3	−3.753	0.341	0.739	0.819	0.855	0.864	0.867	0.869
−2.4	−3.800	0.351	0.725	0.795	0.823	0.830	0.832	0.833
−2.5	−3.845	0.360	0.711	0.771	0.793	0.798	0.799	0.800
−2.6	−3.889	0.368	0.696	0.747	0.764	0.768	0.769	0.769
−2.7	−3.932	0.376	0.681	0.724	0.738	0.740	0.740	0.741
−2.8	−3.973	0.384	0.666	0.702	0.712	0.714	0.714	0.714
−2.9	−4.013	0.390	0.651	0.681	0.683	0.689	0.690	0.690
−3.0	−4.051	0.396	0.636	0.660	0.666	0.666	0.667	0.667

Interagency Advisory Committee on Water Data (1982).

EXAMPLE 3.11

APPLICATION OF GAMMA DISTRIBUTION TO FLOOD FLOWS

What is the magnitude of the 100-yr flood for Cypress Creek ($\bar{Q} =$ 4144 cfs, $S_Q = 3311$ cfs) using the gamma-3 and gamma-2 distributions?

For the gamma-3, $C_s = 1.981$. Linear interpolation in Table 3.4 gives $K = 3.595$. Thus,

$$Q_{100} = 4144 + 3.595 \cdot 3311 = 16{,}050 \text{ cfs.}$$

For the gamma-2, $C_s = 2CV^2 = 2 \cdot 0.799^2 = 1.277$. Linear interpolation in Table 3.4 gives $K = 3.197$. Thus,

$$Q_{100} = 4144 + 3.197 \cdot 3311 = 14{,}730 \text{ cfs.}$$

A plot of these two distributions could be used to solve the inverse problem of determining the probability of the 10,000-cfs flood. It will be illustrated in Example 3.17.

The Log Pearson Type 3 Distribution

When the three-parameter gamma distribution is applied to the logs of the random variables, it is customarily called the log Pearson type 3 (LP3) distribution. It plays an important role in hydrology because it has been

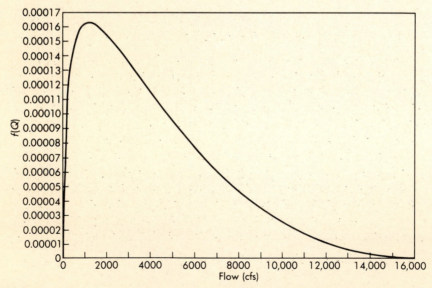

FIGURE 3.16

Log Pearson type 3 probability density function (PDF). Parameters for Example 3.12.

recommended for application to flood flows by the U.S. Interagency Advisory Committee on Water Data (1982). The shape of the LP3 is quite flexible due to its three parameters; an example is shown in Fig. 3.16.

Its use is entirely analogous to the lognormal discussed earlier; however, the moments of the transformed and untransformed variables will not be related here. Instead, the data are transformed by taking logarithms (either base 10 or natural) and the gamma-3 distribution applied exactly as in the preceding section. This means that the magnitudes can readily be computed for return periods shown in Table 3.4, but the inverse problem of determining the return period (or CDF) corresponding to a given magnitude should be done graphically or with the tables mentioned in the preceding section.

EXAMPLE 3.12

APPLICATION OF LP3 DISTRIBUTION TO FLOOD FLOWS

Statistics of the logs of Cypress Creek are given in Example 3.3; the skewness is -1.117. For the 100-yr flood, Table 3.4 gives $K = 1.504$ by linear interpolation. Thus, if $y = \log_{10} Q$,

$$y_{100} = 3.463 + 1.504 \cdot 0.424 = 4.101$$

and the 100-yr flow is

$$Q_{100} = 10^{4.101} = 12{,}610 \text{ cfs.}$$

Magnitudes predicted by the LP3 for this example will be graphed in Example 3.17, and the inverse problem (e.g., determining the return period of a flood of 10,000 cfs) will be considered at that time.

The Gumbel (Extreme Value Type I) Distribution

This PDF arises from the theory of extremes (Gumbel, 1958) and has an appropriate shape (Fig. 3.17) but is unbounded on both the lower and upper ends. However, the possibility of a negative value (e.g., a negative flow) is rarely of concern in its application. Once again, the PDF is unneeded. The CDF has the peculiar double exponential form

$$F(x) = e^{-e^{-\alpha(x-u)}}, \qquad -\infty \le x \le \infty, \tag{3.69}$$

which may be readily evaluated without tables. Parameter α is a scale parameter and parameter u (the mode) is a location parameter. They are related to the mean and variance by

$$\mu = u + \frac{\gamma}{\alpha} = u + \frac{0.5772}{\alpha} \tag{3.70}$$

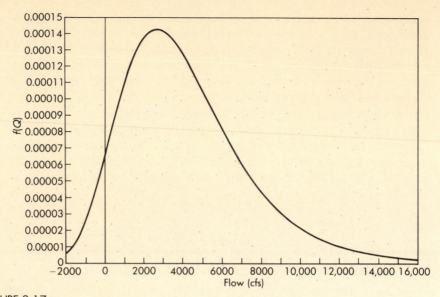

FIGURE 3.17

Gumbel probability density function (PDF). Parameters for Example 3.13.

and

$$\alpha = \frac{\pi}{\sqrt{6}\,\sigma},$$ (3.71)

which provide the method of moments estimates. (γ is called Euler's constant.) Lettenmaier and Burges (1982) have shown that no correction for the number of data points n is necessary in evaluating α and u. The skewness is constant and equal to 1.1396. The argument of the innermost exponential,

$$y \equiv \alpha(x - u),$$ (3.72)

is called the reduced variate. Since y is a unique function of F (and T), it may be used to obtain a linear plot of the distribution, using appropriate scaling, as will be seen later.

If the CDF (Eq. 3.69) is used to solve for x in terms of F or T and is combined with the moment relationships, Eqs. (3.70) and (3.71), a frequency factor may be readily derived (Kite, 1977):

$$K = -0.7797\{0.5772 + \ln[\ln(1/F)]\}$$
$$= -0.7797\{0.5772 + \ln[\ln(T/(T-1))]\},$$ (3.73)

where T is the desired return period. The frequency factor may be used, or Eq. (3.69) may be applied directly to obtain x.

If the CDF is evaluated at the mean, it is found that

$$F(\mu) = 0.57 \tag{3.74}$$

and

$$T(\mu) = \frac{1}{1 - F} = 2.33 \text{ yr.} \tag{3.75}$$

If a plot of flow vs. return period is available, this relationship is some-times used as a crude estimate of the mean of the flows, assuming they obey a Gumbel distribution.

Analogous to the normal and gamma distributions, two log-Gumbel distributions may be defined. These are known as the extreme value type II and III distributions (Gumbel, 1958; Benjamin and Cornell, 1970). The former may be applied to floods (maxima) while the latter (also known as the Weibull distribution) is usually applied to droughts (minima). Space does not permit their further consideration here.

EXAMPLE 3.13

APPLICATION OF GUMBEL DISTRIBUTION TO FLOOD FLOWS

Again, the 100-yr flood will be determined for the data of Example 3.3.

a) What is the 100-yr flood?

From Eq. (3.73), the frequency factor is evaluated as

$$K = -0.7797\{0.5772 + \ln[\ln(100/99)]\} = 3.137.$$

Thus,

$$Q_{100} = 4144 + 3.137 \cdot 3311 = 14{,}530 \text{ cfs.}$$

b) What is the return period of a flood of 10,000 cfs?

Since the CDF may be evaluated analytically, this problem is easily solved, but the parameters α and u must first be found. From Eqs. (3.70) and (3.71),

$$\hat{\alpha} = \pi/\sqrt{6}\, S_Q = 3.87 \cdot 10^{-4} \text{ cfs}^{-1}$$

and

$$\hat{u} = \overline{Q} - 0.5772/\hat{\alpha} = 2654 \text{ cfs.}$$

Then

$$F(10000) = e^{-e^{-3.87(10000-2654)}} = 0.943,$$

and the return period of the 10,000-cfs flow is $1/(1 - 0.943) = 18$ yr.

EXAMPLE 3.14

COMPARISON OF 100-YR FLOODS

The 100-yr floods predicted by the five distributions for flood frequency analysis discussed above are compared in Table 3.5. The wide range is unsettling; which prediction should be used? A preliminary screening can be performed on the basis of skewness. The untransformed data and their logarithms (Example 3.3) both exhibit nonzero skewness, so the normal distribution is suspect. All the other distributions have nonzero skewness, but only the gamma and log-gamma (LP3) theoretically match the data exactly (since they are three-parameter distributions). But the best fit is best determined on the basis of a comparison of the fitted (theoretical) CDF to the empirical CDF. This is done graphically, as explained in the next section.

TABLE 3.5

Comparison of Predicted 100-yr Floods and Measured and Predicted Skewness Values

DISTRIBUTION	PREDICTED 100-YR FLOOD (cfs)	PREDICTED SKEWNESS	MEASURED SKEWNESS
Normal	11,850	0	1.98
Lognormal	28,130	2.91	1.98
		0*	−1.12*
Gamma (Pearson 3)	16,050	1.98	1.98
Log Pearson 3	12,610	−1.12*	−1.12*
Gumbel	14,530	1.14	1.98

* Skewness of logarithms of data.

3.6

GRAPHICAL PRESENTATION OF DATA

Introduction

As in much of statistics, a visual inspection of the fit of the frequency distribution is probably the best aid in determining how well an individual distribution fits a set of data or which distribution fits "best." Two questions must be addressed to plot the data: (1) What kind of graph paper (i.e., type of scaling) should be used? and (2) How should the data points be plotted (the question of **plotting position**)?

Probability Paper

Although ordinary graph paper could be used to plot the CDF vs. the magnitude of the random variable, it is customary to try to use paper scaled such that the theoretical fit is a straight line. That is, it is desirable to plot magnitude x vs. some function of the CDF $h[F(x)]$ such that x vs. h is linear. If this can be done, one ordinate of the paper will be the magnitude and the other the function h, but this axis will be labeled with values of F, not h. Among other advantages, this permits a graphical fit of the distribution to the data simply by drawing a straight line through the plotted data points.

The frequency factor K (Eq. 3.43) serves as the required function $h(F)$ as long as it is not also a function of skewness. For example, for the normal distribution, $K = z$, the standard normal variate. Clearly, a plot of x vs. z is linear (Eq. 3.59), and equal increments of z on one axis of probability paper can be labeled with the corresponding value of F for convenience. This is exactly how normal probability paper is constructed (Fig. 3.18). Lognormal paper is easily constructed also, either by plotting the logs on an arithmetic scale or, more commonly, by providing a log scale instead of an arithmetic scale for the magnitude of the variable (Fig. 3.19). A good

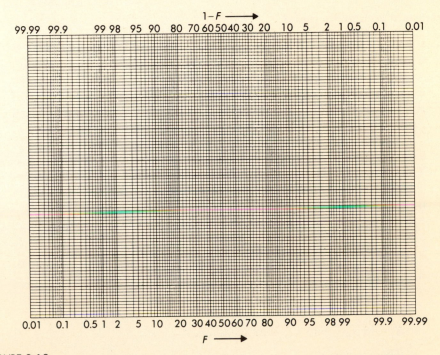

FIGURE 3.18

Normal probability paper.

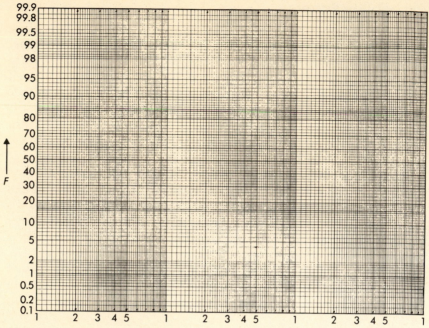

FIGURE 3.19

Lognormal probability paper.

source for normal, lognormal, and many other kinds of graph paper is Craver (1980).

For the normal distribution, the straight fitted line passes through the estimate of the mean of the data at the 50% probability ($z = 0$) and through the location $\bar{x} \pm S_x$ at the 84.1% or 15.9% frequencies (values of $F(x)$), respectively, from which $\bar{x}$ and S_x can be found graphically.

The problem is not quite as simple for the lognormal, since it is the logs, not the untransformed variables, that are distributed linearly vs. z. Thus, the 50% CDF value corresponds to the median of x, not the mean. To estimate the mean and standard deviation of the untransformed data from a graphical lognormal fit, the following procedure must be used.

1. Estimate the mean of the logs and the standard deviation of the logs using the procedure just described for the normal distribution. That is, the logarithm of the value plotted at 50% is the mean of the logs:

$$\bar{y} = \ln x_m = \ln x_{50} \tag{3.76}$$

and

$$S_y = \ln x_{84.1} - \ln x_{50} = \ln(x_{84.1}/x_{50}), \tag{3.77}$$

where $y = \ln x$.

2. Knowing these moments of y, the moments of the untransformed

variable x must be calculated from the relationships given in Eqs. (3.62) and (3.63).

This procedure will be illustrated later in Example 3.15.

Probability paper may also be constructed for the Gumbel distribution by using the reduced variate y (Eq. 3.72). The linear plot is then x vs. y (Fig. 3.20), with corresponding values of F plotted instead of (or in addition to) y using Eq. (3.69). Parameter u of the Gumbel distribution corresponds to $y = 0$ ($F = 0.368$) and parameter α is found graphically from the slope of the fitted line. Example 3.16 will illustrate this procedure.

Unfortunately, no probability paper can be constructed for the gamma or log-gamma distributions since the frequency factor is a function of the skewness. Thus, a probability scale would be needed for every value of C_s, which is not feasible. It is common to plot gamma or LP3 distributions on lognormal paper, on which they appear as smooth curves, with higher curvature for higher skewness. This will be illustrated in Example 3.17.

Finally, it may be mentioned that the exponential distribution has a very simple probability paper. If the exceedance frequency is $G(x) = 1 - F(x)$, then

$$G(x) = e^{-\lambda x} \tag{3.78}$$

and

$$\ln G = -\lambda x. \tag{3.79}$$

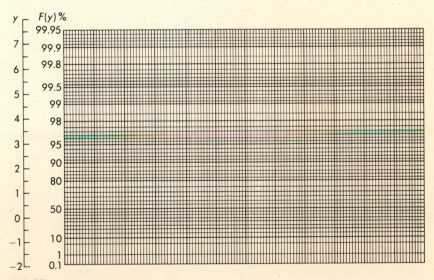

FIGURE 3.20

Gumbel probability paper.

Thus, ordinary semilog paper serves as probability paper for the exponential distribution with G plotted on the log scale.

Plotting Position

To plot the flood (or other hydrologic) data, the data values must first be ranked from 1 to n (the number of years of record) in order of decreasing magnitude. Thus magnitude $m = 1$ is the largest value and $m = n$ is the smallest. The rank m and number of years of record n are then used to compute a plotting position, or empirical estimate of frequency F, or return period T.

In hydrology, the most common plotting position is the Weibull formula:

$$T = \frac{n+1}{m} \tag{3.80}$$

and

$$F = 1 - \frac{m}{(n+1)}, \tag{3.81}$$

which is analyzed in detail by Gumbel (1958). For example, the largest value from a 25-yr record would plot at a return period of 26 yr or a CDF of $25/26 = 0.962$.

However, the venerable Weibull formula has been criticized because it does not provide an estimate of the CDF F such that $E(F)$ equals the theoretical value for the mth largest out of n total samples for any underlying distribution other than the uniform, thus excluding all of the distributions commonly employed for flood frequency and other hydrologic analysis (Cunnane, 1978). Instead, a generalized form first proposed by Gringorten (1963) may be used:

$$T = \frac{n+1-2a}{m-a} \tag{3.82}$$

and

$$F = 1 - \frac{m-a}{n+1-2a} \tag{3.83}$$

The parameter a depends on the distribution and equals 0.375 for the normal (or lognormal) and 0.44 for the Gumbel, with a value of 0.40 suggested as a good compromise for the customary situation in which the exact distribution is unknown. Although the Weibull formula still finds much acceptance in hydrology, Eqs. (3.82) and (3.83) are better from a theoretical standpoint and will be used in the following examples, with parameter $a = 0.4$. For example, the largest value in a 25-yr record would plot with a return period of 42 yr and CDF = 0.976. Hirsch and Stedinger (1987) offer alternative plotting position formulas for instances in which a continuous historical flood record is augmented by estimates of historical flood peaks prior to the period of continuous stream gaging.

Confidence Limits, Outliers, and Zeros

Confidence limits are control curves plotted on either side of the fitted CDF, with the property that, if the data belong to the fitted distribution, a known percentage of the data points should fall between the two curves. Unfortunately, the computation of these limits differs for the different distributions; Kite (1977) summarizes the procedures to be used for each.

However, to illustrate the large uncertainty in frequency estimates, an approximate procedure used by Benjamin and Cornell (1970) will be demonstrated. They use the Kolmogorov-Smirnov (KS) goodness of fit statistic to plot confidence limits for the Gumbel and lognormal distributions, with the implication that it will serve as an approximate procedure for other distributions as well. Let $F(x)$ be the *predicted* value of the CDF. Then a confidence interval on the CDF can be constructed such that

$$\text{Prob}(F \leq F_u) = \text{Prob}(F \leq F + KS) = 1 - \alpha \tag{3.84}$$

and

$$\text{Prob}(F_l \leq F) = \text{Prob}(F - KS \leq F) = 1 - \alpha, \tag{3.85}$$

where KS is the Kolmogorov-Smirnov statistic at confidence level α (not to be confused with the parameter of the Gumbel distribution), and subscripts u and l mean upper and lower, respectively. Values of KS are listed in Table 3.6 as a function of α and the sample size n. The $(1 - 2\alpha)$

TABLE 3.6

Kolmogorov-Smirnov Statistics

SAMPLE SIZE	KS VALUE (fraction) FOR CONFIDENCE LEVEL		
	$\alpha = 10\%$	$\alpha = 5\%$	$\alpha = 1\%$
5	0.51	0.56	0.67
10	0.37	0.41	0.49
15	0.30	0.34	0.40
20	0.26	0.29	0.35
25	0.24	0.26	0.32
30	0.22	0.24	0.29
40	0.19	0.21	0.25
Large n	$1.22/\sqrt{n}$	$1.36/\sqrt{n}$	$1.63/\sqrt{n}$

Data from J. R. Benjamin and C. A. Cornell, *Probability Statistics and Decision for Civil Engineers* (New York: McGraw-Hill Book Company, 1970, p. 667.)

percent confidence limits may be formed on $F(x)$ by

$$\text{Prob}(F - KS \leq F \leq F + KS) = 1 - 2\alpha. \tag{3.86}$$

For example, for a sample size of 40, 90% confidence limits may be placed on the predicted CDF by

$$\text{Prob}(F - 0.21 \leq F \leq F + 0.21) = 90\%.$$

These limits may be plotted on the graph of magnitude vs. CDF and control curves smoothed in for intermediate values; the curves can then be used to estimate confidence limits on magnitudes as well as CDF values. The procedure will be illustrated in Examples 3.15 and 3.16.

The KS statistic is weak inasmuch as it is independent of the actual distribution being plotted. It is also a constant and does not reflect the additional uncertainty in predicted values of F at the extremes of the plotted points (i.e., $m = 1$). Finally, it cannot be used to compute F_u when $F + KS > 1.0$ or to compute F_l when $F - KS < 0$. The best procedure is to use the method appropriate to each distribution. The details are beyond the scope of this text; see Kite (1977).

Outliers are data points that fall an "unusually large" distance away from the fitted CDF and pose another problem in frequency analysis. Should they be included in the analysis or are they anomalies? Statistical tests are available (e.g., Interagency Advisory Group on Water Data, 1982), but in the end the decision about whether or not to retain an outlier is usually made subjectively on the basis of confidence in the individual data value.

The presence of zeros among the data is usually treated by means of a mixed distribution (Fig. 3.7). If there are n_o zeros among n samples, the discrete probability $P(0) = \text{Prob}(Q = 0)$ is estimated as n_o/n. A continuous PDF is then fit to the remainder of the data using the moments of the nonzero data, but scaled such that its total mass is $1 - P(0)$ instead of 1.0. This is done by simply multiplying values of the CDF by $1 - P(0)$. The base value of the CDF on a plot is thus $P(0)$. An example of this procedure is given by Haan (1977).

EXAMPLE 3.15

GRAPHICAL FIT OF LOGNORMAL DISTRIBUTION

The Cypress Creek data from Example 3.3 are to be used to fit the lognormal distribution and to plot the results. The fit has already been performed in Example 3.10 using the method of moments. Hence, the fitted distribution will be plotted using those results.

To plot the data points, the data of Table 3.2 are first ranked, as shown in Table 3.7. (Table 3.7 and Fig. 3.21 were both prepared using microcomputer spreadsheet software, a very convenient tool for frequency

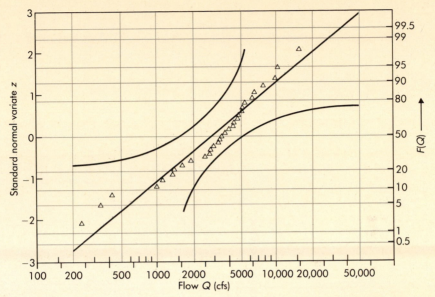

FIGURE 3.21

Lognormal plot of Cypress Creek flows, with 90% confidence intervals.

tabulations.) Return periods and CDF values are then assigned using Eqs. (3.82) and (3.83), also indicated in Table 3.7. To plot 90% confidence intervals, the theoretical (fitted) CDF is computed for each flow and F_u and F_l are computed using $KS \approx 0.24$ for $\alpha = 5\%$ and $n = 31$ yr (Table 3.6). Clearly, F_u cannot be computed for $F \geq 0.76$ (i.e., for $F \geq 1 - KS$) and F_l cannot be computed for $F \leq 0.24$. Return periods corresponding to F_u and F_l are also shown. Finally, the data, fitted distribution, and confidence limits are plotted in Fig. 3.21.

The distribution could be fit by eye by drawing a straight line through the data points. However, here the fit of Example 3.10 is shown. The 50% magnitude (the median) is

$$Q_m = 10^y = 10^{3.463} = 2904 \text{ cfs} \qquad (y = \log_{10} x).$$

A second point for the straight line is found by adding the standard deviation of the logs to the mean of the logs:

$$y_{84.1} = \bar{y} + S_y = 3.463 + 0.424 = 3.887.$$

Thus

$$Q_{84.1} = 10^{3.887} = 7709 \text{ cfs}.$$

Two points are now available with which to draw the straight fitted CDF.

The poor fit of the distribution at the low end is not of concern when the analysis is for flood conditions. In fact, if the fit were to be made by eye,

TABLE 3.7

Tabulated Cypress Creek Data for Lognormal Plot

YEAR	RANK (m)	EMPIRICAL RETURN PERIOD	EMPIRICAL CDF	FLOW (cfs)	FITTED CDF	UPPER CONFIDENCE LEVEL*	LOWER CONFIDENCE LEVEL†
1949	1	52.0	0.981	15600	0.958		0.718
1960	2	19.5	0.949	10300	0.903		0.663
1945	3	12.0	0.917	9840	0.894		0.654
1954	4	8.7	0.885	7760	0.843		0.603
1973	5	6.8	0.853	6560	0.798		0.558
1961	6	5.6	0.821	6260	0.784		0.544
1957	7	4.7	0.788	5440	0.740	0.980	0.500
1968	8	4.1	0.756	5230	0.727	0.967	0.487
1946	9	3.6	0.724	5170	0.723	0.963	0.483
1950	10	3.3	0.692	4740	0.692	0.932	0.452
1974	11	2.9	0.660	4710	0.690	0.930	0.450
1953	12	2.7	0.628	4400	0.665	0.905	0.425
1969	13	2.5	0.596	4300	0.656	0.896	0.416
1972	14	2.3	0.564	3980	0.627	0.867	0.387

Year	n		CDF	Value	Fitted CDF	Fitted CDF + KS*	Fitted CDF − KS†
1959	15	2.1	0.532	3690	0.597	0.837	0.357
1975	16	2.0	0.500	3460	0.571	0.811	0.331
1952	17	1.9	0.468	3310	0.553	0.793	0.313
1966	18	1.8	0.436	3210	0.541	0.781	0.301
1958	19	1.7	0.404	3000	0.513	0.753	0.273
1970	20	1.6	0.372	2820	0.488	0.728	0.248
1964	21	1.5	0.340	2770	0.481	0.721	0.241
1955	22	1.4	0.308	2520	0.442	0.682	0.202
1971	23	1.4	0.276	1900	0.332	0.572	0.092
1947	24	1.3	0.244	1620	0.275	0.515	0.035
1965	25	1.3	0.212	1400	0.227417	0.467417	
1962	26	1.2	0.179	1360	0.218595	0.458595	
1967	27	1.2	0.147	1110	0.162505	0.402505	
1963	28	1.1	0.115	1000	0.137651	0.377651	
1951	29	1.1	0.083	427	0.024584	0.264584	
1956	30	1.1	0.051	340	0.013943	0.253943	
1948	31	1.0	0.019	235	0.005171	0.245171	

* Fitted CDF + KS value of 0.24.
† Fitted CDF − KS value of 0.24.

189

the three low points might be ignored and a somewhat better fit at the upper end obtained. The fit is conservative in the sense that predicted flood magnitudes for rare events are higher at a given CDF (or return period) than the empirical values (not to be construed as a generalization of the lognormal).

What are the mean and standard deviation of the untransformed flows using the graphical fit? This question may make more sense when the fit has been performed by eye instead of analytically as in Example 3.10. Nonetheless, assume that the mean and standard deviations of the base 10 logs of the flows are found to be 3.463 and 0.424, respectively, by an inverse of the procedure just explained in the previous paragraph. Then $\overline{Q}$ and S_Q must be found using the relationships of Eqs. (3.62) and (3.63). The log statistics are first converted to natural logs (if necessary, as it is in this example):

$$\overline{\ln Q} = 2.301 \cdot 3.463 = 7.968$$

and

$$S_{\ln Q} = 2.301 \cdot 0.424 = 0.976.$$

Eqs. (3.62) and (3.63) then give

$$\overline{Q} = e^{7.9568 - (0.976^2/2)} = 1793 \text{ cfs}$$

and

$$S_Q = [(e^{0.976^2} - 1) \cdot 1793^2]^{0.5} = 2263 \text{ cfs}.$$

These values may be compared to 4144 and 3311 cfs, respectively, computed from the data (Example 3.3). Once again: the log of the mean is not the mean of the logs!

The 90% Kolmogorov-Smirnov confidence limits form very wide bands about the fitted distribution. On the one hand, this indicates that the fit is acceptable (at the 90% level) since all of the plotted points fall within the bands. On the other hand, the bands reflect the extreme uncertainty of the probability estimates. For instance, considering a flow of 5000 cfs, from Fig. 3.21 the probability is 90% that the true CDF for this flow lies between 50% and 95%. Put another way, the probability is 90% that the true return period for a flow of 5000 cfs lies between 2 and 20 yr—a very wide interval!

EXAMPLE 3.16

GRAPHICAL FIT OF GUMBEL DISTRIBUTION

A procedure similar to that of Example 3.15 is used here. Computations for the graphical fit of the Gumbel distribution to the Cypress Creek

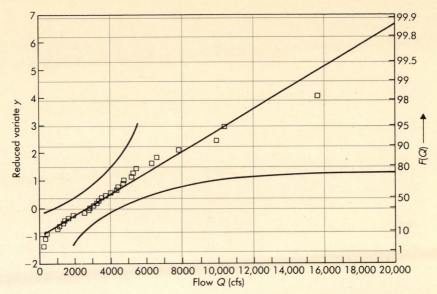

FIGURE 3.22

Gumbel plot of Cypress Creek flows, with 90% confidence intervals.

floods are shown in Table 3.8; all computations are analogous to those of Example 3.15. The fit of the Gumbel distribution is shown on Gumbel paper (extreme-value paper) in Fig. 3.22. (Once again, both Table 3.8 and Fig. 3.22 were prepared using microcomputer spreadsheet software.) The fitted line corresponds to the method of moments fit of Example 3.13. The value at the mode at $F = 0.368$ is the mode $u = 2654$ cfs. A second point with which to draw the straight line fit can be found by solving for the value of F for any other value of Q. This was already done in Example 3.13, in which it was found that $F(10000) = 0.943$.

(If the fitted line has been drawn by eye, the inverse problem of finding parameters α and u is done by simply reversing the procedure. Parameter u is found where $F = 0.368$. A second point along the line is then used to solve for the reduced variate y [Eq. 3.72], which is then used to solve for α, Q and u being known.)

From the graph (or analytically), the theoretical mean, corresponding to $F = 0.57$, is $Q = 4140$ cfs, in agreement with the data.

Although the fitted line is fairly "close" to most of the plotted points, the largest measured flood falls well below the fit. That is, the flood of 15,600 cfs is predicted by the Gumbel distribution to have a much higher CDF (0.993) and return period (143 yr) than its plotted value (0.981 and 52 yr, respectively). This particular fit is nonconservative (not a generalization); the predicted 100-yr flood is 1070 cfs less than one that occurred already during the 31-yr study period. Although superficially the

TABLE 3.8

Tabulated Cypress Creek Data for Gumbel Plot

YEAR	RANK (m)	EMPIRICAL RETURN PERIOD	EMPIRICAL CDF	FLOW (CFS)	FITTED y	FITTED CDF	UPPER CONFIDENCE LEVEL*	LOWER CONFIDENCE LEVEL†
1949	1	52.0	0.981	15600	5.010102	0.993351		0.753351
1960	2	19.5	0.949	10300	2.959002	0.949451		0.709451
1945	3	12.0	0.917	9840	2.780982	0.939903		0.699903
1954	4	8.7	0.885	7760	1.976022	0.870559		0.630559
1973	5	6.8	0.853	6560	1.511622	0.802075		0.562075
1961	6	5.6	0.821	6260	1.395522	0.780591		0.540591
1957	7	4.7	0.788	5440	1.078182	0.711618	0.951618	0.471618
1968	8	4.1	0.756	5230	0.996912	0.691413	0.931413	0.451413
1946	9	3.6	0.724	5170	0.973692	0.685445	0.925445	0.445445
1950	10	3.3	0.692	4740	0.807282	0.640139	0.880139	0.400139
1974	11	2.9	0.660	4710	0.795672	0.636813	0.876813	0.396813
1953	12	2.7	0.628	4400	0.675702	0.601217	0.841217	0.361217
1969	13	2.5	0.596	4300	0.637002	0.589267	0.829267	0.349267
1972	14	2.3	0.564	3980	0.513162	0.549580	0.789580	0.309580

1959	15	2.1	0.532	3690	0.400932	0.511864	0.751864	0.271864
1975	16	2.0	0.500	3460	0.311922	0.480927	0.720927	0.240927
1952	17	1.9	0.468	3310	0.253872	0.460339	0.700339	0.220339
1966	18	1.8	0.436	3210	0.215172	0.446461	0.686461	0.206461
1958	19	1.7	0.404	3000	0.133902	0.416997	0.656997	0.176997
1970	20	1.6	0.372	2820	0.064242	0.391496	0.631496	0.151496
1964	21	1.5	0.340	2770	0.044892	0.384388	0.624388	0.144388
1955	22	1.4	0.308	2520	−0.05185	0.348810	0.588810	0.108810
1971	23	1.4	0.276	1900	−0.29179	0.262151	0.502151	0.022151
1947	24	1.3	0.244	1620	−0.40015	0.224908	0.464908	
1965	25	1.3	0.212	1400	−0.48529	0.196978	0.436978	
1962	26	1.2	0.179	1360	−0.50077	0.192049	0.432049	
1967	27	1.2	0.147	1110	−0.59752	0.162411	0.402411	
1963	28	1.1	0.115	1000	−0.64009	0.150068	0.390068	
1951	29	1.1	0.083	427	−0.86184	0.093711	0.333711	
1956	30	1.1	0.051	340	−0.89551	0.086414	0.326414	
1948	31	1.0	0.019	235	−0.93615	0.078069	0.318069	

* Fitted CDF + KS value of 0.24.

† Fitted CDF − KS value of 0.24.

fit appears to be worse than that of the lognormal (Fig. 3.21), the log scale on the latter is deceptive. The deviation of the largest observed flood from the fitted line is comparable for the two examples. Again, the 90% KS confidence limits produce very wide bands.

Comparison of Fits

The question of which distribution gives the best fit may be addressed quantitatively using measures such as the chi-squared statistic and Kolmogorov-Smirnov test (Benjamin and Cornell, 1970; Haan, 1977), but these tests are seldom helpful in discriminating among different distributions because their confidence limits are so large that both tend to lead to acceptance of the hypothesis that the distribution fits the data. See, for example, the very wide 90% confidence intervals plotted in Figs. 3.21 and 3.22 using the KS statistic. An alternative is to use heuristic measures of goodness of fit, such as the average of absolute values of deviations between the fitted and plotted CDF (Benson, 1968). Another possibility is to include the third and fourth moments of the data in the analysis since these may be used to categorize different distributions (Harr, 1977). But the problems of estimation of higher moments have already been discussed. In the end, the decision is often subjective and based on a preference for the underlying mechanism of one distribution versus another. That is, the engineer may prefer the multiplicative mechanism of the lognormal distribution, the theory of extremes of the Gumbel distribution, or the empiricism of the gamma or log Pearson type 3 distributions.

Graphical comparison of the fitted and plotted CDFs is of considerable use in the decision. King (1971) has shown the form that several CDFs exhibit when plotted on the probability paper of another distribution, and Reich and Renard (1981) apply these shapes to hydrology. The shapes for the lognormal and Gumbel (extreme-value) distributions are shown in Fig. 3.23. For example, if the data are plotted on lognormal paper and a Gumbel distribution fits the data, the data should appear with the curvature in the lower left-hand plot. Unfortunately, the gamma and log Pearson 3 distributions cannot be assessed in this manner since they have no probability paper.

Instead of trying several distributions and attempting to select the best fit, one could simply accept the recommendations of the U.S. Interagency Advisory Committee on Water Data (1982) (formerly the U.S. Water Resources Council; Benson, 1968) that the LP3 distribution is best suited for flood frequency analysis. On the other hand, the log-Gumbel (extreme-value type II) was determined by the Natural Environment Research Council (1975) to be preferable for flood frequency analysis in Great Britain. Individual rivers are likely to vary in their optimal distribution, and the question of which distribution to use for floods and myriad other hydrologic variables will always remain open.

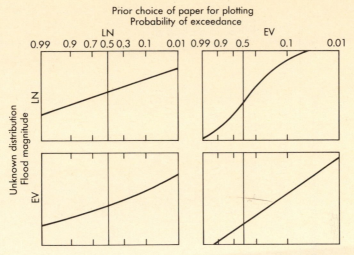

FIGURE 3.23

Effect on cumulative distribution function (CDF) shape of plotting on alternative probability paper. (From Reich and Renard, 1981, Fig. 4.)

EXAMPLE 3.17

GRAPHICAL COMPARISON OF FIVE DISTRIBUTIONS

Which distribution is best for the data for Cypress Creek? There is no right answer to this question, but all the distributions may be plotted together on lognormal paper for a qualitative assessment. Several predicted magnitudes are needed since all but the lognormal will plot as smooth curves. These calculations are shown in Table 3.9 (including more than just the required two points for the lognormal). The resulting plots are shown in Fig. 3.24 along with selected data points.

The curvature criteria discussed earlier (Fig. 3.23) do not appear to help here since the curvature does not clearly favor an alternative distribution and may even be S-shaped (see Figs. 3.21 and 3.22 for a clearer view). The gamma distribution fits best at the upper end of the scale, whereas the Gumbel and LP3 appear to fit best near the middle of the range of data, although differences are slight over much of the range. However, the lognormal has the greatest deviation at the upper end and would probably not be selected as "best," nor would the normal with its tendency to go negative at the low end (Table 3.9) and its large deviation from many of the data points. Since the gamma fits best at the upper end and does not deviate greatly over most of the range of data, it is probably the best choice here.

It is important to realize that, were confidence limits shown (as on Figs. 3.21 and 3.22), they would be very wide at the upper limit of the

TABLE 3.9

Computation of Predicted Flood Magnitudes, Cypress Creek, near Houston, Tex., 1945 – 1975

RETURN PERIOD	CDF	NORMAL		LOGNORMAL		GAMMA		LOG PEARSON 3		GUMBEL	
		$K = z$	Q (cfs)	$K = z$	Q (cfs)	K	Q (cfs)	K	Q (cfs)	K	Q (cfs)
100	0.99	2.326	11845	2.326	28134	3.197	14729	1.504	12609	3.137	14530
50	0.98	2.054	10945	2.054	21572	2.906	13766	1.424	11662	2.592	12727
25	0.96	1.751	9942	1.751	16048	2.217	11484	1.316	10495	2.044	10911
10	0.90	1.282	8389	1.282	10152	1.304	8462	1.103	8525	1.305	8463
5	0.80	0.842	6932	0.842	6607	0.613	6174	0.847	6639	0.719	6526
2	0.50	0	4144	0	2904	−0.304	3137	0.183	3472	−0.164	3600
1.0101	0.01	−2.326	−3557	−2.326	300	−0.999	836	−3.099	141	−1.641	−1289

The table was computed using microcomputer spreadsheet software.

Mean = 4144 cfs

Standard deviation = 3311 cfs

Skewness = 1.981

Mean (log 10) = 3.463

Standard deviation (logs) = 0.424

Skewness (logs) = −1.117

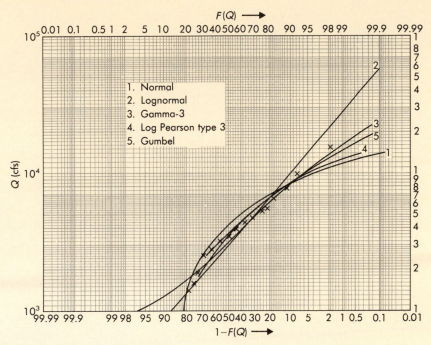

FIGURE 3.24

Comparison of five fitted cumulative density functions (CDFs) for Cypress Creek flows.

CDF, reflecting the uncertainty of predicting the magnitude of, say, the 100-yr flood from only 31 yr of data. The uncertainty can be quantified by computing the standard deviation of the estimates at the extremes, using a different procedure for each distribution (Kite, 1977). Unfortunately, about the only way to reduce the uncertainty is to use a longer period of record (not possible) or a regional analysis, in which regional streamflow data are aggregated and used in a regression-type analysis (Haan, 1977; Kottegoda, 1980).

These plots can be used to interpolate for flood frequencies for the gamma and LP3 distributions, as suggested in Examples 3.11 and 3.12.

SUMMARY

This chapter contains only a sampling of possible topics in frequency analysis. Other topics of particular importance are improved estimation procedures by the method of maximum likelihood (see Section 3.3) and confidence limits and uncertainty (see Section 3.6). The reader may refer to the cited references for additional information.

In this chapter, introductory definitions related to description of data and probability theory were followed by methods for fitting theoretical frequency distributions to measured data. One discrete and six continuous distributions were described and applied. The emphasis has been on flood frequency analysis, but the procedures can be applied to any set of independent hydrologic data, such as rainfall, stage, annual flow volumes, temperatures, water quality parameters, etc. There is no consensus regarding distributions most suited to any particular variable; however, for storm runoff quality, studies have shown that storm runoff event mean concentrations (i.e., flow-weighted average storm event concentrations) are almost universally distributed lognormally (Driscoll, 1986).

Although most of the distributions discussed can also be applied to minima (e.g., droughts) the parameter estimates may vary from those given for the maxima to accommodate the possible negative skewness of the droughts. For use of any distribution other than the normal for minima, reference should be made to Gumbel (1958), Benjamin and Cornell (1970), or Haan (1977) for parameter estimation.

Hydrologic analysis often wanders further into statistics and time series analysis. It is assumed that the reader is familiar with common tools of statistics such as regression and correlation analysis, which are described in almost any statistics text. Time series analysis deals with the treatment of data while retaining their temporal sequence, with possible serial correlation and other underlying relationships (e.g., periodic cycles) among the individual data points. Chatfield (1984) provides an introduction to the topic, and hydrologic applications may be found in texts such as Kottegoda (1980), Salas et al. (1980), and Bras and Rodriguez-Iturbe (1985).

PROBLEMS

3.1. Data for Cypress Creek for the period 1945–1984 are listed in the following table. Using these data, develop a relative frequency histogram and a cumulative frequency histogram for Cypress Creek, 1945–1984. Use a class interval of 2000 cfs.

UNRANKED DATA		RANKED DATA	
Year	Flow (cfs)	Rank	Flow (cfs)
1945	9840	1	15,600
1946	5170	2	10,300
1947	1620	3	9840
1948	235	4	7760
1949	15,600	5	6560
1950	4740	6	6260

UNRANKED DATA		RANKED DATA	
Year	Flow (cfs)	Rank	Flow (cfs)
1951	427	7	5730
1952	3310	8	5440
1953	4400	9	5230
1954	7760	10	5170
1955	2520	11	5060
1956	340	12	4740
1957	5440	13	4710
1958	3000	14	4590
1959	3690	15	4400
1960	10,300	16	4300
1961	6260	17	4210
1962	1360	18	3980
1963	1000	19	3860
1964	2770	20	3830
1965	1400	21	3690
1966	3210	22	3460
1967	1110	23	3310
1968	5230	24	3210
1969	4300	25	3150
1970	2820	26	3080
1971	1900	27	3000
1972	3980	28	2820
1973	6560	29	2770
1974	4710	30	2730
1975	3460	31	2520
1976	3080	32	1900
1977	2730	33	1620
1978	3860	34	1400
1979	4210	35	1360
1980	3150	36	1110
1981	5730	37	1000
1982	3830	38	427
1983	5060	39	340
1984	4590	40	235

3.2. a) Use the data found in Problem 3.1 to calculate the mean, standard deviation, and skew coefficient (Eqs. 3.37, 3.38, and 3.41) of the Cypress Creek data (1945–1984).

b) Repeat part (a) using the log (base 10) of the Cypress Creek data.

3.3. A temporary cofferdam is being designed to protect a 5-yr construction project from the 25-yr flood. What is the risk that the cofferdam will be overtopped

a) at least once during the 5-yr project,

b) not at all during the project,

    c) in the first year only,

    d) in the fourth year and fifth year exactly?

3.4. A recreational park is built near Buffalo Creek. The stream channel can carry 200 m³/s, which is the peak flow of the 5-yr storm of the watershed. Find the following.

    a) The probability that the park will flood next year

    b) The probability that the park will flood at least once in the next 10 yr

    c) The probability that the park will flood 3 times in the next 10 yr

    d) The probability that the park will flood 10 times in the next 10 yr

Problems 3.5 – 3.8 refer to the Cypress Creek data found in Problem 3.1.

3.5. Assume that the Cypress Creek data for the period 1945 – 1984 are normally distributed. Find the following.

    a) Peak flow of the 50-yr flood

    b) Peak flow of the 25-yr flood

    c) Probability that a flood will be less than or equal to 2000 cfs

    d) Return period of the 2000-cfs flood

3.6. Assume that the Cypress Creek data for 1945 – 1984 are lognormally distributed. Find the following.

    a) Peak flow of the 100-yr flood

    b) Peak flow of the 50-yr flood

    c) Probability that a flood will be less than or equal to 2000 cfs

    d) Return period of the 2000-cfs flood

3.7. Assume that the Cypress Creek data for 1945 – 1984 fit a log Pearson 3 distribution. Find the following.

    a) Peak flow of the 100-yr flood

    b) Peak flow of the 25-year flood

    c) Return period of the 2000-cfs flood

3.8. Assume that the Cypress Creek data for 1965 – 1984 fit a log Pearson 3 distribution (statistics of base 10 logs are: $C_s = -1.15$, mean = 3.5375, Var = 0.03849, and standard deviation = 0.1962). Find the peak flow of the 100-yr flood and compare it with the value found in part (a) of Problem 3.7. Explain the difference knowing that 95% of residential development along Cypress Creek occurred after 1965.

3.9. Assume that the Cypress Creek data for 1945 – 1984 fit a 3-parameter gamma distribution. Find the following.

    a) Peak flow of the 100-yr flood

    b) Peak flow of the 50-yr flood

    c) Probability that a flood will be less than or equal to 2000 cfs

    d) Return period of the 2000-cfs flood

3.10. Assume that the Cypress Creek data for 1945 – 1984 fit a Gumbel distribution. Find the following.

a) Peak flow of the 100-yr flood
b) Peak flow of the 50-yr flood
c) Probability that a flood will be less than or equal to 2000 cfs
d) Return period of the 2000-cfs flood

3.11. Match the letters on the right with the numbers on the left to complete
the mathematical statements about PDF properties. Assume that x is
a normally distributed annual occurrence.

1. $\int_{\mu}^{\mu} f(x)dx = \square$

2. $\int_{\square}^{\infty} f(x)dx = 0.02$

3. $\int_{\mu}^{\mu+\square} f(x)dx = 0.34$

4. $\int_{-\infty}^{\square} f(x)dx = 0.5$

5. $\int_{m_1}^{m_2} f(x)dx = \square$

a) Standard deviation
b) Median
c) 0
d) $P(m_1 \leq x \leq m_2)$
e) Value expected every 50 yr
f) Variance
g) $F(x)$

3.12. Knowing that the Gumbel extreme-value distribution is defined as

$$F(y) = e^{-e^{-y}} = \text{Prob}\{Y \leq y\}$$

and that the return period $T(y)$ is

$$T(y) = 1/[1 - F(y)]$$

a) Prove that for floods with large T values, the magnitude of y can be
approximated by $\ln T$. Use a ln series expansion:

$$\ln(1 + x) = x - \frac{x^2}{2} + \frac{x^3}{3} - \cdots, \qquad |x| < 1.$$

b) An old design standard was to build a structure to withstand twice
the magnitude of the largest flood that has occurred historically.
Based on this design, determine the recurrence interval T_d of the
design storm if the largest flood on record has $T = 50$ yr.
c) Compare the relative magnitude of a 100-yr event to a 50-yr event
using the result from part (a). Is the design standard in part (b)
reasonable?

3.13. The total annual runoff for a small watershed was determined to be
approximately normal with a mean of 36 cm and a variance of
29 cm^2. Determine the probability that the total runoff from the basin
will exceed 25 cm in all 4 of the next consecutive 4 yr.

Problems 3.14–3.17 refer to the following data on Spring Creek.

YEAR	FLOW (cfs)	RANK	FLOW (cfs)
1940	3420	1	42,700
1941	42,700	2	31,100
1942	14,200	3	20,700
1943	8000	4	19,300
1944	5260	5	19,300
1945	31,100	6	14,200
1946	12,200	7	12,200
1947	10,000	8	12,100
1948	1430	9	10,700
1949	3850	10	10,300
1950	19,300	11	10,000
1951	0*	12	8760
1952	4130	13	8000
1953	8760	14	7560
1954	1400	15	7340
1955	3570	16	7340
1956	0*	17	6720
1957	4600	18	5260
1958	5260	19	5260
1959	6720	20	4660
1960	20,700	21	4600
1061	10,700	22	4130
1962	0*	23	3850
1963	1590	24	3570
1964	1770	25	3420
1965	2430	26	2430
1966	4660	27	1770
1967	1010	28	1590
1968	12,100	29	1430
1969	10,300	30	1400
1970	1400	31	1400
1971	1300	32	1300
1972	7560	33	1010
1973	19,300		
1974	7340		
1975	7340		

*The years where flow = 0 cfs should not be used in computations since they are outliers.

3.14. a) Find the mean, standard deviation, and skew coefficient (Eqs. 3.37, 3.38, and 3.41) for the Spring Creek data for 1940–1975.
b) Repeat part (a) for the log (base 10) of the Spring Creek data for 1940–1975.
c) Use the Weibull formula (Eqs. 3.80 and 3.81) to determine the plotting positions of the Spring Creek data. Plot the data on normal probability paper. Graphically fit a normal distribution.
d) Repeat part (c) for the log (base 10) of the data or use lognormal paper. Graphically fit a lognormal distribution.

3.15. Assume that the Spring Creek data for 1940–1975 are lognormally distributed. What is the
a) peak flow of the 100-yr flood,
b) peak flow of the 25-yr flood,
c) peak flow of the 10-yr flood,
d) probability that a flood will be less than or equal to 6000 cfs,
e) return period of the 6000-cfs flood,
f) return period of the 15,000-cfs flood?

3.16. a) Assume that the Spring Creek data for 1940–1975 fit the log Pearson 3 distribution. Repeat Problem 3.15.
b) Repeat part (a) for the 2-parameter gamma distribution.
c) Repeat part (a) for the 3-parameter gamma distribution.

3.17. Assume that the Spring Creek data for 1940–1975 fit the Gumbel distribution.
a) Plot the data on extreme value paper using the Gringorten plotting position (Eqs. 3.82 and 3.83) with parameter $a = 0.4$.
b) Find parameters α and u of the distribution, and plot the fitted distribution on the graph.
c) Repeat Problem 3.15.

3.18. The following parameters were computed for a stream near Dallas, Tex., for 1940–1959, inclusive. The data were transformed to $\log_{10} Q = y$.

$$\bar{y} = 3.52 \quad \text{(mean)}$$
$$S_y = 0.50 \quad \text{(standard deviation)}$$
$$C_s = 0.50 \quad \text{(skewness coefficient)}$$

Find the magnitude of the 25-yr flood assuming that the annual peak flow follows (a) log Pearson 3 distribution and (b) lognormal distribution.

3.19. A probability plot of 66 yr of peak discharges for the Kentucky River near Salvisa, Kentucky, is shown in Fig. P3.19.

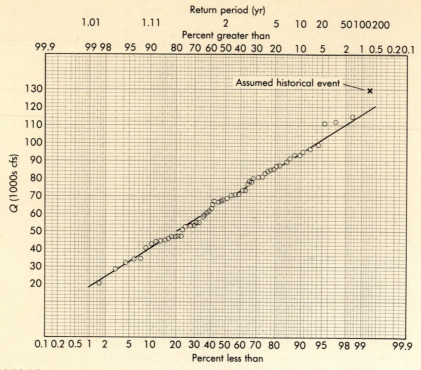

FIGURE P3.19

Normal probability plot for Kentucky River data. (From Haan, 1977, p. 137.)

    a) What probability distribution is being used?
    b) What are the mean and standard deviation of the peak discharges?
    c) If the distribution has other parameters, what are their values?
    d) What is the 25-yr flow?
    e) What is the 100-yr flow?
    f) What is the probability that the annual peak flow will be greater than or equal to 50,000 cfs for all of the next consecutive 3 yr?
    g) What is the probability that at least one 100-yr event will occur in the next 33 yr? In the next 100 yr?
    h) Which plotting position has been used to plot the data points?
    i) Do the Kentucky River data appear to be skewed?

3.20. A probability plot of 19 yr of peak discharges for the West Branch of the Mahoning River near Newton Falls, Ohio, is shown in Fig. P3.20.
    a) Repeat parts (a) through (e) of Problem 3.19.
    b) What is the probability that the annual peak discharge will fall between 5000 and 7000 cfs?

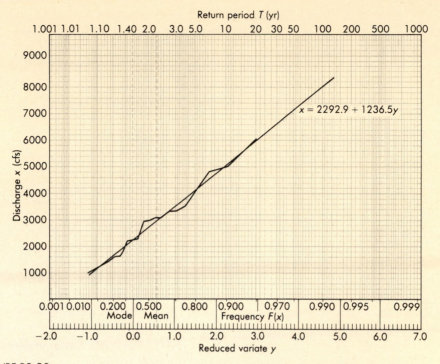

FIGURE P3.20

Annual floods of the West Branch Mahoning River near Newton Falls, Ohio, 1927–1945. (From National Bureau of Standards, 1953, Probability Tables for the Analysis of Extreme Value Data, Applied Mathematics Series 22, U.S. Government Printing Office, Washington, D.C.)

REFERENCES

Abramowitz, M., and I. A. Stegun, 1964, *Handbook of Mathematical Functions,* National Bureau of Standards, U.S. Govt. Printing Office, Washington, D.C. (also published by Dover Publications).

Benjamin, J. R., and C. A. Cornell, 1970, *Probability Statistics and Decision for Civil Engineers,* McGraw-Hill Book Company, New York.

Benson, M. A., 1968, "Uniform Flood-Frequency Estimating Methods for Federal Agencies," *Water Resources Research,* vol. 4, no. 5, October, pp. 891–908.

Bobee, B. B., and R. Robitaille, 1975, "Correction of Bias in Estimation of the Coefficient of Skewness," *Water Resources Research,* vol. 11, no. 6, December, pp. 851–854.

Bras, R. L., and I. Rodriguez-Iturbe, 1985, *Random Functions and Hydrology,* Addison-Wesley Publishing Company, Reading, Massachusetts.

Chatfield, C., 1984, *The Analysis of Time Series: An Introduction,* 3rd edition, Chapman and Hall, New York.

Chemical Rubber Company, *Standard Mathematical Tables,* Cleveland, Ohio, updated periodically.

Chow, V. T. (editor), 1964, *Handbook of Applied Hydrology,* Chapter 8, "Statistical and Probability Analysis of Hydrologic Data," McGraw-Hill Book Company, New York.

Craver, J. S., 1980, *Graph Paper from Your Copier,* H. P. Books, Tucson, Arizona.

Cunnane, C., 1978, "Unbiased Plotting Positions: A Review," *J. Hydrology,* vol. 37, pp. 205–222.

Driscoll, E. D., 1986, "Lognormality of Point and Non-Point Source Pollutant Concentrations," *Proceedings of Stormwater and Water Quality Model Users Group Meeting,* Orlando, Florida, EPA/600/9-86/023, Environmental Protection Agency, Athens, Georgia, March, pp. 157–176.

Fiering, M. B., and B. B. Jackson, 1971, *Synthetic Streamflows,* Water Resources Monograph 1, American Geophysical Union, Washington, D.C.

Gringorten, I. I., 1963, "A Plotting Rule for Extreme Probability Paper," *J. Geophysical Research,* vol. 68, no. 3, pp. 813–814.

Gumbel, E. J., 1958, *Statistics of Extremes,* Columbia University Press, New York.

Haan, C. T., 1977, *Statistical Methods in Hydrology,* Iowa State University Press, Ames.

Harr, M. E., 1977, *Mechanics of Particulate Media: A Probabilistic Approach,* McGraw-Hill Book Company, New York.

Hirsch, R. M., and J. R. Stedinger, 1987, "Plotting Positions for Historical Floods and Their Precision," *Water Resources Research,* vol. 23, vol. 4, April, pp. 715–727.

Interagency Advisory Committee on Water Data, 1982, "Guidelines for Determining Flood Flow Frequency," *Bulletin #17B of the Hydrology Subcommittee,* OWDC, U.S. Geological Survey, Reston, Virginia.

King, J. R., 1971, *Probability Charts for Decision Making,* Industrial Press, New Jersey.

Kite, G. W., 1977, *Frequency and Risk Analyses in Hydrology,* Water Resources Publications, Littleton, Colorado.

Kottegoda, N. T., 1980, *Stochastic Water Resources Technology,* Halstead Press (John Wiley and Sons), New York.

Lettenmaier, D. P., and S. J. Burges, 1982, "Gumbel's Extreme Value I Distribution: A New Look," *J. Hyd. Div., ASCE,* vol. 108, no. HY4, April, pp. 502–514.

National Bureau of Standards, 1950, *Tables of the Binomial Probability Distribution,* Applied Mathematics Series 6, U.S. Government Printing Office, Washington, D.C.

Natural Environment Research Council, 1975, *Flood Studies Report,* 5 vols., Institute of Hydrology, Wallingford, U.K.

Panofsky, H. A., and G. W. Brier, 1968, *Some Applications of Statistics to Meteorology,* Pennsylvania State University Press, University Park.

Parzen, E., 1960, *Modern Probability Theory and Its Applications,* John Wiley and Sons, New York.

Reich, B. M., and K. G. Renard, 1981, "Applications of Advances in Flood Frequency Analysis," *Water Resources Bulletin,* vol. 17, no. 1, February, pp. 67–74.

Salas, J. D., J. W. Delleur, V. Yevjevich, and W. L. Lane, 1980, *Applied Modeling of Hydrologic Time Series,* Water Resources Publications, Littleton, Colorado.

Sangal, B. P., and A. K. Biswas, 1970, "The 3-Parameter Lognormal Distribution and Its Applications in Hydrology," *Water Resources Research,* vol. 6, no. 2, April, pp. 505–515.

Searcy, J. K., 1959, *Flow Duration Curves,* USGS Water Supply Paper 1542-A, Washington, D.C.

Tasker, G. D., and J. R. Stedinger, 1986, "Regional Skew with Weighted LS Regression," *J. Water Resources Planning and Management, ASCE,* vol. 112, no. 2, April, pp. 225–237.

Flood Routing

4.1
HYDROLOGIC AND HYDRAULIC ROUTING

The movement of a flood wave down a channel or through a reservoir and the associated change in timing or attenuation of the wave constitute an important topic in floodplain hydrology. It is essential to understand the theoretical and practical aspects of flood routing to predict the temporal and spatial variations of a flood wave through a river reach or reservoir. Flood routing methods can also be used to predict the outflow hydrograph from a watershed subjected to a known amount of precipitation.

The **storage routing** concept is most easily understood by referring to Fig. 4.1. Inflow and outflow hydrographs for a small level-surface reservoir have been plotted on the same graph. Area A represents the volume of water that fills available storage up to time t_1. Inflow exceeds outflow and the reservoir is filling. At time t_1, inflow and outflow are equal and the maximum storage is reached. For times exceeding t_1, outflow exceeds inflow and the reservoir empties. Area C represents the volume of water that flows out of the reservoir and must equal area A if the reservoir begins and ends at the same level. The peak of the outflow from a reservoir should intersect the inflow hydrograph as shown in Fig. 4.1 because outflow is uniquely determined by reservoir storage or level.

(a)

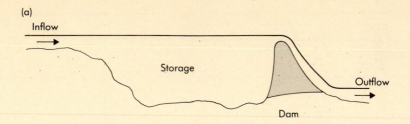

(b)

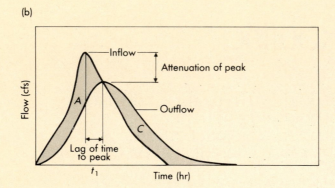

(c)

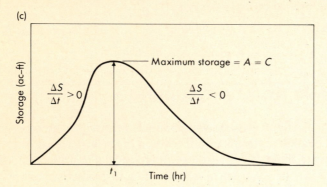

FIGURE 4.1

Reservoir concepts. (a) Reservoir storage. (b) Inflow to and outflow from the reservoir. (c) Storage in the reservoir.

We will see that storage routing through a reservoir will generally attenuate the peak outflow and lag the time to peak for the outflow hydrograph. The rate of change of storage can be written as the continuity equation:

$$I - O = \frac{\Delta S}{\Delta t}, \qquad (4.1)$$

where

I = inflow,

O = outflow,

ΔS = change in storage,

Δt = change in time.

Example 4.1 is a detailed illustration of reservoir storage concepts.

EXAMPLE 4.1

STORAGE COMPUTATIONS

Inflow and outflow hydrographs for a reservoir are depicted in Fig. E4.1(a).

a) Determine the average storage for each one day period ($\Delta t = 1$ day). Graph storage vs. time for the reservoir for the event. Assume that $S_0 = 0$ (the reservoir is initially empty).

b) What is the (approximate) maximum storage reached during this storm event?

SOLUTION

a) The rate of change in storage is equal to inflow minus outflow. First, we tabulate values of I and Q and take their difference. Storage is equal to the area between the inflow and outflow curves, or

$S = \int (I - Q)\, dt.$

This integral can be simply approximated by

$S = \Sigma (I - Q)\, \Delta t,$

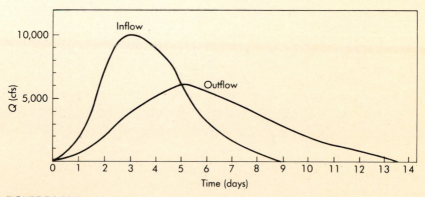

FIGURE E4.1(a)

a method used in Chapter 2 to determine volumes under hydrographs. To minimize error, I and Q values are read at noon each day.

TIME (days)	I (cfs)	Q (cfs)	$\Delta S/\Delta t$ (cfs)
0.5	500	250	250
1.5	3500	1000	2500
2.5	9000	3000	6000
3.5	9750	4500	5250
4.5	8000	5750	2250
5.5	4500	6000	−1500
6.5	2250	5250	−3000
7.5	1250	4250	−3000
8.5	250	3250	−3000
9.5	0	2500	−2500
10.5	0	1500	−1500
11.5	0	1000	−1000
12.5	0	750	−750
13.5	0	0	0

Using $\Delta t = 1$ day, storage for the first day, S_1, is

$$S_1 = S_0 + (I_1 - Q_1)\,\Delta t$$

$$= 0 + (250 \text{ cfs})(1 \text{ day}) \left(\frac{24 \text{ hr}}{\text{day}}\right)\left(\frac{3600 \text{ s}}{\text{hr}}\right)\left(\frac{\text{ac}}{43{,}560 \text{ ft}^2}\right)$$

$$= \boxed{496 \text{ ac-ft}}$$

For day 2,

$$S_2 = S_0 + S_1 + (I_2 - Q_2)\,\Delta t,$$

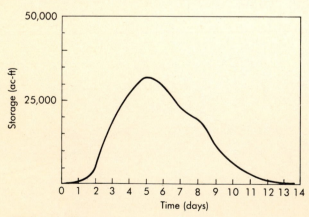

FIGURE E4.1(b)

$$S_2 = 0 + 496 + (2500)(24)(3600) \left(\frac{1}{43,560}\right) \text{ ac-ft}$$

$$= 5455 \text{ ac-ft}$$

The procedure is shown completed in the following table and the hydrograph in Fig. E4.1(b).

TIME (day)	STORAGE (ac-ft)
1	496
2	5455
3	17,356
4	27,769
5	32,232
6	29,256
7	23,306
8	17,356
9	11,405
10	6,446
11	3,471
12	1,488
13	0
14	0

b) The maximum storage, as seen from the table and figure, is 32,232 ac-ft. This occurs at day 5 for this event, as seen from the equation

$$\frac{dS}{dt} = I - Q.$$

S_{max} will occur when dS/dt equals zero. At this point, $I = Q$, which occurs at day 5 on the inflow-outflow hydrographs.

River routing differs from reservoir routing in that storage in a river reach of length L depends on more than just outflow. The peak of the outflow hydrograph from a reach is usually attenuated and delayed compared with that of the inflow hydrograph. Because storage in a river reach is a function of whether stages are rising or falling, storage in this case is a function of both outflow and inflow for the routing reach (see Section 4.3). Also, as river stages rise high enough to inundate a floodplain beyond the banks of the channel, significant velocity reductions are observed in the floodplain compared with the main channel. Example 4.2 presents differences between river and reservoir routing.

EXAMPLE 4.2

RIVER AND RESERVOIR ROUTING CONCEPTS

Figure E4.2(a) illustrates some differences between river and reservoir routing. Prove that, for a level-pool reservoir, the peak of the outflow hydrograph must intersect the inflow hydrograph.

SOLUTION

Storage in a reservoir may be determined from the height of water in the reservoir (see Fig. E4.2b). For instance, in a vertical-walled reservoir,

$V = A_r \cdot H = S$ of the reservoir.

Generally, A_r is a function of depth such that

$S = f(H) = \int A_r(H) \, dH.$

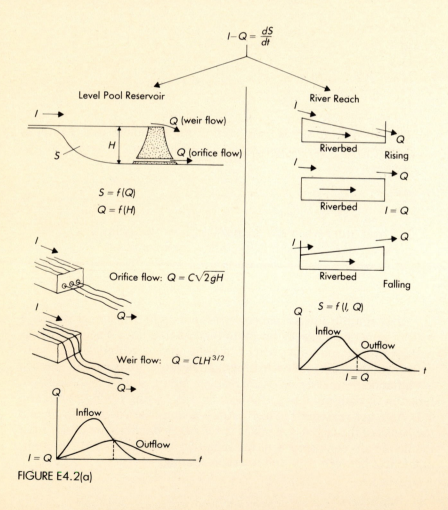

FIGURE E4.2(a)

FIGURE E4.2(b)

While inflow increases storage in a reservoir, outflow may be determined, whether or not inflow is known, if the storage is known:

$$Q = f(S)$$

and, therefore,

$$Q = f(H).$$

From the equation of continuity,

$$I - Q = \frac{dS}{dt} = A_r \frac{dH}{dt},$$

$$I = Q \quad \text{when} \quad \frac{dS}{dt} = 0.$$

Since S is directly proportional to H,

$$\frac{dH}{dt} = 0 \quad \text{when} \quad \frac{dS}{dt} = 0$$

and since Q is also directly proportional to H,

$$\frac{dQ}{dt} = 0 \quad \text{when} \quad \frac{dH}{dt} = 0.$$

Thus,

$$\frac{dQ}{dt} = 0 \quad \text{when} \quad \frac{dS}{dt} = 0.$$

This occurs when Q is at a maximum or a minimum. A minimum occurs when $Q = 0$, the beginning and end of the outflow hydrograph. A maximum occurs at the peak outflow; thus

$$I = Q.$$

Hydrologic Routing Methods

Routing techniques may be classified in two major categories: simple **hydrologic routing** and more complex **hydraulic routing.** Hydrologic routing involves the balancing of inflow, outflow, and volume of storage through use of the continuity equation. A second relationship, the storage-discharge relation, is also required between outflow rate and storage in the system. Applications of hydrologic routing techniques to problems of flood prediction, flood control measures, reservoir design and operation, watershed simulation, and urban design are numerous. Many computer models are available that take input rainfall, convert it to outflow hydrographs, and then route the hydrographs through complex river or reservoir networks using hydrologic routing methods. These applications are presented in detail in Chapters 5 and 6.

Hydraulic Routing Methods

Hydraulic routing is more complex and accurate than hydrologic routing and is based on the solution of the continuity equation and the momentum equation for unsteady flow in open channels. These differential equations are usually solved by explicit or implicit numerical methods on a computer and are known as the St. Venant equations, first derived in 1871, for which no closed-form solutions exist.

Unsteady flow in rivers, reservoirs, and estuaries is caused by motion of long waves due to tides, flood waves, storm surges, and dynamic reservoir releases. These types of wave forms can be adequately described only by the one-dimensional St. Venant equations, which are presented in detail in Section 4.4. In many cases, the governing equations can be simplified to a one-dimensional continuity equation and a uniform flow relationship, referred to as **kinematic wave routing,** which implies that discharge can be computed as a simple function of depth alone. Recently, kinematic wave routing has been added to the HEC-1 flood hydrograph package.

Uniform flow implies a balance between gravitational and frictional forces in the channel. This assumption cannot always be justified, especially on very flat slopes where effects of water surface slope cannot be ignored. Cases where other terms in the momentum equation for hydraulic routing must be retained include (1) upstream movement of tides and storm surges, (2) backwater effects from downstream reservoirs and tributary inflows, (3) flood waves in channels of very flat slope (2 – 3 ft/mi), and (4) abrupt waves caused by sudden releases from reservoirs or dam failures. For these cases, the complete solution of the St. Venant equations should be used. Only a few computer models exist to solve these equations.

4.2

HYDROLOGIC RIVER ROUTING

As a flood wave passes through a river reach, the peak of the outflow hydrograph is usually attenuated and delayed due to channel resistance and storage capacity. Considering a lumped storage approach for the reach, the difference between the ordinates of the inflow and outflow hydrographs, represented by shaded areas in Fig. 4.2, is equal to the rate of change of storage in the reach, as shown in Eq. 4.1. The value of $\Delta S/\Delta t$ in the continuity equation is positive when storage is increasing and negative

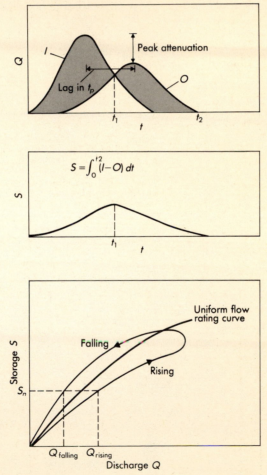

FIGURE 4.2

Storage in a river reach.

when storage is decreasing, and S can be plotted as a function of time. Equation (4.1) can be written in finite-difference form as Eq (4.2), where Δt is referred to as the routing time period and subscripts 1 and 2 denote the beginning and end of the time period, respectively:

$$\frac{1}{2}(I_1 + I_2) - \frac{1}{2}(O_1 + O_2) = \frac{S_2 - S_1}{\Delta t}. \tag{4.2}$$

If storage is plotted against outflow for a river reach, the resulting curve will generally take the form of a loop, as shown in Fig. 4.2. This loop effect implies greater storage for a given outflow during falling stages than during rising stages. If one considers water surface profiles at various times during passage of the flood wave, the concept of prism and wedge storage is useful. This is shown in Fig. 4.3. A large volume of wedge storage may

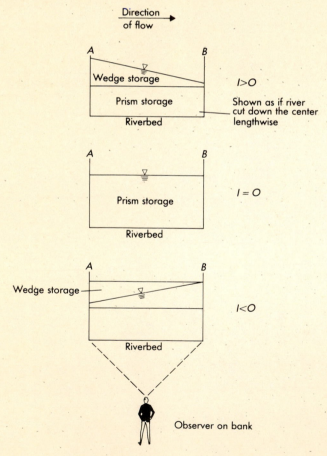

FIGURE 4.3

Prism and wedge storage concepts.

exist during rising stages before outflows have increased. During falling stages, inflow drops more rapidly than outflow, and the wedge storage becomes negative. Hydrologic routing in rivers and channels thus requires a storage relationship that allows for wedge storage. This is accomplished by allowing storage to be a function of both inflow and outflow as in the Muskingum method of flood routing (McCarthy, 1938). The method suffers the disadvantage of assuming the uniform flow rating curve in place of the loop curve shown in Fig. 4.2.

Muskingum Method

The Muskingum method was developed by McCarthy (1938) and utilizes the continuity equation (Eq. 4.2) and a storage relationship that depends on both inflow and outflow. The storage within the reach at a given time can be expressed by (Chow, 1959)

$$S = \frac{b[xI^{m/n} + (1 - x)O^{m/n}]}{a^{m/n}}, \tag{4.3}$$

where inflow and outflow are related to ay^n from Manning's equation, where a and n are constants. Storage in the reach is related to by^m, where b and m are constants. The parameter x defines the relative weighting of inflow and outflow in determining storage volume in the reach.

The Muskingum method assumes that $m/n = 1$ and $b/a = K$, resulting in a linear relationship of the form

$$S = K[xI + (1 - x)O], \tag{4.4}$$

where

$K =$ travel time constant for the reach,

$x =$ weighting factor, which varies from 0 to 0.5 for a given reach.

For the case of linear reservoir routing where S depends only on outflow, $x = 0$ in Eq. (4.4). In smooth uniform channels, $x = 0.5$ yields equal weight to inflow and outflow, which theoretically results in pure translation of the wave. A typical value for most natural streams is $x = 0.2$.

The routing procedure uses the finite-difference form of the continuity Eq. (4.2) combined with Eq. (4.4) in the form

$$S_2 - S_1 = K[x(I_2 - I_1) + (1 - x)(O_2 - O_1)] \tag{4.5}$$

to produce the Muskingum routing equation for a river reach:

$$O_2 = C_0 I_2 + C_1 I_1 + C_2 O_1, \tag{4.6}$$

where

$$C_0 = \frac{-Kx + 0.5 \, \Delta t}{D}, \tag{4.7}$$

$$C_1 = \frac{Kx + 0.5\,\Delta t}{D},$$ (4.8)

$$C_2 = \frac{K - Kx - 0.5\,\Delta t}{D},$$ (4.9)

$$D = K - Kx + 0.5\,\Delta t.$$ (4.10)

This procedure is ideally set up for a calculator or a personal computer. Note that K and Δt must have the same units and that the coefficients C_0, C_1, and C_2 sum to 1.0. With K, x, and Δt known, values C_0, C_1, and C_2 are computed. The routing operation is accomplished by solving Eq. (4.6) for successive time increments, with O_2 of one routing period becoming O_1 of the succeeding period. Example 4.3 illustrates the row-by-row computation, and a computer program is listed in Appendix E.

EXAMPLE 4.3

MUSKINGUM ROUTING

Route the inflow hydrograph tabulated in the following table through a river reach for which $x = 0.2$ and $K = 2$ days. Use a routing period $\Delta t = 1$ day and assume that inflow equals outflow for the first day.

TIME (day)	INFLOW (cfs)
1	4,000
2	7,000
3	11,000
4	17,000
5	22,000
6	27,000
7	30,000
8	28,000
9	25,000
10	23,000
11	20,000
12	17,000
13	14,000
14	11,000
15	8,000
16	5,000
17	4,000
18	4,000
19	4,000
20	4,000

SOLUTION

First, we determine the coefficients C_0, C_1, and C_2 for the reach (Eq. 4.9):

$$C_0 = \frac{-Kx + 0.5\,\Delta t}{D},$$

$$C_1 = \frac{Kx + 0.5\,\Delta t}{D},$$

$$C_2 = \frac{K - Kx - 0.5\,\Delta t}{D},$$

$$D = K - Kx + 0.5\,\Delta t$$

For $K = 2$ days, $\Delta t = 1$ day, and $x = 0.2$,

$$D = 2 - 2(0.2) + 0.5(1)$$

$$= \boxed{2.1,}$$

$$C_0 = \frac{-(2)(0.2) + (0.5)(1)}{(2.1)}$$

$$= \boxed{0.0476,}$$

$$C_1 = \frac{(2)(0.2) + (0.5)(1)}{(2.1)}$$

$$= \boxed{0.4286,}$$

$$C_2 = \frac{2 - (2)(0.2) - (0.5)(1)}{(2.1)}$$

$$= \boxed{0.5238.}$$

We may check our computations by seeing if the coefficients sum to 1:

$$(0.0476) + (0.4286) + (0.5238) = 1.0000.$$

We substitute these values into Eq. (4.6) to obtain

$$O_2 = (0.0476)I_2 + (0.4286)I_1 + (0.5238)O_1.$$

For $t = 1$ day,

$$O_1 = I_1 = \boxed{4000\ \text{cfs.}}$$

For $t = 2$ days,

$$O_2 = (0.0476)(7000) + (0.4286)(4000) + (0.5238)(4000)$$

$$= \boxed{4143\ \text{cfs.}}$$

For $t = 3$ days,

$$O_3 = (0.0476)(11,000) + (0.4286)(7000) + (0.5238)(4143)$$

$$= \boxed{5694\ \text{cfs.}}$$

This procedure is shown completed for $t = 1$ to $t = 20$ days in the following table.

TIME	INFLOW (cfs)	OUTFLOW (cfs)
1	4,000	4,000
2	7,000	4,143
3	11,000	5,694
4	17,000	8,506
5	22,000	12,789
6	27,000	17,413
7	$I_p \rightarrow$ 30,000	22,121
8	28,000	25,778
9	25,000	26,693 $\leftarrow Q_p$
10	23,000	25,792
11	20,000	24,319
12	17,000	22,120
13	14,000	19,539
14	11,000	16,758
15	8,000	13,873
16	5,000	10,934
17	4,000	8,061
18	4,000	6,127
19	4,000	5,114
20	4,000	4,583

Note: Q_p lags I_p by 2 days, i.e., approximately by K.

It is possible to compute outflow at any time if the preceding inflows are known: $I_1, I_2, I_3, \ldots , I_n$. Equation (4.6) can be rewritten as

$$O_n = C_0 I_n + C_1 I_{n-1} + C_2 O_{n-1} \qquad \text{and}$$
$$O_{n-1} = C_0 I_{n-1} + C_1 I_{n-2} + C_2 O_{n-2}. \qquad (4.11)$$

Repeated calculations for $O_{n-2}, O_{n-3}, \ldots$, can be performed so that the following equation for O_n can be derived:

$$O_n = K_1 I_n + K_2 I_{n-1} + K_3 I_{n-2} + \cdots + K_n I_1, \qquad (4.12)$$

where

$$K_1 = C_0; \quad K_2 = C_0 C_2 + C_1; \quad K_3 = K_2 C_2,$$
$$K_i = K_{i-1} C_2 \quad \text{for} \quad i > 2.$$

Determination of Storage Constants

The Muskingum K is usually estimated from the travel time for a flood wave through the reach, and x averages 0.2 for a natural stream. However,

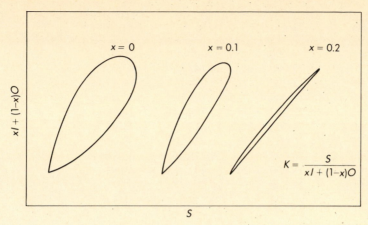

FIGURE 4.4

Selection of Muskingum coefficients.

if both inflow and outflow hydrograph records are available, better estimates for K and x can be made through graphical methods. Storage S is
plotted vs. weighted discharge, $xI + (1 - x)O$, for several selected values
of x, and the plot that yields the most linear single-valued curve provides
the best value for x. The Muskingum method assumes that this curve is a
straight line with reciprocal slope K. Figure 4.4 and Example 4.4 illustrate
the concept of selecting x and K. Thus, the Muskingum method assumes
that storage is a single-valued function of weighted inflow and outflow.
Normally a river must be divided into several reaches for application of
the Muskingum routing method. This requires that flow changes slowly
with time. This method has been shown to work quite well for ordinary
streams with small slopes where the storage-discharge curve is approximately linear. However, in cases involving very steep or mild slopes,
backwater effects, or abrupt waves, dynamic effects of flow may be pro-
nounced and hydraulic routing methods should be used rather than hydrologic methods. Alternatively, the Muskingum-Cunge method may be
used (see Section 4.7).

EXAMPLE 4.4

DETERMINATION OF THE MUSKINGUM ROUTING COEFFICIENTS

The values listed in the following table for inflow, outflow, and storage
were measured for a particular reach of a river. Determine the coefficients
K and x for use in the Muskingum routing equations for this reach.

TIME (hr)	INFLOW (cfs)	OUTFLOW (cfs)	STORAGE (ac-ft)
10	4,000	4,000	6,610
20	7,000	4,143	7,790
30	11,000	5,694	11,170
40	17,000	8,506	16,870
50	22,000	12,789	24,180
60	27,000	17,413	31,950
70	30,000	22,121	39,170
80	28,000	25,778	43,340
90	25,000	26,693	43,560
100	23,000	25,792	41,710
110	20,000	24,319	38,780
120	17,000	22,120	34,870
130	14,000	19,539	30,460
140	11,000	16,758	25,800
150	8,000	13,873	20,990
160	5,000	10,934	16,110
170	4,000	8,061	11,980
180	4,000	6,127	9,420
190	4,000	5,114	8,080
200	4,000	4,583	7,380

SOLUTION

To determine Muskingum coefficients, we guess a value of x and then plot $[xI + (1 - x)Q]$ vs. S. The plot that comes closest to being a straight line is chosen to determine the coefficient values. The average value is $x = 0.2$

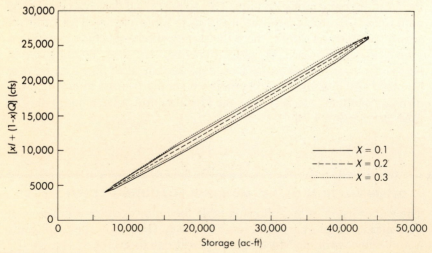

FIGURE E4.4

for a natural stream. Therefore, we assume that x must lie between 0.1 and 0.3. Plots of $[xI + (1 - x)Q]$ vs. S are made for $x = 0.1$, $x = 0.2$, and $x = 0.3$ using the values listed in the following table (see Fig. E4.4).

STORAGE (ac-ft)	$[xI + (1 - x)Q]$ (cfs)		
	$x = 0.1$	$x = 0.2$	$x = 0.3$
6,610	4,000	4,000	4,000
7,790	4,429	4,715	5,000
11,170	6,225	6,755	7,286
16,870	9,355	10,205	11,054
24,180	13,710	14,631	15,552
31,950	18,372	19,331	20,289
39,170	22,909	23,697	24,485
43,340	26,000	26,223	26,445
43,560	26,523	26,355	26,185
41,710	25,513	25,234	24,954
38,780	23,887	23,455	23,023
34,870	21,608	21,096	20,584
30,460	18,985	18,431	17,877
25,800	16,182	15,607	15,031
20,990	13,286	12,699	12,111
16,110	10,341	9,747	9,154
11,980	7,655	7,249	6,843
9,420	5,914	5,702	5,489
8,080	5,003	4,891	4,780
7,380	4,525	4,467	4,408

It can easily be seen that the plot for $x = 0.2$ is the straightest line. K is calculated as the inverse slope of the line:

$$1/K = \frac{4000 \text{ cfs} - 26{,}445 \text{ cfs}}{6610 \text{ ac-ft} - 42{,}240 \text{ ac-ft}} \cdot \frac{1 \text{ ac}}{43{,}560 \text{ ft}^2} \cdot \frac{3600 \text{ s}}{1 \text{ hr}}$$

$$K = \boxed{20 \text{ hr} = 0.8 \text{ day.}}$$

For most streams, there will be a larger looping effect for all values of x. The most linear relationship is chosen to determine K and x.

4.3
HYDROLOGIC RESERVOIR ROUTING

Storage Indication Method

Reservoir or detention basin routing is generally easier to perform than river routing because storage-discharge relations for pipes, weirs, and spillways are single-valued functions independent of inflow. Thus, a simple **storage indication method** or **Puls method** uses the finite-difference form of the continuity equation combined with a storage indication curve ($2S/\Delta t + O$ vs. O). Equation (4.2) can be generalized to the following finite-difference equation for two points in time:

$$(I_n + I_{n+1}) + \left(\frac{2S_n}{\Delta t} - O_n\right) = \left(\frac{2S_{n+1}}{\Delta t} + O_{n+1}\right), \tag{4.13}$$

in which the only unknowns are S_{n+1} and O_{n+1} on the right-hand side. I is known for all n, and S_n and O_n are known for the initial time step; therefore the right-hand side of Eq. (4.13) can be calculated. Values of S_{n+1} and O_{n+1} are then used as input on the left-hand side and the calculation is repeated for the second time interval, and so on. The storage indication curve is a plot of $2S/\Delta t + O$ vs. O, as shown in Example 4.5. Thus, once the right-hand side of Eq. (4.13) has been determined, one can read values of O directly from the curve. Values for $2S/\Delta t - O$ for the left-hand side of Eq. (4.13) are calculated by subtracting $2(O)$ from the right-hand side values. The detailed computations are shown in Example 4.5.

EXAMPLE 4.5

STORAGE INDICATION ROUTING

The design inflow hydrograph shown in Fig. E4.5(a), developed for a commercial area, is to be routed through a reservoir. Assume that initially the reservoir is empty ($S_0 = 0$) and there is no initial outflow ($Q_0 = 0$). Using the depth, storage, and outflow relationships given in the table, route the hydrograph through the reservoir. What is the maximum height reached in the reservoir for this inflow? Use $\Delta t = 10$ min.

DEPTH (ft)	STORAGE (ac-ft)	OUTFLOW (cfs)
0	0	0
1.0	1.0	15
2.0	2.0	32
3.0	3.0	55
4.0	4.0	90
5.0	5.0	125
6.0	6.0	158
7.0	7.5	185
8.0	10.5	210
9.0	12.0	230
10.0	13.5	250
11.0	20.0	270
12.0	22.0	290

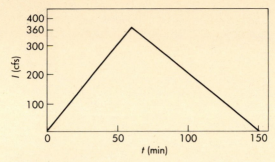

FIGURE E4.5(a)

SOLUTION

First, we develop a storage indication curve for the reservoir. This is a plot of $(2S/\Delta t) + Q$ vs. Q. For instance, at $Q = 90$ cfs, $S = 4.0$ ac-ft and

$$\frac{2S}{\Delta t} + Q = \frac{2(4.0 \text{ ac-ft})\left(\dfrac{43{,}560 \text{ ft}^2}{\text{ac}}\right)}{(10 \text{ min})(60 \text{ s/min})} + 90 \text{ cfs} = 671 \text{ cfs.}$$

Graphical and tabulated results for the storage indication curve are given in Fig. E4.5(b) and the following table.

Q (cfs)	$2S/\Delta t + Q$ (cfs)
0	0
15	160
32	322
55	491
90	671
125	851
158	1,029
185	1,274
210	1,735
230	1,972
250	2,210
270	3,174
290	3,484

Equation (4.13) states:

$$(I_n + I_{n+1}) + \left(\frac{2S_n}{\Delta t} - Q_n\right) = \left(\frac{2S_{n+1}}{\Delta t} + Q_{n+1}\right).$$

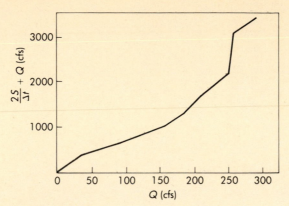

FIGURE E4.5(b)

For $t_n = 0$, $I_n = 0$. At $t_{n+1} = 10$ min, $I_{n+1} = 60$ cfs, as seen in Fig. E4.5(a), and

$$I_n + I_{n+1} = 60 \text{ cfs.}$$

S_n is storage at $t_n = 0$ or S_0, and $S_0 = 0$. $Q_n = Q_0 = 0$. So, referring to Eq. (4.13),

$$60 + (0) = \left(\frac{2S_{n+1}}{\Delta t} + Q_{n+1} \right),$$

$$\left(\frac{2S_{n+1}}{\Delta t} + Q_{n+1} \right) = 60.$$

From the storage indication curve, for

$$\left(\frac{2S}{\Delta t} + Q \right) = 60,$$

$$\boxed{Q = 5 \text{ cfs.}}$$

For $t_n = 10$ min and $t_{n+1} = 20$ min, $I_n = 60$ cfs and $I_{n+1} = 120$ cfs:

$$\frac{2S_n}{\Delta t} - Q_n = \left(\frac{2S_n}{\Delta t} + Q_n \right) - 2Q_n$$
$$= 60 \text{ cfs} - 2(5 \text{ cfs})$$
$$= 50 \text{ cfs,}$$

$$I_n + I_{n+1} + \left(\frac{2S_n}{\Delta t} - Q_n \right) = \left(\frac{2S_{n+1}}{\Delta t} + Q_{n+1} \right),$$

$$60 + 120 + 50 = 230$$

From the storage indication curve, for

$$\left(\frac{2S}{\Delta t} + Q\right) = 230,$$

$$\boxed{Q = 22 \text{ cfs.}}$$

The procedure is shown completed in the following table.

TIME (min)	I_n (cfs)	$(I_n + I_{n+1})$ (cfs)	$\left(\frac{2S_n}{\Delta t} - Q_n\right)$ (cfs)	$\left(\frac{2S_{n+1}}{\Delta t} + Q_{n+1}\right)$ (cfs)	Q_{n+1} (cfs)
0	0	60	0		0
10	60	180	50	60	5
20	120	300	186	230	22
30	180	420	378	486	54
40	240	540	568	798	115
50	300	660	774	1108	167
60	360	680	1046	1434	194
70	320	600	1306	1726	210
80	280	520	1458	1906	224
90	240	440	1516	1978	231
100	200	360	1498	1956	229
110	160	280	1418	1858	220
120	120	200	1282	1698	208
130	80	120	1090	1482	196
140	40	40	854	1210	178
150	0	0	628	894	133
160	0	0	464	628	82
170	0	0	362	462	51
180	0	0	288	362	37
190	0	0	232	288	28
200	0	0	186	232	23

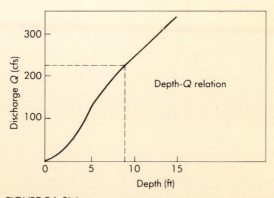

FIGURE E4.5(c)

To determine the maximum height of the reservoir during this inflow, a depth vs. discharge curve is developed using the given data. See Fig. E4.5(c) on page 229.

The maximum outflow found in routing is 231 cfs. The depth corresponding to this, the maximum stage reached by the reservoir, is about 9.0 ft.

Detention Basin Routing

The purpose of flood routing for detention basin design is to determine how the outflow from a detention basin and the storage in the basin vary with time for a known inflow hydrograph. A more accurate numerical routing scheme for solving the continuity and storage equations is the Runge-Kutta technique (Chapra and Canale, 1985). The **Runge-Kutta (R-K) method** for solving ordinary differential equations can be developed to solve the following equations with various orders of accuracy (first, second, third, fourth). The continuity equation is expressed

$$\frac{dV}{dt} = Q_{in}(t) - Q_{out}(H), \tag{4.14}$$

where

$$V = \text{volume of water in storage in the basin,}$$
$$Q_{in}(t) = \text{inflow into the detention basin as a function of time,}$$
$$Q_{out}(H) = \text{outflow from a detention basin as a function of head } (H) \text{ in the basin.}$$

The change in volume dV due to a change in depth dH can be expressed as

$$dV = A_r(H)\, dH, \tag{4.15}$$

where $A_r(H)$ is the surface area related to H. The continuity equation is then expressed as

$$\frac{dH}{dt} = \frac{Q_{in}(t) - Q_{out}(H)}{A_r(H)}. \tag{4.16}$$

Figure 4.5(a) illustrates the first-order R-K technique. Equation (4.16) can be represented by

$$\frac{dH}{dt} = f(H_n, t_n),$$

where

$$H = \text{the dependent variable,}$$
$$t = \text{the independent variable.}$$

In first-order R-K solutions, a finite time increment Δt is chosen. Then

$$\Delta H = f(H_n, t_n) \, \Delta t,$$
$$H_{n+1} = H_n + \Delta H, \tag{4.17}$$

assuming that the initial head H_n is known. However, since ΔH is not constant but is continually changing with time, error is introduced as shown in Fig. 4.5(a). Thus, first-order R-K solutions are relatively accurate only for very small time increments.

The second-order R-K technique alleviates some of this error, as shown in Fig. 4.5(b), where ΔH is calculated at the beginning and end of the chosen time increment Δt, and the two values of ΔH are averaged.

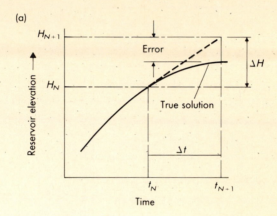

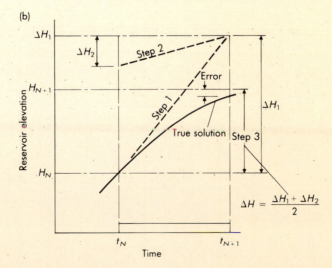

FIGURE 4.5

(a) First-order Runge-Kutta technique. (b) Second-order Runge-Kutta technique.

(Note that the line for step 1 is tangent to $f(H_n, t_n)$ at t_n. The line for step 2 is tangent to $f(H_n, t_n)$ at t_{n+1}.) First, ΔH_1 is found from

$$\Delta H_1 = f(H_n, t_n) \Delta t.$$

Since H_{n+1} has not yet been calculated, ΔH_2 cannot be found at that point. But ΔH_2 is estimated by evaluating Eq. (4.17) at $H_n + \Delta H_1$ and $t_n + \Delta t$:

$$\Delta H_2 = f(H_n + \Delta H_1, t_n + \Delta t).$$

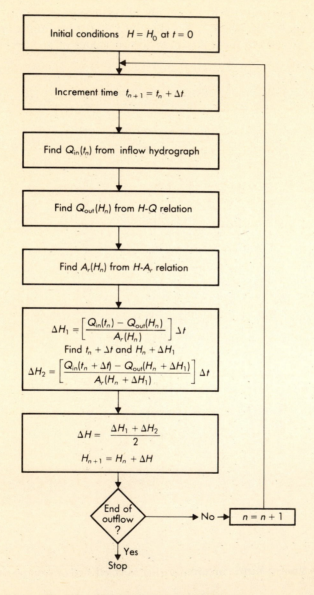

FIGURE 4.6

Flow chart for second-order Runge-Kutta solutions.

Then

$$\Delta H = \frac{\Delta H_1 + \Delta H_2}{2},$$

$$H_{n+1} = H_n + \Delta H. \tag{4.18}$$

Applying this technique to Eq. (4.16), we have

$$\Delta H_1 = \frac{Q_{in}(t_n) - Q_{out}(H_n)}{A_r(H_n)} \Delta t,$$

$$\Delta H_2 = \frac{Q_{in}(t_n + \Delta t) - Q_{out}(H_n + \Delta H_1)}{A_r(H_n + \Delta H_1)} \Delta t. \tag{4.19}$$

The second-order R-K solution of Eq. (4.16) is further illustrated by the flow chart of Fig. 4.6.

By reasoning similar to that used in developing the second-order R-K solution, third- and fourth-order R-K solutions attempt to improve the accuracy of the estimate of ΔH. The third-order technique estimates H_{n+1} by

$$H_{n+1} = H_n + \tfrac{1}{6} [k_1 + 4k_2 + k_3] \Delta t, \tag{4.20}$$

where

$$k_1 = f(t_n, H_n),$$
$$k_2 = f(t_n + \Delta t/2, H_n + \tfrac{1}{2}k_1\Delta t),$$
$$k_3 = f(t_n + \Delta t, H_n - k_1\Delta t + 2k_2\Delta t).$$

In the classical fourth-order R-K technique,

$$H_{n+1} = H_n + \tfrac{1}{6}[k_1 + 2k_2 + 2k_3 + k_4] \Delta t, \tag{4.21}$$

where

$$k_1 = f(t_n, H_n)$$
$$k_2 = f(t_n + \Delta t/2, H_n + \tfrac{1}{2}k_1\Delta t),$$
$$k_3 = f(t_n + \Delta t/2, H_n + \tfrac{1}{2}k_2\Delta t),$$
$$k_4 = f(t_n + \Delta t, H_n + k_3\Delta t). \tag{4.22}$$

The fourth-order R-K scheme is considered to be more accurate than the first, second, or third order. Example 4.6 illustrates the use of the fourth-order R-K solution, and a computer program is included in Appendix E.

EXAMPLE 4.6

FOURTH-ORDER RUNGE-KUTTA METHODS

Bull Creek Watershed has a reservoir with storage relationship

$$S = AH,$$

where A is the area (300 ac) and H is the depth or head of the reservoir in ft.

The area is assumed to be constant. The outflow is governed by the equation

$$Q = 56.25 \, H^{3/2},$$

where Q is in cfs. Route the storm hydrograph represented in the following table through the reservoir using fourth-order Runge-Kutta methods.

TIME (hr)	INFLOW (cfs)
12	40
24	35
36	37
48	125
60	340
72	575
84	722
96	740
108	673
120	456
132	250
144	140
156	10

SOLUTION

The governing equation is

$$\frac{\Delta H}{\Delta t} = f(t, H) = \frac{Q_{in}(t) - Q_{out}(H)}{A_r(H)}.$$

The fourth-order R-K equation will be used as shown in Eqs. (4.21) and (4.22).

The assumption is made that $H_0 = 0$ ft and $Q_{out}(H_0) = 0$ cfs. Values of Q_{in} are interpolated as necessary, and values of Q_{out} are found from the equation $Q = 56.25 H^{3/2}$. The computations are shown for $n = 0$:

$\Delta t = 12$ hr,

$t_0 = 12$ hr,

$H_0 = 0$ ft,

$k_1 = f(t_0, H_0)$

$\quad = [Q_{in}(12) - Q_{out}(0)]/A_r(0)$

$\quad = (40 \text{ cfs} - 0 \text{ cfs})/300 \text{ ac}$

$\quad = (0.1333 \text{ cfs/ac})(1 \text{ ac-in.}/1 \text{ cfs-hr}),$

$\boxed{k_1 = 0.1333 \text{ in./hr},}$

$k_2 = f[(12 + 6), (0 + 0.5(0.1333 \text{ in./hr})(12 \text{ hr})(1 \text{ ft}/12 \text{ in.})]$

$\quad = [Q_{in}(18) - Q_{out}(0.0667)]/300 \text{ ac}$

$\quad = (37.5 \text{ cfs} - 0.9679 \text{ cfs})/300 \text{ ac},$

$$k_2 = 0.1218 \text{ in./hr,}$$

$$k_3 = f[(12 + 6), (0 + (0.1218 \text{ in./hr})(12 \text{ hr})(1 \text{ ft/12 in.})/2]$$
$$= [Q_{in}(18) - Q_{out}(0.0609)]/300 \text{ ac}$$
$$= (37.5 \text{ cfs} - 0.8454 \text{ cfs})/300 \text{ ac,}$$

$$k_3 = 0.1222 \text{ in./hr,}$$

$$k_4 = f[(12 + 12), (0 + (0.1222 \text{ in./hr})(12 \text{ hr})(1 \text{ ft/12 in.})]$$
$$= [Q_{in}(24) - Q_{out}(0.1222)]/300 \text{ ac}$$
$$= (35 \text{ cfs} - 2.4029 \text{ cfs})/300 \text{ ac}$$

$$k_4 = 0.1087 \text{ in./hr.}$$

Then

$$H_1 = H_0 + (1/6)(k_1 + 2k_2 + 2k_3 + k_4) \Delta t$$
$$= 0 \text{ ft} + 1/6[0.1333 \text{ in./hr} + 2(0.1218 \text{ in./hr})$$
$$+ 2(0.1222 \text{ in./hr}) + 0.1087 \text{ in./hr}](1 \text{ ft/12 in.})(12 \text{ hr}),$$

$$H_1 = 0.1217 \text{ ft,}$$

and

$$Q_{out}(H_1) = 2.39 \text{ cfs.}$$

Using the computer program listed in Appendix E, the values in the following table were derived. The last column shows the values obtained for outflow from this reservoir using storage indication methods. It can be seen that the peak values differ by less than 1% and that the time to peak is the same for both methods. The Runge-Kutta program relies on a table of H vs. Q values. The more precisely these values are entered into the program, the more accurate the results will be.

n	TIME (hr)	INFLOW (cfs)	HEAD (ft)	AREA (ac)	ΔH (ft)	OUTFLOW (cfs)	$S-I$ OUTFLOW (cfs)
0	12	40	.00	300	.00	0	0
1	24	35	.12	300	.12	3	2
2	36	37	.22	300	.10	6	6
3	48	125	.45	300	.23	18	17
4	60	340	1.10	300	.65	65	63
5	72	575	2.22	300	1.12	186	181
6	84	722	3.46	300	1.24	363	357
7	96	740	4.40	300	.94	520	518
8	108	673	4.86	300	.45	603	604 ← Q_p
9	120	456	4.74	300	.12	580	585
10	132	250	4.14	300	.59	475	479

4.4

GOVERNING EQUATIONS FOR HYDRAULIC RIVER ROUTING

Hydraulic routing differs from hydrologic routing in that both the equation of continuity and the momentum equation are solved simultaneously rather than through the use of an empirical storage-discharge relation. Since closed-form solutions do not exist for the St. Venant equations, various numerical methods have been developed for the computer. **Explicit methods** calculate values of velocity and depth over a grid system based on previously known data for the river reach. **Implicit methods** set up a series of simultaneous numerical equations over a grid system for the entire river, and the equations are solved at each time step. **Characteristic methods** employ the concept of characteristic curves in the xt-plane, produced by converting partial differential equations into ordinary differential equations. The equations can be simplified under certain conditions to allow for the use of a uniform flow equation in place of the full momentum equation. This method is referred to as the **kinematic wave model.**

The general equation of continuity states that inflow minus outflow equals rate of change of storage. For the river element shown in Fig. 4.7:

$$\text{Inflow} = (Q - \frac{\partial Q}{\partial x}\frac{\Delta x}{2})dt + q\,\Delta x\,\Delta t,$$

$$\text{Outflow} = (Q + \frac{\partial Q}{\partial x}\frac{\Delta x}{2})dt,$$

$$\text{Storage change} = \frac{\partial A}{\partial t}\,\Delta x\,\Delta t,$$

where

$q = $ rate of lateral flow per unit length of channel,

$A = $ cross-sectional area.

The equation of continuity becomes

$$\frac{\partial A}{\partial t} + \frac{\partial Q}{\partial x} = q. \tag{4.23}$$

For a unit width b of channel with $v = $ average velocity, continuity becomes

$$y\frac{\partial v}{\partial x} + v\frac{\partial y}{\partial x} + \frac{\partial y}{\partial t} = q/b. \tag{4.24}$$

The momentum equation in the x-direction is produced from a force balance on the river element according to Newton's second law of motion.

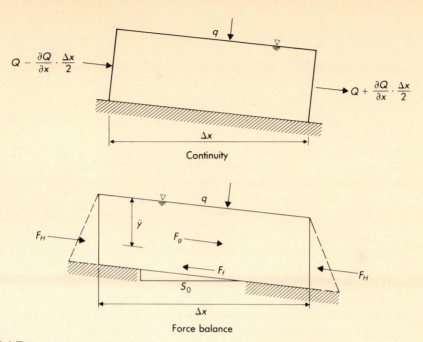

FIGURE 4.7

Continuity and momentum elements for a river reach.

The following external forces are acting on area A (Fig. 4.7):

Hydrostatic: $F_H = -\gamma \dfrac{\partial(\bar{y}A)}{\partial x}\Delta x,$

Gravity: $F_G = \gamma A S_0\,\Delta x,$

Friction: $F_f = -\gamma A S_f\,\Delta x,$

where

  γ = specific weight of water (ρg),

  $\bar{y}$ = distance from the water surface to the centroid of the pressure prism,

  S_f = friction slope,

  S_0 = bed slope.

The rate of change of momentum is expressed from fluid mechanics as

$$F = \frac{d}{dt}(mv),$$

$$\frac{d(mv)}{dt} = m\frac{dv}{dt} + v\frac{dm}{dt} = \rho A\,\Delta x\,\frac{dv}{dt} + \rho vq\,\Delta x, \qquad (4.25)$$

where

$$\frac{dv}{dt} = \frac{\partial v}{\partial t} + v\frac{\partial v}{\partial x}.$$

Equating Eq. (4.25) to the sum of external forces results in

$$\frac{\partial v}{\partial t} + v\frac{\partial v}{\partial x} + \frac{g}{A}\frac{\partial(\bar{y}A)}{\partial x} + \frac{vq}{A} = g(S_0 - S_f). \tag{4.26}$$

For negligible lateral inflow and a wide channel, the equation can be rearranged to yield (Henderson, 1966)

$$S_f = S_0 - \frac{\partial y}{\partial x} - \frac{v}{g}\frac{\partial v}{\partial x} - \frac{1}{g}\frac{\partial v}{\partial t}. \tag{4.27}$$

Frictional losses are usually evaluated by assuming they are the same as for steady, uniform flow conditions, and Manning's equation could be used (Eq. 4.30).

For overland flow and many channel flow situations, some of the terms in Eq. (4.27) can be neglected (Eagleson, 1970). In a typical shallow stream, if the bed slope is 0.01, the rate of change of water depth (dy/dx) will probably not exceed 0.001; the longitudinal velocity gradient term $(v/g)(\partial v/\partial x)$ and the time rate of change of velocity term $(1/g)(\partial v/\partial t)$ will typically be less than 0.001. Thus, the last three terms on the right-hand side of Eq. (4.27) can often be neglected. In overland flow conditions, the three terms will be two orders of magnitude less than those for the bed slope. Various flow routing methods will result depending on the terms neglected (Table 4.1).

The full dynamic wave equations (St. Venant equations), presented as Eqs. (4.23) and (4.27), require numerical techniques for solution and large quantities of measured hydraulic data. These drawbacks can be overcome by neglecting some of the terms as mentioned above. Two approaches that have found wide application in engineering practice in-

TABLE 4.1

Forms of the Momentum Equation

TYPE OF FLOW	MOMENTUM EQUATION
Kinematic Wave (steady uniform)	$S_f = S_0$
Diffusion (Noninertia) Model	$S_f = S_0 - \partial y/\partial x$
Steady Nonuniform	$S_f = S_0 - \partial y/\partial x - (v/g)\partial v/\partial x$
Unsteady Nonuniform	$S_f = S_0 - \partial y/\partial x - (v/g)\partial v/\partial x - (1/g)\partial v/\partial t$

clude the **diffusion** or **noninertia** model and the **kinematic wave** model. The diffusion analogy results in the continuity equation (Eq. 4.23) and the simplified form of the momentum equation (Eq. 4.27) as

$$\frac{dy}{dx} = S_0 - S_f.$$

The diffusion model is described in more detail in Section 4.7.

The kinematic wave model further assumes that the pressure term is negligible, resulting in

$$S_0 = S_f,$$

which means that a simple uniform flow formula such as Manning's equation can be used.

4.5

MOVEMENT OF A FLOOD WAVE

One of the simplest of all wave forms is the monoclinal flood wave shown in Fig. 4.8(a). It is simply a step increase in discharge that moves downstream at wave celerity $u = c$ and with no change in shape. Application of the continuity equation (inflow minus outflow equals change in storage) within the reach and neglecting changes in wave shape results in

$$u(A_2 - A_1) = A_2 v_2 - A_1 v_1, \tag{4.28}$$

where u and v are velocities of the wave and water, respectively, and A is cross-sectional area of the channel. Thus,

$$u = \frac{A_1 v_1 - A_2 v_2}{A_1 - A_2} = \frac{Q_1 - Q_2}{A_1 - A_2},$$

where Q is measured relative to the bank. For the wave of small height in a wide rectangular channel, velocity is called **celerity** and is equal to

$$c = \frac{dQ}{dA} = \frac{1}{B} \frac{dQ}{dy}, \tag{4.29}$$

where B is channel top width. Seddon (1900) first developed Eq. (4.29) for the Mississippi River.

The argument leading to Eq. (4.29) implies that the wave profile is permanent without any change in shape and that Q is a single-valued function of depth or area.

The Manning formula describes flow in an open channel:

$$v = \frac{1.49}{n} R^{2/3} \sqrt{S}, \tag{4.30}$$

(a)

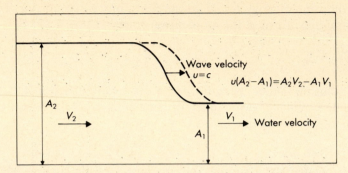

(b)

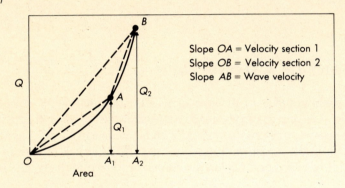

FIGURE 4.8

(a) Monoclinal rising flood wave. (b) Area-discharge relation for streams.

where

$R = A/P$, the hydraulic radius (ft),
$P =$ wetted perimeter (ft),
$A =$ cross-sectional area (ft²),
$S =$ energy slope (ft/ft),
$n =$ roughness coefficient (Manning's),
$1.49 =$ factor to convert from SI units.

Then

$$Q = \frac{1.49}{n} By^{5/3} \sqrt{S} \qquad\qquad (4.31)$$

for a wide rectangular channel where $R \approx y$. Differentiating, we have

$$\frac{dQ}{dy} = \frac{5}{3} \frac{1.49}{n} By^{2/3} \sqrt{S} = \frac{5}{3} Bv, \qquad\qquad (4.32)$$

and substituting into Eq. (4.29) gives

$$c = 1.667v. \tag{4.33}$$

The ratio of c to v is always greater than unity; i.e., wave celerity exceeds velocity for the monoclinal flood wave (Fig. 4.8b).

It can be shown (Chow, 1959) that abrupt dynamic waves such as tidal bores have wave velocities measured relative to the bank of

$$c = v \pm \sqrt{gy} \tag{4.34}$$

and an observer moving at velocity c will see a steady profile. This type of dynamic wave has a different velocity than the one presented in Eq. (4.28), and it can propagate in either direction. A single-valued relation between Q and A is not assumed, and the momentum equation must also be solved for the dynamic wave. Waves of the form presented in Eq. (4.29) are termed **kinematic waves** because they are based on the continuity equation and imply a unique function between Q and y (Lighthill and Whitman, 1955). Kinematic waves imply that $S_f = S_0$ and that all other terms in the governing momentum equation (Eq. 4.27) are negligible.

An alternative derivation of Eq. (4.29) comes directly from the continuity equation for a prismatic channel:

$$\frac{\partial Q}{\partial x} + \frac{B\partial y}{\partial t} = 0, \tag{4.35}$$

or

$$\frac{1}{B}\frac{dQ}{dy}\frac{\partial y}{\partial x} + \frac{\partial y}{\partial t} = 0, \tag{4.36}$$

but when moving with the wave speed c,

$$\frac{dy}{dt} = \frac{dx}{dt}\frac{\partial y}{\partial x} + \frac{\partial y}{\partial t} = 0. \tag{4.37}$$

Thus

$$\frac{dx}{dt} = c = \frac{1}{B}\frac{dQ}{dy} \tag{4.38}$$

since Eqs. (4.36) and (4.37) are both equal to zero. It follows from Eq. (4.37) that to an observer moving with velocity c given by Eq. (4.38), y and Q will appear to be constant (Fig. 4.9).

In a natural flood wave, both kinematic and dynamic waves may be present. The speed of the main flood wave is approximately that of a kinematic wave, with dynamic waves moving ahead of and behind at speeds $v \pm \sqrt{gy}$. The speed of the kinematic wave is the same as the monoclinal wave, Eq. (4.29).

If the kinematic wave moves according to Eq. (4.33), c will be less than the leading dynamic wave speed, provided that the Froude number is

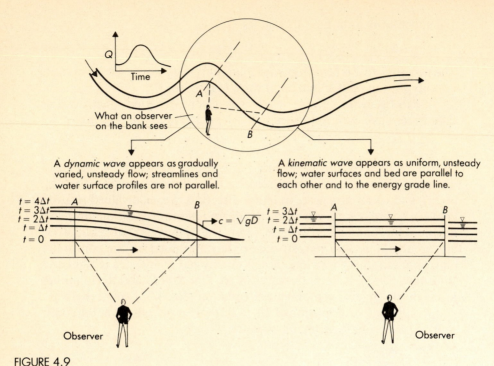

FIGURE 4.9

Visualization of dynamic and kinematic waves.

less than 2, where $\text{Fr} = v/\sqrt{gy}$. This condition occurs in most natural rivers except extremely steep mountain torrents. With a wave celerity of 10 ft/s and a rate of rise of 5 ft/hr, $\partial y/\partial x = 1/7200$. This implies that only for extremely flat slopes or extremely large rates of rise (as for a dam break) are the kinematic assumptions violated. It turns out that, for normal floods in natural rivers, the dynamic wave fronts attenuate very rapidly as long as $\text{Fr} < 2$, and kinematic waves dominate the flood response (Henderson, 1966).

4.6
KINEMATIC WAVE ROUTING

The kinematic wave assumption is that inertial and pressure effects are unimportant and that the weight or gravity force of fluid is approximately balanced by the resistive forces of bed friction (Eq. 4.30 and Table 4.1). Kinematic waves will not accelerate appreciably and can flow only in a downstream direction without crest subsidence. The flood wave will be observed as a uniform rise and fall in water surface over a relatively long period of time. Thus kinematic waves represent the characteristic changes in discharge, velocity, and water surface elevation with time at any one location on an overland flow plane or along a stream channel. Kinematic waves are often classified as uniform, unsteady flows.

Recently, the kinematic wave method of routing overland and river flows has been added to the HEC-1 Flood Hydrograph package, as described in detail in Chapter 5. The following discussion and numerical solutions are based on this addition because of the wide acceptance and availability of the HEC-1 package (Hydrologic Engineering Center, 1980).

The concept incorporated into HEC-1 uses various elements such as overland flow planes, collector channels, and main channels to route kinematic waves (Fig. 4.10). These various elements are combined to describe basin and subbasin responses to storm events. Overland flow is handled separately from open channel flow because of the assumptions inherent in developing the kinematic flow equation for overland flow planes.

Overland flow in the model is distributed over a wide area and at very shallow average depths until it reaches a well-defined collector channel. Pervious and impervious flow surfaces are allowed in HEC-1 with unique slopes, flow lengths, roughnesses, and loss rates. After overland runoff is routed down the length of the overland flow strip, it is then routed along the collector system and eventually into a main channel. Runoff moves through the collector system, picking up additional lateral inflow from adjacent strips uniformly distributed along the system. Collector and main channel kinematic wave routing are similar in theory and differ only in the shape of the collector.

Governing Equation for Kinematic Overland Flow Routing

For the conditions of kinematic flow, and with no appreciable backwater effect, the discharge can be described as a function of depth only, for all x and t:

$$Q = \alpha y^m, \tag{4.39}$$

where

$\qquad Q = $ discharge in cfs,

$\qquad \alpha, m = $ kinematic wave routing parameters.

Henderson (1966) presents the normalized momentum equation (Eq. 4.27) in the form

$$Q = Q_0 \left(1 - \left(\frac{1}{S_0} \frac{\partial y}{\partial x} + \frac{v}{g} \frac{\partial v}{\partial x} + \frac{1}{g} \frac{\partial v}{\partial t} + \frac{qv}{gy} \right) \right)^{1/2}, \tag{4.40}$$

where Q_0 is the flow under uniform conditions. This equation describes

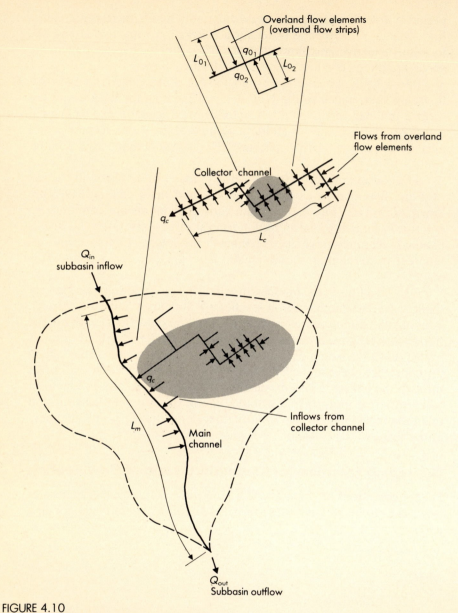

FIGURE 4.10

Relationships between flow elements.

the condition of kinematic flow if the sum of terms to the right of the minus sign is much less than one. Then

$$Q \approx Q_0. \tag{4.41}$$

It has been shown previously that the kinematic wave form dominates only if Fr < 2. Woolhiser and Liggett (1967) analyzed characteristics of the rising overland flow hydrograph and found that the dynamic terms can generally be neglected if

$$k = \frac{S_0 L}{y\text{Fr}^2} > 10, \tag{4.42}$$

where

L = the length of the plane,

Fr = $v/\sqrt{gy}$ with v = overland velocity,

y = the depth at the end of the plane,

S_0 = the slope,

k = the dimensionless kinematic flow number.

These results are best summarized in Fig. 4.11, where Q_* (dimensionless flow) is plotted versus t_* (dimensionless time) for various values of k in Eq. (4.42). It can be seen that for $k \leq 10$, large errors in calculated Q_* result by deleting dynamic terms from the momentum equation for overland flow.

The kinematic wave equation for an overland flow segment on a wide plane with shallow flows can be derived from Eq. (4.39) and Manning's equation for overland flow:

$$q = \frac{1.49}{N} \sqrt{S_0}\, y_0^{5/3} \tag{4.43}$$

Values of N for overland flow are typically greater than Manning's n for channels and are presented in Table 4.2, based on field and laboratory investigations.

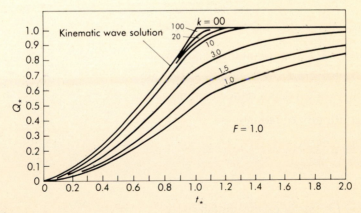

FIGURE 4.11

Effect of kinematic wave number k on the rising hydrograph. (From Woolhiser and Liggett, 1967.)

TABLE 4.2

Effective Roughness Parameters for Overland Flow

SURFACE	N
Dense Growth*	0.4–0.5
Pasture*	0.3–0.4
Lawns*	0.2–0.3
Bluegrass sod**	0.2–0.5
Short-grass prairie**	0.1–0.2
Sparse vegetation**	0.05–0.13
Bare clay-loam soil (eroded)**	0.01–0.03

Concrete/asphalt
  Very Shallow Depths*
  (depths less than 1/4 inch)
	0.10–0.15

Small Depths*
  (depths on the order of 1/4 inch
  to several inches)
	0.05–0.10

* From Crawford and Linsley (1966)
** From Woolhiser (1975)

Rewriting equations in terms of flow per unit width for overland flow, we have

$$q_0 = \alpha_0 y_0{}^{m_0}, \tag{4.44}$$

where

$$\alpha_0 = \frac{1.49}{N} \sqrt{S_0} = \text{conveyance factor,}$$

$m_0 = 5/3$ from Manning's equation,

S_0 = average overland flow slope,

y_0 = mean depth of overland flow.

The continuity equation is

$$\frac{\partial y_0}{\partial t} + \frac{\partial q_0}{\partial x} = i - f, \tag{4.45}$$

where

$i - f$ = rate of excess rainfall (ft/s),

q_0 = flow rate per unit width (cfs/ft),

y_0 = mean depth of overland flow (ft).

Finally, by substitution of Eq. (4.45) in Eq. (4.44), we have

$$\frac{\partial y_0}{\partial t} + \alpha_0 m_0 y_0^{m_0-1} \frac{\partial y_0}{\partial x} = i - f, \tag{4.46}$$

which can be solved numerically for $y_0 = f(x, t, i - f)$. Once y_0 is found, it is substituted into Eq. (4.44) to give a value for q_0. Equations (4.44) and (4.46) form the complete kinematic wave equations for overland flow. Analytical solutions for runoff from an impermeable plane surface are presented in a later section. Numerical techniques for solving these equations are described in more detail in the section on finite differences.

Kinematic Channel Routing

Simple cross-sectional shapes such as triangles, trapezoids, and circles are used as representative collectors or stream channels. These are completely characterized by slope, length, cross-sectional dimensions, shape, and Manning's n value. The basic forms of the equations are similar to the overland flow Eqs. (4.44) and (4.45). For stream channels or collectors,

$$\frac{\partial A_c}{\partial t} + \frac{\partial Q_c}{\partial x} = q_0, \tag{4.47}$$

$$Q_c = \alpha_c A_c^{m_c}, \tag{4.48}$$

where

A_c = cross-sectional flow area (ft²),

Q_c = discharge (cfs),

q_0 = overland inflow per unit length (cfs/ft),

α_c, m_c = kinematic wave parameters for the particular channel.

The values of α_c and m_c are derived for a simple triangular section in Example 4.7 and results are presented only for rectangular, trapezoidal, and circular shapes in Table 4.3. Figure 4.12 presents shape parameters.

EXAMPLE 4.7

KINEMATIC CHANNEL PARAMETERS

Determine α_c and m_c for the case of a triangular prismatic channel.

SOLUTION

The geometry of the channel cross section is shown in Fig. E4.7.

$$\text{Area} = A_c = zy_c^2$$

$$\text{Wetted perimeter} = P_c = 2y_c \sqrt{1 + z^2}$$

$$\text{Hydraulic radius} = R = A_c/P_c$$

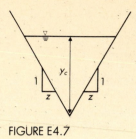

FIGURE E4.7

Substituting these into Manning's equation (Eq. 4.30), we have

$$Q_c = \frac{1.49}{n} \sqrt{S} \frac{A_c^{5/3}}{P_c^{2/3}} = \frac{1.49 \sqrt{S}}{n} \frac{(z^{5/3} y_c^{10/3})}{1.59 \, y_c^{2/3}(1 + z^2)^{1/3}}$$

$$= \frac{0.94 \sqrt{S}}{n} \left(\frac{z}{1 + z^2}\right)^{1/3} (zy_c^2)^{4/3} = \frac{0.94 \sqrt{S}}{n} \left(\frac{z}{1 + z^2}\right)^{1/3} A_c^{4/3}.$$

From Eq. (4.48), $Q_c = \alpha_c A_c^{m_c}$. Therefore,

$$\boxed{\alpha_c = \frac{0.94 \sqrt{S}}{n} \left(\frac{z}{1 + z^2}\right)^{1/3}, \qquad m_c = 4/3.}$$

TABLE 4.3

Kinematic Channel Parameters*

SHAPE	α_c	m_c
Rectangular	$\dfrac{0.72 \sqrt{S_c}}{n}$	$\dfrac{4}{3}$
Trapezoidal	Variable†	$\dfrac{4}{3}$ to $\dfrac{5}{3}$
Circular	$\dfrac{0.804 \sqrt{S_c}}{n} (D_c^{1/6})$	1.25
	D_c = diam (ft)	

* From Hydrologic Engineering Center, 1979.

† For trapezoidal

$$Q_c = \frac{1.49}{n} \sqrt{S_c} (A_c')^{5/3} \left(\frac{1}{w + zy_c \sqrt{1 + z^2}}\right)^{2/3},$$

where A_c' = effective cross-sectional area at depth y_c.

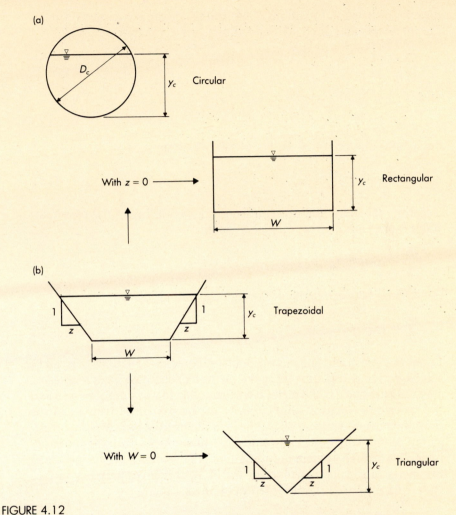

FIGURE 4.12

Two basic channel shapes and their variations used by the HEC-1 flood hydrograph package for kinematic wave stream routing.

Analytical Solutions for an Impermeable Plane†

The kinematic wave equations have an important advantage over the more complete St. Venant equations in that analytical solutions are possible for simple watershed conditions. For the case of an impermeable plane, $y = A/b$, $q = Q/b$, and $R = y$, where y is depth, b is width, and q is

† This section may be omitted without loss of continuity in the text.

flow per unit width. Equations (4.44) and (4.45) can be written, dropping the subscripts for simplicity and letting $i_e = i - f$ = excess rainfall,

$$\frac{\partial y}{\partial t} + \frac{\partial q}{\partial x} = i_e \tag{4.49}$$

and

$$q = \alpha y^m. \tag{4.50}$$

or, substituting Eq. (4.50) in Eq. (4.49), we obtain

$$\frac{\partial y}{\partial t} + \alpha m y^{m-1} \frac{\partial y}{\partial x} = i_e. \tag{4.51}$$

Recalling the definition of dy/dt from Eq. (4.37), Eq. (4.51) essentially states that to an observer moving at speed

$$\frac{dx}{dt} = \alpha m y^{m-1}, \tag{4.52}$$

the relation between depth of flow and rainfall excess is

$$\frac{dy}{dt} = i_e. \tag{4.53}$$

Equations (4.52) and (4.53) are ordinary differential equations that can be solved for y and x vs. time using the method of characteristics (Eagleson, 1970). The essence of the method when applied to wave motions is to find the space-time locus of discontinuity in the partial derivatives of the important dependent variables. The locus defines the path of wave propagation along which one may study the phenomena in terms of easily integrable ordinary differential equations. The solutions apply only along the characteristic curves defined by Eq. (4.52). For an initially dry surface ($y_0(t = 0) = 0$), we can integrate Eq. (4.53) to obtain

$$y = y_0 + i_e t = i_e t. \tag{4.54}$$

Substituting into Eq. (4.52) yields

$$\frac{dx}{dt} = \alpha m (i_e t)^{m-1}, \tag{4.55}$$

which integrates to

$$x = x_0 + \alpha i_e^{m-1} t^m, \tag{4.56}$$

or, simplifying,

$$x = x_0 + \alpha y^{m-1} t. \tag{4.57}$$

Equation (4.57) relates x to y and t, where x_0 is the point from which the forward characteristics begin (at $t = 0$), measured from the upslope end of the plane. Discharge at any point along the characteristic is given by Eq. (4.50), with y substituted as

$$q = \alpha(i_e t)^m. \tag{4.58}$$

Typical characteristic curves for the kinematic wave are sketched in Fig. 4.13. The curve ABC represents the equilibrium depth profile, which occurs at times $t \geq t_c$, the time of concentration for the plane. At $t = t_c$, the outflow reaches its maximum value, which must be equal to inflow. The equilibrium depth profile persists regardless of how long the rainfall lasts, as long as $t = D \geq t_c$, where D is rainfall duration.

Another characteristic might emanate from a point interior to the plane x_0 and travel the distance $L - x_0$ during time t_0. Equations (4.54), (4.57), and (4.58) can be used to calculate depth and discharge at each point (x, t) along this characteristic, until it arrives at point F. We can use these same equations to explore simple relationships for time of concentration, equilibrium depth profile, and the rising outflow hydrograph at $x = L$ on the plane.

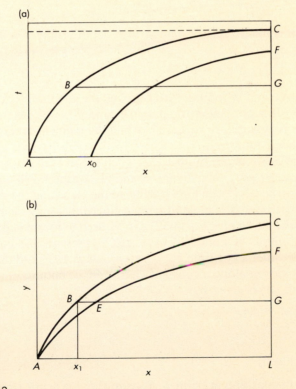

FIGURE 4.13

Kinematic wave characteristic curves. (a) Characteristics for kinematic wave. (b) Water depth profiles.

Time of Concentration

From Eq. (4.56), assuming that $t = t_c$ for $x - x_0 = L$, we can solve for the time of concentration (equilibrium time):

$$t_c = (L/\alpha i_e^{m-1})^{1/m} \tag{4.59}$$

Note that this is the time for a kinematic wave to travel the distance L, not the travel time for a parcel of water. For Manning's equation, t_c in minutes is

$$t_c = (6.9/i_e^{0.4})(nL\sqrt{S_0})^{0.6} \tag{4.60}$$

for i_e in mm/hr and L in m. The factor 6.9 changes to 0.928 for i_e in in./hr and L in ft. One must be extremely careful to make sure that consistent units are used in Eqs. (4.59) and (4.60).

Equilibrium Depth Profile

Solving Eqs. (4.54) and (4.57) simultaneously with the boundary condition that at $x_0 = 0$, $y_0 = 0$, results in

$$y(x) = (i_e x/\alpha)^{1/m}, \tag{4.61}$$

which is the equation for the equilibrium depth profile. For Manning's equation in SI units, Eq. (4.61) becomes

$$y(x) = (ni_e x/\sqrt{S_0})^{0.6}. \tag{4.62}$$

Henderson and Wooding (1964) derived the kinematic equations for the falling hydrograph under two possible conditions: (1) when the rising hydrograph is at equilibrium and (2) when the rising hydrograph is at a point less than equilibrium.

For case 1, Eq. (4.53) implies that when the rainfall stops and duration D is greater than t_c, then $dy/dt = 0$ and y is constant. From Eq. (4.52), the characteristic curves are lines parallel to the plane and the wave speed dx/dt, the depth, and discharge all remain constant along a characteristic. Figure 4.13(b) shows how Eq. (4.52) can be used to locate depth profiles after the cessation of rainfall. Curve ABC represents the equilibrium depth profile. After some time, the depth profile is AEF, and the depth at point B has moved along a constant characteristic path to point E. The new x-coordinate is given by

$$\Delta x = \alpha m y^{m-1} \Delta t, \tag{4.63}$$

or

$$x = x_1 + \alpha m y^{m-1}(t - D), \tag{4.64}$$

where x_1 is the position at point B. Substituting for depth y gives:

$$x = \frac{\alpha y^m}{i_e} + \alpha m y^{m-1}(t - D) \tag{4.65}$$

Finally, for the point $x = L$ and $q_L = \alpha y_L{}^m$, we obtain the equation for the recession hydrograph:

$$L = \frac{q}{i_e} + mq^{(1-1/m)}\alpha^{(1/m)}(t - D), \tag{4.66}$$

which must be solved iteratively. Note that the equilibrium depth profile remains constant as long as rainfall continues at a steady rate i_e.

For case 2, the duration D is less than t_c and the depth profile resembles ABG in Fig. 4.13(b). The depth at point B will move at a constant rate and reach the end of the plane at time t^*, evaluated from

$$t^* = D + \frac{L - x_1}{dx/dt}, \tag{4.67}$$

which can be rearranged to

$$t^* = D(1 + [1/m(t_c/D)^m - 1]). \tag{4.68}$$

The discharge at the end of the plane will remain constant between $D \leq t \leq t^*$ and will be equal to

$$q = \alpha(i_e D)^m. \tag{4.69}$$

For $t > t^*$, the recession follows Eq. (4.65) for case 1.

Figure 4.14 shows typical outflow hydrographs for several possible

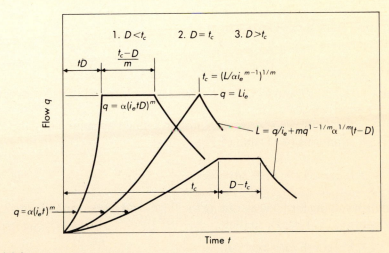

FIGURE 4.14

Outflow hydrograph shape for different storm durations ($D = t_d$) but similar total excess rain i_e.

relationships among D and t_c. The discharge at the mouth prior to D or t_c is given by

$$q = \alpha(i_e t)^m. \tag{4.70}$$

If $D > t_c$, then the hydrograph top is horizontal as indicated by case 3 in Fig. 4.14. If $D = t_c$, then the hydrograph resembles case 2 and a single peak occurs. Finally, if $D < t_c$ for a short storm event, then the hydrograph remains constant until the influence of the upstream end reaches the outlet (case 1).

Figure 4.15 from Wooding (1965) depicts the effect of steady infiltration rate f on watershed discharge due to a steady rainfall rate. Note that it is generally necessary to use numerical models to get hydrograph shapes for the case of infiltration losses.

More detailed treatment of kinematic waves can be found in Eagleson (1970), Overton and Meadows (1976), Raudkivi (1979), and Stephenson and Meadows (1986).

Finite-Difference Approximations

Finite-difference approximations must be used to solve the kinematic wave equations for actual watersheds and variable rainfalls and can be derived by considering a continuous function $y(x)$ and its derivatives (Chapra and Canale, 1985). A Taylor series expansion of $y(x)$ at $x + h$, where $h = \Delta x$, is

$$y(x + h) = y(x) + hy'(x) + \frac{h^2}{2!} y''(x) + \frac{h^3}{3!} y'''(x) + \cdots, \tag{4.71}$$

and the Taylor series expansion at $x - h$ is

$$y(x - h) = y(x) - hy'(x) + \frac{h^2}{2!} y''(x) - \frac{h^3}{3!} y'''(x) + \cdots, \tag{4.72}$$

where $y'(x)$, $y''(x)$, and $y'''(x)$ are the first, second, and third derivatives of $y(x)$, respectively.

A central difference approximation is defined by subtracting the expansion equations (4.71) and (4.72)

$$y(x + h) - y(x - h) = 2hy'(x) - \frac{2h^3}{3!} y'''(x) + \cdots \tag{4.73}$$

and then solving for $y'(x)$ assuming the third- and higher-order forms are negligible ($y'''(x) \ll 1$), so that

$$y'(x) = \frac{y(x + h) - y(x - h)}{2h}, \tag{4.74}$$

which has an error of approximation (truncation error) of order h^2.

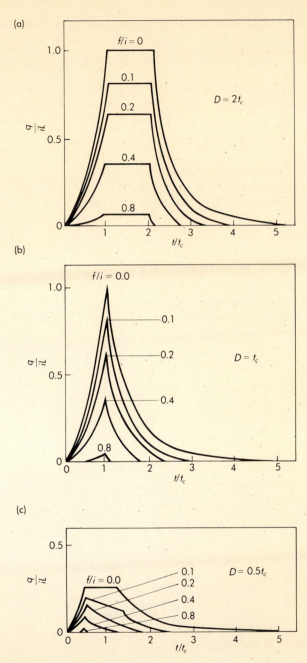

FIGURE 4.15

Effects of infiltration on catchment discharge. (From Wooding, 1965.)

A forward difference is defined by considering the expansion, Eq. (4.71), as

$$y(x + h) = y(x) + hy'(x) + \cdots. \tag{4.75}$$

Solving for $y'(x)$ and assuming the second- and higher-order terms are negligible, we have

$$y'(x) = \frac{y(x + h) - y(x)}{h}, \tag{4.76}$$

which has an error of approximation of order h.

The backward-difference approximation can be determined considering the expansion, Eq. (4.72), and solving for $y'(x)$:

$$y'(x) = \frac{y(x) - y(x - h)}{h}. \tag{4.77}$$

An xt-plane (Fig. 4.16) consisting of a rectangular net of discrete points is used to represent the continuous solution domain defined by the independent variables (x and t) in the governing flow equations (Eqs. 4.44 and 4.46). The lines parallel to the x-axis are spaced by the time increment

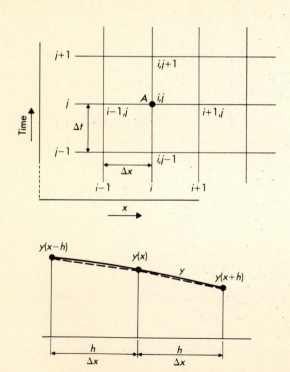

FIGURE 4.16

Illustration of finite-difference grid and method.

Δt. The discrete points in the solution domain are defined by the subscript i for the x position and the subscript j for the time line.

Explicit Finite-Difference Solution for Kinematic Wave

The governing kinematic wave equation to be solved is of the form

$$\frac{\partial A_c}{\partial t} + \alpha_c m_c A_c^{m_c-1} \frac{\partial A_c}{\partial x} = q_0, \tag{4.78}$$

where Eq. (4.48) has been substituted into Eq. (4.47). The finite-difference numerical techniques to solve this equation have been developed over a period of years by Harley (1975), Resource Analysis, Inc. (1975), and Bras (1973). These methods have been incorporated into the HEC-1 Flood Hydrograph package (Hydrologic Engineering Center, 1979), and coding information and boundary conditions are presented there. Further details on numerical methods for kinematic wave routing are contained in Overton and Meadows (1976) and Stephenson and Meadows (1986).

The governing equations are similar in form for overland flow, collector elements, and main channels; therefore, solution details are presented only for the collector elements since the others are basically the same. Finite-difference approximations are made for the various derivatives in Eq. (4.78) over a grid space in x and t. Explicit computations advance along the x-dimension downstream for each time step Δt until all the flows and stages are calculated along the entire distance L. The time increment is advanced by Δt, and the procedure is repeated once again. Figure 4.16 shows the grid space in x and t over which the solution is found.

The standard form of the equation solves for the values of A_c and Q_0 at point A in Fig. 4.16 as a function of known values at previous points in x and t. Thus, using backward differences and an explicit solution results in

$$\frac{A(i, j) - A(i, j-1)}{\Delta t} + \alpha_c m_c \left[\frac{A(i, j-1) + A(i-1, j-1)}{2} \right]^{m_c-1}$$

$$\times \left[\frac{A(i, j-1) - A(i-1, j-1)}{\Delta x} \right]$$

$$= \frac{q(i, j) + q(i, j-1)}{2}, \tag{4.79}$$

where two of the terms are averaged over two points in space or time. We solve for $A(i, j)$ directly as a function of other known values and then compute $Q(i, j)$ from

$$Q(i, j) = \alpha_c (A(i, j))^{m_c}. \tag{4.80}$$

We can also compute y for a given flow, time, and location knowing Q and A.

This method requires $c\Delta t/\Delta x \leq 1$ to maintain numerical stability (Hydrologic Engineering Center, 1979). However, if $c > \Delta x/\Delta t$, the numerical scheme must be modified to maintain numerical stability. The conservative form of Eq. (4.78) solves first for $Q(i, j)$ as

$$Q(i, j) = Q(i - 1, j) + \bar{q}\Delta x$$
$$- [\Delta x/\Delta t][A(i - 1, j) - A(i - 1, j - 1)],$$

where

$$\bar{q} = \frac{q(i, j) + q(i, j - 1)}{2}. \tag{4.81}$$

Knowing $Q(i, j)$, $A(i, j)$ is found by solving

$$A(i, j) = \left(\frac{Q(i, j)}{\alpha_c}\right)^{1/m_c}. \tag{4.82}$$

Some numerical instability can occur under certain conditions using the above explicit method.

4.7

HYDRAULIC RIVER ROUTING

The application of the St. Venant equations (Eqs. 4.23 and 4.27) in flood routing involves solving them simultaneously down the length of a stream channel. An excellent review of methods is given in Price (1974), who compares four of the more important numerical methods for hydraulic flood routing with exact analytical solutions for the monoclinal wave (Section 4.5). Price found that the four-point implicit numerical scheme of Amein and Fang (1970) was the most efficient of the methods and is most accurate when $\Delta x/\Delta t$ is approximately equal to the speed of the monoclinal wave, the kinematic wave speed. The other methods examined included (1) the leap-frog explicit method, (2) the two-step Lax Wendroff explicit method, and (3) the fixed mesh characteristic method. In the numerical schemes, the stage was defined at the upstream boundary and rating curve conditions were specified at the downstream boundary.

The objective of the numerical flood routing method is to simulate the propagation of a flood downstream in a river, given a stage or discharge hydrograph at an upstream section. The preceding methods provide the most accurate solution available for predicting floods downstream in a large river or for considering effects of channel alteration or reservoir storage. Because the computational effort is relatively large for any numerical solution, careful consideration should be given to selecting one of the available hydraulic methods versus one of the simpler, approximate methods such as kinematic wave or the Muskingum method.

Characteristic Methods

The early approaches to numerical flood routing were based on the characteristic form of the governing equations. This form basically replaces the partial differential equations with four ordinary differential equations, which are then solved numerically along the characteristic curves. Henderson (1966) and Overton and Meadows (1976) present details of the conversion. However, the characteristic method is cumbersome because of required interpolation over the variable grid mesh in space and time. The method involves integration along two sets of characteristic curves, and a network of variable points is located in the xt-plane by the intersection of the forward and backward characteristic curves. With the advent of the more efficient implicit methods described in the following sections, the characteristic method offers no additional accuracy with relatively complex interpolation problems.

Explicit Methods

Explicit methods are primarily the outcome of the pioneering work of Stoker (1957) and Issacson, et al. (1956). A network of nodes as shown in Fig. 4.17 is defined for solving the governing equations in x and t. The variables are known at points L, M, and R. The centered difference solution is used to solve Eqs. (4.23) and (4.27) for velocity $V(P)$ and depth $y(P)$ at point P. The following numerical approximations are made for the various parameters and their derivatives over the grid at point m:

$$V(m) = [V(R) + V(L)]/2, \tag{4.83}$$

$$\frac{\partial V(m)}{\partial x} = \frac{V(R) - V(L)}{2\Delta x}, \tag{4.84}$$

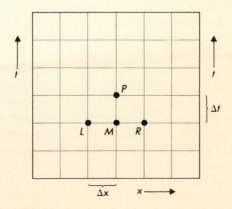

FIGURE 4.17
Grid used for explicit solution.

$$\frac{\partial V(m)}{\partial t} = \frac{V(P) - V(m)}{\Delta t}.$$ (4.85)

When inserted into the governing St. Venant equations, $V(P)$ and $y(P)$ are computed, where B = channel width:

$$V(P) = \tfrac{1}{2}[V(R) + V(L)]$$

$$- \frac{\Delta t}{2\Delta x}\{\tfrac{1}{2}[V(R)^2 - V(L)^2] + g[y(R) - y(L)]\}$$

$$+ g\frac{\Delta t}{2}[2S_0 - S_f(R) - S_f(L)]$$

$$- \frac{1}{B}\frac{\Delta t}{2}\left\{\frac{[V(R) + V(L)][q(R) + q(L)]}{y(R) + y(L)}\right\},$$ (4.86)

$$y(P) = \tfrac{1}{2}[y(R) + y(L)]$$

$$+ \frac{\Delta t}{2\Delta x}\left\{\frac{\Delta x}{B}[q(R) + q(L)]\right.$$

$$\left. - [y(R)V(R) - y(L)V(L)]\right\}.$$ (4.87)

The solution of the explicit equations proceeds by solving for $y(P)$ first, substituting into Eq. (4.86), and solving for $V(P)$ as a function of the known values of S_f, $V(L)$, and $V(R)$. The initial condition and boundary conditions (upstream and downstream) must be known from inflow hydrographs and stage-outflow relations. The major disadvantage of the explicit approach is the requirement of using small time steps, known as the Courant condition, to avoid stability problems ($\Delta t \leq \Delta x/c$, where c is defined in Eq. 4.34).

Implicit Methods

Implicit methods have received the most interest in recent years because they overcome the stability problems of explicit methods and allow fewer time steps for flood routing. The four-point implicit method of Amein and Fang (1970) has the feature of solving the nonlinear simultaneous finite-difference equations using the Newton iteration technique. In this method, a set of N equations with N unknowns results from writing the equations in space and time and incorporating upstream and downstream boundary conditions. The details of the iteration process are given in Amein and Fang (1970) with relevant boundary conditions. The method was applied to several actual flood predictions in North Carolina and was found to be accurate and more efficient than the explicit or characteristic methods. Computer time can be further reduced using the Gaussian elimination procedure suggested by Fread (1971). More recent comparisons by Price (1974) with exact analytical solutions for the monoclinal wave also found the four-point implicit method to be most efficient of the ones tested.

The following approximations to the derivative terms are made for the grid points shown in Fig. 4.18. Hydraulic variables are known at points 1, 2, and 3 from boundary and initial conditions, and the unknowns are $Q(4)$, $v(4)$, $A(4)$, and $S_f(4)$.

$$\frac{\partial Q}{\partial x} = \frac{Q(4) + Q(3) - Q(2) - Q(1)}{2\Delta x} \tag{4.88}$$

$$\frac{\partial A}{\partial t} = \frac{A(4) + A(2) - A(3) - A(1)}{2\Delta t} \tag{4.89}$$

$$\frac{\partial y}{\partial x} = \frac{y(4) + y(3) - y(2) - y(1)}{2\Delta x} \tag{4.90}$$

$$\frac{\partial v}{\partial x} = \frac{v(4) + v(3) - v(2) - v(1)}{2\Delta x} \tag{4.91}$$

$$\frac{\partial v}{\partial t} = \frac{v(4) + v(2) - v(3) - v(1)}{2\Delta t} \tag{4.92}$$

$$S_f = \tfrac{1}{4}[S_f(1) + S_f(2) + S_f(3) + S_f(4)] \tag{4.93}$$

$$q = \tfrac{1}{4}[q(1) + q(2) + q(3) + q(4)] \tag{4.94}$$

Hydraulic variables at nodes 1, 2, and 3 are known from initial and boundary conditions, and all unknowns are evaluated at point 4 at the first time step. The resulting set of algebraic finite-difference equations is nonlinear and must be solved using an iterative scheme. The solution procedure solves for all unknowns at one advanced time level simultaneously, and larger Δx and Δt values can be used than in the explicit method.

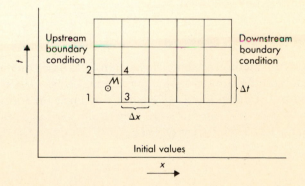

FIGURE 4.18
Network of points for implicit method.

Diffusion Model

The diffusion wave model (also called the noninertial model) assumes that the inertia terms in Eq. (4.27) are negligible compared with the pressure, friction, and gravity terms. The momentum equation becomes

$$\frac{\partial y}{\partial x} = S_0 - S_f \tag{4.95}$$

and when combined with the continuity equation (4.23), the resulting equation resembles a form of the classic diffusion equation in mathematics:

$$\frac{\partial Q}{\partial t} + c \frac{\partial Q}{\partial x} = D_1 \frac{\partial^2 Q}{\partial x^2}, \tag{4.96}$$

where c is wave celerity and D_1 is a diffusion coefficient. The effect of the diffusion term, compared with the kinematic wave, is to make y decrease in the view of an observer moving at velocity c. Thus, attenuation of the flood wave is included in the diffusion model, but not for the kinematic wave. Attenuation may be important in long rivers or for rivers with large floodplains.

Ponce et al. (1978), Raudkivi (1979), and Henderson (1966) attempted to evaluate criteria for using kinematic, diffusion, or fully dynamic flood routing models. Ponce et al. offered the first useful criteria for application of models:

$$T_B S_0(v_0/y_0) \quad > 171 \qquad \text{Kinematic,} \tag{4.97}$$
$$T_B S_0(g/y_0)^{1/2} > \ 30 \qquad \text{Diffusion,} \tag{4.98}$$

where T_B is duration of the flood wave, S_0 is bed slope, v_0 and y_0 are initial velocity and depth, and g is gravity. Thus, the kinematic model applies to shallow flow on steep slopes or to long-duration flood waves. The diffusion model applies to a wider range of conditions, and the full dynamic model applies to any condition.

Muskingum-Cunge Method

The Muskingum method presented in Section 4.2 is based on the continuity equation and the storage-discharge relation. Cunge (1969) extended the method into a finite-difference scheme that takes the form

$$K \frac{d}{dt} [\epsilon Q_j + (1 - \epsilon)Q_{j+1}] = Q_j - Q_{j+1}, \tag{4.99}$$

where

$$Q_j = \text{inflow to reach,}$$
$$Q_{j+1} = \text{outflow from reach,}$$

K = travel time parameter,

ϵ = weighting factor ($0 \leq \epsilon \leq 0.5$).

In finite-difference form,

$$\frac{K}{\Delta t}[\epsilon Q_j^{n+1} + (1 - \epsilon)Q_{j+1}^{n+1} - \epsilon Q_j^n - (1 - \epsilon)Q_{j+1}^n]$$

$$= \tfrac{1}{2}(Q_j^{n+1} - Q_{j+1}^{n+1} + Q_j^n - Q_{j+1}^n). \qquad (4.100)$$

Cunge then showed that if $K = \Delta x/c$, then Eq. (4.100) is the finite-difference form of the kinematic wave equation

$$\frac{\partial Q}{\partial t} + c\frac{\partial Q}{\partial x} = 0, \qquad (4.101)$$

and if $D_1 = (1/2 - \epsilon)c\Delta x$, then the diffusion equation results in

$$\frac{\partial Q}{\partial t} + c\frac{\partial Q}{\partial x} = D_1\frac{\partial^2 Q}{\partial x^2}. \qquad (4.102)$$

If the diffusion coefficient is defined as

$$D_1 = \alpha \overline{Q}/L,$$

then

$$\epsilon = \frac{1}{2} - \frac{\alpha \overline{Q}}{Lc\Delta x}, \qquad (4.103)$$

where L is the length of the whole reach, Δx is the subreach length, $\overline{Q}$ is the average of the peak flows at the upper and lower ends, and α is an attenuation parameter.

According to the Cunge method, the outflow hydrograph at the downstream end is calculated similarly to Eq. (4.6) using

$$Q_{j+1}^{n+1} = C_1 Q_j^n + C_2 Q_j^{n+1} + C_3 Q_{j+1}^n + C_4, \qquad (4.104)$$

where

$$C_1 = \frac{K\epsilon + \Delta t/2}{D},$$

$$C_2 = \frac{\Delta t/2 - K\epsilon}{D},$$

$$C_3 = \frac{K(1 - \epsilon) - \Delta t/2}{D},$$

$$C_4 = \frac{q\Delta t\Delta x}{D}, \qquad (4.105)$$

$$D = K(1 - \epsilon) + \Delta t/2. \qquad (4.106)$$

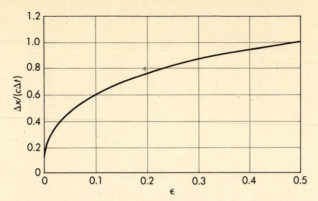

FIGURE 4.19

Curve for $\Delta x/(c\Delta t)$. (From Cunge, 1969, Journal of Hydraulic Research.)

With Δt an integral number of hours, the length Δx is chosen so that $\Delta x/c\Delta t$ lies below the curve in Fig. 4.19. A complete example of the Muskingum-Cunge routing procedure is presented in Example 4.8, as adapted from Raudkivi (1979).

EXAMPLE 4.8

MUSKINGUM-CUNGE METHOD

The hydrograph at the upstream end of a river is given in the following table. The reach of interest is 18 km long. Using a subreach length Δx of 6 km, determine the hydrograph at the end of the reach using the Muskingum-Cunge method. Assume $c = 2$ m/s, $\alpha = 3.6 \times 10^5$, and no lateral inflow.

TIME (hr)	FLOW (m³/s)	TIME (hr)	FLOW (m³/s)
0	10	10	146
1	12	11	129
2	18	12	105
3	28.5	13	78
4	50	14	59
5	78	15	45
6	107	16	33
7	134.5	17	24
8	147	18	17
9	150	19	12
		20	10

SOLUTION

Set

$$K = \Delta x / c$$
$$= \frac{6 \text{ km} \cdot 1000 \text{ m/km}}{2 \text{ m/s}}$$

$$\boxed{K = 3000 \text{ s (50 min)}.}$$

Then Eq. (4.103) gives

$$\epsilon = \frac{1}{2} - \frac{\alpha \overline{Q}}{Lc\Delta x}, \qquad \text{where } \overline{Q} = 148 \text{ m}^2/\text{s}$$
$$= \frac{1}{2} - \frac{3.6 \times 10^5 \cdot 148 \text{ m}^3/\text{s} \cdot 1 \text{ km}^2/1}{18 \text{ km} \cdot 2 \text{ m/s} \cdot 6 \text{ km} \cdot 10^6 \text{ m}^2}$$

$$\boxed{\epsilon = 0.253.}$$

No lateral inflow means $q = 0$. Δt is found from Fig. 4.19:

$$\Delta x / (c\Delta t) \leq 0.82,$$
$$\Delta t > \Delta x / (c)(0.82),$$
$$\Delta t > \frac{6 \text{ km} \cdot 1000 \text{ m/km}}{2 \text{ m/s} \cdot 0.82}$$
$$\Delta t > 3658 \text{ s}.$$

Use

$$\boxed{\Delta t = 7200 \text{ s (120 min)}.}$$

Then Eqs. (4.105) and (4.106) give

$$D = K(1 - \epsilon) + \tfrac{1}{2}\Delta t$$
$$= 50(1 - 0.253) + 0.5(120)$$

$$\boxed{D = 97.33 \text{ min},}$$

$$C_1 = \frac{K\epsilon + \Delta t/2}{D}$$
$$= [50(0.253) + 0.5(120)]/97.33$$

$$\boxed{C_1 = 0.7466,}$$

$$C_2 = \frac{\Delta t/2 - K\epsilon}{D}$$
$$= [0.5(120) - 50(0.253)]/97.33$$

$$\boxed{C_2 = 0.4863,}$$

$$C_3 = \frac{K(1 - \epsilon) - \Delta t/2}{D}$$

$$= [50(1 - 0.253) - 0.5(120)]/97.33$$

$$\boxed{C_3 = -0.2329,}$$

$$C_4 = \frac{q\Delta t\Delta x}{D} = 0.$$

Check

$$\Sigma C_i = 0.7466 + 0.4863 - 0.2329 = 1.0.$$

Then

$$Q_{j+1}^{n+1} = C_1 Q_j^n + C_2 Q_j^{n+1} + C_3 Q_{j+1}^n + C_4. \tag{4.104}$$

For $j = 0$, $\Delta x = 0$ and $\Delta t = 2$ hr. Thus the hydrograph for $j = 0$ is given in column 3 of the following table. For any j and $n = 0$ (time 0) the flow is assumed to be 10 m³/s. For $j = 0$, $n = 0$:

$$Q_1^1 = C_1 Q_0^0 + C_2 Q_0^1 + C_3 Q_1^0,$$
$$= (0.7466)(10) + (0.4863)(18) + (-0.2329)(10)$$
$$Q_1^1 = 13.89 \text{ m}^3/\text{s}.$$

For $j = 0$, $n = 1$:

$$Q_1^2 = C_1 Q_0^1 + C_2 Q_0^2 + C_3 Q_1^1$$
$$= (0.7466)(18) + (0.4863)(50) + (0.2329)(13.89)$$
$$Q_1^2 = 34.51 \text{ m}^3/\text{s}.$$

The process is continued through $j = 3$, $n = 14$ to obtain the following values, where $\Delta t = 2$ hr and $\Delta x = 6$ km.

$n\,\Delta t$ (hr)	$j\,\Delta x$			
	0	6	12	18 km
0	10	10	10	10
2	18	13.89	11.89	10.92
4	50	34.51	24.38	18.19
6	107	81.32	59.63	42.96
8	147	132.44	111.23	88.60
10	146	149.91	145.88	133.35
12	105	125.16	138.82	145.37
14	59	77.93	99.01	117.94
16	33	41.94	55.52	73.45
18	17	23.14	29.63	38.75
20	10	12.17	16.29	21.02
22	10	9.49	9.91	12.09
24	10	10.12	9.70	9.30
26	10	9.97	10.15	10.01
28	10	10.01	9.95	10.08

Dynamic Wave Model (DWOPER)

Hydrologic routing techniques are not adequate for modeling unsteady flows subject to backwater or tidal effects and channels with very mild bottom slopes. For this reason, the NWS Hydrologic Research Laboratory developed a hydrodynamic model known as DWOPER (Dynamic Wave Operational Model) in the 1970s (Fread, 1978). DWOPER is being implemented on a number of major river systems including the Mississippi, Ohio, and Red Rivers.

DWOPER is based on the complete solution of the one-dimensional St. Venant equations. The equations are solved using a Newton-Raphson iterative technique for a weighted four-point nonlinear implicit finite difference scheme. The model is applicable to rivers of varying physical features such as irregular geometry, lateral inflow, flow diversions, variable roughness parameters, local head losses (such as bridge piers), and wind effects. The implicit finite-difference technique is numerically stable enough to allow large time steps to be used with no loss of accuracy.

DWOPER was applied to the lower Mississippi River between Red River Landing and Venice (Fread, 1978). This reach is about 290 miles long, has an average bottom slope of 0.034 ft/mi, depth variations from 30 ft to 200 ft, and flow is contained between levees along most of its length. The model was calibrated for a 1969 flood event and the simulated hydrograph had an average root mean square (RMS) error of 0.25 ft when compared with the actual values. Several other floods were then simulated and resulted in an average RMS error of less than 0.50 ft. Similar levels of accuracy have been achieved on the Mississippi-Ohio-Cumberland-Tennessee River system and the Columbia-Willamette River system. This shows that DWOPER is a powerful and accurate model for flood forecasts as well as day-to-day river forecasts for water supply, irrigation, power, recreation, and water quality interests. More interested readers should consult original references by Fread.

Rating Curves

Rating curves are plots of water level (stage) vs. discharge. These are developed from continuous records of water level coupled with discrete discharge measurements made at times of various river stages to produce a unique relationship between the two for a particular location and cross section. However, unless flow can be approximated as uniform ($S_f = S_0$), discharge will be greater during the rising stage of a flood hydrograph than during the falling stage, leading to a loop rating curve, as shown in Fig. 4.20. If the hydrograph has multiple peaks, the curve can be very complex, as shown in Fig. 4.20(a). A theoretical shape for a smooth, single-peak hydrograph is shown in Fig. 4.20(b).

The physical explanation for the hysteresis present in the rating curve lies primarily with the water surface slope. It is the next most significant

(a)

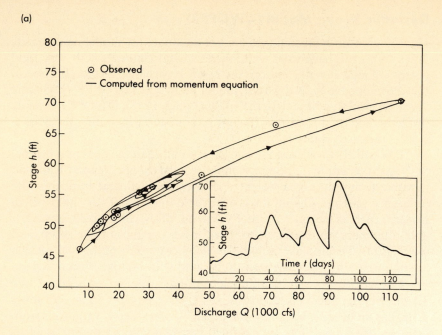

(b)

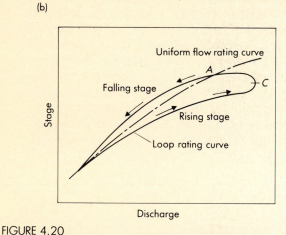

FIGURE 4.20

**Loop rating curves. (a) Stage-discharge relation for Red River, Alex-
andria, Louisiana. (From Fread, 1978.) (b) Theoretical rating curve.
(From Henderson, 1966, p. 392.)**

term in the momentum equation (Eq. 4.27); if it is retained, it leads to
diffusion or noninertial routing (Eq. 4.95). When the momentum equation is rearranged into the form of Eq. (4.40), it is seen that the flow equals
uniform flow multiplied by a factor that depends on the water surface
slope, a lateral inflow term (the last term), and two inertial terms of lesser

significance. During the rising portion of the flood wave, the water surface slope is in the same direction as the slope of the bed S_0 and acts to accelerate the flow (the slope appears as wedge storage in Fig. 4.3a). During this time, the slope $\partial y/\partial x$ is negative and adds to the flow in Eq. (4.40). After the flood crest has passed, the water surface slope is positive (Fig. 4.3c) and acts to decrease the flow in Eq. (4.40). Physically, the positive water surface slope acts to move water in the reverse direction, thus decelerating the positive flow.

The water surface slope effect can be approximated using depth vs. time data in the following manner. Retaining only the slope term in Eq. (4.40) gives

$$\frac{Q}{Q_0} = \sqrt{1 - \frac{1}{S_0}\frac{\partial y}{\partial x}}. \tag{4.107}$$

The kinematic wave equation (Eq. 4.37) can be used to approximate the slope as

$$\frac{\partial y}{\partial x} = -\frac{1}{c}\frac{\partial y}{\partial t}, \tag{4.108}$$

where $c = $ kinematic wave celerity. Substituting Eq. (4.108) into Eq. (4.107) leads to the Jones formula (Henderson, 1966):

$$\frac{Q}{Q_0} = \sqrt{1 + \frac{1}{S_0 c}\frac{\partial y}{\partial t}}. \tag{4.109}$$

Here it can be seen that flow is greater than uniform during periods of rising stage ($\partial y/\partial t > 0$) and less than uniform during periods of falling stage, leading to hysteresis and the loop rating curve. The advantage of the Jones formula is that it involves only terms that can be measured at a single gaging location, whereas computation of $\partial y/\partial x$ must involve at least two locations. Additional corrections to Eq. (4.109) can be made to account for subsidence of the flood wave (Henderson, 1966; Stephenson and Meadows, 1986).

Interesting features of the loop rating curve are described by Henderson (1966). Point C in Fig. 4.20(b) is the point of maximum flow and point A is the point of maximum stage. Thus an observer on the bank watching a flood will experience the maximum flow rate before experiencing the maximum stage. The kinematic wave method uses the uniform flow relationship shown in Fig. 4.20(b) and thus cannot simulate the loop rating curve.

The USGS computes rating curves for all of its gaging stations; these are usually plotted on logarithmic scales, which tend to "straighten" the fit (Kennedy, 1983). Fortunately, most gaging data tend to produce rating curves that can be represented as single-valued, thus avoiding the complex problems of hysteresis just discussed.

SUMMARY

Chapter 4 covered hydrologic (storage) and hydraulic routing methods to predict flood movement for rivers and reservoirs. Hydrologic routing requires solution of the lumped continuity equation and a storage-outflow relation. The relation may be a function of both inflow and outflow for a river, but for a reservoir, storage is related only to outflow. Simple numerical methods are presented for solving the flood routing equations through time.

Hydraulic river routing is more complex than hydrologic routing in that both the one-dimensional continuity and momentum equations must be solved. In many cases, the momentum equation may be simplified to Manning's equation; the procedure is called kinematic wave routing. Most practical kinematic wave solutions require numerical methods to compute actual hydrographs, such as those that are currently included in the HEC-1 Flood Hydrograph package.

Full hydraulic (dynamic) routing requires solution of the St. Venant equations, for which no analytical solution exists. Advanced numerical schemes must be employed to maintain mathematical stability and provide accurate solutions. Explicit and implicit finite-difference methods are the most commonly used techniques. One of the most accurate computer models for dynamic routing is DWOPER, which uses the four-point implicit method. DWOPER has been extensively tested for floods on the Mississippi, Ohio, and Columbia Rivers with excellent results. However, large amounts of data are required for full hydraulic routing as compared to simpler hydrologic methods.

PROBLEMS

4.1. An inflow hydrograph is measured for a cross section of a stream. Compute the outflow hydrograph at a point five miles downstream using the Muskingum method. Assume $K = 12$ hr, $x = 0.15$, and outflow equals inflow initially. Plot the inflow and outflow hydrographs.

TIME	INFLOW (cfs)
9:00 A.M.	50
3:00 P.M.	75
9:00 P.M.	150
3:00 A.M.	450
9:00 A.M.	1000
3:00 P.M.	840
9:00 P.M.	750

TIME	INFLOW (cfs)
3:00 A.M.	600
9:00 A.M.	300
3:00 P.M.	100
9:00 P.M.	50

4.2. Using the Muskingum method, route the following inflow hydrograph assuming (a) $K = 4$ hr, $x = 0.12$, and (b) $K = 4$ hr, $x = 0.0$. Plot the inflow and outflow hydrographs for each case, assuming initial outflow equals initial inflow.

TIME (hr)	INFLOW (cfs)
0	0
2	5
4	25
6	50
8	35
10	21
12	13
14	7.5
16	2.5
18	0

4.3. A detention pond needs to be designed with a total capacity of 30 ac-in. of storage. The inflow hydrograph for the pond is given in Fig. P4.3. Assume that the pond is initially 50% full and outflow will cease when the pond is again 50% full. Graphically determine the peak outflow from the pond assuming a linear rise and fall in the outflow hydrograph. Draw the outflow hydrograph. At what time does outflow cease?

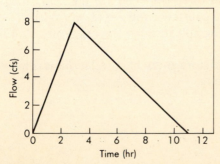

FIGURE P4.3

4.4. An inflow hydrograph is given for a reservoir that has a weir-spillway outflow structure. The flow through the spillway is governed by the equation

$$Q = 3.75Ly^{3/2} \text{ (cfs)},$$

where L is the length of the weir and y is the height of the water above the spillway crest. The storage in the reservoir is governed by

$$S = 300y \text{ (ac-ft)}.$$

Using $\Delta t = 12$ hr, $L = 15$ ft, and $S_0 = Q_0 = 0$, route the inflow hydrograph through the reservoir using the storage indication method.

TIME (hr)	INFLOW (cfs)
12	40
24	35
36	37
48	125
60	340
72	575
84	722
96	740
108	673
120	456
132	250
144	140
156	10

4.5. A reservoir has a linear S-Q relationship of

$$S = KQ,$$

where $K = 1.21$ hr. The inflow hydrograph for a storm event is given in the table.

a) Develop a simple recursive relation using the continuity equation and S-Q relationship for the linear reservoir [i.e., $aQ_2 = bQ_1 + c\bar{I}$, where a, b, and c are constants and $\bar{I} = (I_1 + I_2)/2$].

b) Storage route the hydrograph through the reservoir using $\Delta t = 1$ hr.

c) Explain why the shape of storage-discharge relations is usually not linear for actual reservoirs.

TIME (hr)	INFLOW (m³/s)
0	0
1	100
2	200

TIME (hr)	INFLOW (m³/s)
3	400
4	300
5	200
6	100
7	50
8	0

4.6. Given a reservoir with a storage-discharge relationship governed by the equation

$$S = KQ^{3/2},$$

route the inflow hydrograph for Problem 4.5 using storage routing techniques and a value of $K = 0.75$ for Q in cfs and S in cfs-hr. Discuss the differences in the outflow hydrograph for this reservoir and for the reservoir of Problem 4.5. Use $\Delta t = 1$ hr.

4.7. Blue Hole Lake has a spillway structure with the following relationship:

$$Q = 3.75Ly^{3/2},$$

where $L = 10$ m is the length of the weir and y is the height of the water above the spillway crest. The storage in the reservoir is governed by the relation

$$S = 1.5 \times 10^6 y.$$

Flow is measured in m³/s, length and depth are measured in m, and storage is measured in m³. Using a time step of 1 hr and assuming that the initial outflow is equal to zero, route the following inflow hydrograph through the reservoir using the Runge-Kutta FORTRAN program listed in Appendix E. Initial storage is 0.

TIME (hr)	INFLOW (m³/s)
0	0
1	200
2	300
3	500
4	450
5	400
6	300
7	200
8	100
9	50
10	0

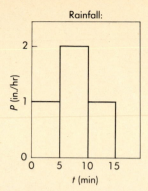

FIGURE P4.8

4.8. A small rectangular parking lot drains into a rectangular channel along its lower edge. Using the kinematic wave program listed in Appendix E, determine the flow into the channel from the parking lot. Assume that the parking lot may be modeled in two reaches of the same length with a time step of 5 min. The rainfall data and pertinent characteristics are listed in Fig. P4.8 and in the following table.

Overland flow characteristics:

Dimensions of lot: 50 ft × 4000 ft

Overland flow length: 50 ft

Overland slope: 0.06 ft/ft

Manning's *n* value $= 0.3$

Imperviousness $= 100\%$

$\Delta x = 25$ ft

$\Delta t = 5$ min

4.9. The parking lot of Problem 4.8 drains into a rectangular channel with the characteristics listed in the following table. Using the kinematic wave program, determine the outflow from the channel due to overland flow from the parking lot.

Channel characteristics:

Length $= 4000$ ft

Slope $= 0.003$ ft/ft

Manning's *n* value $= 0.025$

Bottom width $= 2$ ft

$\Delta x = 80$ ft

$\Delta t = 5$ min

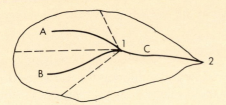

FIGURE P4.10

4.10. A storm event occurred on Falls Creek Watershed that produced a
rainfall pattern of 5 cm/hr for the first 10 min, 10 cm/hr in the second 10 min, and 5 cm/hr in the next 10 min. The watershed is divided into three subbasins (see Fig. P4.10) with the unit hydrographs
as given in the table. Subbasins A and B had a loss rate of 2.5 cm/hr
for the first 10 min and 1.0 cm/hr thereafter. Subbasin C had a loss
rate of 1.0 cm/hr for the first 10 min and 0 cm/hr thereafter. Using a
simple lag routing method, determine the storm hydrograph at point
2. Assume a lag time of 20 min.

UNIT HYDROGRAPHS

Subbasin A		Subbasin B		Subbasin C	
Time (min)	Q (m³/s)	Time (min)	Q (m³/s)	Time (min)	Q (m³/s)
0	0	0	0	0	0
10	5	10	5	10	16.7
20	10	20	10	20	33.4
30	15	30	15	30	50.0
40	20	40	20	40	33.4
50	25	50	25	50	16.7
60	20	60	20	60	0
70	15	70	15		
80	10	80	10		
90	5	90	5		
100	0	100	0		

4.11. Repeat Problem 4.10 using Muskingum routing methods instead of
simple lag routing. Discuss the differences between the two storm
hydrographs. Use $x = 0.2$, $K = 20$ min, and $\triangle t = 10$ min.

4.12. The storage equation has been given as

$$S_i = K[xI_i + (1 - x)O_i]$$

and the continuity equation as

$$\bar{I} = \bar{O} + \Delta S/\Delta t.$$

Given that I and O are the average inflow and outflow within the time period and ΔS is the change in the storage, derive the Muskingum river routing equation:

$$O_2 = C_0 I_2 + C_1 I_1 + C_2 O_1,$$

where I_2 and O_2 refer to the inflow and outflow at the end of the time period and I_1 and O_1 refer to those values at the beginning of the time period. Verify the equations for C_0, C_1, and C_2 given in Section 4.2.

4.13. Using the storage routing program listed in Appendix E, repeat Problem 4.5. Compare the results.

4.14. Using the Muskingum routing program listed in Appendix E, repeat Problem 4.2. Compare the results.

4.15. A storm event yielded the inflow and outflow hydrographs listed in the table. Derive the constants K and x needed to apply Muskingum routing methods to this reach. Use these constants to route the inflow hydrograph, either by hand or with the computer program of Problem 4.14, and compare the computed outflow hydrograph to the observed outflow hydrograph.

TIME (hr)	INFLOW (m³/s)	OUTFLOW (m³/s)
0	11	11
12	17.5	10.5
24	51.5	17
36	54.5	27.5
48	43	37.5
60	29.5	42.5
72	19.5	40
84	14	32
96	11	22
108	10	15
120	9.5	11
132	9	10

4.16. Repeat Problem 4.4 using the storage indication routing program listed in Appendix E.

4.17. Develop a new flood routing method for rectangular cross sections based on the Muskingum and storage indication techniques. Instead of using the Muskingum storage equation $S = f(K, x, I, Q)$, use Manning's equation in the form

$$Q = ay^m,$$

where a and m are constants. Assume prismatic channel conditions at each time step, and derive the necessary equation for flood routing in a given rectangular channel with length L, inflows I_1 and I_2, channel width B, and outflows Q_1 and Q_2.

4.18. How does the method in Problem 4.17 compare with the kinematic wave routing method described in this chapter? What hydraulic conditions are necessary for kinematic wave assumptions to be valid for overland flow? What are advantages and disadvantages of using kinematic channel routing compared with solving the St. Venant equations?

REFERENCES

Amein, M. and C. S. Fang, 1970, "Implicit Flood Routing in Natural Channels," *ASCE, J. Hyd. Div.,* vol. 96, no. HY12, pp. 2481–2500.

Bras, R. L., 1973, *Simulation of the Effects of Urbanization on Catchment Response,* thesis presented to MIT, Cambridge, Massachusetts.

Carnahan, B., H. A. Luther, and J. O. Wilkes, 1969, *Applied Numerical Methods,* John Wiley & Sons, New York.

Chapra, S. C., and R. P. Canale, 1985, *Numerical Methods for Engineers with Personal Computer Applications,* McGraw-Hill Book Company, New York.

Chow, V. T., 1959, *Open Channel Hydraulics,* McGraw-Hill Book Company, New York.

Cunge, K. A., 1969, "On the Subject of a Flood Propagation Method (Muskingum Method)," *J. Hyd. Res.,* vol. 7, no. 2, pp. 205–230.

Crawford, N. H., and R. K. Linsley, 1966, *Digital Simulation in Hydrology, Stanford Watershed Model IV,* Tech. Rep. 39, Civil Engineering Dept., Stanford University, Stanford, California.

Eagleson, P. S., 1970, *Dynamic Hydrology,* McGraw-Hill, New York.

Fread, D. L., 1971, "Flood Routing in Meandering Rivers with Flood Plains," *Rivers '76,* vol. 1, Symp. Inland Waterways for Navigation, Flood Control, and Water Diversions, ASCE, pp. 16–35.

Fread, D. L., 1973, *A Dynamic Model of Stage-Discharge Relations Affected by Changing Discharge,* NOAA Technical Memorandum NWS HYDRO-16, U.S. Department of Commerce, National Oceanic and Atmospheric Administration, National Weather Service, Office of Hydrology, Silver Spring, Maryland.

Fread, D. L., 1978, "National Weather Service Operational Dynamic Wave Model," *Verification of Math. and Physical Models in Hydraulic Engr.,* Proc. 26th Annual Hydr. Div., Special Conf., ASCE, College Park, Maryland.

Harley, B. M., 1975, *Use of the MITCAT Model for Urban Hydrologic Studies,* presented at the Urban Hydrology Training Course at the Hydrologic Engineering Center, Davis, California.

Henderson, F. M., 1966, *Open Channel Flow,* Macmillan, New York.

Henderson, F. M., and R. A. Wooding, 1964, "Overland Flow and Groundwater from a Steady Rainfall of Finite Duration," *J. Geophys. Res.,* vol. 69, no. 8, pp. 39–67.

Hydrologic Engineering Center, 1979, *Introduction and Application of Kinematic Wave Routing Techniques Using HEC-1,* Training Document no. 10, United States Army Corps of Engineers, Davis, California.

Hydrologic Engineering Center, 1980, *HEC-1 Flood Hydrograph Package: User's Manual and Programmer's Manual,* updated 1981, United States Army Corps of Engineers, Davis, California.

Issacson, E., J. J. Stoker, and B. A. Troesch, 1956, "Numerical Solution of Flood Prediction and River Regulation Problems," *Inst. Math. Sci.,* Report no. IMM-235, New York University, New York.

Kennedy, E. J., 1983, *Computation of Continuous Records of Streamflow,* Book 3, Chapter A13, Techniques of Water Resources Investigations of the United States Geological Survey, Distribution Branch, USGS, Alexandria, Virginia.

Lai, C., 1986, "Numerical Modeling of Unsteady Open-Channel Flow," *Advances in Hydroscience,* B. C. Yen (editor), vol. 14, Academic Press, New York, pp. 161–333.

Lighthill, M. J., and G. B. Whitham, 1955, "I: Flood Movement in Long Rivers," *Proc. R. Sci.,* ser A., vol. 229, pp. 281–316.

McCarthy, G. T., 1938, "The Unit Hydrograph and Flood Routing," unpublished paper presented at a conference of the North Atlantic Div., United States Army Corps of Engineers.

Overton, D. E., and M. E. Meadows, 1976, *Stormwater Modeling,* Academic Press, New York.

Ponce, V. M., H. Indlekofer, and D. B. Simons, 1978, "Convergence of Four-Point Implicit Water Wave Models," *ASCE, J. Hyd. Div.,* vol. 104, no. HY7, pp. 947–958.

Price, R. K., 1974, "Comparison of Four Numerical Methods for Flood Routing," *ASCE Proc., J Hyd. Div.,* vol. 100, no. HY7, pp. 879–899.

Raudkivi, A. J., 1979, *Hydrology,* Pergamon Press, Elmsford, New York.

Resource Analysis, Inc., 1975, *MITCAT Catchment Simulation Model, Description and Users Manual, Version 6,* Cambridge, Massachusetts.

Seddon, J., 1900, "River Hydraulics," *Trans. ASCE,* vol. 43, pp. 217–229.

Stephenson, D., and M. E. Meadows, 1986, *Kinematic Hydrology and Modeling,* Elsevier Science Publishing Company, New York.

Stoker, J. J., 1957, *Water Waves,* Interscience Press, New York.

Wooding, R. A., 1965, "A Hydraulic Model for the Catchment-Stream Problem: I: Kinematic-Wave Theory," *J. Hydrol.,* vol. 3, pp. 254–267.

Woolhiser, D. A., 1975, "Simulation of Unsteady Overland Flow," in *Unsteady Flow in Open Channels,* K. Mahmod and V. Yehjevich, editors, Water Resources Publications, Colorado.

Woolhiser, D. A., and J. A. Liggett, 1967, "Unsteady One Dimensional Flow over a Plane: The Rising Hydrograph," *Water Resources Res.,* vol. 3, no. 3, pp. 753–771.

5

Hydrologic Simulation
Models and Watershed
Analysis

5.1
INTRODUCTION TO SIMULATION MODELS

The rainfall-runoff process was presented in detail in Chapter 2, where we
showed how input rainfall is distributed into various components of evaporation, infiltration, detention or depression storage, overland flow, and
eventually streamflow. The actual shape and timing of the response hydrograph for a particular watershed have been shown to be a function of
many physiographic, land use, and climatic variables, which are listed in
Table 5.1. Rainfall intensity and duration are the major driving forces of
the rainfall-runoff process, followed by watershed characteristics that
translate the rainfall input into an output hydrograph at the outlet of the
basin. Size, slope, shape, soils, and storage capacity are all important
parameters in watershed geomorphology. Land use and land cover parameters can significantly alter the natural hydrologic response through

TABLE 5.1

Factors Affecting the Hydrograph

1. Rainfall intensity
2. Rainfall duration
3. Watershed size
4. Watershed slope
5. Watershed shape
6. Watershed storage
7. Watershed morphology
8. Channel type
9. Land use/land cover
10. Soil type
11. Percent impervious

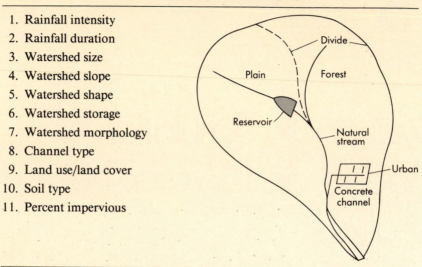

increases in impervious cover, altered slopes, and improved drainage channel networks.

Advances in computer methods over the past two decades combined with larger and more extensive data-monitoring efforts have allowed for the development and application of simulation models in hydrology. Such models incorporate various equations to describe hydrologic transport processes and account for water balances through time. Chapters 1, 2, and 4 have presented a number of computational methods routinely used to convert rainfall into a storm hydrograph at the basin outlet. Computer models allow for parameter variation in space and time through the use of well-known numerical methods. Complex rainfall patterns and heterogeneous watersheds can be simulated with relative ease if hydrologic data are sufficient, and various design and control schemes can be tested with the models.

Major research programs by Harvard University, Stanford University, and the U.S. Army Corps of Engineers during the 1960s were pioneering efforts in an attempt to use digital computers to simulate watershed behavior. The Stanford Watershed Model (Crawford and Linsley, 1966), which later evolved into the Hydrologic Simulation Program-FORTRAN (Johanson, et al., 1980), was the first available major watershed model. From this early effort, a range of modeling approaches was developed and applied during the 1960s and 1970s for urban stormwater, flood hydrology, agricultural drainage, reservoir design, and river basin management.

Hydrologic simulation models can be classified according to a wide range of characteristics, as shown in Table 5.2. For watershed analysis, the major categories of interest include lumped parameter versus distributed parameter, event versus continuous, and stochastic versus deterministic. Lumped parameter models transform actual rainfall input into runoff output by conceptualizing that all watershed processes occur at one spatial point (as in a "black box"). Model parameters may or may not have a direct physical definition in the system. Synthetic unit hydrographs are a widely used example. Distributed parameter models attempt to describe physical processes and mechanisms in space, as evidenced by certain classes of hydrologic simulation models. While distributed models are theoretically more satisfying, data are often lacking to calibrate and verify the simulation results.

Event models such as the HEC-1 Flood Hydrograph Package (Hydrologic Engineering Center, 1981) or the EPA Storm Water Management Model (SWMM) (Huber et al., 1981) or the SCS TR20 (Soil Conservation Service, 1975a) simulate single-storm responses for given rainfall input data. Unit hydrograph or kinematic wave methods are used to generate storm hydrographs, which are then routed within stream channels. Continuous models such as the Stanford Watershed Model, SWMM, and STORM are based on long-term water balance equations and thus account directly for effects of antecedent conditions. These models are probably most useful in watersheds with large areas of pervious land.

Some models include a random or stochastic component to represent input rainfall, which is then used to generate time series of streamflow. The time series can then be statistically evaluated using flood frequency analysis. Hydrologic synthesis allows hydrologists to extend short historical records, such as streamflow, to longer sequences based on statistical

TABLE 5.2
Hydrologic Models

MODEL TYPE	EXAMPLE OF MODEL
Lumped parameter	Snyder Unit Hydrograph
Distributed	Kinematic wave
Event	HEC-1, SWMM
Continuous	Stanford Watershed Model, SWMM, HSPF, STORM
Physically based	HEC-1, SWMM, HSPF
Stochastic	Synthetic streamflows
Numerical	Explicit kinematic wave
Analytical	Nash IUH

methods. Synthetic sequences either preserve the statistical character of the historical record or follow a prescribed probability distribution, such as a lognormal or log Pearson type 3 (Chapter 3). Mass curve analysis assumes that historical records will repeat exactly. Random generations assume that successive flows are independent and distributed according to a known probability distribution. Finally, Markov techniques assume that the next flow in a sequence is related to a subset of previous flows and is usually restricted to shorter time intervals of analysis. Details on stochastic models can be found in Bras and Rodriguez-Iturbe (1984).

A major advantage of simulation models is the insight gained by gathering and organizing data required as input to the mathematical algorithms that comprise the overall model system. This exercise can often guide the collection of additional data or direct the improvement of mathematical formulations to better represent watershed behavior. Another advantage is that many alternative schemes for development or flood control can be quickly tested and compared with simulation models.

The major limitation of simulation models is the inability to properly calibrate and verify applications in which input data are lacking. An overreliance on very sophisticated computer models that failed to perform in the 1970s and 1980s has led to a more skeptical approach to modeling in recent years. Current practice assumes that the simplest model that will satisfactorily describe the system for the given input data should be used. Model accuracy is largely determined by available input data and observed input and output time series at various locations in a watershed.

Despite their limitations, simulation models still provide the most logical and scientifically advanced approach to understanding the hydrologic behavior of complex watershed and water resources systems. Model developments and applications have opened a new era in the science of hydrology and have led to many new design and operating policies never before realized or tested. In recent years, several excellent reviews of models in hydrology have been published, including Fleming (1975), Delleur and Dendrou (1980), McPherson (1975), Feldman (1981), Kibler (1982), and Whipple et al. (1983). The reader is referred to these references for details beyond the scope of this chapter.

5.2

ORGANIZATION FOR HYDROLOGIC WATERSHED ANALYSIS

With many hydrologic models available to the hydrologist or engineer, very little new model development is currently being supported. Rather, the engineer must select one of the available simulation models based on characteristics of the system to be studied, the objectives to be met, and the available budget for data collection and computer expense. Once the model is selected, the steps involved in watershed simulation analysis generally follow the sequence of Table 5.3.

Steps 1 and 5 are the most important in the overall sequence. The selection of a model is a very difficult and important decision since the success of the analysis hinges on accuracy of the results. Table 5.2 should be considered along with characteristics of the watershed, study objectives, and so on, to develop a modeling plan. In general, unless watershed data are extensive in space and time, the usual approach to watershed analysis is to use a deterministic event model with lumped parameter concepts for developing hydrographs and flood routing. HEC-1 (Hydrologic Engineering Center, 1981) and TR-20 (Soil Conservation Service, 1975a) are two of the most widely used models for typical watershed analysis.

If a watershed has extensive data available on rainfall, infiltration, base flow, streamflow, and soils and land use, then either the Stanford Watershed Model or Hydrological Simulation Program–FORTRAN (HSPF) can be applied for calculating continuous long-term water balances and outflow hydrographs. For the case of a well-defined urban drainage network, a distributed event model such as SWMM or ILLUDAS (Terstriep and Stall, 1974) can be applied to define hydrologic

TABLE 5.3
Steps in Watershed Simulation Analysis

1. Select model based on study objectives and watershed characteristics, availability of data, and project budget.

2. Obtain all necessary input data — rainfall, infiltration, physiography, land use, channel characteristics, streamflow, design floods, and reservoir data.

3. Evaluate and refine study objectives in terms of simulations to be performed under various watershed conditions.

4. Choose methods for determining subbasin hydrographs and channel routing.

5. Calibrate model using historical rainfall, streamflow, and existing watershed conditions. Verify model using other events under different conditions while maintaining same calibration parameters.

6. Perform model simulations using historical or design rainfall; various conditions of land use; various control schemes for reservoirs, channels, or diversions.

7. Perform sensitivity analysis on input rainfall, routing parameters, and hydrograph parameters as required.

8. Evaluate usefulness of the model and comment on needed changes or modifications.

response for components throughout the system. Urban stormwater models are discussed in more detail in Section 5.3 and Chapter 6.

Step 5 in Table 5.3, model calibration and verification, is extremely important in fitting the model parameters and producing accurate and reliable results in steps 6 and 7. Model calibration involves selecting a measured set of input data (rainfall, channel routing, land use, etc.) and measured output hydrographs for model application. The controlling parameters in the model are adjusted until a "best fit" is obtained for this set of data. The model should then be "verified" by simulating a second or third event (different rainfall) and keeping all other parameters unchanged to produce a comparison of predicted and measured hydrographs. A detailed example of a calibrated and verified model is described in Section 5.6, where an actual case study is highlighted to indicate difficulties and complexities that are often encountered in watershed analyses.

5.3

DESCRIPTION OF MAJOR HYDROLOGIC SIMULATION MODELS

A selected number of the most popular event, continuous, and urban runoff models for hydrologic simulation are listed in Table 5.4. Universities or federal agencies supported the development of most of these models. Some of the models are very well documented and, as a result, have seen wide application to watersheds, especially in the United States. Others have been applied to only specific areas in the country. Extreme care and judgment must be exercised in applying any one of these models to areas where data exist to define unit hydrograph and routing parameters.

Several of the models in Table 5.4 have been extensively reviewed by Delleur and Dendrou (1980) and will only be briefly described below. The Stanford Watershed Model (Crawford and Linsley, 1966) was one of the first and most comprehensive hydrologic models. The Stanford Watershed Model (SWM-IV) is made up of a sequence of routines for each process in the hydrologic cycle on a continuous basis. The continuity equation is balanced for each time step, and precipitation, interception, evapotranspiration, infiltration, soil moisture storage, overland flow, interflow, and channel routing are all incorporated. The model requires a large amount of input data for a watershed and consequently is not used as often as other available models.

Several versions of the model in FORTRAN have evolved over the years, e.g. at the University of Kentucky, Ohio State University, and the University of Texas. The Hydrocomp Simulation Program (HSP) was the commercial successor to SWM-IV and was modified to include water quality considerations, kinematic wave routing, and variable time steps.

TABLE 5.4

Selected Simulation Models in Hydrology

MODEL	AUTHOR	DATE	DESCRIPTION
1. Stanford	Crawford and Linsley	1966	Stanford Watershed Model
2. HEC-1	HEC	1973	Flood Hydrograph Package
3. HEC-2	HEC	1976	Water Surface Profiles
4. HEC-3	HEC	1973	Reservoir System Analysis for Conservation
5. HEC-4	HEC	1971	Monthly Streamflow Simulation
6. HEC-5	HEC	1979	Simulation of Flood Control Systems
7. SCS-TR20	USDA SCS	1974	Hydrologic simulation model
8. USDA HL-74	USDA ARS	1975	Model of watershed hydrology
9. HSPF	Hydrocomp, EPA	1980	Hydrological Simulation Program-FORTRAN
10. MITCAT	Eagleson et al.	1970	Event rainfall-runoff urban model
11. SWMM	Metcalf & Eddy; Camp, Dresser, and McKee; University of Florida	1971, 1981	Storm Water Management Model
12. ILLUDAS	Illinois State Water Survey	1974	Illinois Urban Drainage Area Simulator
13. STORM	HEC	1974	Storage, Treatment, Overflow, Runoff Model
14. USGS	USGS, Dawdy	1975	Flood routing in urban areas
15. Penn St.	Aron, Lakatos	1976	Urban Runoff Model
16. DWOPER	NWS, Fread	1978	NWS Operational Dynamic Wave Model

The current version is HSPF, which is a redesigned modular version that performs simulations of a variety of hydrologic and water quality processes on or under the land surface, in channels, and in reservoirs (Johanson et al., 1980).

One of the most useful modules in HSPF is the nonpoint source model (NPS), which provides for continuous simulation of nonpoint pollutants from urban and undeveloped land surfaces. It is available as a separate model from EPA and has been extensively tested with HSP by the North Virginia Planning Commission. Donigian and Crawford (1976, 1979) developed the model and user's manual for EPA.

MITCAT was developed by Bravo et al. (1970) at MIT for single-event rainfall. Overland flow was treated by kinematic wave methods for the first time, and channel flows were routed by means of a linearized St. Venant equation. Several infiltration equations can be used along with a variable time step. A modularized version of the model is described by Harley et al. (1970).

The EPA funded the development of the Storm Water Management Model (SWMM) by Metcalf and Eddy, Inc., University of Florida, and Water Resources, Inc. (now Camp, Dresser, and McKee) (1971). SWMM is the most comprehensive urban runoff model; it provides for continuous and event simulation for a variety of catchments, conveyance, storage, treatment, and receiving streams. Both water quantity and quality can be simulated and flow routing can be performed by nonlinear reservoir methods, kinematic wave methods, or with the full St. Venant equations in the SWMM EXTRAN Block. The model has gone through a number of revisions; version 3 (Huber et al., 1981; Roesner et al., 1981) is currently available, with version 4 expected in 1987. SWMM is described in more detail with an example in Chapter 6.

The STORM model was developed by the Hydrologic Engineering Center (1975; Roesner et al., 1974) to estimate quantity and quality of urban stormwater runoff. The program performs continuous simulation using years of continuous hourly rainfall data, but individual storms may also be analyzed. Infiltration is estimated using a weighted runoff coefficient or SCS methods. Downstream storage and treatment processes may be simulated but there is no channel routing. The model provides a much simpler alternative to SWMM.

The Illinois Urban Drainage Area Simulator (ILLUDAS) was developed by Terstriep and Stall (1974) and is a single-storm event model that can handle a detailed storm-sewer system. The model can be run in a design mode or in a simulation mode for a particular watershed. ILLUDAS requires input rainfall, pipe configurations, and subbasin areas to calculate flows at various points in the system. Runoff is generated using a time-area curve. The model has recently been updated to include a pressurized or surcharged condition in the pipe system (Chiang and Bedient, 1986) and is called the PIBS model. The method compares well with the SWMM EXTRAN Block and runs on a personal computer.

The Penn State runoff model (Aron and Lakatos, 1976) was developed as a simple and concise model that emphasizes the timing of subarea flow contributions to peak rates at various points in a watershed. The model provides a practical approach for the analysis of runoff for small to midsize communities. The watershed is subdivided into pervious and impervious areas, water conveyance, and storage systems. A study of the parameter sensitivity of the model and of SWMM was performed by Kibler and Aron (1978).

The Soil Conservation Service (SCS) has developed a number of methods and models that have been described in detail by McCuen (1982). Graphical, tabular, and chart methods are available for the SCS TR-55, entitled "Urban Hydrology for Small Watersheds." These methods are primarily based on the runoff curve number method described in Chapter 2 and will not be presented here.

The SCS TR-20 is a computerized method for solving hydrologic problems using SCS procedures. The program develops runoff hydrographs, routes these through channel reaches and reservoirs, and combines or separates hydrographs at confluences. For each subarea, the area, runoff curve number, and time of concentration are required. Routing procedures are somewhat simpler than those used in HEC-1 and include the convex routing method or a simple routing coefficient. The overall model produces results somewhat similar to those of HEC-1 and is particularly suited to examining the effects of detention storage in a watershed.

The preceding discussion has provided a review of many of the popular hydrologic simulation models used for watershed analysis. The reader should consult user documentation directly from authors before attempting to apply the models to an actual watershed. The next three sections are devoted to a detailed presentation of the HEC-1 model along with examples and case studies. It is considered by many to be the most versatile model and is the most often used of the available models described here.

5.4

HEC-1 FLOOD HYDROGRAPH PACKAGE

The main purpose of HEC-1 is to simulate hydrologic processes during flood events. It was developed over a number of years by the U.S. Army Corps of Engineers, Hydrologic Engineering Center. The process of converting precipitation to direct runoff can be simulated by HEC-1 for small subbasins or large complex watersheds, as shown in Fig. 5.1. HEC-1 is a general flood hydrograph package with the following capabilities:

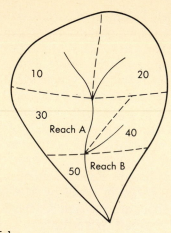

FIGURE 5.1

HEC-1 model configuration.

1. Optimal estimation of unit hydrograph, loss rate, and streamflow routing parameters from measured data
2. Simulation of watershed runoff and streamflow from historical or design rainfall
3. Computation of damage frequency curves and expected annual damages for various locations and multiple flood control plans
4. Simulation of reservoir outflow for dam safety analysis

HEC-1 is one of the most widely used methods for watershed simulation and flood event analysis. The following discussion will describe the features of the model in more detail followed by a case study. Consult the user's manuals and associated documents from the Hydrologic Engineering Center (1981) for more complete coverage.

Watershed Delineation

HEC-1 uses parameters averaged in space and time to simulate the runoff process. The size of subbasins, routing reaches, or computation interval is selected based on the basin physiography, available rainfall data, available streamflow data, and required accuracy. As shown in Fig. 5.1, a complex watershed is subdivided into small and relatively homogeneous subbasins according to drainage divides. Routing reaches are identified and the overall order of the runoff computation is thus defined for input to HEC-1.

The size of a subbasin should generally be in the range of 1–10 mi^2 because of limitations of unit hydrograph theory. Routing reaches should be long enough so that a flood wave will not travel faster than the computation time step. Otherwise, numerical errors in flood calculations will occur.

Precipitation

Precipitation P is computed for each subbasin in Fig. 5.1 using historical data or synthetic design storms. The model allows for (1) incremental P (depth of P for each time interval) for each subbasin, (2) total P and a time distribution, or (3) historical gage data together with areal weighting coefficients for each subbasin. Standard design storms can be used in the form of (1) depth vs. duration data, (2) probable maximum precipitation, or (3) standard project precipitation. The precipitation data must be input at a constant time interval, but this interval may be different from the computational time interval in the model. HEC-1 also allows for the incorporation of snowfall and snowmelt, as discussed in the next section.

Snowfall and Snowmelt

HEC-1 is capable of simulating snowfall and snowmelt. Up to ten elevation zones of equal increments may be specified in each subbasin. Usually, an elevation increment of 1000 ft is used, but any increment may be specified as long as the air temperature lapse rate corresponds to the change in elevation within the zone. Temperature data are input for the bottom of the lowest elevation zone. Temperatures are then reduced by the lapse rate (°C or °F per change in elevation). The temperature at which melting will occur is input in the data, which are then used to determine whether precipitation falls as snow (melt temperature $+2$°C or °F) or as rain. Snowmelt occurs when the temperature is equal to or greater than the melt temperature and is calculated by either the degree-day method or an energy budget method (see Section 2.8). For more detail on snowfall and snowmelt calculations in HEC-1, see the HEC-1 user's manual (Hydrologic Engineering Center, 1981) and *Runoff from Snowmelt* (U.S. Army Corps of Engineers, 1960).

Loss Rate Analysis

HEC-1 contains four methods for computing the loss of precipitation to interception and infiltration as shown in Table 5.5. The simplest is the initial and constant loss rate function, in which the initial loss volume is satisfied before the constant loss rate begins. This method is the same as the ϕ index for infiltration (see Section 1.5) if initial loss is zero and is often used in design storm analysis or where data are insufficient to allow calculation with more sophisticated methods.

   The HEC exponential loss rate function uses loss rate as a function of accumulated soil moisture losses (Fig. 5.2). The equation computes loss rate L by the function AP^E, where A is a combined effect of interception

TABLE 5.5

HEC-1 Loss Methods

METHOD	DESCRIPTION
1. Initial and Constant	Initial loss volume is satisfied and then constant loss rate begins (Section 1.5).
2. HEC exponential	Loss function is related to antecedent soil moisture condition and is a continuous function of soil wetness (see Fig. 5.2).
3. SCS curve number	Initial loss is satisfied before calculating cumulative runoff as a function of cumulative rainfall.
4. Holtan method	Infiltration rate is computed as an exponential function of available soil moisture storage from Holtan's equation.

and infiltration rates as a function of accumulated soil moisture losses C, P is precipitation intensity, and E is an exponent for precipitation effects. This function is difficult to apply in actual watersheds because the parameters are not readily determined from measureable quantities.

The SCS method employs the SCS curve number, which is directly related to land use and soil properties of a watershed. The SCS approach (see Section 2.4) is popular because of its application to ungaged areas and its large empirical data base.

The Holtan loss rate equation (Holtan et al., 1975) is related to

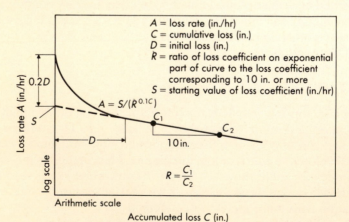

FIGURE 5.2

HEC-1 loss rate function ($L = AP^E$).

watershed characteristics, but the parameters must still be calibrated, which is impossible for ungaged areas. This method is similar to the HEC exponential method but does not consider precipitation intensity.

Subbasin Runoff Calculation

Several methods for surface runoff computations in HEC-1 are available in the model and can be selected by the user. These are presented in Table 5.6 and include the unit hydrograph methods of Clark (1945), Snyder (1938), and the Soil Conservation Service (1975b). Known unit hydrographs can also be directly input. The kinematic wave overland flow computation was added to HEC-1 in 1979 and allows more accurate representation of urbanized areas for UH calculations (Hydrologic Engineering Center, 1979).

Clark's method is based on the time-area curve method described in Section 2.2. The time-area histogram, determined from isochrones of the watershed, is convoluted with a unit design storm hyetograph to yield the hydrograph, which is then routed through linear reservoir storage to allow for peak attenuation.

Snyder's method, described in Section 2.4, provides time to peak t_p and peaking coefficient C_p, which are insufficient to produce the curve of the hydrograph. Clark's method is therefore used to smooth and define the shape of the hydrograph corresponding to Snyder's coefficients. The direct runoff ordinates are calculated by convolution of the unit hydrograph with the net precipitation for the subbasin (Fig. 5.3).

The kinematic wave process was adapted from the MIT catchment model (Harley, 1975) and is described in more detail in Section 4.6. It represents a nonlinear runoff response compared with linear unit hydrograph methods. The kinematic wave method relies on parameters that are generally measureable from a basin such as slope, land use, lengths, channel shape, roughness, and area. Overland flow, collector, and main channel elements are used to represent the characteristics of a drainage basin (Fig. 5.4).

TABLE 5.6
Surface Runoff Methods in HEC-1

1. Unit hydrograph input directly

2. Clark hydrograph method (time-area method)

3. Snyder unit hydrograph method (see Section 2.4)

4. SCS method (curve number method + SCS unit hydrograph)

5. Kinematic wave for overland hydrograph (see Section 4.6)

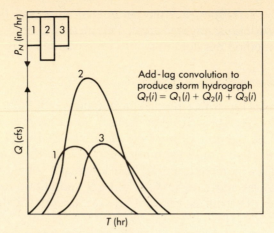

FIGURE 5.3

Snyder's unit hydrograph method and convolution.

In the kinematic wave method, the main equations are Manning's equation of flow and continuity (Eqs. 4.47 and 4.48):

$$Q_c = \alpha_c A_c^{m_c} \quad \text{(flow)},$$

$$\frac{\partial A_c}{\partial t} + \frac{\partial Q_c}{\partial x} = q_0 \quad \text{(continuity)}. \tag{5.1}$$

These are solved numerically to produce Q as a function of time and space.

Once overland flow has been calculated, it is directed to one or two levels of collector channels before entering the main channel. Collector and main channels can have various shapes and roughnesses. A subbasin's total contribution to the main channel is determined at any time by

$$QSB = QDA \left(\frac{\text{Subbasin area}}{\text{Typical drainage area}} \right), \tag{5.2}$$

where

QSB = HEC-1 subbasin outflow,

QDA = typical drainage area outflow.

QSB may be lumped at either end of the main channel or laterally distributed.

Base Flow Calculation

The base flow part of HEC-1 uses a logarithmic decay function. Equation (5.3) defines the parameters that are used in the model, in which the recession flow threshold Q_R and the decay constant RTIOR must be specified by the user. Figure 5.5 defines the relation between the stream-flow hydrograph and user-defined parameters. Equation (5.3) computes recession flow Q as

$$Q = Q_R(RTIOR)^{-n\,\Delta t}, \qquad\qquad (5.3)$$

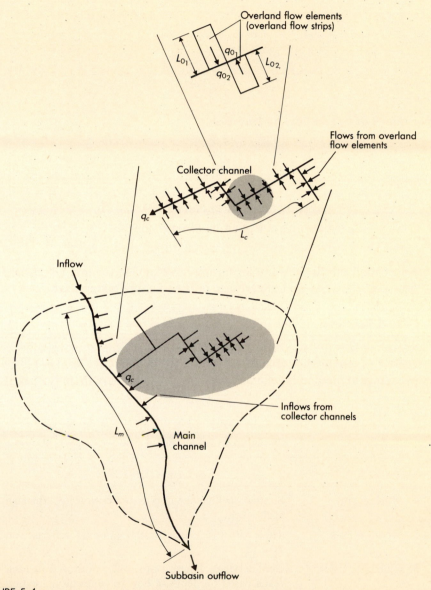

FIGURE 5.4

Kinematic wave flow elements. (From Hydrologic Engineering Center, 1981.)

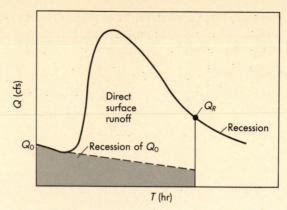

FIGURE 5.5
Components of streamflow hydrograph.

where

$$Q = \text{recession flow rate at end of } n\Delta t,$$
$$Q_R = \text{flow rate at beginning of recession,}$$
$$\text{RTIOR} = \text{ratio of recession flows } n\Delta t \text{ increments apart.}$$

Many applications of HEC-1 are for major urban flood events, where base flow is a relatively small percentage of the total hydrograph flow.

Flood Routing

Table 5.7 lists the major methods of flood routing that are available in HEC-1. These methods have been presented in detail in Chapter 4. The Muskingum method is generally used for streams where there is a linear

TABLE 5.7

Flood Routing in HEC-1

METHOD	DESCRIPTION
1. Muskingum	Storage coefficient (x) plus travel time (K) through each reach
2. Modified Puls	Table of storage versus outflow for each reach based on HEC-2
3. Kinematic wave	Channel shape, length, slope, and n; outflow from each reach based on depth of flow in continuity equation and Manning's equation (Eqs. 5.1)

relation between storage and weighted discharge. The modified Puls uses a storage-discharge (S-Q) relation that must be determined from field surveys and can be calculated for a reach using the HEC-2 water surface profile model (see Chapter 7).

The kinematic wave channel routing method does not allow flood peak attenuation, which is a problem where the floodplain is extensive. It has the advantage of allowing for laterally distributed inflow along the main channel (Hydrologic Engineering Center, 1979). Manning's equation forms the basis for the kinematic wave method as described in detail in Section 4.6.

Other HEC-1 Capabilities

HEC-1 can be calibrated for observed storm events using a parameter optimization algorithm. A number of gaged events must be available for comparison with model predictions. Parameter selection is often accomplished by systematic trial and error: parameter values are selected, the model is executed with historical rainfall, and the resulting runoff hydrograph is compared with the observed hydrograph. The process is repeated until a desirable best fit occurs.

An automatic optimization procedure in HEC-1 is a useful alternative to the trial-and-error parameter selection process. The optimization involves an objective function for which a minimum value is sought, subject to certain constraints (ranges) on the parameters. The function takes the form

$$\text{STDER} = \sum_{i=1}^{N} (\text{QOBS}_i - \text{QCOMP}_i)^2 \left(\frac{WT}{N}\right), \tag{5.4}$$

where STDER is the error index; QOBS_i is the observed hydrograph ordinate for time i; QCOMP_i is the computed ordinate for time i from HEC-1; N is the total number of hydrograph ordinates; and WT is a weighting function that emphasizes accurate reproduction of peak flows. The parameter values are bounded by upper and lower values. A more detailed description of this method is available in the user's manual (Hydrologic Engineering Center, 1981).

A second major capability of HEC-1 is for ungaged areas where parameter selection is particularly difficult. The HEC (1980) describes a complete program for modeling ungaged areas using frequency analysis and regionalization of hydrologic parameters. The experience of the analyst with the parameters of the model is extremely important in the application to ungaged areas. Three detailed case studies are presented in the HEC report.

Finally, HEC-1 can be used in detailed evaluation of flood control measures. Depth-area-duration relations can be simulated for a river system, multiple storms can be simultaneously executed for detailed comparison, and several alternative flood control plans can be analyzed in a single run. Economic calculations for expected annual flood damage can also be performed for any point of interest in a river basin. Thus, HEC-1 capability is very extensive, incorporating large amounts of data manipulation required for watershed simulation.

The best way for the student or practicing hydrologist to obtain a working knowledge of a model such as HEC-1 is, of course, to apply it to an actual watershed. But without some careful guidance on model setup, parameter selection, and calibration efforts, it becomes a very complicated assignment. To help overcome these stumbling blocks and not lose sight of the usefulness of HEC-1, a simple example and a fairly complete case study are presented in the following sections.

5.5
INPUT AND OUTPUT DATA FOR HEC-1

Input Data Overview

The major advantage of using HEC-1 compared with other models is the simple organization of input data. HEC-1 uses "card images" for input

TABLE 5.8

Basic Data Categories

Job control	
I__	Job initialization
V__	Variable output summary
O__	Optimization
J__	Job type
Hydrology and hydraulics	
K__	Job step control
H__	Hydrograph transformation
Q__	Hydrograph data
B__	Basin data
P__	Precipitation data
L__	Loss (infiltration) data
U__	Unitgraph data
M__	Melt data
R__	Routing data
S__	Storage data
D__	Diversion data
W__	Pump withdrawal data
Miscellaneous	
E__	Economics data
ZZ	End of job

that indicate the format of the river basin data and determine output formats. Each card image consists of a terminal input line of 80 columns, in either fixed or free format. Fixed format consists of ten 8-column fields, and free format allows data values to be separated by commas. In the following, the word *card* is used to refer to a single line of terminal input.

The watershed simulation data are identified on each card by a unique two-character code in columns 1 and 2. These codes identify data to be read or they activate various program options. The first character of the code identifies the general category (i.e., P for precipitation data), and the second character specifies a certain type of data (i.e., PB for basin average total precipitation). A summary of the basic data categories is shown in Table 5.8.

For watershed simulation, the actual sequence of the input data prescribes how the hydrologic flow and routing work. The example data organization for a watershed is depicted in Table 5.9, where it can be seen that each KK card is used as the beginning of a computational block. For example, subbasin runoff, hydrograph combination, and a routing step all begin with a KK card, giving the input a "block" structure.

TABLE 5.9
Example of Input Data Organization

SEGMENT	CODE	DESCRIPTION OF DATA
Job control	ID	Title and description of model
	IT/IN	Time interval and beginning time
	IO	Output control for entire job
Runoff from first subbasin	KK	Station name
	KM	Comment card (optional)
	BA	Basin area
	P__	Precipitation method
	L__	Loss rate method
	U__	Rainfall transformation method
Route hydrograph from subbasin to next point	KK	Station name
	KM	Comment card (optional)
	R__	Routing method
Combine routed hydrographs	KK	Station name
	KM	Comment card (optional)
	HC	Number of hydrographs to be combined
Runoff from next next subbasin	KK	
	KM	
	etc.	

Process continues until the mouth of the watershed is reached.

	ZZ	Indicates end of job

Each subbasin runoff group includes a BA card for basin area, a P__ card for precipitation input, an L__ card for loss rate to infiltration, and a U__ card for unit hydrograph transformation of rainfall. A base flow (BF) card may also be included in this group.

A number of methods are available for specifying rainfall hyetographs or total subbasin rainfall (Fig. 5.6). The PI or PC cards allow incremental or cumulative rainfall to be input as a time series. The PG

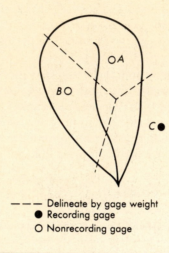

--- Delineate by gage weight
● Recording gage
○ Nonrecording gage

DESCRIPTION		DATA LINE			
Comments	ID	Title of data set			
Time interval information	IT				
Rainfall data by gage, both	PG	A	5.7		
recording and nonrecording	PG	B	6.5		
	PG	C			
	PC	0.02	0.09	0.17	etc.
		(data from recording gage)			
Basin data	KK	Sample basin runoff block			
	BA	1.5			
Weightings for gages (basin	PT	A	B	C	
average)	PW	0.40	0.45	0.15	
Weightings for temporal	PR	C			
distribution	PW	1			
Loss rate	L__				
Unit hydrograph	U__				

FIGURE 5.6

Typical rainfall input methods.

card is used to specify the total rainfall for a nonrecording gage or to indicate that recorded data follow on PI or PC cards, as shown in Fig. 5.6. If precipitation does not change from one subbasin to another, then the data may be input only once, with the first KK group. The (PR, PW) and (PT, PW) cards specify which gages and corresponding weights are to be used to calculate a subbasin's average precipitation. A recording gage can be used as both a storm total and a recording gage station.

Loss rates can be computed using any of four methods contained in HEC-1 (Table 5.5). The LE, LU, LS, and LH cards provide the necessary codes for loss rate calculations. In the following example the LS or SCS method has been employed because of its well-established data base. If loss rate data do not change across the watershed, they need to be specified only once in the first KK group.

Unit hydrograph methods were presented in Chapter 2, and Table 5.6 indicates the methods available in HEC-1. The UI, UC, US, UD, and UK cards are used to specify the particular method for each subbasin. It is recommended that specific data (i.e., TC and R coefficients) be derived for each different subbasin for a more accurate solution. The kinematic overland flow method (UK) was introduced in HEC-1 in 1981 and is described in detail in Chapter 4.

Once a storm hydrograph has been computed or combined from upstream areas, channel and reservoir routing is performed using methods indicated on the R__ cards (see Example 5.1). The most popular methods include Muskingum (RM), storage-discharge or modified Puls (RS), and kinematic wave (RK) (see Chapter 4). The SV/SQ cards are used to input storage-discharge relations, and the SQ/SE cards are used to input a rating curve of discharge vs. elevation. For reservoir routing, a number of special card groups can be used to fully characterize outflow structures. The SA, SV, and SE cards indicate reservoir area-volume-elevation tables, and SS or SL cards specify spillway and low-level outlet data. These options are described in more detail in the example in Section 5.6.

The combination card HC allows for the summation of two or more hydrographs at a particular station in the watershed. The HC step will use the most recently calculated hydrographs in the system. Thus, the progression of hydrologic calculation is in the order provided by the KK card groupings. In Fig. 5.7, runoff from subbasin 1 is routed from *A* to *B* and is combined at point *B* with subbasin 2. After routing to point *C*, runoff from subbasin 3 is added to produce a single hydrograph at point *C*. The process continues until the outlet of the watershed has been reached. If a measured gage exists at any point, the QO card can be used to compare computed and observed flows, and a statistical summary is provided.

Reservoirs, channelization, or diversion options can be evaluated with HEC-1 for a particular watershed by adjusting the appropriate subbasin runoff or routing steps to reflect the change. Effects of land use

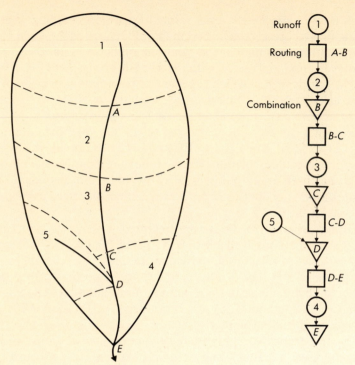

FIGURE 5.7

Schematic of HEC-1 computational order.

change through time can be represented by adjusting subbasin parameters
and channelization routing parameters. Different rainfall intensities and
patterns and different loss rates can easily be integrated into the watershed
analysis. HEC-1 provides a convenient and well-documented tool for
hydrologic assessment. Selection of parameters and the overall data organization in HEC-1 are highlighted in Example 5.1 and the detailed example is presented in Section 5.6.

EXAMPLE 5.1

A small watershed has the parameters listed in the following tables. A unit
hydrograph and Muskingum routing coefficients are known for subbasin
3, shown in Fig. E5.1(a). A rainfall hyetograph is shown in Fig. E5.1(b) for
a storm event that occurred on June 19, 1983. Assume that the rain fell
uniformly over the watershed. Use the information given to develop a

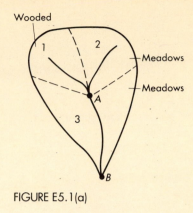

FIGURE E5.1(a)

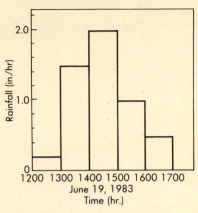

FIGURE E5.1(b)

HEC-1 input data set to model this storm. Run the model to determine the predicted outflow at point B.

SUBBASIN NUMBER	TC (hr)	R (hr)	SCS CURVE NUMBER	% IMPERVIOUS	AREA (mi²)
1	2.5	5.5	66	0	2.5
2	2.8	7.5	58	0	2.7
3	–	–	58	0	3.3

	TIME (hr)	0	1	2	3	4	5	6	7
UH FOR SUBBASIN 3:	U (cfs)	0	200	400	600	450	300	150	0

Muskingum coefficients: $x = 0.15$, $K = 3$ hr

SOLUTION

The input data set is as follows:

```
ID ***              EXAMPLE 5.1
ID ***
ID ***      HEC-1 INPUT DATA SET
ID ***
IT    60 19JUN83      1200      100
IO     4
KK  SUB1
KM       RUNOFF FROM SUBBASIN 1
PI   0.2      1.5      2.0      1.0      0.5
BA   2.5
LS           66        0
```

```
UC    2.5      5.5
KK    SUB2
KM        RUNOFF FROM SUBBASIN 2
BA    2.7
LS              58        0
UC    2.8      7.5
KK     A
KM        COMBINE RUNOFF FROM SUB 1 WITH RUNOFF FROM SUB 2 AT A
HC     2
KKA TO B
KM        MUSKINGUM ROUTING FROM A TO B
RM     1       3      0.15
KK    SUB3
KM        RUNOFF FROM SUBBASIN 3
BA    3.3
LS              58        0
UI     0      200     400     600     450     300     150        0
KK     B
KM        COMBINE FLOW FROM SUB3 AND ROUTED POINT B
HC     2
ZZ
```

DISCUSSION

The ID lines (card image) are used to identify the data set. The IT line sets the computation interval to 60 min, the starting date and time to June 19, 1983, 12:00 noon, and the maximum number of hydrograph ordinates computed to 100. The IO line sets the program to produce print of both the input data and a master summary of the output data. No hydrographs will be plotted.

The KK line signals the beginning of a computational block. KM lines are used to indicate the type of computation that will be performed in that particular group.

A PI line is used to input the incremental rainfall values. The time increment between the rainfall values is 60 min (see IT line). Since the rainfall is uniform over the watershed, only one PI line is necessary or the PI lines may be repeated for each subbasin runoff computational block. In either case, the same rainfall pattern is used to compute runoff from each subbasin.

For each subbasin runoff block, BA lines are used to input the subbasin area in mi^2. The LS line indicates that the SCS loss method will be used and the SCS curve number and percent imperviousness are entered here. In subbasins 1 and 2, UC lines are used to indicate that the Clark unit hydrograph method will be used. Values for TC and R are entered on the UC line. The unit hydrograph for subbasin 3 is provided in the given data and is input on the UI line.

HC lines indicate that a hydrograph combination step is being performed and the number of hydrographs to be combined is entered. An

RM line is used to enter the Muskingum routing coefficients for the route step between points A and B. The ZZ line (no data) indicates that the end of the data set has been reached.

The HEC-1 program was run using the preceding data set. The information in the following table was taken from the master summary output by the program.

OPERATION	STATION	PEAK FLOW	TIME OF PEAK	BASIN AREA
HYDROGRAPH AT				
	SUB1	365.	6.00	2.50
HYDROGRAPH AT				
	SUB2	215.	7.00	2.70
2 COMBINED AT				
	A	574.	6.00	5.20
ROUTED TO				
	A TO B	434.	9.00	5.20
HYDROGRAPH AT				
	SUB3	632.	7.00	3.30
2 COMBINED AT				
	B	972.	7.00	8.50

Output Data Overview

Most of the HEC-1 output can be controlled by the user, and a variety of summary outputs can be printed easily. The data used in each hydrograph computation (KK group) can be printed along with hydrographs, rainfall, storage, and other data as needed. Output is generally controlled by the IO card, which can be overridden by the KO card for a specific KK group. The output control provides an echo of input data, which should be used to check actual input data.

Hydrographs may be printed as tables or graphed as a printer plot. Rainfall, losses, and net rainfall are included in the table and the plot. Inflow and outflow hydrographs are printed and plotted along with storage for each routing step. A list of error messages may be printed that may or may not stop execution of the program. Output should always be checked for possible errors or warnings.

The program generates hydrologic summaries of the calculations throughout a watershed. Special summaries can also be generated using the VS or VV cards. The standard program summary shows the peak flow and accumulated drainage area for each hydrograph computation in the simulation. Time of peak and averaged flow values are presented along with stage information for reservoir routing. A sample summary is presented in Example 5.1 for a watershed with three subbasins. The user can choose time series data at selected stations to be displayed in tables, each containing up to ten columns. Rainfall, losses, stages, storages, and hydrographs can be printed in any desired order using the VS and VV cards. In general, the output format for HEC-1 is very easy to understand and can be made as detailed as necessary.

5.6
HEC-1 WATERSHED ANALYSIS: CASE STUDY

Watershed Description

The watershed selected for detailed study in this section is Keegans Bayou, a tributary to Brays Bayou in Houston, Texas. Brays Bayou watershed was chosen for a number of reasons. First, urban development since the mid-1970s has been significant and has created a severe flood problem in the downstream portion of the watershed. Second, Brays Bayou is one of the best-monitored hydrologic systems in the country, with four U.S. Geological Survey (USGS) stream gages and ten precipitation gages. The overall watershed area is 129 mi^2 and includes areas that are undeveloped as well as areas of full urbanization. Third, Brays Bayou has been the subject of several detailed hydrologic studies over the past ten years, representing a unique case study for the application of the HEC-1 model.

Keegans Bayou is located in the far upper end of Brays Bayou, as shown in Fig. 5.8, and drains an area of 17.9 mi^2. The watershed consists of five subareas, each with its own particular set of hydrologic parameters. The subarea divisions are based on topographic and land use data. Keegans Bayou ranges in percent development from 15% in subarea 1 to 70% in subarea 3. The subareas range in size from 2.48 to 4.68 mi^2.

Physical Parameters

Physical parameters for Keegans Bayou are presented in Table 5.10 for each of the subareas. These data were obtained from a detailed analysis of USGS topographic maps and current land use maps, with the watershed area delineated on each. Parameters of main interest include channel slopes, channel lengths, percent development, and percent conveyance, which is the ratio of flow in the channel to total flow. Figure 5.9 shows a map from which parameters are determined for a particular subarea.

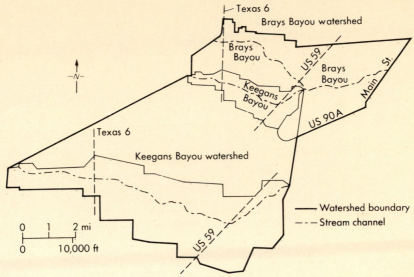

FIGURE 5.8

Location of Keegans Bayou within Brays Bayou watershed.

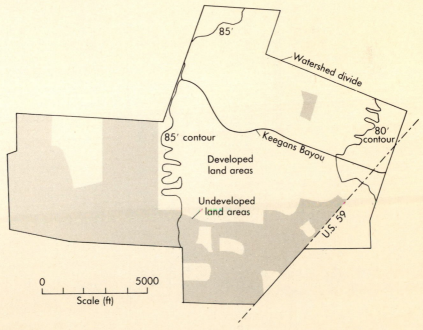

FIGURE 5.9

Subarea land use and topographic data.

TABLE 5.10

Physical Parameters for Keegans Bayou

SUBBASIN NUMBER	AREA (mi²)	SLOPE (ft/mi)	CHANNEL SLOPE (ft/mi)	CHANNEL LENGTH (mi)	LENGTH TO CENTRAL AREA (mi)	PERCENT DEVELOPMENT (%)	PERCENT CONVEYANCE (%)
1	3.50	2.35	10.0	2.936	1.586	15	100
2	4.68	2.35	10.0	3.220	1.776	60	100
3	4.00	4.30	10.0	2.214	1.468	70	100
4	2.48	3.30	10.0	2.379	1.361	60	80
5	3.26	3.30	10.0	2.178	0.923	60	80

Theoretically, the preceding physical parameters could be determined for any watershed given the existence of topographic maps and land use data.

Rainfall Data

The HEC-1 model can accept rainfall data in a number of ways: historical data from recording or nonrecording gages or design storm data. For the Keegans Bayou example, historical data from an April 1979 storm event is presented in Table 5.11 for the three recording rain gages and two stream gages shown in Fig. 5.10. The Thiessen polygons for rainfall weighting are also shown in Fig. 5.10 and Table 5.12 for the five subareas.

 The 10-yr and 100-yr synthetic design storms for the Houston area are depicted in Fig. 5.11. These distributions were derived using the National Weather Service distributions (TP-40) for 1-hr to 24-hr durations. The rainfalls were applied to Keegans Bayou uniformly for each subarea using the PH card in HEC-1. Thus, one can input any design rainfall into HEC-1, calculate the outflow, and compare the result with a given historical rainfall. In this way, one can determine the approximate frequency of occurrence of the historical storm. In practice, the 2-, 5-, 10-, 25-, 50-, 100-, and 500-yr rainfalls are evaluated in most flood studies. An alternative, continuous simulation approach for determination of flood frequency is shown in Chapter 6.

TABLE 5.11
April 1979 Cumulative Rainfall Data

TIME	GAGE NUMBERS		
(hr)	303R	4780	4800
1600	0.93	1.00	0.87
1630	0.95	1.00	0.88
1700	1.03	1.10	0.93
1730	1.06	1.10	0.95
1800	1.29	1.20	1.05
1830	1.94	1.40	1.13
1900	4.18	2.50	1.38
1930	4.32	2.90	1.88
2000	4.38	3.00	1.96
2030	4.54	3.10	2.25
2100	4.70	3.20	2.36
2130	4.76	3.40	2.58
2200	4.84	3.50	2.65
2230	4.88	3.50	2.69
2300	4.90	3.50	2.71

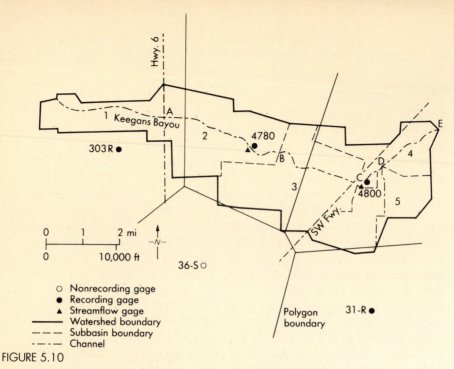

FIGURE 5.10

Location of gages and Thiessen polygons for Keegans Bayou.

Unit Hydrograph Data

Unit hydrograph methods were presented in Chapter 2. In this example the Clark ($TC \& R$) method and the kinematic wave method will be used to illustrate their applications to an urban watershed such as Keegans Bayou. The $TC \& R$ method is briefly summarized in Table 5.13 with empirical equations that have been developed for the Houston area.

TABLE 5.12

Precipitation Gage Weights

	GAGE NUMBERS		
SUBBASIN	303R	4780	4800
1	1.00	0.00	0.00
2	0.33	0.67	0.00
3	0.00	0.61	0.39
4	0.00	0.01	0.99
5	0.00	0.00	1.00

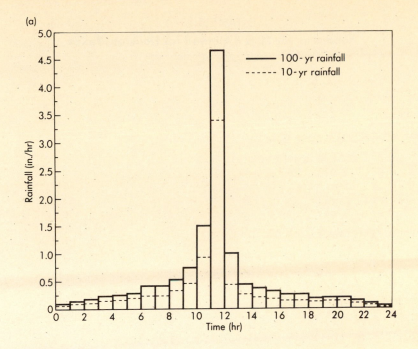

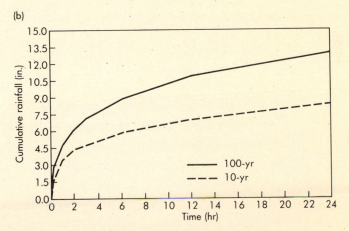

FIGURE 5.11

Design hyetographs for Houston, Texas.

These equations are based on the analysis of a number of watersheds along the Texas Gulf Coast (Harris County Flood Control District, 1983). Many similar equations have been generated for other regions in the country, and the user of HEC-1 is strongly urged to evaluate the $TC \& R$ method carefully before applying it to ungaged watersheds.

TABLE 5.13

Equations for TC & R Method for Houston, Texas

$$TC + R = C\left(\frac{L}{\sqrt{S}}\right)^{0.706},$$

where

$C = 4295[\% \text{ development}]^{-0.678}[\% \text{ Conveyance}]^{-0.967}$,

$C = 7.25$ if % development is ≤ 18.

$$TC = C'\left(\frac{L_{ca}}{\sqrt{S}}\right)^{1.06},$$

where C' is taken from the following:

S_0(ft/mi)	% DEVELOPMENT	C'
> 40	0	5.12
$20 < S \leq 40$	0	3.79
≤ 20	0	2.46
> 40	100	1.95
≤ 20	100	0.94

(If the percent development is between 0 and 100, C' is found by linear interpolation.)

$$R = (TC + R) - TC,$$

where

$$L = \text{length of channel (mi)},$$
$$L_{ca} = \text{length along channel to centroid of area (mi)},$$
$$S = \text{channel slope (ft/mi)},$$
$$S_0 = \text{representative overland slope (ft/mi)},$$
$$\% \text{ development} = \text{percent of area that is developed (\%)},$$
$$\% \text{ conveyance} = \text{ratio of flow in channel to total flow (\%)}$$
$$TC = \text{time of concentration (hr)},$$
$$R = \text{routing constant (hr)}.$$

The physical parameters presented in Table 5.10 for Keegans Bayou were used to develop the values for TC and R shown in Table 5.14. Generally, the higher the percent development and percent conveyance in a subarea, the lower the value of TC and R. Thus, subarea 1 has the highest value and subarea 3 has the lowest value of $TC + R$. Individual values for

TABLE 5.14

Clark Unit Hydrograph (*TC & R*) Modeling Parameters

SUBBASIN	$TC + R$	TC	R	% IMPERVIOUS	SCS CURVE NUMBER
1	10.272	2.314	7.958	0	80
2	7.126	1.809	5.317	16	80
3	3.891	0.968	2.923	14	80
4	4.675	1.140	3.535	13	80
5	4.393	0.755	3.638	13	80

each subarea are input to the HEC-1 model using the UC card group. Note also that the SCS curve number method was selected for infiltration losses and those parameters are input to the HEC-1 model on the LS card.

The input data for the kinematic wave method are somewhat more difficult since the method involves more parameters. The method for overland flow and collector channels was presented in detail in Chapter 4. Table 5.15 presents the overland length, slope, roughness, and contributing area for each subarea, divided into impervious and pervious sections. Note that the impervious channel lengths and roughnesses are much lower than their pervious counterparts. Table 5.15 also presents the data for collector channels and main channels for each subarea. Both trapezoidal and circular collectors are represented, with roughness coefficients ranging from 0.03 to 0.06. These low *n* values indicate the effects of channelization. Thus, the kinematic wave method requires the input of two tables of data compared with one table for the *TC & R* method. The resulting outflow hydrographs from the two methods will be compared to measured values later in this section.

Channel Storage Routing Data

The HEC-1 model converts unit hydrographs for each subarea into storm hydrographs through the process of hydrograph convolution as described in Chapter 2. For Keegans Bayou, the flood hydrograph from subarea 1 is computed for the outlet of that subarea, point *A* in Fig. 5.12. The storm hydrograph from subarea 2 is assumed to occur at point *B*, and the hydrograph from subarea 3 is assumed to occur at point *C*. It is therefore required that the hydrograph from subarea 1 at point *A* be routed through reach *AB* and then combined with the hydrograph from subarea 2 at point *B*. Likewise, the combined hydrograph at point *B* must be routed to point *C* and then combined with the hydrograph from subarea 3. Larger watersheds with more complicated channel networks can be represented in the HEC-1 model in a similar fashion.

Storage routing data for a watershed such as Keegans Bayou are sometimes difficult to obtain, especially if hydraulic calculations have not

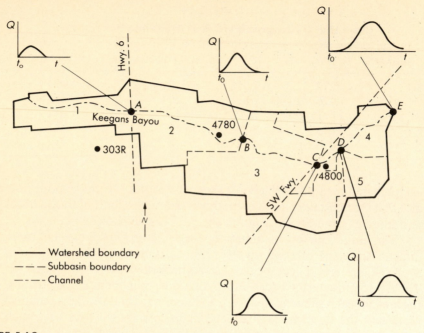

FIGURE 5.12

Computational points for hydrograph progression.

TABLE 5.15

Kinematic Wave Modeling Parameters

OVERLAND FLOW					
Subbasin Number	Type of Cover	Length (ft)	Slope (ft/ft)	Manning's n	Area (%)
1	Impervious	200	0.0005	0.17	19.4
	Pervious	3000	0.0005	0.40	80.6
2	Impervious	500	0.0007	0.26	62.1
	Pervious	3000	0.0007	0.40	37.9
3	Impervious	400	0.0007	0.19	69.9
	Pervious	4000	0.0007	0.40	30.1
4	Impervious	300	0.0010	0.19	61.0
	Pervious	2000	0.0010	0.35	39.0
5	Impervious	300	0.0010	0.19	60.0
	Pervious	2000	0.0010	0.35	40.0

(continued)

TABLE 5.15 *(continued)*

COLLECTOR/MAIN CHANNELS

Subbasin Number	Type Channel	Length (ft)	Slope (ft/ft)	n	Area (mi²)	Shape	BW (ft)	Side Slopes
1	Collector	1000	0.0017	0.060	0.17	TRAP	0	1
	Main	15500	0.0004	0.060		TRAP	22	5
2	Collector	7000	0.0017	0.060	0.38	TRAP	0	1
	Main	17000	0.0004	0.060		TRAP	27	4
3	Collector	6000	0.0008	0.030	0.35	TRAP	0	1
	Main	11700	0.0008	0.060		TRAP	25	4
4	Collector	1500	0.0006	0.030	0.50	CIRC	2	
	Collector	2000	0.0006	0.030	0.50	TRAP	8	2
	Main	13000	0.0006	0.060		TRAP	30	4
5	Collector	3500	0.0006	0.030	0.20	CIRC	4	
	Main	11500	0.0006	0.030		TRAP	12	2.5

been completed for the main channel. Often in well-studied basins, the HEC-2 model has been applied to develop water surface and hydraulic information (refer to Chapter 7). Computations with HEC-2 or a similar model allow the development of storage-discharge relations for each reach of the main channel *(AB, BC, CD, DE)* in Keegans Bayou. This information is then used in HEC-1 to route the hydrographs from one point to another in the main channel. Table 5.16 presents storage in ac-ft vs. discharge in cfs for the four reaches in Keegans Bayou. The routing data are entered into HEC-1 using the SV and SQ cards to develop the storage-discharge tables. The routing method is called the modified Puls (Chow, 1964), which is a simple storage routing method applicable to both channel and reservoir routing (Chapter 4).

Resulting Hydrographs

The HEC-1 model was run using data presented in Tables 5.10–5.16 and using two different unit hydrograph methods, the Clark *(TC & R)* and kinematic wave. Three different rainfalls were tested: the April 1979 storm, the 10-yr frequency design storm, and the 100-yr frequency design storm. The HEC-1 input data sets used in the April 1979 runs are shown for both methods in Tables 5.17 and 5.18. Resulting peak outflows for all three storms are listed in Table 5.19.

Figures 5.13 and 5.14 depict the hydrographs for the April 1979 storm at Roark Road (point *C*, Fig. 5.12) for both the Clark and kinematic wave methods. The measured hydrograph is plotted along with that computed by HEC-1 for the storm event. It can be seen that the Clark method tends to overpredict peak flows in this watershed for this storm while the kinematic wave method comes much closer to measured values. Both

TABLE 5.16

Channel Storage Routing Data*

| REACH $(A - B)$ | | REACH $(B - C)$ | | REACH $(C - D)$ | | REACH $(D - E)$ | |
S	Q	S	Q	S	Q	S	Q
0	0	0	0	0	0	0	0
202	2580	202	3580	67	4000	202	4000
359	3580	359	4580	73	5000	218	5000
541	4580	541	5580	104	6000	313	6000
763	5580	763	6580	146	7000	439	7000
978	6580	978	7580	233	8500	700	8500
1202	7975	1202	8975	362	10500	1086	10500

* *S* is in ac-ft and *Q* is in cfs.

TABLE 5.17

```
ID***                    HEC-1 EXAMPLE NUMBER 1
ID***                    KEEGANS BAYOU WATERSHED
ID***
ID***                    TC&R METHOD
ID***                    APRIL 1979 RECORDED STORM RAINFALL
ID***
IT      30 19APR79     1600        33
IN      30 19APR79     1600
IO       1
PG    4800
PC     .87      .88      .93      .95     1.05     1.13     1.38     1.88     1.96     2.25
PC    2.36     2.58     2.65     2.69     2.71
PG    303R
PC     .93      .95     1.03     1.06     1.29     1.94     4.18     4.32     4.38     4.54
PC    4.70     4.76     4.84     4.88     4.90
PG    4780
PC    1.00     1.00     1.10     1.10     1.20     1.40     2.50     2.90     3.00     3.10
PC    3.20     3.40     3.50     3.50     3.50
KK    SUB1
KM       RUNOFF FROM SUBBASIN 1
PT    303R
PW       1
PR    303R
PW       1
BA    3.50
LS              80
UC    2.314    7.958
KK    A - B
KM       CHANNEL STORAGE ROUTING FROM A TO B
RS       1      STOR      -1
SV       0      202      359      541      763      978     1202
SQ       0     2580     3580     4580     5580     6580     7975
KK    SUB2
KM       RUNOFF FROM SUBBASIN 2
PT    4780     303R
PW     .67      .33
PR    4780
PW     .67      .33
BA    4.68
LS              80       16
UC    1.809    5.317
KK    B
KM       COMBINE SUB2 AND ROUTED POINT B
HC       2
KK    B - C
KM       CHANNEL STORAGE ROUTING FROM B TO C
RS       1      STOR      -1
SV       0      202      359      541      763      978     1202
SQ       0     3580     4580     5580     6580     7580     8975
KK    SUB3
KM       RUNOFF FROM SUBBASIN 3
PT    4780     4800
PW     .61      .39
PR    4780     4800
```

(continued)

TABLE 5.17 *(continued)*

```
PW    .61      .39
BA   4.00
LS             80        14
UC  0.968    2.923
KK    C
KM        COMBINE FLOW FROM SUB3 AND ROUTED POINT B
HC     2
KK  ROARK
KO     1
KM         MEASURED AT ROARK
IN    30  19APR79     1600
QO    10       9        9       11       14       20       29       59      199      433
QO   717     865      992     1039     1070     1020      938      767      700      642
QO   510     450      400      380      375      360      345      330      315      303
QO   300     295      295
KK  C - D
KM        CHANNEL STORAGE ROUTING FROM C TO D
RS     2     STOR      -1
SV     0      67       73      104      146      233      362
SQ     0     4000     5000     6000     7000     8500    10500
KK  SUB5
KM       RUNOFF FROM SUBBASIN 5
PT  4800
PW     1
PR  4800
PW     1
BA   3.26
LS             80        13
UC  0.755    3.638
KK    D
KM        COMBINE SUB5 AND ROUTED POINT D
HC     2
KK  D - E
KM        CHANNEL STORAGE ROUTING FROM D TO E
RS     2     STOR      -1
SV     0     202      218      313      439      700     1086
SQ     0     4000     5000     6000     7000     8500    10500
KK  SUB4
KM       RUNOFF FROM SUBBASIN 4
PT  4780     4800
PW   .01      .99
PR  4780     4800
PW   .01      .99
BA   2.48
LS             80        13
UC  1.140    3.535
KK  MOUTH
KM        COMBINE FLOW FROM SUB4 AND ROUTED POINT E AT MOUTH
HC     2
ZZ
```

TABLE 5.18

```
ID***                       HEC-1 EXAMPLE NUMBER 1
ID***                       KEEGANS BAYOU WATERSHED
ID***
ID***                       KINEMATIC WAVE METHOD
ID***                       APRIL 1979 RECORDED STORM RAINFALL
ID***
IT      30 19APR79     1600        33
IN      30 19APR79     1600
IO       O         2
PG    4800
PC     .87      .88      .93      .95     1.05     1.13     1.38     1.88     1.96     2.25
PC    2.36     2.58     2.65     2.69     2.71     2.71     2.71     2.71     2.71     2.71
PG    303R
PC     .93      .95     1.03     1.06     1.29     1.94     4.18     4.32     4.38     4.54
PC    4.70     4.76     4.84     4.88     4.90     4.90     4.90     4.90     4.90     4.90
PG    4780
PC    1.00     1.00     1.10     1.10     1.20     1.40     2.50     2.90     3.00     3.10
PC    3.20     3.40     3.50     3.50     3.50     3.50     3.50     3.50     3.50     3.50
KK    SUB1
KM          RUNOFF FROM SUBBASIN 1
PT    303R
PW       1
PR    303R
PW       1
BA    3.50
LS              90                        80
UK     200  0.0005      .17     19.4
UK    3000  0.0005      .40     80.6
RK    1000  0.0017     .060      .17     TRAP        O        1
RK   15500  0.0004     .060              TRAP       22        5
KK    A - B
KM          CHANNEL STORAGE ROUTING FROM A TO B
RS       1      STOR       -1
SV       O       202      359      541      763      978     1202
SQ       O      2580     3580     4580     5580     6580     7975
KK    SUB2
KM          RUNOFF FROM SUBBASIN 2
PT    4780      303R
PW     .67       .33
PR    4780
PW     .67       .33
BA    4.68
LS              88                        80
UK     500  0.0005      .26     62.1
UK    3000  0.0005      .40     37.9
RK    7000  0.0017     .060      .38     TRAP        O        1
RK   17000  0.0004     .060              TRAP       27        4
KK       B
KM          COMBINE SUB2 AND ROUTED POINT B
HC       2
KK    B - C
KM          CHANNEL STORAGE ROUTING FROM B TO C
RS       1      STOR       -1
SV       O       202      359      541      763      978     1202
```

(continued)

TABLE 5.18 (continued)

```
SQ      O     3580    4580    5580    6580    7580    8975
KK    SUB3
KM         RUNOFF FROM SUBBASIN 3
PT    4780    4800
PW     .61     .39
PR    4780    4800
PW     .61     .39
BA    4.00
LS             88                    80
UK    400   0.0007    .19    69.9
UK    4000  0.0007    .40    30.1
RK    6000  0.0008   .030    .35    TRAP      O       1
RK   11700  0.0008   .060           TRAP     25       4
KK    C
KM         COMBINE FLOW FROM SUB3 AND ROUTED POINT C
HC      2
KKROARK
KM         MEASURED AT ROARK
IN    30  19APR79   1600
QO    10       9       9      11      14      20      29      59     199     433
QO    717    865     992    1039    1070    1020     938     767     700     642
QO    510    450     400     380     375     360     345     330     315     303
QO    300    295     295
KK  C - D
KM         CHANNEL STORAGE ROUTING FROM C TO D
RS      1     STOR     -1
SV      O      67      73     104     146     233     362
SQ      O    4000    5000    6000    7000    8500   10500
KK    SUB5
KM         RUNOFF FROM SUBBASIN 5
PT    4800
PW      1
PR    4800
PW      1
BA    3.26
LS             91                    80
UK    300   0.0010    .19    60.0
UK    2000  0.0010    .35    40.0
RK    3500  0.0006   .030    0.20   CIRC      4
RK   11500  0.0006   .030           TRAP     12     2.5
KK    D
KM         COMBINE SUB5 AND ROUTED POINT D
HC      2
KK  D - E
KM         CHANNEL STORAGE ROUTING FROM D TO E
RS      2     STOR     -1
SV      O     202     218     313     439     700    1086
SQ      O    4000    5000    6000    7000    8500   10500
KK    SUB4
KM         RUNOFF FROM SUBBASIN 4
PT    4780    4800
PW     .01     .99
PR    4780    4800
PW     .01     .99
```

(continued)

TABLE 5.18 *(continued)*

BA	2.48						
LS		91			80		
UK	300	0.0010	.19	61.0			
UK	2000	0.0010	.35	39.0			
RK	1500	0.0006	.030	0.50	CIRC	2	
RK	2000	0.0006	.030	0.50	TRAP	8	2
RK	12560	0.0006	.060		TRAP	30	4
KK	MOUTH						
KM		COMBINE FLOW FROM SUB4 AND ROUTED POINT E AT MOUTH					
HC	2						
ZZ							

methods predict time to peak t_p very well, while the KW method better matches the volume of runoff (area under the hydrograph). Table 5.20 shows a detailed HEC-1 comparison of the computed vs. measured hydrographs at Roark Road for the kinematic wave method for the April 1979 storm. In general, it is always useful to check three parameters against the measured gage data in calibrating HEC-1 input: peak flow, volume under the hydrograph, and time to peak. Some adjustment of parameters may be necessary to calibrate the model fully.

Once the HEC-1 input data are calibrated for several measured storm events, predictions for the 10-yr, 100-yr, or any other design storm can be made by changing rainfall input (see Chapter 3). The resulting 10-yr and 100-yr design storm hydrographs then provide an estimate of peak flows at various points along the main channel. These flows can be used with hydraulic computations to derive elevations for the 10-yr and 100-yr floodplain as described in Chapter 7.

TABLE 5.19

Comparison of Peak Flows (cfs)

		PEAK FLOW AT	
STORM	MODEL	Roark Road (Point C)	Mouth (Point E)
April 1979	Kinematic wave	976	1,394
	TC & R	1,223	1,553
10-yr	Kinematic wave	5,412	8,219
	TC & R	4,772	7,383
100-yr	Kinematic wave	10,184	11,227
	TC & R	7,475	10,611

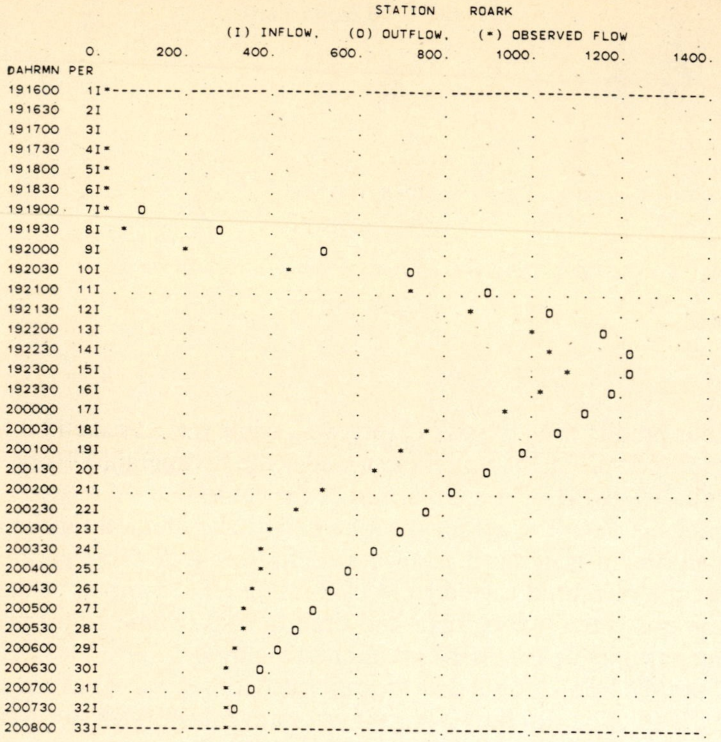

```
                                         STATION      ROARK
                              (I) INFLOW,    (O) OUTFLOW,   (*) OBSERVED FLOW
                  0.        200.       400.        600.       800.       1000.       1200.      1400.
       DAHRMN PER
       191600   1I*--------- ---------- ---------- ---------- ---------- ---------- ----------.
       191630   2I           .                .         .           .          .          .
       191700   3I           .                .         .           .          .          .
       191730   4I*          .                .         .           .          .          .
       191800   5I*          . .              . .       .           .          .          .
       191830   6I*          .                .         .           .          .          .
       191900   7I*   O      .                .         .           .          .          .
       191930   8I   *       .    O           .         .           .          .          .
       192000   9I           *            .   O         .           .          .          .
       192030  10I           .            . *           O  .        .          .          .
       192100  11I  . . . . . . . . . . . . . . . . . * . . . .O. . . . . . . . . . . . . . .
       192130  12I           .                .         .   *       .  O       .          .
       192200  13I           .                .         .           *          O  .       .
       192230  14I           .                .         .           . *          .O        .
       192300  15I           .                .         .           .   *        .O        .
       192330  16I           .                .         .           .  *        O.         .
       200000  17I           .                .         .         *          O          .
       200030  18I           .                .         .      *          . . O          .
       200100  19I           .                .       . *          O.          .          .
       200130  20I           .                .      . *        O          .          .
       200200  21I  . . . . . . . . . . . . . * . . . . . .O. . . . . . . . . . . . . . . . .
       200230  22I           .            . *        .     O     .          .          .
       200300  23I           .            *          .   O       .          .          .
       200330  24I           .          *.         . O         .          .          .
       200400  25I           .          *.       O.          .          .          .
       200430  26I           .       *. .        O  .         .          .          .
       200500  27I           .     *.         O   .          .          .          .
       200530  28I           .     *.  . O        .          .          .          .
       200600  29I           .   *.   .O         .          .          .          .
       200630  30I           .  *    O.          .          .          .          .
       200700  31I  . . . . . . . . *. . O . . . . . . . . . . . . . . . . . . . . . . . . . .
       200730  32I           .    *O    .         .          .          .          .
       200800  33I--------- ----*----- ---------- ---------- ---------- ---------- ----------.
```

FIGURE 5.13

**Computed vs. measured hydrographs, April 1979 storm: *TC & R*
method.**

Sources of error in the calibration of HEC-1 to a watershed include
the following:

1. Spatially variable input rainfall with too few rain gages across the
   watershed
2. Variable channel conditions and inaccurate estimates of routing coefficients
3. Infiltration or loss inaccuracies
4. Nonlinear hydrographs that cannot be properly represented by linear
   unit hydrograph theory.

Once a watershed has been evaluated under existing conditions of
land use and channel physiography, the HEC-1 model can be used to test
flood control plans that might include a reservoir or detention pond or
some diversion of part of the flow to another watershed.

TABLE 5.20

COMPARISON OF COMPUTED AND OBSERVED HYDROGRAPHS

	SUM OF FLOWS	EQUIV DEPTH	MEAN FLOW	TIME TO CENTER OF MASS	LAG C.M. TO C.M.	PEAK FLOW	TIME OF PEAK
PRECIPITATION EXCESS		1.029		4.66			
COMPUTED HYDROGRAPH	13551.	0.862	411.	9.46	4.80	976.	7.50
OBSERVED HYDROGRAPH	14201.	0.903	430.	9.29	4.64	1070.	7.50
DIFFERENCE	-650.	-0.041	-20.	0.16	0.16	-94.	0.0
PERCENT DIFFERENCE	-4.58				3.52	-8.76	

	STANDARD ERROR	56.	44.	AVERAGE ABSOLUTE ERROR
	OBJECTIVE FUNCTION	64.	28.74	AVERAGE PERCENT ABSOLUTE ERROR

HYDROGRAPH AT STATION   ROARK

DA MON HRMN ORD	COMP Q	OBS Q	RESIDUL
19 APR 1600 1	0.	10.	-10.
19 APR 1630 2	0.	9.	-9.
19 APR 1700 3	0.	9.	-9.
19 APR 1730 4	0.	11.	-11.
19 APR 1800 5	0.	14.	-14.
19 APR 1830 6	0.	20.	-20.
19 APR 1900 7	7.	29.	-22.
19 APR 1930 8	51.	59.	-8.
19 APR 2000 9	168.	199.	-31.
19 APR 2030 10	355.	433.	-78.
19 APR 2100 11	571.	717.	-146.

DA MON HRMN ORD	COMP Q	OBS Q	RESIDUL
19 APR 2130 12	764.	865.	-101.
19 APR 2200 13	902.	992.	-90.
19 APR 2230 14	970.	1039.	-69.
19 APR 2300 15	976.	1070.	-94.
19 APR 2330 16	940.	1020.	-80.
20 APR 0000 17	877.	938.	-61.
20 APR 0030 18	804.	767.	37.
20 APR 0100 19	728.	700.	28.
20 APR 0130 20	655.	642.	13.
20 APR 0200 21	589.	510.	79.
20 APR 0230 22	531.	450.	81.

DA MON HRMN ORD	COMP Q	OBS Q	RESIDUL
20 APR 0300 23	480.	400.	80.
20 APR 0330 24	436.	380.	56.
20 APR 0400 25	398.	375.	23.
20 APR 0430 26	366.	360.	6.
20 APR 0500 27	339.	345.	-6.
20 APR 0530 28	315.	330.	-15.
20 APR 0600 29	296.	315.	-20.
20 APR 0630 30	278.	303.	-25.
20 APR 0700 31	264.	300.	-36.
20 APR 0730 32	251.	295.	-44.
20 APR 0800 33	240.	295.	-55.

```
                                    STATION    ROARK
                           (I) INFLOW,    (O) OUTFLOW,    (*) OBSERVED FLOW
                0.          200.       400.       600.       800.      1000.      1200.
        DAHRMN PER
        191600    1I*--------.---------.---------.---------.---------.---------.
        191630    2I         .         .         .         .         .         .
        191700    3I         .         .         .         .         .         .
        191730    4I*        .         .         .         .         .         .
        191800    5I*        .         .         .         .         .         .
        191830    6I*        .         .         .         .         .         .
        191900    7I*        .         .         .         .         .         .
        191930    8I   *     .         .         .         .         .         .
        192000    9I        O  *       .         .         .         .         .
        192030   10I         .   O . * .         .         .         .         .
        192100   11I  . . . . . . . . ..O. . . * .         .         .         .
        192130   12I         .         .         .O  .  *  .         .         .
        192200   13I         .         .         .   O . * .         .         .
        192230   14I         .         .         .      O . *        .         .
        192300   15I         .         .         .        O.     *   .         .
        192330   16I         .         .         .        O  .*      .         .
        200000   17I         .         .         .      O  *  .      .         .
        200030   18I         .         .         .    *  O     .     .         .
        200100   19I         .         .         . *O       .        .         .
        200130   20I         .         .       .*O         .         .         .
        200200   21I  . . . . . . . . . .*  .O. . . . . . .         .         .
        200230   22I         .         .    *   O .         .         .         .
        200300   23I         .         .  *    O  .         .         .         .
        200330   24I         .         .*.  O   .         .         .         .
        200400   25I         .         *O       .         .         .         .
        200430   26I         .       * .         .         .         .         .
        200500   27I         .         *         .         .         .         .
        200530   28I         .    O*   .         .         .         .         .
        200600   29I         .   O*    .         .         .         .         .
        200630   30I         .  O*     .         .         .         .         .
        200700   31I  . . . . .O.*. . .         .         .         .         .
        200730   32I         . O *     .         .         .         .         .
        200800   33I---------.-O--*----.---------.---------.---------.---------.
```

FIGURE 5.14

Computed vs. measured hydrographs, April 1979 storm: kinematic wave method.

Flood Control Alternatives

One of the most powerful features of the HEC-1 model is its ability to represent changes in watershed physiography or land use that may occur as a watershed develops. For Keegans Bayou, we will consider two examples: (1) a large detention pond for the storage of flood waters in subarea 5 and (2) the diversion of a portion of the flow at point C into a channel that leads to another watershed. In both cases, the watershed will be assumed to have existing land use.

To input a detention pond into HEC-1, the storage routing data must be changed to represent the storage-discharge relation for the particular detention pond. The pond data are located in the input through logical placement so that all contributing hydrographs route through the pond. The starting water elevation in the pond depends on the operation of the system. The discharge characteristics can be considered in detail through options available in the model, but generally pipe flow is assumed for low flows, and weir flow is assumed for flood levels. Details on equations that govern such flows are contained in Chapter 7.

TABLE 5.21

**S-Q Relation for Ponds
Used in Keegans Bayou**

STORAGE (ac-ft)	DISCHARGE (cfs)
0	0
330	481
1,761	1,107
2,493	1,317
3,269	1,507
5,409	1,997
5,591	2,336
5,771	2,815
6,000	3,569
6,000	13,000

The characteristics of the detention pond for Keegans Bayou are depicted in Table 5.21. The storage-discharge data are based on pond geometry, which is a function of available land area and available outfall depth. The existing 100-yr peak flow in Keegans Bayou is 13,240 cfs at point D and 11,227 cfs at the mouth. To control the peak flows at the mouth of the watershed, one pond was located at point B and a second pond was located just upstream of point D along the stream (Fig. 5.15).

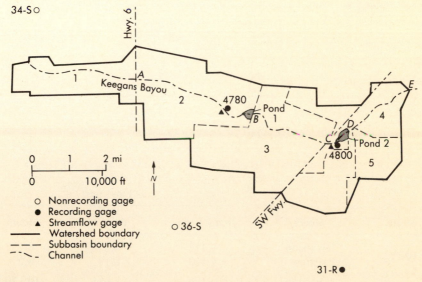

FIGURE 5.15

Location of detention ponds in Keegans Bayou.

Table 5.22 shows the relative effects of these two ponds compared with existing conditions (case 1). Case 2 shows the effect of moving the second pond downstream of point D. Case 1 yields a 13% reduction in peak flow while case 2 produces a 52% reduction in peak flow (Fig. 5.16), indicating the importance of proper pond location.

TABLE 5.22

Comparison of Peak Flows for Various Conditions

		PEAK FLOW IN CFS AT			PERCENT REDUCTION (%)
MODEL	CONDITIONS	Point D	Pond 2	Mouth	
Existing	No ponds	13,240	–	11,227	–
Case 1	Pond 1 at B Pond 2 before D	6,461	1,238	9,783	13
Case 2	Pond 1 at B Pond 2 after D	12,385	1,547	5,346	52

TABLE 5.23

Diversion at Roark Road

INFLOW (cfs)	DIVERSION (cfs)
0	0
500	375
1,000	750
2,500	1,875
5,000	3,750
10,000	7,500

COMPARISON OF RESULTS USING THE KINEMATIC WAVE METHOD

		Peak Flows (cfs)		
Storm	Model	Roark Road	Point D	Mouth
10-yr	Diversion	5,412	4,876	6,457
	Existing	5,412	8,197	8,219
100-yr	Diversion	10,184	8,081	10,170
	Existing	10,184	13,240	11,227

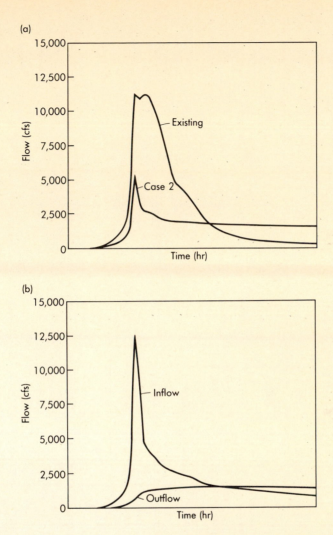

FIGURE 5.16

(a) Effect of ponds on peak flow at mouth for 100-yr design rainfall. (b) Inflow and outflow for pond 2, 100-yr design rainfall.

A second alternative for flood control is diversion of flood runoff to another watershed. This is accomplished in HEC-1 via the divert cards, which allow the user to define a flow level above which a certain percentage will be diverted from the main channel. While this option implies that there must be an adjacent watershed able and willing to receive the excess flood water, it can be a viable alternative in some cases. Application to Keegans Bayou involved diverting from point *C* and observing the effect on the resulting hydrograph at the watershed outlet. Table 5.23 presents

TABLE 5.24
Fully Developed Conditions: Kinematic Wave Model

RUNOFF SUMMARY
FLOW IN CUBIC FEET PER SECOND
TIME IN HOURS, AREA IN SQUARE MILES

OPERATION	STATION	PEAK FLOW	TIME OF PEAK	AVERAGE FLOW FOR MAXIMUM PERIOD			BASIN AREA	MAXIMUM STAGE
				6-HOUR	24-HOUR	72-HOUR		
HYDROGRAPH AT	SUB1	8689.	13.00	3018.	1040.	511.	3.50	
ROUTED TO	A - B	4400.	13.50	2933.	1039.	511.	3.50	
HYDROGRAPH AT	SUB2	8442.	13.50	3836.	1333.	660.	4.68	
2 COMBINED AT	B	12842.	13.50	6769.	2369.	1171.	8.18	
ROUTED TO	B - C	7799.	15.00	6288.	2367.	1171.	8.18	
HYDROGRAPH AT	SUB3	10391.	13.00	3498.	1182.	580.	4.00	
2 COMBINED AT	C	15124.	13.00	9473.	3545.	1751.	12.18	
ROUTED TO	C - D	12046.	13.50	9397.	3545.	1751.	12.18	
HYDROGRAPH AT	SUB5	8686.	12.50	2953.	1014.	496.	3.26	
2 COMBINED AT	D	18436.	13.00	12066.	4545.	2247.	15.44	
ROUTED TO	D - E	12782.	15.50	11254.	4544.	2247.	15.44	
HYDROGRAPH AT	SUB4	6660.	13.00	2215.	757.	372.	2.48	
2 COMBINED AT	MOUTH	14357.	13.00	12862.	5298.	2619.	17.92	

*** NORMAL END OF JOB ***

these results along with the divert inflow-outflow relationship. Once again, if the resulting peak flows are too high at the outlet, more water must be diverted.

One final exercise with the HEC-1 model application to Keegans Bayou is a consideration of full development conditions in the watershed. We assume full urbanization, which will require significant changes in unit hydrograph and loss rate parameters. There will also be associated channelization of laterals and the main stream, which will require changes in the routing parameters. As an illustration of these changes, Table 5.24 presents results of the fully developed computer run for Keegans Bayou, and it should be obvious that such a condition will further worsen the flood problem downstream. Peak flows have been increased by a factor of 28% at the mouth for the 100-yr storm. Thus, some measure of flood control will probably be required before the watershed is allowed to fully develop.

SUMMARY

Simulation models in hydrology have been developed and applied over the last two decades with remarkable success. These models incorporate various equations to describe hydrologic processes in space and time. Complex rainfall patterns affecting large watersheds can be simulated, and computer models can be used to test various design and control schemes. A selected number of the most popular rainfall-runoff models for hydrologic simulation are contained in Table 5.4. The Stanford Watershed Model, HEC-1, SCS TR-20, HSPF, SWMM, and ILLUDAS are some of the most popular and comprehensive models for watershed analysis.

The second half of the chapter is devoted to a detailed presentation of HEC-1 by the U.S. Army Corps of Engineers, considered by many to be the most versatile and most often used hydrologic model. HEC-1 is a general flood hydrograph package for handling watershed runoff given historical or design rainfall.

Chapter 5 concludes with a detailed case study and application of HEC-1 to an urban watershed. Two different models are developed, one using Clark unit hydrograph methods and one using kinematic wave methods, and the results from each are compared to observed flow data. Then several flood control alternatives are modeled, including flow diversion and detention ponds, to indicate the overall capabilities of the HEC-1 program.

PROBLEMS

Problems 5.1–5.9 deal with the hypothetical watershed with the following parameters. Problems 5.1–5.5 involve input data setup only, while Problems 5.6–5.9 require actual HEC-1 program computations.

SUBBASIN NUMBER	AREA (ac)	L (ft)	L_{ca} (ft)	S (ft/mi)	PERCENT DEVELOPMENT (%)	PERCENT CONVEYANCE (%)
1	666	5,400	3,000	20	30	80
2	1,485	9,000	5,050	10	28	80
3	1,517	8,750	4,975	10	75	100
4	2,752	10,000	6,340	10	80	100

Assume $S = S_0$ in each subbasin. The infiltration data are as follows:

SUBBASIN NUMBER	PERCENT IMPERVIOUS (%)	SCS CURVE NUMBER
1	20	78
2	18	80
3	52	85
4	60	90

A schematic representation of the watershed is shown in Fig. P5.1.

5.1. Calculate the TC and R coefficients for each subbasin in the given watershed. Assume that the equations listed in Table 5.13 are applicable and that $S = S_0$.

5.2. Using the data shown in Fig. 5.11, set up the input data lines for the 24-hr hypothetical storms for the 100-yr frequency and the 10-yr frequency.

5.3. Set up a HEC-1 input data set for the watershed using Clark's unit

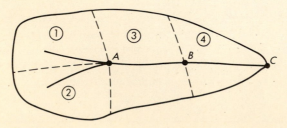

FIGURE P5.1

hydrograph method (*TC* & *R*) and the 100-yr design storm. Intermediate summaries and hydrographs should be printed as well as the
master summary. Use Muskingum routing methods where necessary.
Assume that $K = 0.60$ hr and $x = 0.4$ in reaches AB and BC, with
only one subreach (field 1 of the RM card).

5.4. Revise the HEC-1 input data set of Problem 5.3 to reflect a detention
pond near point B. Assume that the pond can be modeled in such a
way that only the runoff from subbasin 3 should be routed through it.
The storage-discharge relation is given as follows:

S (ac-ft)	0	55	294	416	550	550
Q (cfs)	0	120	277	329	377	5000

5.5. The watershed is predicted to grow in the next 20 to 30 yr and perhaps
reach 100% development in that time. If this occurs, certain parameters will change, as shown in the following table. Compute the new *TC*
and *R* coefficients and change the input data set of Problem 5.3 to
reflect the fully developed conditions. Assume that the Muskingum
values used in Problem 5.3 will still be correct.

SUBBASIN	PERCENT DEVELOPMENT (%)	PERCENT CONVEYANCE (%)	PERCENT IMPERVIOUS (%)	SCS CURVE NUMBER
1	75	100	38	83
2	75	100	30	81
3	100	100	85	94
4	100	100	85	94

5.6. Run the HEC-1 model to reflect the existing conditions, the effect of
the pond with existing conditions, fully developed conditions, and
the effect of the pond with fully developed conditions for both the
100-yr and the 10-yr design storms. Compare the peak flows at points
B and C with the values listed below.

CONDITIONS MODELED	STORM MODELED	PEAK FLOW IN cfs AT	
		Point B	Point C
Existing	100-yr	6,715	12,643
	10-yr	4,304	8,271
Existing with pond	100-yr	6,715	10,995
	10-yr	2,325	5,918
Fully developed	100-yr	9,493	15,185
	10-yr	6,141	9,921
Fully developed with pond	100-yr	9,493	13,672
	10-yr	3,820	7,369

5.7. Develop the necessary storage-discharge relation for a single pond to be placed at point A so that the 100-yr flow at point C is reduced to 75% of the existing 100-yr flow. Assume existing development conditions and place the pond so that all of the flow from upstream of point A must flow through the pond.

5.8. Repeat Problem 5.7, placing the pond so that all the flow above point B, including runoff from subbasin 3, is accepted into the pond. Compare the resulting peak flows at points B and C and discuss the differences. Use storage in the pond of Problem (5.4) times 5.

5.9. For the Muskingum coefficients given in Problem 5.3, determine the dimensions of a wide rectangular channel needed to carry the 100-yr flow under existing development conditions. Use a Manning's n value of 0.04.

REFERENCES

Aron, G., and D. F. Lakatos, 1976, *Penn State Urban Runoff Model: User's Manual,* Res. Pub. 96, Institute for Research on Land and Water Resources, Pennsylvania State University, University Park, Pennsylvania.

Bras, R. L., and I. Rodriguez-Iturbe, 1984, *Random Functions and Hydrology,* Addison-Wesley Publishing Co., Reading, Massachusetts.

Bravo, C. A., M. H. Harley, F. E. Perkins, and P. S. Eagleson, 1970, *A Linear Distributed Model of Catchment Runoff,* Rep. No. 123, Hydrodynamics Laboratory, Massachusetts Institute of Technology, Cambridge, Massachusetts.

Chiang, C. Y., and P. B. Bedient, 1986, "PIBS Model for Surcharged Pipe Flow," *ASCE J. Hyd. Eng. Div.,* vol. 112, no. 3, pp. 181–192.

Chow, V. T. (editor), 1964, *Handbook of Applied Hydrology,* McGraw-Hill Book Company, New York.

Clark, C. O., 1945, "Storage and the Unit Hydrograph," *Trans. ASCE,* vol. 110, pp. 1419–1446.

Crawford, N. H., and R. K. Linsley, 1966, *Digital Simulation in Hydrology, Stanford Watershed Model IV,* Tech. Rep. 39, Civil Engineering Dept., Stanford University, Stanford, California.

Delleur, J. W., and S. A. Dendrou, 1980, "Modeling the Runoff Process in Urban Areas," *CRC Crit. Rev. Environ. Control,* vol. 10, pp. 1–64.

Donigian, A. S., Jr., and N. H. Crawford, 1976, *Modeling Nonpoint Pollution from the Land Surface,* EPA-600/3-76-083, U.S. Environmental Protection Agency, Athens, Georgia.

▲ Kissimmee River Floodplain in
South Florida
◄ Athabasca River in the Canadian
Rockies (Morton Photo/Graphics)

▲ Martin's Brook near
   N. Reading, Massachusetts
   (Morton Photo/Graphics)
◄ Guadalupe River in Central
   Texas

▲ Control Gates for Large Water
  Supply and Recreational
  Reservoir
▶ Mill Stream in Massachusetts
  (Morton Photo/Graphics)

▲ Concrete-lined Channel for Flood Control
▼ Irrigation Canal in Texas

► Stormwater Inlet in
Residential Area
▼ Roadside Swale and Culvert,
Woodlands, Texas

▲ Ipswich Floodplain in N. Reading, Massachusetts (Morton Photo/Graphics)
▼ Flooded Texas Medical Center Parking Garage (1976)

Texas
Medical
Center   Garage No 3

▲ Flooded Commuter Parking Lot along Brays Bayou in Houston, Texas (1983)
▼ Residential Houses Located in a Floodplain

▲ U.S. Geological Survey Stream Gaging Station with Telemetry
▼ U.S. EPA R. S. Kerr Environmental Research Laboratory Drilling Rig for Ground Water Monitoring Wells

Donigian, A. S., Jr., and N. H. Crawford, 1979, *User's Manual for the Nonpoint Source (NPS) Model*, Environ. Res. Info. Center, U.S. Environmental Protection Agency, Cincinnati, Ohio.

Feldman, A. D., 1981, "HEC Models for Water Resources System Simulation: Theory and Experience," *Advances in Hydroscience*, vol. 12, pp. 297–423, Academic Press, New York.

Fleming, G., 1975, *Computer Simulation Techniques in Hydrology*, American Elsevier Publishing Co., New York.

Harley, M. H., F. E. Perkins, and P. S. Eagleson, 1970, *A Modular Distributed Model of Catchment Dynamics*, Rep. No. 133, R. M. Parsons Laboratory, Massachusetts Institute of Technology, Cambridge, Massachusetts.

Harley, M. H., 1975, *MITCAT Catchment Simulation Model: Description and User's Manual Version 6*, Resource Analysis Corporation, Massachusetts.

Harris County Flood Control District, 1983, *Flood Hazard Study of Harris County Hydrologic Methodology*, Houston, Texas.

Holtan, H. N., G. J. Stiltner, W. H. Henson, and N. C. Lopez, 1975, *USDAHL-74 Revised Model of Watershed Hydrology*, Tech. Bull. No. 1518, Agricultural Research Service, U.S. Department of Agriculture, Washington, D.C.

Huber, W. C., J. P. Heaney, S. J. Nix, R. E. Dickinson, and D. J. Polmann, 1981, *Storm Water Management Model User's Manual, Version III*, EPA-600/2-84-109a (NTIS PB84-198423), U.S. Environmental Protection Agency, Athens, Georgia.

Hydrologic Engineering Center, 1975, *Urban Storm Water Runoff: STORM, Generalized Computer Program 723-58-L2520*, U.S. Army Corps of Engineers, Davis, California.

Hydrologic Engineering Center, 1979, *Introduction and Application of Kinematic Wave Routing Techniques Using HEC-1*, Training Doc. No. 10, U.S. Army Corps of Engineers, Davis, California.

Hydrologic Engineering Center, 1980, *Hydrologic Analysis of Ungaged Watersheds with HEC-1 (preliminary)*, U.S. Army Corps of Engineers, Davis, California.

Hydrologic Engineering Center, 1981, *HEC-1 Flood Hydrograph Package: User's Manual* and *Programmer's Manual*, updated 1985, U.S. Army Corps of Engineers, Davis, California.

Johanson, R. C., J. C. Imhoff, and H. H. Davis, 1980, *User's Manual for Hydrological Simulation Program-FORTRAN (HSPF)*, EPA-600/9-80-015, U.S. EPA, Athens, Georgia.

Kibler, D. F. (editor), 1982, *Urban Stormwater Hydrology*, Water Resources Monograph 7, American Geophys. Union, Washington, D.C.

Kibler, D. F., and G. Aron, 1978, "Effect of Parameter Sensitivity and Model Structure in Urban Runoff Simulation," *Proc. Int. Symp. Storm Water Management,* University of Kentucky, Lexington, Kentucky.

McCuen, R. H., 1982, *A Guide to Hydrologic Analysis Using SCS Methods,* Prentice-Hall, Englewood Cliffs, New Jersey.

McPherson, M. B., 1975, *Urban Mathematical Modeling and Catchment Research in the U.S.A.,* ASCE Urban Water Resources Research Program Tech. Memo. No. IHP-1, NTIS No. PB 260 685.

Metcalf & Eddy, Inc., University of Florida, Gainesville, and Water Resources Engineers, Inc., 1971, *Storm Water Management Model, for Environmental Protection Agency,* 4 Volumes, EPA Rep. Nos. 11024DOC07/71, 11024DOC08/71, 11024DOC09/71, and 11024DOC10/71.

Roesner, L. A., H. M. Nichandros, R. P. Shubinski, A. D. Feldman, J. W. Abbott, and A. O. Friedland, 1974, *A Model for Evaluating Runoff Quality in Metropolitan Master Planning,* ASCE Urban Water Resources Research Program Technical Memorandum No. 23, NTIS PB-234312, ASCE, New York.

Roesner, L. A., R. P. Shubinski, and J. A. Aldrich, 1981, *Storm Water Management Model User's Manual Version III: Addendum I, EX-TRAN,* EPA-600/2-84-109b (NTIS PB84-198431), U.S. EPA, Cincinnati, Ohio.

Snyder, F. F., 1938, "Synthetic Unit Hydrographs," *Trans. Am. Geophys. Union,* vol. 19, part 1, pp. 447–454.

Soil Conservation Service, 1975a, *Computer Program for Project Formulation, Hydrology,* Tech. Release No. 20, U.S. Department of Agriculture, Washington, D.C.

Soil Conservation Service, 1975b, *Urban Hydrology for Small Watersheds,* Tech. Release No. 55, United States Department of Agriculture, Washington, D.C.

Terstriep, M. L., and J. B. Stall, 1974, *The Illinois Urban Drainage Area Simulator, ILLUDAS,* Bull. 58, Illinois State Water Survey, Urbana, Illinois.

U.S. Army Corps of Engineers, 1960, *Runoff from Snowmelt,* Eng. Man. 1110-2-1406, U.S. Army, Washington, D.C.

Whipple, W., N. S. Grigg, T. Grizzard, C. W. Randall, R. P. Shubinski, and L. S. Tucker, 1983, *Stormwater Management in Urbanizing Areas,* Prentice-Hall, Englewood Cliffs, New Jersey.

6

Urban Hydrology

6.1
CHARACTERISTICS OF URBAN HYDROLOGY

Scope of This Chapter

This chapter describes techniques commonly applied in urban hydrology, emphasizing techniques not already discussed earlier in the text and modifying conventional techniques specifically for the urban situation. Modifications include unique loss estimates for calculation of rainfall excess, decreased lag times in unit hydrograph procedures, and emphasis on kinematic wave techniques for overland flow routing. The choice of input rainfall for construction of a design hyetograph is of special importance in urban hydrology because impervious urban surfaces rapidly convert the rainfall to runoff, such that short time increment variations in the rainfall hyetograph usually create similar variations in the runoff hydrograph. Sewer hydraulics and options for control of urban flooding are discussed.

Several comprehensive computer models also are available for application in urban situations; the chapter culminates in a review of such models plus a case study. A considerable volume of literature is available on urban hydrology, including texts by Delleur and Dendrou (1980), Kibler (1982), and Whipple et al. (1983).

Introduction

Although the same physical principles hold as for elsewhere in the hydrologic cycle, the hydrology of urban areas is dominated by two distinct
characteristics: (1) the preponderance of impervious surfaces (e.g., pavement, roofs) and (2) the presence of man-made or hydraulically "improved" drainage systems (e.g., a sewer system). Thus the response of an
urban catchment to rainfall is much faster than that of a rural catchment
of equivalent area, slope, and soils. In addition, the runoff volume from
an urban catchment is larger because there is less pervious area available
for infiltration. These characteristics are well documented and are illustrated in Fig. 6.1. While the control problem may be aggravated by the
faster response and greater runoff volumes, these same characteristics of
urbanization also tend to make the engineering analysis problem somewhat easier because estimation of losses is simplified and channel characteristics of shape, slope, and roughness are much better known.

Drainage systems in urban areas may rely on natural channels, but
most cities have a sewer network for removal of stormwater. If the system
is exclusively for stormwater removal, it is called a **storm sewer**. If the
same conduit also carries domestic sewage, it is called a **combined sewer**. A

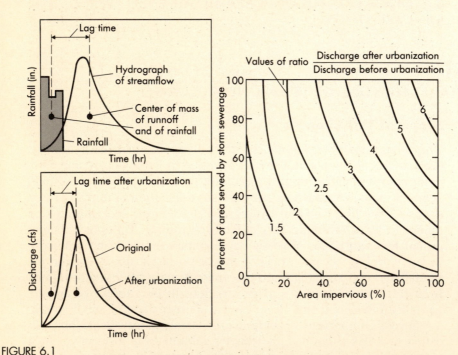

FIGURE 6.1

**Effect of urbanization on urban runoff hydrograph. (a) Shape. (b) Peak
flows. (From Leopold, 1968.)**

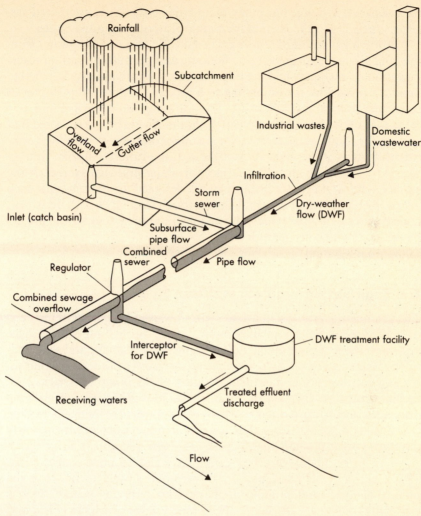

FIGURE 6.2

Urban drainage system. (From Metcalf and Eddy et al., 1971.)

combined sewer usually has a **regulator** (hydraulic control structure such as a weir or orifice) at its downstream end that diverts the dry-weather flow (domestic sewage) into an **interceptor,** which carries it to a treatment plant (Fig. 6.2). During wet weather, when the hydraulic capacities of the interceptor and the regulator are exceeded, the combined sewer overflow (a mixture of stormwater and sewage) is released directly to the receiving water body, with the potential for pollution problems. When the stormwater and domestic sewage are carried in separate conduits, the drainage system is said to be **separated,** as opposed to combined. Most newer cities have separate systems, but many older cities retain their combined systems, especially in the northeastern and midwestern United States.

Storm and combined sewers are installed to remove stormwater from the land surface, thus preventing flooding and permitting normal transportation on highways and the like. As such, they are usually designed to handle a peak flow corresponding to a given return period (Chapter 3) according to local regulations (2–5 yr for suburban drainage, and 10–50 yr for major highways is typical). It should always be remembered that a storm (or combined) sewer system is a **minor drainage system** for stormwater. That is, if the capacity is exceeded, the runoff will then find an alternative pathway along the surface, or **major drainage system.** If the surface system (streets, for instance) is not designed to accommodate such a flow, then the surface flow may flow through structures, with disastrous consequences. In other words, the major drainage system should always be designed with consideration of the possibility of failure or exceedance of the minor system.

The Engineering Problem in Urban Hydrology

The engineering problem in urban hydrology usually consists of the need to control peak flows and maximum depths throughout the drainage system. If the hydraulic grade line is too high, sewers may **surcharge,** that is, the water level may rise above the crown (top) of the sewer conduit, leading occasionally to basement flooding or discharge to streets. New facilities must be designed to minimize such occurrences, and existing drainage systems must often be modified to correct for them. Exceeding the capacity of an existing system is a problem that often occurs in newly developed areas that are served by an old sewer system.

The water quality of urban runoff may also be poor (Lager et al., 1977), and special measures may be required simply to improve the quality of the runoff prior to discharge into receiving waters. This is especially true for the discharge of combined sewer overflows (Field and Struzeski, 1972).

Alternatives for control of urban runoff quantity are many; a representative list is given in Table 6.1 and discussed later. In new developments it may be possible simply to design the drainage system with conduits or channels with sufficient hydraulic capacity to pass all expected flows, but this is seldom an economical alternative in older systems. Attenuation of flood peaks using storage, whether on the surface or within the drainage system, is the most common control alternative. Storage also has the secondary benefit of encouraging infiltration (if the storage basin is on pervious soil) and evaporation, thus reducing the runoff volume as well as the peak. In addition, storage facilities act as sedimentation basins for water quality control. Many additional options are available for quality control (Field, 1984).

TABLE 6.1

Alternatives for Control of Urban Runoff Quantity

MEASURE	DESCRIPTION
Storage	
Detention	Water detained and released after attenuation by reservoir effect
Retention	Water retained in facility, not released downstream; removed only by infiltration and evaporation
Examples	Basins, concrete tanks, available storage within sewer system, swales, parks, roofs, parking lots
Advantages	Inexpensive if land is available; can promote water conservation and flow equalization
Disadvantages	Costly if land unavailable; storage on private property (e.g., roofs, parking lots) subject to maintenance problems
Increased Infiltration	Reduction of impervious area
Examples	Porous pavement and parking lots
Advantages	Promotes water conservation
Disadvantages	Unsuitable for old cities; unable to reduce volumes of large storms; occasional structural problems
Traditional Flood Control Measures	Measures used in urban and nonurban settings
Examples	Channelization, floodplain zoning, floodproofing
Advantages	Well studied; effective in new areas
Disadvantages	Zoning, floodproofing may be unsuitable for old cities
Other	
Examples	Inlet restrictions, improved maintenance
Comment	Options for older systems

Control options may often be actively traded off against each other during the design phase; when a mixture of controls is used, the cost for one will usually decrease as the use of the other is increased (Heaney and Nix, 1977). Economic optimization is a necessary part of any design.

Data and information needed for analysis of problems in urban hydrology are similar to those needed in other areas of hydrology — for example, information on rainfall, surface catchment characteristics, and characteristics of the drainage system, natural or man-made. When analytical methods are used to calculate hydrographs, it is highly desirable to have measurements available with which to calibrate and verify predictions.

Modern analytical tools range from simple methods for predicting peak flows and runoff volumes to sophisticated computer models for predicting the complete hydrograph at any point in the drainage system. Such methods usually include a means for converting rainfall to runoff from the land surface, followed by a method for routing flows through the drainage system. Many standard computer models are available for both mainframe computers and microcomputers, and most techniques can easily be programmed in FORTRAN, BASIC (Sharp and Sawden, 1984), or other languages.

Design Objectives

The engineering objective when dealing with urban hydrology is to provide for control of peak flows and maximum depths at all locations within the drainage system. In essence it is the same flood control problem dealt with in Chapter 4, but with the added considerations associated with urbanization. Secondary problems include basement flooding, surcharging, and water quality control.

The analytical problems that must be solved to address these objectives are the predictions of runoff peaks, volumes, and complete hydrographs anywhere in the drainage system. These problems are often separated into those involving the surface drainage system, for which rainfall must be converted into an overland flow hydrograph (or inlet hydrograph, for flow into the inlet to the sewer system), and those involving the channel or sewer system, which often may be handled through conventional flow routing techniques (Chapter 4). However, there is a tendency in urban hydrology to lump the two systems together and produce one aggregate hydrograph (or peak flow or runoff volume) at the system outlet, such as by a unit hydrograph.

The distinction between the needs to predict peak flows, runoff volumes, and complete hydrographs is important since the three objectives may require very different analytical methods. In general, predictions of peaks and volumes lend themselves to simplified techniques, while hydrographs require a more comprehensive analysis.

Estimation of base flow in urban drainage systems also requires special consideration since water may enter the channels both as **infiltration** (seepage into a conduit from ground water) and as domestic sewage. The total of the two is often evaluated from monitoring at downstream treatment plants or from special infiltration/inflow (I/I) studies (Environmental Protection Agency, 1977).

A further distinction will be made regarding the application of analytical techniques, namely, the design of new systems versus the alleviation of drainage problems in old ones. More options are usually available for the former, whereas existing systems may impose severe constraints on control options for the latter.

6.2

REVIEW OF PHYSICAL PROCESSES

Rainfall-Runoff

Conversion of rainfall into runoff in urban areas is usually somewhat simplified because of the relatively high imperviousness of such areas, although in residential and open-land districts the calculation of infiltration into pervious surfaces may still represent a critical factor in the analysis. Once computed, however, rainfall excess may be converted into a runoff hydrograph using almost any of the methods used for natural catchments.

When hydrographs are to be computed, special effort is required to obtain adequate rainfall data. This is because urban areas respond quickly to rainfall transients, in contrast to natural catchments, which dampen out short-term fluctuations. Thus rainfall data should be available at 5-min or shorter increments to adequately predict the runoff hydrograph. A further consideration is the storm direction; as for natural catchments, the hydrograph can have a considerably higher peak if the storm moves down the catchment toward the outlet. This means that one rain gage is seldom sufficient; at least three, plus a wind measurement, may be necessary to describe the dynamics of moving storms (James and Scheckenberger, 1984).

Catchment Description

The urban catchment is characterized by its area, shape, slope, soils, land use, imperviousness, roughness, and storage. The area and imperviousness are the two most important parameters for a good prediction of hydrograph volumes. Although it is a seemingly straightforward parameter, estimation of the percent imperviousness can be subtle. In particular, it is usually necessary to distinguish between **hydraulically connected impervious areas** and those that are not. Hydraulically connected impervious areas are those that drain directly into the drainage system, such as a street surface with curbs and gutters that direct the runoff into a storm sewer inlet. **Nonhydraulically connected areas** include rooftops or driveways that drain onto pervious areas. Runoff from such areas does not enter the storm drainage system unless the pervious areas become saturated.

Estimates of imperviousness can be made by measuring such areas on aerial photographs or by considering land use; typical values are shown in Table 6.2. For large urban areas, imperviousness can be estimated on the basis of population density (e.g., Stankowski, 1974):

$$I = 9.6 \mathrm{PD}^{(0.573 - 0.017 \ln \mathrm{PD})}, \tag{6.1}$$

TABLE 6.2

**Imperviousness by Type of Land
Use for Nine Ontario Cities
(Sullivan et al., 1978)**

LAND USE	PERCENT IMPERVIOUSNESS	
	Average	Range
Residential	30	22–44
Commercial	81	52–90
Industrial	40	11–57
Institutional	30	17–38
Open	5	1–14

where

I = percent imperviousness,

PD = population density (persons/ac).

Equation (6.1) is based on a regression analysis of 567 communities in New Jersey, so it should be used with caution elsewhere! Imperviousness is often used as a calibration parameter in models.

Superimposed on the catchment is the drainage system, either natural or (more likely) **improved.** Improved drainage usually consists of a network of channels and conduits that form a sewer system that normally flows with a free surface, i.e., open channel flow, as opposed to pressure flow in a pipeline. The drainage system thus has its own set of parameters to describe its geometry and hydraulic characteristics. For a new system, all of these parameters are design parameters and may be varied according to the engineer's needs. For an existing system, a major effort at collecting data may be required to accurately describe the as-built system configuration. This is especially difficult in old systems in which *ad hoc* repairs and alterations may have been made over many decades and for which a field survey may be required if accurate data, such as elevations of conduit **inverts** (bottoms), are to be obtained.

Calculation of Losses

Rainfall excess is computed as rainfall minus losses (Chapter 2). Losses result from depression storage (or "initial abstraction") on vegetation and other surfaces, infiltration into pervious surfaces, and evaporation. For an individual storm, evaporation is relatively unimportant, but for a long-term analysis of the urban water budget (or for long-term simulation),

TABLE 6.3

Depression Storage Estimates in Urban Areas

CITY	COVER	DEPRESSION STORAGE (in.)	REFERENCE
Chicago	Pervious	0.25	Tholin and Kiefer (1960)
	Impervious	0.0625	
Los Angeles	Sand	0.20	Hicks (1944)
	Loam	0.15	
	Clay	0.10	

evaporation is just as important as it is in natural catchments. Infiltration calculations can also proceed in the same manner as for natural areas (Chapter 1).

Depression storage is difficult to separate from infiltration over pervious areas; rough estimates for large urban areas include those shown in Table 6.3. For highly impervious areas, measurements have been made to relate depression storage to the slope of the catchment (Fig. 6.3). Depres-

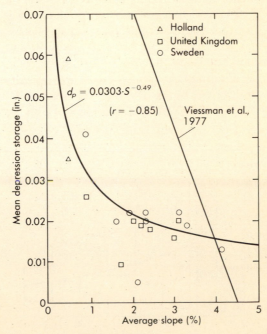

FIGURE 6.3

Depression storage vs. catchment slope. (After Kidd, 1978; Viessman et al., 1977.)

sion storage is often used as a calibration parameter in models since, although representative of a real loss, its estimation is very difficult *a priori.*

EXAMPLE 6.1

COMPUTATION OF RAINFALL EXCESS

The rainfall hyetograph shown in Fig. 6.4 and listed in Table 6.4 is subject to a depression storage loss of 0.15 in. and Horton infiltration (Eq. 1.18) with parameters $f_0 = 0.45$ in./hr, $f_c = 0.05$ in./hr, and $k = 1$ hr^{-1}. Calculate the hyetograph of rainfall excess.

SOLUTION

Depression storage is customarily removed "off the top," prior to the deduction of any infiltration losses. The volume of the first 10-min rainfall increment is 0.3 in./hr $\times$ 1/6 hr = 0.05 in. This leaves $0.15 - 0.05 = 0.10$ in. of depression storage to be filled during the next rainfall increment(s). This occurs at a time

$$t = 1/6 \text{ hr} + (0.10 \text{ in.}/0.8 \text{ in./hr}) = 1/6 \text{ hr} + 1/8 \text{ hr} = 17.5 \text{ min.}$$

Thus infiltration begins at 17.5 min, as sketched in Fig. 6.4. The infiltration calculations are summarized in Table 6.4 and the rainfall excess is sketched in Fig. 6.4. Note that a negative rainfall excess is not possible. Note also that for this example, infiltration capacity is assumed to continue to diminish at the same rate between times 40 and 50 min even though the rainfall rate is less than the infiltration capacity.

The volumes of rainfall and rainfall excess are found by summing the ordinates and multiplying by the time interval (1/6 hr). Thus the rainfall volume is 3.5 $\times$ 1/6 = 0.583 in. The volume of rainfall excess is com-

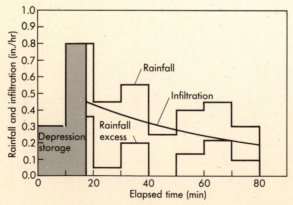

FIGURE 6.4

Rainfall and rainfall excess hyetographs.

TABLE 6.4

Computation of Rainfall Excess

TIME INTERVAL (min)	RAINFALL (in./hr)	INFILTRATION CAPACITY AT START OF INTERVAL (in./hr)	AVERAGE INFILTRATION CAPACITY (in./hr)	AVERAGE RAINFALL EXCESS (in./hr)
0–10	0.30			
10–17.5	0.80			
17.5–20	0.80	0.450	0.442	0.358
20–30	0.45	0.434	0.404	0.046
30–40	0.55	0.375	0.350	0.200
40–50	0.25	0.325	0.304	0*
50–60	0.40	0.283	0.265	0.135
60–70	0.45	0.247	0.232	0.218
70–80	0.30	0.217	0.204	0.096
80–90	0.00	0.191		0.000

* Negative value set to zero.

puted similarly, except that the first time increment is only 2.5 min instead of 10 min. Thus the volume of rainfall excess is $0.695 \times 1/6 + 0.358 \times 2.5/60 = 0.131$ in. Since the depression storage volume was 0.15 in., the actual infiltration volume is $0.583 - 0.15 - 0.131 = 0.302$ in. Note that this is *not* equal to the area under the infiltration curve since not all of the infiltration capacity was used during the interval between 40 and 50 min. A runoff coefficient can be defined simply as the ratio of runoff (rainfall excess) to rainfall. Here the ratio is $0.131/0.583 = 0.22$.

When rainfall excess is computed, it is customary to leave it tabulated in the original hyetograph time increments (10 min for this example). In this example there is a complication for the first interval, since it begins at 17.5 min. There are at least two options: (1) The starting time could remain at 17.5 min and if equal time intervals are needed, an interval of 2.5 min could be used; (2) the first, short interval of rainfall excess could be averaged over the time interval from 10 to 20 min. This would yield a rainfall excess value of $0.358 \times 2.5/10 = 0.090$ in./hr for the interval from 10 to 20 min (and zero for the first interval).

Time of Concentration

The time of concentration t_c was introduced in Chapter 4 as part of the development of kinematic wave theory. It is worth a brief review of the definitions of t_c since it is fundamental to much of the analysis in urban hydrology. There are two related definitions:

1. The time of concentration is the travel time of a *wave* to move from the hydraulically most distant point in the catchment to the outlet.
2. The time of concentration is the time to equilibrium of the catchment under a steady rainfall excess (i.e., when the outflow from the catchment equals the rainfall excess onto the catchment).

Note especially that t_c is *not* the travel time taken by a parcel of water to move down the catchment, as is so often cited in texts. The catchment is in equilibrium when the time t_c is reached because the outlet then "feels" the inflow from every portion of the catchment. Since a wave moves faster than a parcel of water does, the time of concentration (and equilibrium) occurs sooner than if based on overland flow (or channel) water velocities (Overton and Meadows, 1976). For overland flow, the wave speed (celerity) is usually given by the kinematic wave equation

$$c = mV = \alpha m y^{m-1}, \tag{6.2}$$

where

$\qquad c =$ wave speed,

$\qquad V =$ average velocity of water,

$\qquad y =$ water depth,

$\qquad \alpha, m =$ kinematic wave parameters in the uniform flow momentum equation, and

$$q = yV = \alpha y^m \tag{6.3}$$

where $q =$ flow per unit width of the catchment.

Using Manning's equation, the parameter m is 5/3 for turbulent flow and 3 for laminar flow. Thus the wave can travel at a rate of from 1.7 to 3 times as fast as does a parcel of water. Corresponding values of the parameter α for turbulent and laminar flow, respectively, are

$$\alpha = \frac{k_m \sqrt{S}}{n} \qquad (m = 5/3) \qquad \text{turbulent flow}, \tag{6.4}$$

$$\alpha = gS/3v \qquad (m = 3) \qquad \text{laminar flow}, \tag{6.5}$$

where

$\qquad S =$ slope (small, such that $\sin S \approx \tan S \approx S$),

$\qquad n =$ Manning's roughness,

$\qquad g =$ acceleration of gravity,

$\qquad v =$ kinematic viscosity, and

$\qquad k_m = 1.49$ for units of ft and s in Eq. (6.4),

$\qquad\qquad = 1.0$ for units of m and s.

The time of concentration for overland flow by kinematic wave theory is

$$t_c = \left(\frac{L}{\alpha i_e^{m-1}}\right)^{1/m},$$

(6.6)

where

L = length of overland flow plane,

i_e = rainfall excess.

Care should be taken to use consistent units in Eq. (6.6). It can be seen that t_c depends inversely on the rainfall excess; thus higher-intensity rainfall excess will reduce t_c. Alternative equations for t_c based on water parcel travel times seldom include this inherent physical property of the catchment (McCuen et al., 1984; Huber, 1987).

EXAMPLE 6.2

KINEMATIC WAVE FLOW COMPUTATION

Water flows down an asphalt parking lot that has a 1% slope. What are the values of α and m (Eq. 6.3) for the following two cases:

a) laminar flow ($T = 20°C$),

b) turbulent flow ($n = 0.013$)?

What is the flow per unit width q (cfs/ft) for each of the two cases, for a water depth of 0.3 in.?

SOLUTION

Laminar flow has an exponent $m = 3$ and a coefficient α, given by Eq. (6.5). At a temperature of 20°C, $v = 1.003 \times 10^{-2}$ cm²/s $= 1.080 \times 10^{-5}$ ft²/s (Table C.1). Thus

$$\alpha = gS/3v = \frac{32.2 \times 0.01}{3 \times 1.080 \times 10^{-5}} = 9938 \text{ ft}^{-1}\text{s}^{-1}.$$

For turbulent flow, the exponent $m = 5/3$, and from Eq. (6.4),

$$\alpha = \frac{1.49S^{0.5}}{n} = \frac{1.49 \times 0.01^{0.5}}{0.013} = 11.46 \text{ ft}^{1/3}/\text{s}.$$

A depth of 0.3 in. = 0.025 ft. The flow per unit width is given by Eq. (6.3):

Laminar: $q = 9938 \times 0.025^3 = 0.155$ cfs/ft,

Turbulent: $q = 11.46 \times 0.025^{5/3} = 0.024$ cfs/ft.

For the same depth, the laminar flow rate is higher, but in most actual cases turbulent flow is observed.

In free-surface flow, waves in channels and conduits move with the dynamic wave speed

$$c = V \pm \sqrt{gA/B}, \tag{6.7}$$

where

A = cross-sectional area in the channel,

B = surface width.

Downstream wave speeds (positive sign) given by Eq. (6.7) are obviously greater than the water velocity V. Wave speeds calculated using the negative sign are downstream for supercritical flow and upstream for subcritical flow. If flow through closed conduits (pressure flow) occurs in the catchment, waves move with the speed of sound (Wylie and Streeter, 1978), very fast indeed.

To summarize: any catchment, urban or otherwise, responds to an input of rainfall through the passage of *waves* downstream along all water pathways—overland flow, open channels and conduits, and closed conduits. The net effect is that the time of concentration (or time to equilibrium) is reached well in advance of the travel time required by a parcel of water to move the complete distance downstream. An accurate value of t_c is essential in the use of the rational method, presented in Section 6.4, for estimation of peak flows as well as for evaluation of the catchment response to changes in the rainfall regime.

Lag Times

Other timing parameters of the urban hydrograph include the various possible lag times discussed in Chapter 2 (Fig. 2.7). These lag parameters, such as time to peak t_p, have the distinct advantage of being possible to measure, whereas the time of concentration is very difficult to measure (Overton and Meadows, 1976) and is more of a conceptual parameter. Lag times are often incorporated into unit hydrograph theory and conceptual models.

Flow Routing

Flow routing through the urban drainage system may be accomplished using almost any of the methods discussed in Chapter 4. The task is made somewhat easier because of the presence of uniform channel sections (e.g., circular pipes) in the sewer system. But at the same time, difficulties may arise at the many junctions and hydraulic structures encountered in a sewer system. Most of these difficulties may be overcome through the use of more sophisticated hydraulic methods discussed in Section 6.5 (such as the St. Venant equations) if desired (Yen, 1986); the level of the analysis should be tailored to the desired accuracy of the predictions, and all predictions should be verified against measured flows wherever possible.

6.3

RAINFALL ANALYSIS

Data Sources

Rainfall is the driving force in most hydrologic analysis. Two types of rainfall data are commonly required in urban hydrology: (1) "raw" point precipitation data (i.e., actual hyetographs) and (2) processed data, usually in the form of frequency information. Computerized point precipitation data are available from the National Weather Service (NWS) through its National Climatic Data Center (Asheville, North Carolina 28801-2696) in the form of hourly or 15-min values from recording rain gages. The hourly data are also given in the monthly "Climatological Data," published for each state by the NWS. Photocopies of the actual rain gage charts are available for desired dates from the National Climatic Data Center if more detailed time resolution is needed. The NWS network of weather stations is the prime source for all meteorological data, but their data may often be supplemented by other sources that collect data continuously or as part of special projects. These sources include the U.S. Geological Survey (USGS), the Soil Conservation Service (SCS), and the Army Corps of Engineers as well as state agencies, local utilities, and universities. It is often necessary to contact several groups to obtain all available data for a catchment.

It is useful to note that the rainfall data published in the NWS's printed publications are often slightly different from the computerized data obtained from the National Climatic Data Center because the values come from different rain gages. These gages may be within a few feet of each other, but still the recording rain gage data (which are computerized) almost always differ from the nonrecording gage data (which are printed as daily totals in official publications). For example, monthly totals from two such gages at St. Leo, Florida (about 30 miles north of Tampa) are compared in Table 6.5. Neither gage consistently yields larger or smaller values than the other. The nonrecording (printed) data are traditionally regarded as "official."

Processed data include many varieties of statistical summaries, but the most common type of processed data used in urban hydrology is in the form of intensity-duration-frequency (IDF) curves, discussed in Chapter 1 and later in this chapter. IDF curves are available from the NWS (Hershfield, 1961), the SCS, and local sources such as state departments of transportation (e.g., Weldon, 1985). The best information usually comes from local sources.

TABLE 6.5

Comparison of Recording and Nonrecording Rainfall Totals, St. Leo, Florida

YEAR	MONTH	NONRECORDING GAGE (in.)	RECORDING GAGE (in.)	DIFFERENCE (in.)	PERCENT DIFFERENCE*
1977	Jan.	3.09	2.9	0.19	6.1
	Feb.	4.26	4.0	0.26	6.1
	Mar.	1.55	1.7	−0.15	−9.7
	Apr.	0.39	0.3	0.09	23.1
	May	1.38	1.2	0.18	13.0
	Jun.	4.63	4.6	0.03	0.6
	Jul.	6.78	6.7	0.08	1.2
	Aug.	12.25	11.8	0.45	3.7
	Sep.	8.66	8.9	−0.24	−2.8
	Oct.	1.31	1.3	0.01	0.8
	Nov.	1.88	1.7	0.18	9.6
	Dec.	3.48	3.3	0.18	5.2
	Annual	49.66	48.4	1.26	2.5
1978	Jan.	4.72	4.8	−0.08	−1.7
	Feb.	6.20	5.8	0.40	6.5
	Mar.	4.01	4.0	0.01	0.2
	Apr.	1.59	1.5	0.09	5.7
	May	7.83	8.1	−0.27	−3.4
	Jun.	6.12	6.0	0.12	2.0
	Jul.	9.52	11.3	−1.78	−18.7
	Aug.	4.70	3.5	1.20	25.5
	Sep.	1.21	1.2	0.01	0.8
	Oct.	0.29	0.3	−0.01	−3.4
	Nov.	0.02	0.0	0.02	100.0
	Dec.	4.54	4.3	0.24	5.3
	Annual	50.75	50.8	−0.05	−0.1

* $100 \times$ Difference/Nonrecording

Rainfall Measurement

As mentioned earlier, a unique aspect of urban hydrology is the fast response of an urban catchment to the rainfall input. This means that high-frequency information in the rainfall signal is transformed into high-frequency pulses in the runoff hydrograph since the highly impervious urban catchment does not significantly dampen such fluctuations. In most cases a tipping bucket rain gage will provide an adequate resolution of high frequencies, in contrast to weighing bucket gages, which are also commonly employed. Tipping bucket gages are much easier to install in a network because the electrical pulses can be transmitted over a telephone

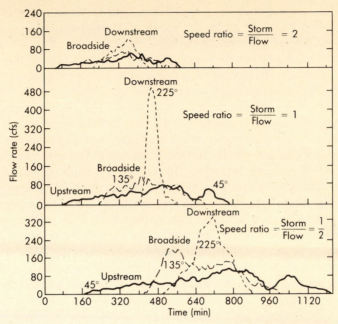

FIGURE 6.5

Effect of storm moving upstream, downstream, and broadside to an urban catchment. (From Surkan, 1974.)

or other communication device or recorded on a data logger at the site. In most urban modeling work, it is necessary to have hyetographs with 5-min or better resolution, although this may be relaxed somewhat for larger basins.

Unless the catchment is very small (such that a storm completely "covers" the basin spatially), the outfall hydrograph will also be sensitive to the storm direction (James and Shtifter, 1981). To simulate the effects of a moving (or "kinematic") storm, it is necessary to have more than one rain gage so that the longitudinal movement (up or down the basin) can be defined. Movement of the storm *across* the flow direction does not affect the hydrograph as much as longitudinal movement. These effects can be seen in Fig. 6.5.

Intensity-Duration-Frequency Curves: Use and Misuse

Intensity-duration-frequency (IDF) curves summarize *conditional* probabilities (frequencies) of rainfall depths or average intensities. Specifically, IDF curves are graphical representations of the probability that a certain average rainfall intensity will occur, given a duration; their derivation is discussed by McPherson (1978). For example, IDF curves for Tallahassee, Florida, are shown in Fig. 6.6 (Weldon, 1985). The return period of an average intensity of 3.6 in./hr for a duration of 40 min is 5 yr. (As described in Chapter 3, the return period is the reciprocal of the probability that an event will be equaled or exceeded in any one year or other time unit.) Conversely, if one wanted to know what average intensity could be expected for a duration of 20 min once, on the average, every 10 yr, Fig. 6.6 gives a value of 5.6 in./hr. These values are conditional on the duration — the duration must always be specified by some means to use IDF curves — and the duration is *not* the duration of an actual rainstorm.

A critical characteristic of IDF curves is that the intensities are indeed averages over the specified duration and do not represent actual time histories of rainfall. The contour for a given return period could represent the smoothed results of several different storms. Moreover, the duration is not the actual length of a storm; rather, it is merely a 20-min period, say, within a longer storm of any duration, during which the average intensity happened to be the specified value. In fact, given that the IDF curves are really smoothed contours, unless a data point falls on a contour, they are completely hypothetical! A detailed analysis of the method of construct-

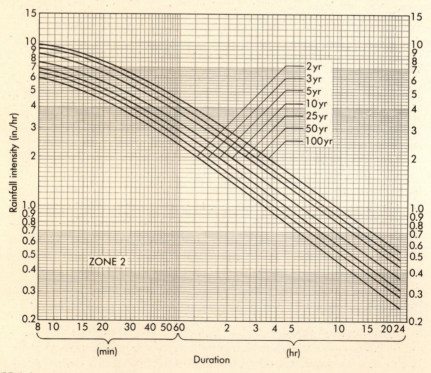

FIGURE 6.6

Intensity-duration-frequency curves for the Tallahassee, Florida, region. (From Weldon, 1985.)

ing IDF curves and the perils of improper interpretations is given by McPherson (1978).

The most common problem with the IDF curves is that they are often misused to assign a return period to a storm event depth or average intensity or vice versa. But what duration should be used for such an analysis? After all, a given depth can result from infinitely many combinations of average intensity and duration, and as the duration increases, the return period for a given depth decreases. Since the durations on IDF curves do not represent actual storm durations in any sense, it is only a crude approximation to use the curves to define, say, the 10-yr rainfall event on the basis of total depth. Instead, this must be done by a frequency analysis of storm depths, after having separated the time series of rainfall values into independent storm events (described later in this chapter).

The preponderance of IDF information in the hydrologic literature is primarily a result of the need for IDF curves for use in the rational method (described in Section 6.4). As will be seen, the IDF data are properly applied in this case. But to reiterate: (1) IDF curves do *not* represent time histories of real storms—the intensities are *averages* over the indicated duration; (2) a single curve represents data from several different storms; (3) the duration is *not* the duration of an actual storm and most likely represents a shorter period of a longer storm; and (4) it is incorrect to use IDF curves to obtain a storm event volume because the duration must be arbitrarily assigned.

Definition of a Storm Event

Return periods (or frequencies) can be assigned to rainfall events on the basis of several different parameters, but most commonly on the basis of total volume, average intensity, peak intensity, duration, or interevent time. To work with any of these parameters, the rainfall time series must first be separated into a series of discrete, independent "events." When this is done they may be ranked by volume or any desired parameter and a conventional frequency analysis performed (Chapter 3).

For ease of computation, a statistical measure is usually employed to separate independent storm events. A minimum interevent time (MIT) is defined such that rainfall pulses separated by a time less than this value are considered part of the same event. The process can be illustrated by an example.

EXAMPLE 6.3

SEPARATION OF RAINFALL INTO EVENTS

Given the hypothetical rainfall sequence shown in Fig. 6.7, determine the number of events corresponding to various values of the minimum interevent time (MIT).

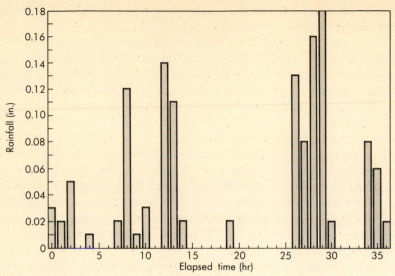

FIGURE 6.7

Hypothetical hourly rainfall hyetograph for illustration of event separation.

SOLUTION

The results are summarized in the following table:

MIT	NUMBER OF EVENTS
0	20
1	7
2	5
3	4
4	3
5	2
6	2
≥7	1

When an MIT value of zero is used, all rainfall hours are considered to be separate events; hence the number of events is equal to the number of rainfall hours. For an MIT value of 1 hr, at least 1 dry hour is required to separate events, leading to the 7 events (clusters of contiguous wet hours) given in the table. Since there are no rainfall hours separated by more than 6 hr, all rainfall is considered to be a single event for an MIT value greater than or equal to 7 hr. Note that the magnitudes of the hourly rainfall values play no role in this analysis.

Several possibilities exist for a statistical measure with which to define independence between rainfall values, including examination of the correlation between rainfall values at different lags (Tavares, 1975; Heaney et al., 1977). However, the easiest definition (Hydroscience, 1979; Restrepo-Posada and Eagleson, 1982) is based on the observation that interevent times are usually well described by an exponential probability density function (Section 3.5), for which the standard deviation equals the mean (or their ratio, the coefficient of variation CV, equals 1.0). The procedure is to use trial values of MIT until the CV of the interevent times is closest to 1.0. Typical values of the resulting MIT may range from 5 to 50 hr for a time series of hourly rainfall values, depending on location and season. The analysis is facilitated by the use of a computer program such as SYNOP (Environmental Protection Agency, 1976; Hydroscience, 1979) that performs the event separation and frequency analysis for a given MIT value.

Having completed the event separation, the events may be ranked by the parameter of interest for any desired frequency analysis. For example, the seven highest total storm volumes for a 25-yr record of hourly rainfalls in Tallahassee, Florida, are shown in Table 6.6, for which the MIT is 5 hr. It is clear that these volumes can arise from actual storms of greatly differing durations, certainly not fixed as on an IDF curve. For comparison, 24-hr depths are also shown, taken from IDF data for the Tallahassee region (Weldon, 1985). Notice that the 24-hr depths are in the correct range of return periods for total storm depths, but this would not be true if other durations were used.

TABLE 6.6

Comparison of 24-hr IDF Storm Depths with SYNOP Frequency Analysis of Historical Storms

STORM DATE	RETURN PERIOD* (yr)	DURATION (hr)	DEPTH (in.)	INTENSITY (in./hr)
9/20/69	26.0	54	13.41	0.25
IDF	25	24	10.08	0.42
7/17/64	13.0	33	9.76	0.30
IDF	10	24	8.64	0.36
12/13/64	8.7	36	9.73	0.27
7/28/75	6.5	53	8.84	0.17
7/21/70	5.2	20	8.18	0.41
IDF	5	24	7.32	0.31
8/30/50	4.3	36	7.34	0.20
6/18/72	3.7	46	7.17	0.16

* By Weibul formula (Eq. 3.80). Actual return period will be different if based on a fitted distribution (Chapter 3).

Choice of Design Rainfall

Often in an urban analysis it is necessary to provide a "design storm" for a model with which to evaluate the effectiveness of a drainage system. Sometimes these rainfall inputs consist of measured hyetographs obtained during a monitoring program of the catchment and are used for calibration and verification of a model. In this case, there is little question about the choice of rainfall input, but more often than not an agency will require a design storm corresponding to a specified return period, such as a "25-yr storm." There are several ways to define such an event, and the method of constructing and applying a design storm can be very controversial (McPherson, 1978; Patry and McPherson, 1979; Harremöes, 1983; Adams and Howard, 1985).

The first question is To what storm parameter does the return period apply? That is, is it a 25-yr storm based on runoff volume, peak flow, average flow, or rainfall volume, rainfall average intensity, etc.? Equally important, given that antecedent conditions in the catchment can be highly variable and that the catchment will alter the nature of the rainfall input in any case, the return period of a storm based on rainfall characteristics will not, in general, be the same as the return period of the same storm based on runoff characteristics. For example, the 25-yr storm based on rainfall volume may be only the 5-yr storm based on runoff volume if the catchment is dry before the storm occurs. Thus it is an error to assign a return period to a runoff characteristic based on a frequency analysis of rainfall.

Synthetic Design Storms

The use of IDF curves is intimately linked to most of the methods for constructing **synthetic** (hypothetical) **design storms.** A typical synthetic design storm is constructed in the following manner. First, a duration is specified, often arbitrarily by an agency, say 24 hr. Second, the 24-hr depth is obtained from an IDF curve for the desired return period. Third, the rainfall must be distributed in time to construct the synthetic hyetograph. There are major differences in the literature in this last step (Arnell, 1982). The shape varies mainly depending on the type of storm to be simulated: a cyclonic storm usually has the highest intensities near the middle, and a convective storm usually has the highest intensities near the beginning.

EXAMPLE 6.4

CONSTRUCTION OF SCS TYPE II, 24-HR DESIGN STORM

Construct a 5-yr, 24-hr design storm for Tallahassee, using the SCS type II distribution.

SOLUTION

The distribution (percentage mass curve) is tabulated in Table 6.7. The duration of the design storm must be 24 hr to use the SCS type II distribution, which is specifically given for a 24-hr duration. From the Tallahassee IDF curve shown in Fig. 6.6 the average intensity for 24 hr and a 5-yr return period is 0.305 in./hr. This leads to a total depth for 24 hr of 7.32 in., as indicated in Table 6.6. The total depth is allocated over the hourly increments in Table 6.7, and the average intensities for each hour are plotted in Fig. 6.8. Notice the very high intensity during hour 12, reflecting the fact that 42.5% (i.e., $66 - 23.5\%$, from Table 6.7) of the total rainfall is assumed to occur during this interval. In the concluding case study to this chapter, it will be seen that this high intensity can lead to excessively high peak flows when used for modeling purposes.

TABLE 6.7

24-hr Design Storm from SCS Type II Distribution
(Total depth = 7.32 in.)

HOUR	PERCENT TOTAL DEPTH	CUMULATIVE DEPTH (in.)	AVERAGE INTENSITY (in./hr)
0	0.00	0.00	
1	1.00	0.07	0.07
2	2.20	0.16	0.09
3	3.55	0.26	0.10
4	4.91	0.36	0.10
5	6.20	0.45	0.09
6	8.10	0.59	0.14
7	10.00	0.73	0.14
8	12.10	0.89	0.15
9	14.70	1.08	0.19
10	18.60	1.36	0.29
11	23.50	1.72	0.36
12	66.00	4.83	3.11
13	77.40	5.67	0.83
14	82.10	6.01	0.34
15	85.30	6.24	0.23
16	88.10	6.45	0.20
17	90.10	6.60	0.15
18	92.20	6.75	0.15
19	93.50	6.84	0.10
20	94.80	6.94	0.10
21	96.10	7.03	0.10
22	97.40	7.13	0.10
23	98.70	7.22	0.10
24	100.00	7.32	0.10

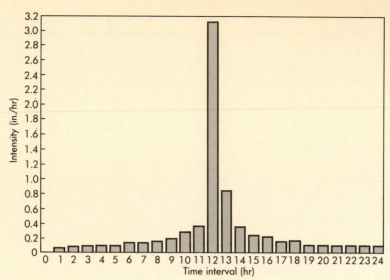

FIGURE 6.8

SCS type II, 5-yr, 24-hr design storm for Tallahassee, Florida.

McPherson (1978) compares a 5-yr, 3-hr synthetic storm for Chicago with historic storms based on total storm event depth. As shown in Fig. 6.9, not only is the depth of the synthetic storm different from the two historic storms that bracket a 5-yr return period, but the hyetograph shapes bear no relation to the assumed shape for the synthetic storm. Thus, although the synthetic storm may certainly be applied in a study, the shape of the storm is so unusual and its duration so arbitrary that the true return period of the storm, based on any criterion, is totally unknown. Additional comparisons will be seen in the case study at the end of this chapter.

Alternatives to Synthetic Design Storms

Synthetic design storms are very popular because they are relatively easy to construct and use. They require only IDF curves and an assumed shape for the hyetograph, not extensive historic rainfall data. What are the alternatives? The best method is to analyze historical runoff data for the catchment to identify storms of interest based on a frequency analysis of peak flows, flood depths, runoff volumes, etc. The storms that caused, say, the 5-yr peak flow could then be used for design purposes (given adequate information on antecedent conditions). This would involve several storms, with very different temporal distributions, each with a return period close to 5 yr, based on some criterion. The case study at the conclusion of this chapter illustrates a collection of such storms, based on simulation results, not monitoring results.

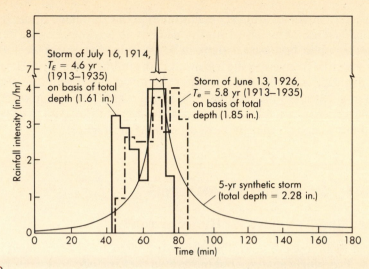

FIGURE 6.9
Actual vs. synthetic storm patterns, Chicago. (From McPherson, 1978.)

Selection of storms from a monitoring program is seldom possible in urban areas because flow monitoring programs are typically short, and there may be extensive development under way in the catchment that alters the rainfall-runoff mechanisms, thus destroying the homogeneity of the runoff record. Probably the best alternative is to apply a continuous simulation model to simulate runoff hydrographs for a period of, say, 25 yr based on historic rainfall records available from the National Climatic Data Center. The model must first be calibrated using measured rainfall and runoff data. A frequency analysis can then be performed on the simulated hydrographs — on the parameters of interest — from which historical design storms can be identified for more detailed analysis (Huber et al., 1986). (The detailed analysis must also have a means for establishing the correct antecedent conditions, a problem automatically handled during the continuous simulation.) Both of these options (frequency analysis of measured or simulated hydrographs) have the advantage of an analysis of runoff, not of rainfall. This continuous simulation approach is routinely used by the U.S. Geological Survey and is demonstrated in the case study at the end of this chapter.

A third alternative is a frequency analysis on rainfall *events,* similar to that shown in Table 6.6. If historical events from such an analysis are subsequently used for design storms, the disadvantage remains of basing return periods on rainfall characteristics instead of on runoff characteristics, but at least there would not be an assumption of a duration.

To summarize: the use of historical storms for design purposes is preferred because the frequency analysis can be directed to the specific parameter of interest without an assumed duration and because there is

no need to make an arbitrary assumption about the shape of the hyetograph. The use of historical storms also has an advantage when dealing
with the public, because a design can be presented to prevent the flooding
that occurred from a specific real event in the public's memory. The main
disadvantage is simply the extra effort involved in a continuous simulation or a frequency analysis based on storm events rather than on conditional frequencies readily available from published IDF data. As a result,
synthetic design storms are most often used in practice.

6.4
METHODS FOR QUANTITY ANALYSIS

Peak Flow, Volume, or Hydrograph?

Recall that there are several possible parameters to be determined in an
urban hydrologic analysis, but most often they include peak flow, runoff
volume, or the complete runoff hydrograph. Related parameters might be
the hydraulic grade line or flooding depths. If hydrographs are predicted,
then peaks and volumes are an implicit part of the analysis, but some of
the simpler methods will not necessarily provide all parameters of possible interest.

Peak Flows by the Rational Method

The rational method dates from the 1850s in Ireland (Dooge, 1973) and is
called the Lloyd-Davies method in Great Britain. It is one of the simplest
and best-known methods routinely applied in urban hydrology, although
it contains subtleties that are not always appreciated. Peak flows are
predicted by the simple product

$$Q_p = k_c CiA, \tag{6.8}$$

where

Q_p = peak flow (cfs or m³/s),
C = runoff coefficient,
i = rainfall intensity (in./hr or mm/hr),
A = catchment area (ac or ha),
k_c = conversion factor.

When U.S. customary units are used, the conversion factor $k_c = 1.008$ to
convert ac-in./hr to cfs and is routinely ignored; this conversion is the
basis for the term "rational" in the rational method. (The approximate
equivalence of ac-in./hr to cfs is worth remembering for convenient rough
calculations.) For the alternative metric units given with Eq. (6.8), the
conversion factor $k_c = 0.00278$ to convert ha-mm/hr to m³/s.

Due to assumptions regarding the homogeneity of rainfall and equilibrium conditions at the time of peak flow, the rational method should not be used on areas larger than about 1 mi^2 (2.5 km^2) without subdividing the overall catchment into subcatchments and including the effect of routing through drainage channels. Since actual rainfall is not homogeneous in space and time, the rational method becomes more conservative (i.e., it overpredicts peak flows) as the area becomes larger.

All catchment losses are incorporated into the runoff coefficient C, which is usually given as a function of land use (Table 6.8). When multiple land uses are found within the catchment, it is customary to use an area-weighted runoff coefficient in Eq. (6.8). A better estimate would be obtained from measurements since it is often assumed (but is not necessarily true) that

$$C = \frac{\text{Runoff volume}}{\text{Rainfall volume}}. \tag{6.9}$$

When tabulated for actual storm data, there is usually a considerable variation in C values, but the calculation is useful since it gives a rough idea of catchment losses. One of the principal fallacies of the rational method is the assumption that a real catchment will experience a constant fractional loss, regardless of the total rainfall volume or antecedent conditions. The approximation is better as the imperviousness of the catchment increases; Eq. (6.8) is probably most nearly exact for a rooftop.

The intensity i is obtained from an IDF curve (e.g., Fig. 6.6) for a specified return period under the assumption that the duration t_r equals the time of concentration t_c. This is physically realistic because the time of concentration also is the time to equilibrium, at which time the whole catchment contributes to flow at the outfall. Thus, if $t_r < t_c$, then equilibrium would not be reached and it would be wrong to use the total area A. If $t_r > t_c$, then equilibrium would have been reached earlier and a higher intensity should be used. The key is a proper evaluation of t_c. Since it is inversely related to the rainfall intensity, the kinematic wave equation should be used. Since t_c by Eq. (6.6) is proportional to the (unknown) rainfall intensity, an iterative solution is necessary, combining Eq. (6.6) with the IDF curves. Such an iterative solution is easily accomplished, as will be illustrated in Example 6.6.

IDF curves may sometimes be approximated by the following functional form (Meyer, 1928):

$$i = \frac{a}{b + t_r}, \tag{6.10}$$

where a and b are regression coefficients obtained graphically or by least squares. When substituted into Eq. (6.6), the equation can rapidly be solved for $t_r = t_c$ by Newton-Raphson iteration (Hornbeck, 1982).

TABLE 6.8

Typical C Coefficients for 5- to 10-yr Frequency Design*

DESCRIPTION OF AREA	RUNOFF COEFFICIENTS
Business	
Downtown areas	0.70–0.95
Neighborhood areas	0.50–0.70
Residential	
Single-family areas	0.30–0.50
Multiunits, detached	0.40–0.60
Multiunits, attached	0.60–0.75
Residential (suburban)	0.25–0.40
Apartment dwelling areas	0.50–0.70
Industrial	
Light areas	0.50–0.80
Heavy areas	0.60–0.90
Parks, cemeteries	0.10–0.25
Playgrounds	0.20–0.35
Railroad yard areas	0.20–0.40
Unimproved areas	0.10–0.30
Streets	
Asphalt	0.70–0.95
Concrete	0.80–0.95
Brick	0.70–0.85
Drives and walks	0.75–0.85
Roofs	0.75–0.95
Lawns, Sandy Soil	
Flat, 2%	0.05–0.10
Average, 2–7%	0.10–0.15
Steep, 7%	0.15–0.20
Lawns, Heavy Soil	
Flat, 2%	0.13–0.17
Average, 2–7%	0.18–0.22
Steep, 7%	0.25–0.35

* Viessman et al., 1977.

To avoid the complications of an iterative process, constant overland flow inlet times are often used to approximate the time of concentration. These vary from 5 to 30 min with 5 to 15 min most commonly used (American Society of Civil Engineers and Water Pollution Control Federation, 1969). A value of 5 min is appropriate for highly developed, impervious urban areas with closely spaced stormwater inlets, increasing to 10 to 15 min for flat developed urban areas of somewhat lesser density. Values of 20 to 30 min are appropriate for residential districts with widely spaced inlets, with longer values for flatter slopes and larger areas. Due to the ease with which iterative computations can be performed on the computer (Example 6.6), there should be little need for simplifications such as inlet times.

In large catchments, the drainage channels may also need to be considered in the estimate of t_c. For this purpose, the wave travel times along the flow pathways may be added to yield an overall time of concentration. Travel times in channels and conduits should be based on the wave speed given by Eq. (6.7), using an estimate of the channel size, depth, and velocity. Since one objective may be the design of this channel, it may again be necessary to iterate to find the matching values of t_c and t_r. This possibility is examined in Example 6.6.

EXAMPLE 6.5

RATIONAL METHOD DESIGN WITH INLET TIMES

The 10-yr peak flow at a stormwater inlet in Tallahassee, Florida, is to be determined for a 40-ha area in rolling terrain. An inlet time of 20 min may be assumed. Land use is as follows.

LAND USE	AREA (ha)	RUNOFF COEFFICIENT*
Single-family residential	30	0.40
Commercial	3	0.60
Park	7	0.15

* From Table 6.8.

SOLUTION

We find an area-weighted runoff coefficient:

$$\overline{C} = \frac{\Sigma A_i C_i}{\Sigma A_i} = \frac{30 \cdot 0.4 + 3 \cdot 0.6 + 7 \cdot 0.15}{30 + 3 + 7} = 0.37.$$

From Fig. 6.6, for an inlet time of 20 min and return period of 10 yr, the average intensity is 5.6 in./hr (142 mm/hr). The peak flow from Eq. (6.8) is

$$Q_p = 0.00278 \cdot 0.37 \cdot 142 \cdot 40$$
$$= 5.8 \text{ m}^3/\text{s}.$$

EXAMPLE 6.6

RATIONAL METHOD DESIGN WITH KINEMATIC WAVE

Drainage design is to be accomplished for a 4-ac asphalt parking lot in Tallahassee for a 5-yr return period. The dimensions are such that the overland flow length is 1000 ft down a 1% slope.

a) What will be the peak runoff rate?

A runoff coefficient of 0.95 will be assumed (see Table 6.8), as will a Manning roughness coefficient of 0.013 for asphalt (Chow, 1959). From Example 6.2, the kinematic wave parameters for this problem are $\alpha = 11.46$ and $m = 5/3$ for turbulent flow. The IDF curves of Fig. 6.6 may be used to obtain intensity i as a function of duration t_r. However, an iterative procedure must be used since the time of concentration t_c is also a function of i.

Two options for the iterative procedure have been explained in the text. Use of the Meyer equation to approximate an IDF curve is a good alternative if extrapolation of the IDF curve is required, as could occur if the necessary time were less than the minimum value of 8 min given in Fig. 6.6. Alternatively, values can simply be read from the graph and inserted into Eq. (6.6) on a trial basis until $t_r = t_c$. This second method is used here and is easily facilitated using a spreadsheet program on a microcomputer, as indicated in Table 6.9.

For the calculations, a value of t_r is assumed and a value of i is read from Fig. 6.6. Rainfall excess is computed by multiplying intensity from the IDF curve by the runoff coefficient, followed by computation of t_c and Q_p. (These last three steps are done automatically by the spreadsheet program.) Although only the final value of Q_p is of use, the last column shows the relative insensitivity of the peak flow to the changes in t_r due to the "flatness" of the IDF curve in this region. Caution must be taken to use consistent units in Eq. (6.6). Thus, when evaluating the equation, rainfall intensity must be converted to ft/s and the resulting value of t_c converted from seconds to minutes. For example, the first trial for $t_r = 10$ min gives:

$$t_c = \frac{\left[\dfrac{1000}{11.46 \cdot (6.4 \cdot 0.95/43,200)^{2/3}}\right]^{0.6}}{60} = 8.45 \text{ min,}$$

where the factor of 43,200 converts in./hr to ft/s. Because of the "flat" response of t_c, convergence is easily obtained in this example, to the accuracy of the IDF curves. (Remember that the last two columns are automatically calculated by the spreadsheet software.)

b) What size concrete pipe ($n = 0.015$) is required for drainage at the downstream end of the parking lot, for an available slope of 2%?

TABLE 6.9

Iterative Calculations for Rational Method

TRIAL t_r (min)	INTENSITY i (in./hr)	CALCULATION OF t_c (min)	$Q_p = CiA$ (cfs)
10	6.4	8.45	24.3
8	6.6	8.35	25.1
8.4	6.5	8.40	24.7

Standard pipe sizes (in the United States) come in 6-in. increments to a 60-in. diameter, followed by 12-in. increments. Trial computations using Manning's equation show that a 2-ft-diameter pipe is adequate. The full-flow capacity (about 8% less than maximum capacity) is

$$Q_f = \left(\frac{1.49}{n}\right) A_f R^{2/3} S^{1/2}$$

$$= \left(\frac{1.49}{0.015}\right) 3.14 \cdot 0.5^{2/3} \cdot 0.02^{1/2} = 27.8 \text{ cfs.}$$

c) For the calculation of the time of concentration, should a correction be made for the wave travel time in the drainage pipe?

The length of the pipe is equal to half the width of the 4-ac parking lot. The width is

$$W = 4 \text{ ac} \times 43{,}560 \text{ ft}^2/\text{ac}/1000 \text{ ft} = 174 \text{ ft,}$$

so the half-width is 87 ft. During the approach to equilibrium, let average conditions in the pipe be represented by a half-full state. Using Manning's equation, we find the velocity

$$V = \left(\frac{1.49}{0.015}\right) 0.5^{2/3} \cdot 0.02^{1/2} = 8.85 \text{ ft/s.}$$

The wave speed is, from Eq. (6.7),

$$c = 8.85 + \left(\frac{32.2 \times 1.57}{2}\right)^{1/2} = 13.9 \text{ ft/s,}$$

where the half-full area is 1.57 ft^2 and the half-full top width is equal to the diameter of 2 ft.

The wave travel time is $87/13.9 = 6.3$ s. If added to the overland flow value for t_c this would amount to such a small correction that it could not be seen while working with the graphical IDF curves. Hence, this correction can be neglected.

In summary, the rational method is based on the false assumption that losses may be treated as a constant fraction of rainfall regardless of the amount of rainfall or of the antecedent conditions. Fortunately, this assumption gets better as the degree of urbanization increases (i.e., as imperviousness increases), and intelligent use of the rational method (i.e., in conjunction with the kinematic wave equation for time of concentration) should yield a reasonable approximation to the desired peak flow.

Coefficient and Regression Methods

Regression techniques are often incorporated into other methods such as synthetic unit hydrographs (Chapter 2), in which parameters of the hydrograph are related to physical parameters of the catchment. Peak flows may also be found by regression analysis (e.g., Bensen, 1962), but this application is usually performed only for large nonurban catchments. The USGS also uses regression techniques to relate frequency analysis results to catchment characteristics. However, the most straightforward application simply relates measured values of desired independent and dependent variables, such as rainfall and runoff.

When measured rainfall and runoff data are available, it is common to regress the runoff against the rainfall. If a linear equation is fit to the data, it will have the form

$$R = CR(P - DS), \tag{6.11}$$

where

R = runoff depth,
P = rainfall depth,
CR = slope of the fitted line (approximate runoff coefficient),
DS = depression storage (depth).

Here depths, not flows, are predicted. Since the depths are equivalent to volumes (when multiplied by the catchment area), depth and volume are often used interchangeably. The form of Eq. (6.11) assumes that there will be no runoff until the depression storage is filled by rainfall. (If DS is negative, the equation can be rewritten to indicate a positive intercept on the runoff axis. This situation implies a base flow, that is, a contribution to runoff even with zero rainfall.) The slope CR is almost equivalent to a runoff coefficient (Eq. 6.09), except that part of the losses are incorporated into the depression storage. Equation (6.11) works best for monthly or annual totals rather than for individual storm events and is not suited for situations in which there is appreciable carryover in the catchment storage (Diskin, 1970).

EXAMPLE 6.7

REGRESSION OF RUNOFF VS. RAINFALL

Rainfall and runoff volumes (depths) for ten monitored storms for the Megginnis Arm catchment in Tallahassee, Florida, are listed in Table 6.10 and plotted in Fig. 6.10. The fit of the linear regression is also shown in the figure. The parameters listed in the figure indicate that the regression is significant at the 95% level (Haan, 1977), although in this instance the regression is heavily influenced by the single "large" data point at 4.55 in. of rain and 1.4 in. of runoff. The fitted line has the equation

$$\text{Runoff} = 0.308(\text{Rainfall} - 0.059),$$

with both parameters measured in inches. Although the depression storage value of about 0.06 in. indicates that at least that much rain must fall before runoff is expected, the parameter is not significantly different from zero in this instance. This is typical of urban areas in which impervious land cover tends to generate some runoff even for small rainfall totals.

TABLE 6.10

Rainfall-runoff data for Megginnis Arm Catchment, Tallahassee, Florida

DATE OF STORM EVENT			RAIN	RUNOFF
Mo	Day	Yr	(in.)	(in.)
6	6	79	1.48	0.316
9	21	79	1.13	0.201
9	25	79	0.54	0.070
9	26	79	2.40	0.586
9	27	79	0.73	0.148
3	9	80	0.92	0.296
4	10	80	1.35	0.721
5	22	80	0.86	0.285
5	23	80	1.43	0.537
2	10	81	4.55	1.400

Regression – Flood Frequency Methods

The U.S. Geological Survey (USGS) extrapolates urban flood frequencies by regression analysis. Peak flow prediction by this method has been performed for urban areas for the United States as a whole (Sauer et al., 1983) and for many individual cities. For example, flood volumes and flood peaks have been analyzed for Tallahassee by Franklin (1984) and Franklin and Losey (1984), respectively. The procedure first assembles all

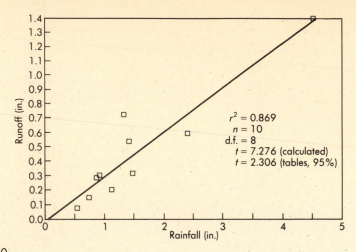

FIGURE 6.10

Runoff depth vs. rainfall depth for ten storm events in Tallahassee, Florida.

data for an area. If the runoff record is long enough, flood frequency estimates are made directly using the log Pearson type 3 distribution (Section 3.5). If the record is insufficient for a frequency analysis of measured peaks (or volumes), the data are used to calibrate a simulation model. A long-term rainfall record is then input to the model and is used to compute a synthesized record of flood peaks (or volumes). A log Pearson type 3 distribution is then fit to the synthesized record. In both cases (measured or synthesized data), the log Pearson type 3 analysis is used to compute flood magnitudes (peaks or volumes) corresponding to return periods of from 2 to 500 yr.

The peaks or volumes at different return periods are then related to basin parameters by multiple regression analysis. Seven statistically significant independent variables in the U.S. nationwide study (Sauer et al., 1983) for prediction of flood peaks were basin size, channel slope, basin rainfall, basin storage, an index of basin development (reflecting man-made changes to the drainage system), impervious area, and the equivalent rural discharge for the same return period. Equivalent rural discharges are available regionally from similar USGS studies of rural areas (e.g., Bridges, 1982, for Florida). Three-parameter equations that performed almost as well used only the basin area, the development index, and the equivalent rural discharge as independent variables. Even further simplification may be possible locally. For example, flood peak estimates of this type in Tallahassee (Franklin and Losey, 1984) used only the basin area and percent impervious area. Since the nationwide equations depend on local estimates of rural peaks, they cannot be demonstrated here.

However, an example of local estimates will be presented with the case study at the end of the chapter.

The regression equations can be used to estimate flows at ungaged sites. Modifications are available to account for extensive basin storage (e.g., due to lakes or wetlands), land use changes, etc. Local offices of the USGS should be contacted for access to the studies for specific regions.

Unit Hydrographs

Unit hydrographs (UH), introduced in Chapter 2, have also been applied in urban areas, with the parameters of the UH related to catchment parameters by regression analysis. This was first performed by Eagleson (1962) and later by several researchers, including Brater and Sherrill (1975) and Espey et al. (1977). An application of an urban UH was presented in Chapter 2.

Time-Area Methods

Time-area methods have also been applied with success in urban areas because of the relative ease of estimation of losses. The origin of the time-area curve was discussed in Chapter 2, but it should be emphasized that, strictly speaking, the isochrones of equal travel time shown in Fig. 2.6 should be travel times of *waves,* not parcels of water, to reach the outlet, since the principle involved is one of time to equilibrium from each contributing area, i.e., the same principle discussed earlier in the context of the kinematic wave. This concept is often ignored in applications of time-area methods, but a relatively easy correction to water parcel travel times can be made using Eq. (6.2) in which water velocities are multiplied by the appropriate kinematic wave exponent to approximate wave velocities. Two widely used urban models that employ time-area methods are the Road Research Laboratory (RRL) model (Watkins, 1962; Stall and Terstriep, 1972; Terstriep and Stall, 1969) and the ILLUDAS model (Terstriep and Stall, 1974). If the time-area diagram is approximated by a triangle and outflows are then routed through a linear reservoir, the technique is sometimes called the Santa Barbara method (Stubchaer, 1975). The time-area method is demonstrated in Example 2.2.

Kinematic Wave

The kinematic wave equations have been discussed earlier in this chapter as well as in Chapters 2 and 4. The numerical techniques shown in Chapter 4 are usually necessary for urban areas with unsteady and nonuniform rainfall excess. However, the analytical solution for the case of steady, uniform rainfall excess (Eagleson, 1970) highlights the concept of wave travel times (Eqs. 6.2 and 6.7) and the time of concentration (Eq. 6.6), both immensely useful in urban hydrology. An example of a watershed model that uses kinematic waves (HEC-1) was discussed in Chapter 5. Urban models that use kinematic waves include the MITCAT model (Harley et al., 1970) and the SWMM model (Huber et al., 1981) for the channel routing component.

Linear and Nonlinear Reservoirs

A catchment surface may be conceptualized as a "reservoir" with inflows due to rainfall (and possible upstream contributions) and outflows due to evaporation, infiltration, and surface runoff (Fig. 6.11a). This has led to several methods for conversion of rainfall into runoff, based on solutions to the reservoir routing equations (Dooge, 1973). For use of reservoir methods, surface storage is spatially lumped; that is, there is no variation with horizontal distance, and the storage is conceptualized as a "tank" with inflows and outflows. Spatial variations may be incorporated by distributing these lumped storages over the catchment to reflect parameter variations and so on (Fig. 6.11b). The distributed storages are then linked by channel routing routines. The equations for reservoir routing can be solved both for conversion of rainfall into runoff and for application to "real" reservoirs, such as detention basins. The latter will be described in Section 6.6.

As shown in Chapter 4, several routing techniques are based on a lumped form of the continuity equation:

$$Q_i - Q = d V / dt \qquad\qquad (6.12)$$

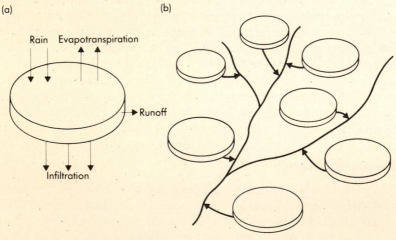

(a)   (b)

FIGURE 6.11

Conceptual reservoir models. (a) Individual catchment. (b) River network. (Note: "reservoirs" are "shapeless." A circular shape is used only for convenience.)

where

Q_i = inflow (cfs),

Q = outflow (cfs),

V = storage (ft³),

t = time (s).

The inflow $Q_i(t)$ may consist of upstream flows or rainfall or both and is assumed to be known. A second equation is thus needed to solve for the two unknowns, $Q(t)$ and $V(t)$; Muskingum and detention basin routing were shown for this purpose in Chapter 4. An additional alternative is to use the relationship

$$Q = KV^b, \qquad (6.13)$$

where K and b are power function parameters that may be fit by regression techniques or through physical relationships. For example, outflow by a weir or orifice or by Manning's equation lends itself naturally to a power function, especially if depth $h(t)$ is used as the dependent variable instead of $V(t)$ by the relationship

$$dV/dt = A(h) \frac{dh}{dt}, \qquad (6.14)$$

where

A = surface area and is a function of depth h.

Then a weir outflow can be represented as

$$Q = C_w L_w (h - h_0)^{1.5}, \qquad (6.15)$$

where

L_w = weir length (perpendicular to flow),

h_o = weir crest elevation,

C_w = weir coefficient.

The weir coefficient is dimensional and depends on several factors, especially the weir geometry (Daugherty et al., 1985). Common values for horizontal weirs perpendicular to the flow direction are $C_w = 3.33$ ft$^{0.5}$/s for U.S. customary units and $C_w = 1.84$ m$^{0.5}$/s for metric units.

An orifice would be included as

$$Q = C_d A_o \sqrt{2g(h - h_o)}, \qquad (6.16)$$

where

C_d = discharge coefficient,

A_o = area of orifice,

h_o = elevation of orifice centerline.

Submerged culverts often behave as orifices with discharge coefficients ranging from 0.62 for a sharp-edged entrance to nearly 1.0 for a well-rounded entrance (Daugherty et al., 1985). Apart from their universal presence near highways, culverts are widely used as outlets from detention ponds in urban areas.

Finally, Manning's equation can be used as the second relationship between storage and outflow. For a wide rectangular channel (as for overland flow) the hydraulic radius is equal to the depth, and Manning's equation has the form

$$Q = W(k_m/n)(h - DS)^{5/3}S^{1/2}, \tag{6.17}$$

where

W = width of (overland) flow,

n = Manning's roughness,

DS = depression storage (depth),

S = slope.

(The constant k_m was discussed in conjunction with Eq. 6.4.) The relationship (6.17) can be coupled with the continuity equation for generation of overland flow, as shown in Example 6.8.

EXAMPLE 6.8

NONLINEAR RESERVOIR MODEL FOR OVERLAND FLOW

Derive a method for generation of overland flow from rainfall by coupling the continuity equation, Eq. (6.12), with Manning's equation, Eq. (6.17).

SOLUTION

Let inflow to the "reservoir" equal the product of rainfall excess i_e and catchment area A. Using U.S. customary units, Manning's equation can be substituted into the continuity equation, yielding

$$i_e A - W(1.49/n)(h - DS)^{5/3}S^{1/2} = A\frac{dh}{dt},$$

in which the surface area A is assumed to be constant. Dividing by the area gives

$$i_e + WCON(h - DS)^{5/3} = \frac{dh}{dt}, \tag{6.18}$$

where

$$WCON \equiv -\frac{1.49\ WS^{1/2}}{An} \tag{6.19}$$

and lumps all Manning's equation parameters into one comprehensive constant.

For use in a simulation model, Eq. (6.18) must be solved numerically since i_e can be an arbitrary function of time and since the equation is nonlinear in the dependent variable h. A simple finite-difference scheme for this purpose is

$$\frac{h_2 - h_1}{\Delta t} = i_e + WCON \left(\frac{h_1 + h_2}{2} - DS\right)^{5/3}, \tag{6.20}$$

where subscripts 1 and 2 denote the beginning and end of a time step, respectively, and the time step is of length Δt. The rainfall excess i_e must be an average value over the time step. Equation (6.20) must be solved iteratively since it is transcendental in the unknown depth h_2. This method is used in the SWMM model (Huber et al., 1981; Roesner, 1982), wherein solution of Eq. (6.20) is accomplished at each time step by a Newton-Raphson iteration (Hornbeck, 1982).

The special case of $b = 1$ in Eq. (6.13) corresponds to a **linear reservoir,** which is often used as a conceptual catchment model (Zoch, 1934; Dooge, 1973) and sometimes as a reservoir or stream routing model. Routing using a linear reservoir is easily accomplished using the Muskingum method shown in Chapter 4 with Muskingum parameter $x = 0$. Analytical solutions are also available that may be evaluated with more or less ease depending on the functional form of the inflow $Q_i(t)$ (Henderson, 1966; Medina et al., 1981). The general case of a nonlinear reservoir ($b \neq 1$ in Eq. 6.13) has received analytical attention (see Dooge, 1973, for a summary) and is readily solved by numerical integration of Eqs. (6.12) and (6.13).

6.5
SEWER SYSTEM HYDRAULICS

Flow Routing

A sewer system is similar to any other network of channels except that the geometry is well known and regular (e.g., a system of circular pipes). It is sometimes necessary, however, to conduct a field survey to obtain invert and ground surface elevations since the system may have been altered or plans may have been lost since construction. In principle, open channel flow through such a system should be readily described by flow routing methods that properly account for the geometry of the sewer cross section. Many shapes are found in various cities (e.g., circular, horseshoe, egg); the geometries of several are described by Chow (1959), Davis (1952), and Metcalf and Eddy (1914). Flow routing is often made more complicated, however, by energy losses at various structures and inflow points, possible surcharging and pressure flow regimes, supercritical flow and transitions in steep areas, and complex behavior at diversions such as pump stations and combined sewer overflow regulators (Fig. 6.2). The energy grade line along a hypothetical sewer with surcharge conditions is shown in Fig. 6.12.

For a network of pipes with no complex structures, flow routing can be accomplished by almost any of the methods discussed in Chapter 4. For example, kinematic wave, Muskingum, nonlinear reservoir, and diffusion methods have all been applied to urban drainage systems, but mostly in the context of urban runoff models since numerical solutions are almost always required. Methods that are suitable for "desktop calculations" are mostly limited to those that combine the runoff generation and routing, such as unit hydrographs.

St. Venant Equations

The "simpler" methods previously discussed (those that use fewer than all the terms in the momentum equation) have the advantage of computational simplicity but are limited to (1) situations in which backwater in the sewer network is unimportant and (2) dendritic networks, in which any branching of the network in the downstream direction does not depend on flow conditions downstream. Weirs and combined sewer regulators satisfy the latter condition as long as they are not submerged due to downstream backwater effects, and for many urban drainage networks backwater may be neglected through most of the system. In areas in which backwater must be considered, it is sometimes due to a "reservoir effect,"

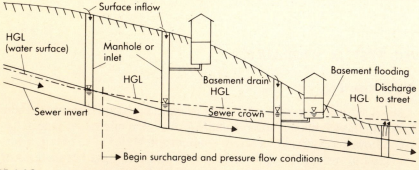

FIGURE 6.12

Hydraulic grade line (HGL) along a surcharged sewer segment. High HGL at downstream end could be caused by high tail water elevation or excessive inflows to sewer system.

in which a horizontal water surface extends upstream in the network due to the ponding action of a weir, for instance. This situation may be treated as an in-line reservoir. But the most general case of a drainage system with low slopes and possible downstream boundary condition of a fixed water level (as at the entrance to a large river, lake, or estuary) requires the complete St. Venant equations (Eqs. 4.23 and 4.26) to simulate the resulting backwater effects. Solution of the complete equations is also necessary in the case of looped networks, in which the flow along each pathway of a loop is a function of downstream conditions. Note that this transient analysis differs from the classical steady-state backwater analysis (illustrated in Chapter 7).

The numerical methods discussed in Chapter 4 for application to the St. Venant equations can be applied equally well in urban drainage systems; a survey of several different approaches is given by Yen (1986). Successful models have been built on the method of characteristics (Sevuk et al., 1973), explicit methods (Roesner et al., 1981) and the four-point implicit method (Cunge et al., 1980). However, the method of characteristics seems to have the most limitations regarding the special structures and flow conditions encountered in sewer systems; in particular, both the explicit and implicit methods are more easily adapted to surcharge conditions.

Surcharging

Surcharging occurs when the hydraulic grade line (HGL) of the flow is above the sewer crown, as in the downstream segment sketched in Fig. 6.12. If the HGL rises above the level of basement drains connected to the sewer system, water will be backed into basements and will cause basement flooding. Health problems may also result if combined sewage enters basements, as sometimes occurs in older cities. The most spectacular effects of surcharging are blown manhole covers and geysers out of the sewer system when the HGL rises well above the ground surface. If the HGL remains below the ground surface and below the level of basement drains, usually no special problem is associated with surcharging, and the flow may even be accelerated by pressurizing the conduit.

Since surcharged conduits are really just a network flowing under pressurized conditions, one way of simulating the surcharge conditions is to treat the network like a water distribution network, in which an iterative solution is used for heads (HGL) at nodes while maintaining continuity and friction losses. This method has been applied on a steady-state basis by Wood and Heitzman (1983) and as part of the more general transient Extran Model by Roesner et al. (1981). Another option is to use the Preissmann slot method, in which a small "imaginary" slot above the sewer allows free-surface flow to be maintained at all times and permits the use of only the St. Venant equations. Models that utilize this scheme include Carredas (SOGREAH, 1977) and S11S (Hoff-Clausen et al., 1981).

The transition from free-surface flow to a surcharged condition is not necessarily smooth and may be accompanied by the release of trapped air, hydraulic jumps, and pressure waves (Song et al., 1983; McCorquodale and Hamam, 1983). No general sewer routing model deals with this problem; special hydraulic model studies or numerical techniques are needed and must be used if it is important.

Routing at Internal Hydraulic Structures

Structures that may be found in a storm or combined sewer include weirs, orifices, pump stations, tide gates, storage tanks, and diversion structures or regulators of various kinds that may or may not behave as a conventional weir or orifice. To the extent that they do, Eq. (6.16) may be used to represent an orifice and Eq. (6.15) a weir. Special allowance must be made for submerged weirs and weirs of other geometries (see Chow, 1959; Henderson, 1966).

Side-flow weirs are often encountered in sewer systems. An exact analysis requires analysis of the gradually varied flow past the weir (Chow, 1959; Henderson, 1966; Hager, 1987). The flow per unit length along the side-flow weir is

$$q = C_s h^{3/2}, \tag{6.21}$$

where

q = discharge per unit length over the side weir (cfs/ft or m²/s),

h = head over the weir crest (ft or m),

C_s = weir coefficient.

The head will vary along the length of the weir, and a numerical solution is required for the head and discharge. The coefficient C_s is a function of the geometry of the weir installation and of the upstream Froude number. As an approximation, a constant value of 4.1 for U.S. customary units and 2.26 for metric units may be used (Ackers, 1957). The full functional relationship for C_s is presented by Hager (1987).

Similarly, a gradually varied flow analysis may be used for the exact analysis of flow over an open **grate inlet** on the street surface. Resource material for the hydraulics of flow over grate inlets on the surface of the ground is described by Ring (1983).

Tide gates are often installed at the outlet of a sewer system to prevent backflow of the receiving water into the sewer conduits; they open only when the head inside the conduits is higher than the head in the receiving water. As such, they are often treated as an orifice, with discharge coefficients available from the manufacturer.

Pumps are usually operated in at least two stages. The first stage is turned on when the water in the wet well rises to a certain level, and the second stage and any additional stages are started when and if the water rises to higher specified levels. The pumps may simply raise the water to a higher level for continued free-surface flow or they may discharge through a **force main** (a pressurized conduit). Outflow for either situation will consist of pulses as the pumps alternately turn on and off. Records of pumping cycles sometimes serve as a means for approximate flow measurements, but the actual discharge depends on the head vs. discharge rating curve for the pumps.

Weirs and other structures often create storage in the system. Routing through such storage can be accomplished by conventional means (Chapter 4), as described below.

6.6

CONTROL OPTIONS

Several options for control of urban runoff quantity were listed in Table 6.1, of which storage in one form or another is dominant. Regulations in many states and localities now require storage for quantity or water quality control in developing urban areas, and most new highways are designed with roadside swales (low depressions, usually vegetated) for mitigation of runoff peaks and volumes. Storage and other measures identified in Table 6.1 will be examined briefly here.

Detention/Retention

Detention storage involves detaining or slowing runoff, as in a reservoir, and then releasing it. In **retention storage,** runoff is not released downstream and is usually removed from storage only by infiltration through a porous bottom or by evaporation. Both types of storage are very common, although designed retention becomes less practical as the size of the drainage area increases. The required retention basin volume should be based on an analysis of storm event volumes, as discussed in Section 6.3.

EXAMPLE 6.9

DETENTION BASIN SIZING

Obtain an estimate for the size of a detention basin required to hold the runoff from a 5-yr (volume) storm for the 2230-ac Megginnis Arm catchment in Tallahassee, Florida.

SOLUTION

The regression relationship developed in Example 6.7 can be used as a very rough method for converting rainfall into runoff. The 5-yr storm event rainfall depth is approximately 8.18 in., from Table 6.6. Thus,

Runoff $= 0.308(8.18 - 0.059) = 2.50$ in. $= 0.21$ ft.

The required volume is the depth times the catchment area:

Volume $= 0.21$ ft $\times$ 2230 ac $\times$ 43560 ft^3/ac $= 2.04 \times 10^7$ ft^3.

If the 5-yr, 24-hr depth of 7.32 in. from the Tallahassee IDF curves were used, a similar calculation would yield a runoff depth of 2.24 in. (0.186 ft) and a volume of 1.81×10^7 ft^3. This value is slightly less than the value obtained from the frequency analysis of storm event depths.

Note that this crude regression analysis ignores important factors of antecedent conditions, variable loss coefficients and spatial and temporal rainfall variations that can be included in the analysis when using certain rainfall-runoff models.

Detention basins (also called retarding basins, or stormwater management basins) are usually designed with a low flow or underflow outlet and with an emergency weir or spillway for high flows. A hypothetical design is shown in Fig. 6.13. Flow routing may be performed by the methods of Chapter 4, as shown in Example 6.10, which illustrates how the detention pond can be treated as a nonlinear reservoir and the equations solved numerically.

EXAMPLE 6.10

FLOW ROUTING THROUGH A DETENTION POND

The hypothetical design of Fig. 6.13 receives as inflow the triangular hydrograph shown in Fig. 6.14. Determine the outflow and stage hydrographs for the conditions given below.

Outflow is only by a culvert of 0.225 m (225 mm) diameter that behaves as an orifice (Eq. 6.16), with discharge coefficient $C_d = 0.9$. The surface area of the basin is given by the power relationship

$A = 400h^{0.7}$,

with A in m^2 and h in m above the culvert centerline (i.e., $h = 0$ at the culvert/orifice centerline). The basin has an initial depth of 0.5 m (above the culvert centerline).

Combining Eqs. (6.12), (6.14), and (6.16) yields the governing differential equation for the depth $h(t)$:

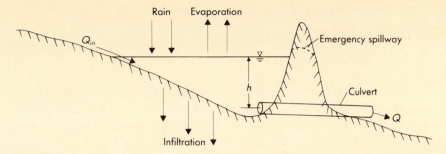

FIGURE 6.13

Hypothetical detention basin.

$$Q_i(t) - C_d A_0 \sqrt{2gh} = 400h^{0.7} \frac{dh}{dt}. \tag{6.22}$$

Placing the derivative on the left-hand side, dividing by the expression for surface area, and substituting for C_d and A_0 puts the equation in a form suitable for numerical solution:

$$\frac{dh}{dt} = \frac{Q_i(t) - 0.0358 \cdot \sqrt{19.6h}}{400h^{0.7}}, \tag{6.23}$$

in which the substitution $g = 9.8$ m/s² has also been made.

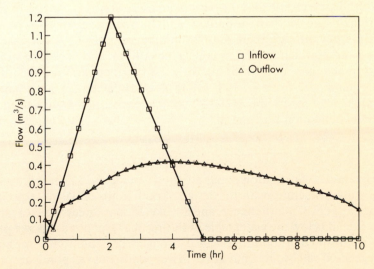

FIGURE 6.14

Inflow and outflow hydrographs for a detention pond.

If the right-hand side is denoted $f(h, t)$, then Eq. (6.23) can be written as

$$\frac{dh}{dt} = f(h, t). \tag{6.24}$$

Many ways exist for the numerical solution of Eq. (6.24) (Runge-Kutta methods were used in Chapter 4), but the Euler method (Hornbeck, 1982) is perhaps the simplest (and also the crudest) and will be illustrated here. For a time step Δt, the derivative is approximated as

$$\frac{dh}{dt} \approx \frac{h(t + \Delta t) - h(t)}{\Delta t}. \tag{6.25}$$

Substituting into Eq. (6.24) and solving for the unknown $h(t + \Delta t)$ gives

$$h(t + \Delta t) = h(t) + \Delta t\, f(h, t). \tag{6.26}$$

It is important to understand that the function $f(h, t)$ in Eq. (6.26) is always evaluated at the "old" time step, that is, at time t, not time $t + \Delta t$. Recalling that $f(h, t)$ equals the right-hand side of Eq. (6.23), the computations can be set up in a tabular fashion, as shown in Table 6.11 for a time step Δt of 15 min or 900 s. A spreadsheet program for a microcomputer is convenient for this purpose, or the solution can be programmed using any desired computer language. (Table 6.11 is the printout of a spreadsheet formulation.) The predicted outflow hydrograph is also plotted in Fig. 6.14. The dip in the hydrographs at the first time step is due to the combination of zero inflow at $t = 0$ and the initial 0.5-m head on the culvert, which means that the basin is initially draining. At $t = 15$ min, the inflow is greater than the outflow, and the water level starts to rise. The maximum depth of 6.95 m and the outflow of 0.418 m³/s occur at the time (4 hr) at which the inflow equals the outflow, and hence the storage must be at a maximum from Eq. (6.12). If the attenuation of the inflow hydrograph peak from 1.2 to 0.418 m³/s is insufficient, it is a relatively simple task to repeat the analysis with alternative parameters until the desired attenuation is achieved.

Examples of detention ponds are shown in Fig. 6.15. There are many practical considerations in the design of both kinds of basins, especially siting, land costs, maintenance, and the like. In some areas the excavation may extend below the water table, resulting in a "wet pond." A wet pond may also result if the bottom is sealed through sedimentation. Some kinds of vegetation can become nuisances in wet ponds, but wet ponds are often more aesthetically pleasing. On the other hand, some dry "ponds" are hardly noticed by the public since they may be implemented as multiple-use facilities, such as parks and recreation areas, and filled only during

TABLE 6.11

Detention Pond Analysis by Euler Method Solution of Equations for Nonlinear Reservoir

TIME (hr)	Q_{in} (cms)	$h(t)$ (m)	Q_{out} (cms)	A_{surf} (m^2)	$f(h, t) = dh/dt$ (m/s)	$h(t + dt)$ (m)
0.00	0.00	0.500	0.112	246.2	−0.00045	0.091
0.25	0.15	0.091	0.048	74.4	0.001374	1.328
0.50	0.30	1.328	0.183	487.8	0.000240	1.544
0.75	0.45	1.544	0.197	542.2	0.000466	1.964
1.00	0.60	1.964	0.222	641.7	0.000588	2.495
1.25	0.75	2.495	0.250	758.5	0.000658	3.088
1.50	0.90	3.088	0.278	880.6	0.000705	3.723
1.75	1.05	3.723	0.306	1003.9	0.000741	4.390
2.00	1.20	4.390	0.332	1126.7	0.000770	5.084
2.25	1.10	5.084	0.357	1248.5	0.000594	5.619
2.50	1.00	5.619	0.376	1339.1	0.000466	6.039
2.75	0.90	6.039	0.389	1408.4	0.000362	6.365
3.00	0.80	6.365	0.400	1461.2	0.000273	6.612
3.25	0.70	6.612	0.407	1500.6	0.000195	6.787
3.50	0.60	6.787	0.413	1528.4	0.000122	6.897
3.75	0.50	6.897	0.416	1545.8	0.000054	6.946
4.00	0.40	6.946	0.418	1553.4	−0.00001	6.936
4.25	0.30	6.936	0.417	1551.8	−0.00007	6.868
4.50	0.20	6.868	0.415	1541.2	−0.00013	6.742
4.75	0.10	6.742	0.411	1521.4	−0.00020	6.558
5.00	0.00	6.558	0.406	1492.2	−0.00027	6.314
5.25	0.00	6.314	0.398	1453.0	−0.00027	6.067
5.50	0.00	6.067	0.390	1413.0	−0.00027	5.818
5.75	0.00	5.818	0.382	1372.2	−0.00027	5.568
6.00	0.00	5.568	0.374	1330.6	−0.00028	5.315
6.25	0.00	5.315	0.365	1288.0	−0.00028	5.060
6.50	0.00	5.060	0.356	1244.4	−0.00028	4.802
6.75	0.00	4.802	0.347	1199.7	−0.00028	4.542
7.00	0.00	4.542	0.338	1153.7	−0.00029	4.278
7.25	0.00	4.278	0.328	1106.5	−0.00029	4.012
7.50	0.00	4.012	0.317	1057.8	−0.00029	3.742
7.75	0.00	3.742	0.306	1007.4	−0.00030	3.468
8.00	0.00	3.468	0.295	955.2	−0.00030	3.190
8.25	0.00	3.190	0.283	901.0	−0.00031	2.907
8.50	0.00	2.907	0.270	844.3	−0.00031	2.619
8.75	0.00	2.619	0.256	784.9	−0.00032	2.325
9.00	0.00	2.325	0.242	722.1	−0.00033	2.024
9.25	0.00	2.024	0.225	655.3	−0.00034	1.715
9.50	0.00	1.715	0.207	583.4	−0.00035	1.395
9.75	0.00	1.395	0.187	504.9	−0.00037	1.061
10.00	0.00	1.061	0.163	417.0	−0.00039	0.709

FIGURE 6.15

Typical detention basin installations.

exceptional storms. Storage is also popular as a multipurpose control device since water quality may be enhanced by sedimentation. Many details and considerations in the design and operation of detention/retention facilities are given by Poertner (1981), DeGroot (1982), and Whipple et al. (1983).

In-system storage can be installed in areas where cheaper surface storage is unavailable. Such storage can consist of concrete tanks or over-sized conduits, or it can simply utilize the sewer system volume itself, sometimes by means of adjustable weirs that are used to detain water in the system.

Surface storage may also be available on rooftops, in parking lots, and on other private lands. For example, if a roof is structurally suited, water may be ponded up to a few inches by using special drain fixtures. Unfortunately, there is little incentive for a private owner to endure problems with leaks and maintenance; storage on private lands often is unsuccessful for this reason. When storage does exist on private lands or adjacent to highways, an allowance for this "loss" must be made in the depression storage value used for the catchment.

Increased Infiltration

Since urbanization creates increased imperviousness, one likely control is to increase the amount of pervious area wherever possible. This has been accomplished with porous parking lots through the use of concrete block or similar shapes laid such that water can infiltrate through the soil-filled center, as shown in Fig. 6.16. The blocks lend strength to the soil, and grass may grow on the soil so that the blocks are not visible. Porous pavement has also been tried experimentally (Field, 1984), and difficul-

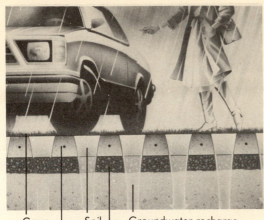

Grass   Soil   Groundwater recharge
Concrete   Sub-base

(a)

(b)

(c)

FIGURE 6.16

Porous pavement for reduction of urban runoff. (a) Pavement structure. (b) Relationship of grass and concrete. (c) Finished parking lot. (Reprinted by permission of Bomanite Corporation, Palo Alto, California.)

ties of possible clogging have been overcome. Although increased infiltration is very useful for small storms and for water quality control (of small storms), it is not of much use for larger storms for which the soil may become saturated.

Other Methods

Inlet restrictions are sometimes used in older cities that suffer from surcharged sewer lines and for which storage is not an option. Such devices (Field, 1984) limit the rate at which surface runoff can enter the sewer system, thus trading off a reduction in the hydraulic grade line in the sewer for increased street surface flooding. However, the latter is usually only temporary, and this measure does reduce basement flooding. Hence inlet restrictions may be the least expensive alternative in older sewer systems.

Nonstructural alternatives include **floodplain zoning** and **floodproofing** of individual structures, as for general floodplain management (Whipple et al., 1983; James and Lee, 1971). Floodplain zoning restricts land use in flood-prone areas (often designated as the 10-yr or 100-yr floodplain) to compatible uses, such as parks, agriculture, and so on. Although some damages may still occur during flooding, major damage to the urban infrastructure is usually avoided. Floodproofing involves raising the foundation of structures above a designated flood elevation, using either fill or pilings or "stilts." The methods may be used in combination, with zoning near the channel and floodproofing farther away (James and Lee, 1971). These two nonstructural alternatives are often not feasible in older cities because of the impossibility of relocating existing structures.

6.7
OPERATIONAL COMPUTER MODELS

Definition

The evolution of hydrologic models was discussed in Chapter 1 and is a natural consequence of computer evolution and availability, theoretical developments, and data collection efforts during the past two decades. Computers and models in fact have revolutionized the manner in which hydrologic computations are now routinely performed. However, many more models have been developed than can be summarized briefly in a text, and a screening method must be employed, leading to the concept of operational models.

In this text, the term *operational* implies three conditions: (1) documentation, especially a user's manual for operation of the program, if any; (2) use and testing by someone other than the model developer; and (3) maintenance and user support. Maintenance most often means support by a government agency; notable examples are the U.S. Army Corps of Engineers, Hydrologic Engineering Center in Davis, California, and the Environmental Protection Agency, Center for Water Quality Modeling in Athens, Georgia. Operational models are generally available and can be acquired and successfully implemented by new users, who would have much more trouble attempting to follow a procedure documented only in a technical paper.

This section discusses a few of the most widely used operational models in urban hydrology. Reviews and comparisons are available from Brandstetter (1976), Delleur and Dendrou (1980), Huber and Heaney (1982), Kibler (1982), Kohlhaas (1982), and Environmental Protection Agency (1983). Water quality processes are not reviewed in this text; a summary of methods is given by Huber (1985). Although most of the models reviewed here were originally developed for mainframe computers, some have also been adapted to microcomputers, as noted.

Soil Conservation Service

The basics of the Soil Conservation Service (SCS) method (1964) have been presented in Chapter 2. The "method" is actually a thick handbook dealing with many aspects of hydrology, from rainfall analysis to flood routing, but the best-known component of the SCS method utilizes information on soil storage, as characterized by curve numbers, to predict the runoff volume resulting from rainfall. A variety of unit hydrograph procedures is then available to distribute the runoff in time. The computerized version of the SCS procedures is called TR 20.

An adaptation of the original method to urban areas is called TR 55 (Soil Conservation Service, 1986). Additional curve numbers are provided in TR 55 as a function of urban land use, and information is presented on travel times and times of concentration for application of the unit hydrograph procedures. The SCS procedures are widely used in engineering practice, and additional source material is available (Viessman et al., 1977; McCuen, 1982). The procedures have been adapted to microcomputers (Soil Conservation Service, 1986). The strength of the SCS method is the enormous data base of soils information; highly site-specific data can be obtained from county soils maps and SCS Soil Survey Interpretation Sheets. (These data are useful as well in applications using methods other than the SCS method.) The weakness of the SCS method is its synthetic design storm philosophy, which was criticized in Section 6.3. With this method it is also difficult to duplicate measured hydrographs in areas with high water tables. This difficulty arises from the sensitivity of the SCS method to the depth of the water table and the method's inability to represent this factor using the three available antecedent conditions (Capece et al., 1984).

ILLUDAS

The Illinois Urban Drainage Area Simulator (ILLUDAS) (Terstriep and Stall, 1974), developed by the Illinois State Water Survey, is an improved version of the British Transport and Road Research Laboratory (RRL) model (Stall and Terstriep, 1972). Based on the time-area method (Chapter 2), the ILLUDAS model also includes pervious area runoff with infiltration estimated by the Horton equation (Eq. 1.18). Site-specific soils data are included for Illinois, but enough information is given in the documentation to adapt the model to other locations. Since it generates complete hydrographs (from design storms), it can be used to estimate volumes as well as peak flows and is perhaps the easiest step upward from a rational method analysis. Output includes routed hydrographs. It also calculates required sizes of circular pipes in the event the input values do not have sufficient capacity. Because of its simplicity, microcomputer availability, and the option of metric units, ILLUDAS has seen extensive application in urban areas and has recently been altered to handle surcharged pipe conditions (Chiang and Bedient, 1986).

Penn State Urban Runoff Model

The Penn State Urban Runoff Model (PSRM) was originally developed for mainframe computers (Lakatos, 1976) for application to the city of Philadelphia and is now supported for microcomputers (Aron, 1987). It is a single-event model developed with the objectives of simplicity and ease of use for design purposes. Infiltration is calculated using SCS methods, overland flow is generated using a nonlinear reservoir method similar to that used in Example 6.8, and pipe routing is by time-of-travel shift without attenuation. Reservoir routing is performed by the storage indication (Puls) method. The PSRM has been applied often (e.g., Kibler and Aron, 1980; Kibler et al., 1981) and is available from the Department of Civil Engineering, Penn State University, University Park, Pennsylvania.

HSPF

The first widely used continuous simulation model in hydrology was the Stanford Watershed Model (Crawford and Linsley, 1966), which was first applied in watersheds in northern California. Input precipitation data for continuous simulation included long-term hourly values from the National Climatic Data Center. These hydrologic roots are now incorporated into the Hydrological Simulation Program — FORTRAN (HSPF), a continuous or single-event hydrologic model that can also be used to simulate non–point source water quality processes (Johanson et al., 1980). Although it has had several urban applications, HSPF has been applied most often in rural situations. The model is supported by the EPA

Center for Water Quality Modeling at Athens, Georgia. A microcomputer version is available from the EPA.

STORM

The first significant use of continuous simulation in urban hydrology came with the Storage, Treatment, Overflow and Runoff Model (STORM), developed by the Army Corps of Engineers for original application to the San Francisco master drainage plan (Roesner et al., 1974; Hydrologic Engineering Center, 1977). The master drainage plan was developed to alleviate combined sewer overflows into San Francisco Bay. The STORM model also uses the National Weather Service hourly rainfall records; 62 yr were analyzed for the STORM simulations of combined sewer overflows in San Francisco. The output was then analyzed to determine the optimum combination of storage and treatment to limit the number of combined sewer overflows to a specified level, using the kind of results shown in Fig. 6.17. As the figure shows, the number of overflows decreases with increasing treatment and storage.

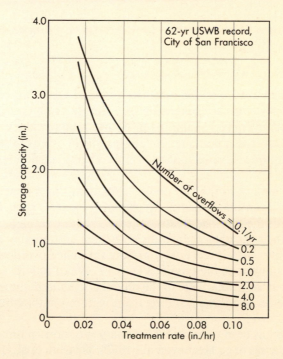

FIGURE 6.17

Number of combined sewer overflows per year as a function of storage capacity and treatment rate, based on a 62-yr continuous simulation. (From Roesner et al., 1974.)

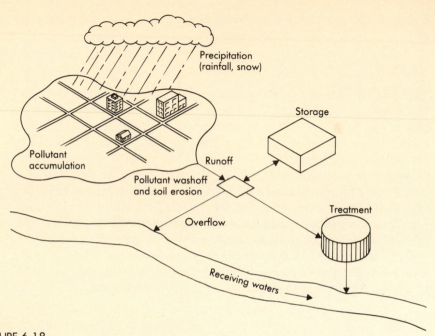

FIGURE 6.18

Conceptualized view of urban system used in STORM. (From Roesner et al., 1974.)

The urban area is schematized in STORM according to Fig. 6.18. Hourly runoff depths are generated very simply using the runoff coefficient method of Eq. (6.11) applied to hourly rainfall values. Depression storage is recovered exponentially between storm events. The SCS curve number technique is available as an option. There is no flow routing of the runoff.

Runoff may be routed through "treatment" at a constant rate to simulated conventional treatment plants. Storage is then filled if the runoff exceeds the specified treatment rate. If both treatment and storage are used to their capacity, there is an "overflow," simulating a combined sewer overflow. A statistical analysis of the simulated overflows is used to prepare an analysis of the type of Fig. 6.17. Water quality may also be simulated, but the primary use of the model is to analyze the frequency of combined sewer overflows resulting from various combinations of treatment and storage.

STORM has been widely used for analysis of non–point source runoff in urban areas (e.g., Heaney et al., 1977; Najarian et al., 1986) and is supported by the Hydrologic Engineering Center at Davis, California. A microcomputer version has been prepared (Bontjie et al., 1984) independently of the HEC.

SWMM

The Storm Water Management Model (SWMM) was developed for the Environmental Protection Agency in 1969–1971 as a single-event model for simulation of quantity and quality processes in combined sewer systems (Metcalf and Eddy et al., 1971). It has since been applied to virtually every aspect of urban drainage, from routine drainage design to sophisticated hydraulic analysis to non–point source runoff quality studies, using both single event and continuous simulation. Through subdividing large catchments and flow routing down the drainage system, SWMM can be applied to catchments of almost any size, from parking lots to subdivisions to cities. A bibliography of SWMM usage is available (Huber et al., 1985).

The current version of SWMM (Huber et al., 1981; Roesner et al., 1981) is segmented into several computational "blocks":

1. The Runoff block generates runoff from rainfall using a nonlinear reservoir method (Example 6.8) and does simple flow routing by the same method. Subsurface flow routing of water infiltrated through the soil surface is optional.

2. The Transport block performs flow routing using a kinematic wave technique.

3. The Extran block performs flow routing by an explicit finite-difference solution of the complete St. Venant equations.

4. The Storage/Treatment block routes through storage units using the Puls (storage indication) method.

5. The Statistics block separates continuous simulation hydrographs and pollutographs (concentration vs. time) into independent storm events, calculates statistical moments, and performs elementary frequency analyses.

Water quality may also be simulated in all blocks except Extran, and the output from continuous simulation may be analyzed by the Statistics block. Metric units are optional. SWMM is supported by the EPA Center for Water Quality Modeling at Athens, Georgia, and is the most prominent urban runoff model. Microcomputer versions are available from EPA and elsewhere (James and Robinson, 1984). A detailed case study of the use of this model is given later in this chapter.

U.S. Geological Survey Models

The U.S. Geological Survey (USGS) is the leading federal agency for research in hydrology and has produced several rainfall-runoff models, including that of Dawdy et al. (1972), which has been updated for urban hydrology by Alley and Smith (1982a, 1982b). The urban model includes generation of runoff and subsequent routing, both by the kinematic wave method. Aids for parameter estimation are included in the models as well. The USGS has conducted studies in many urban areas using the combined modeling-regression approach discussed in Section 6.4.

Model Selection

Although there may be theoretical reasons for preference of one model over another, the choice is often made on much more pragmatic grounds. Considerations include designation of a required model by a permitting agency, model support and documentation, familiarity with techniques used in a model, type of computer required, and availability of data. The last factor is probably the most important. For example, complex flow routing in a sewer cannot be performed without detailed information on invert elevations, ground surface elevations, and hydraulic structures in the sewer system. If such information is not available, a lumped parameter approach using a much simpler method (such as a unit hydrograph) might be appropriate.

When both rainfall and runoff data are available for calibration, most of the methods discussed in this section can be calibrated for the catchment under consideration. In general, the simplest model that is applicable to the problems under study and that can be properly calibrated is probably the best choice. However, simple models are generally able to address fewer hypothetical What if? questions dealing with alternative design formulations because the required physically based procedures may not be incorporated into the models. Hence model choice is an important function of model capability as well as model simplicity.

Finally, model reviews and comparative studies referenced earlier in this section may be consulted for guidance. Since model capabilities continually change and improve, caution should be exercised when reading about models in older references.

6.8
CASE STUDY

Introduction

The calibration and verification of the SWMM model on the Megginnis Arm Catchment in Tallahassee, Florida, will be shown. Results of continuous and single-event SWMM simulations will be illustrated and compared with alternative engineering analyses of the basin for prediction of peak flows, including use of synthetic design storms. The various methodologies will be compared for several return periods, with emphasis on 5-yr peak flows because of the short period of record. Much of the material in this section is extracted from Huber et al. (1986).

The Catchment

The 2230-ac Megginnis Arm Catchment in Tallahassee, Florida (Fig. 6.19) discharges to Lake Jackson north of the city and has experienced both quantity and quality problems in recent years due to increased urbanization. The runoff data start with 1973, and rainfall-runoff data are available since 1979. It is the site of a multimillion-dollar experimental water quality control facility consisting of a detention basin and artificial marsh and has been included by the USGS in its urban runoff studies (Franklin and Losey, 1984). A summary of reports available for the catchment is given by Esry and Bowman (1984), and various data for Tallahassee and the catchment have been illustrated in conjunction with Figs. 6.6 and 6.10, Tables 6.6 and 6.10, and Example 6.7.

Data Input and SWMM Calibration

Rainfall is converted into runoff in the Runoff block of SWMM by the method illustrated in Example 6.8; corresponding input parameters are listed in Table 6.12. The average slope was found by selecting eight points at the edge of the catchment, calculating the path length of each point to the catchment outlet, dividing the path lengths by the change in elevation, and taking an area-weighted average of the eight slopes. Percent impervi-

TABLE 6.12

Runoff Block Parameters for Megginnis Arm Catchment

PARAMETER	VALUE
Area	2230 ac (903 ha)
Width	6000 ft (1830 m)
Percent imperviousness	28.3
Slope	0.0216
Manning's roughness	
Impervious	0.015
Pervious	0.35
Depression storage	
Impervious	0.02 in. (0.5 mm)
Pervious	0.50 in. (13 mm)
Green-Ampt parameters	
Suction	18.13 in. (461 mm)
Hydraulic conductivity	5.76 in./hr (146 mm/hr)
Initial moisture deficit	0.15

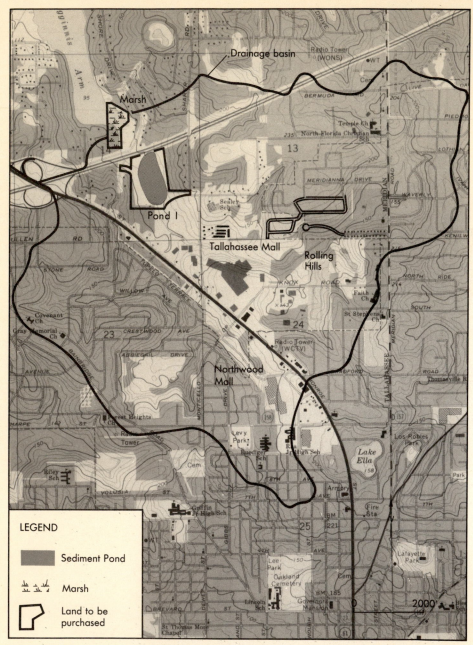

FIGURE 6.19

Megginnis Arm catchment in Tallahassee, Florida. The outlet is at the downstream end of the pond, and rain and flow gages are at the entrance to the pond. (From Huber et al., 1986.)

ousness was obtained from an earlier USGS modeling study by Franklin and Losey (1984).

Green-Ampt infiltration parameters (Chapter 1) were estimated by identifying the soils in the catchment (primarily sandy) from a county soil survey map and then finding the hydraulic conductivity and capillary suction for each soil from data published by Carlisle et al. (1981) and taking a weighted average over the soil types. Manning's n values were selected from charts based on average type of ground cover. Average monthly pan evaporation data were obtained from National Weather Service (NWS) values published by Farnsworth and Thompson (1982), from which actual evapotranspiration (ET) estimates were calculated by multiplying by a pan coefficient of 0.7. Final parameter estimates are shown in Table 6.12. (The subcatchment width is a calibration parameter that is discussed below.) The overall catchment was schematized using only one subcatchment and no channel routing to maintain a reasonable computation time for the continuous simulation.

Rainfall-runoff data for calibration and verification were obtained from available USGS records for the catchment. Ten of the largest storms were selected from the period 1979–1981 and randomly divided into two sets of five storms each, one for calibration and the other for verification. (The ten storms are listed in Table 6.10.) The larger storms were chosen since the ultimate use of the modeling was to be drainage design, for which calibration for large storms is preferable. The model was calibrated for the five storms simultaneously, that is, while maintaining the same parameter values for each (Maalel and Huber, 1984). Calibration of runoff volumes was sufficient using the assumed value of catchment imperviousness (Fig. 6.20); thus calibration for peak flows was achieved only by varying the subcatchment width (a parameter in SWMM that is equivalent to making

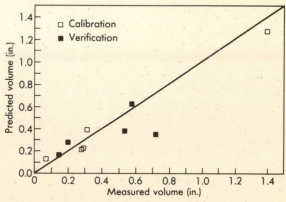

FIGURE 6.20

Goodness of fit of runoff volumes.

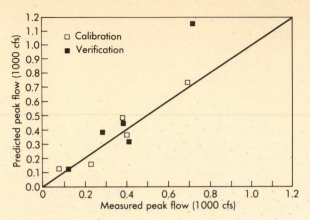

FIGURE 6.21

Goodness of fit of runoff peaks.

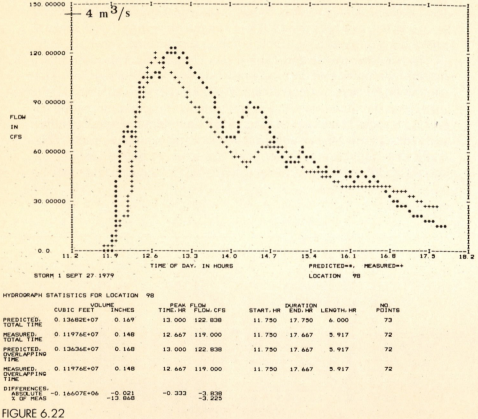

	VOLUME		PEAK FLOW			DURATION		NO.
	CUBIC FEET	INCHES	TIME, HR	FLOW, CFS	START, HR	END, HR	LENGTH, HR	POINTS
PREDICTED, TOTAL TIME	0.13682E+07	0.169	13.000	122.838	11.750	17.750	6.000	73
MEASURED, TOTAL TIME	0.11976E+07	0.148	12.667	119.000	11.750	17.667	5.917	72
PREDICTED, OVERLAPPING TIME	0.13636E+07	0.168	13.000	122.838	11.750	17.667	5.917	72
MEASURED, OVERLAPPING TIME	0.11976E+07	0.148	12.667	119.000	11.750	17.667	5.917	72
DIFFERENCES, ABSOLUTE	-0.16607E+06	-0.021	-0.333	-3.838				
% OF MEAS		-13.868		-3.225				

HYDROGRAPH STATISTICS FOR LOCATION 98

FIGURE 6.22

Predicted and measured hydrograph, September 27, 1979 (verification run).

changes in the slope or roughness; see Eq. 6.19). The results for peak flows are shown in Fig. 6.21. Further efforts might result in a refined calibration, but the agreement between measured and predicted volumes and peaks shown in Figs. 6.20 and 6.21 was considered adequate for this study.

Verification was accomplished by running data for five different storms using the same parameters as in the calibration runs. Results of the verification runs were comparable to the calibration runs and are also shown in Figs. 6.20 and 6.21. A comparison of a predicted and computed hydrograph from the verification runs is shown in Fig. 6.22.

Frequency Analysis of Continuous Results

When the verification was completed, a continuous run of 21.6 yr (259 months: June 1958 to December 1979) was made using hourly rainfall data from the Tallahassee Airport NWS rain gage. Statistical analysis of the predicted flows was performed using the Statistics block of SWMM. The time series of hourly runoff values was separated into 1485 independent storm events by varying the minimum interevent time (MIT; see Section 6.3), until the coefficient of variation of interevent times equaled 1.0, yielding a value of MIT = 19 hr.

The SWMM Statistics block performs a frequency analysis on any or all of the following parameters: runoff volume, average flow, peak flow, event duration, and interevent duration. (If pollutants were being simulated, frequency analyses could also be performed on total load, average load, peak load, flow weighted average concentration, and peak concentration.) Storm events are sorted and ranked by magnitude for each parameter of interest and are assigned an empirical return period in months according to the Weibul formula (Eq. 3.80). The largest-magnitude event for this 259-month simulation was thus assigned a return period of 260 months. Hence, for this simulation, a 5-yr event is bracketed by return periods of 52 and 65 months, both of which were selected as "design events."

On this basis, four storms from the historical record were selected for the "5-yr" storm: two based on peak flow and two based on total flow (runoff volume). The hyetographs of these storms are shown in Fig. 6.23 and their characteristics are listed in Table 6.13.

Of course, the return periods of the individual storms are different when ranked by another parameter. When the four storms of interest are ranked by the other parameters, the corresponding return periods are shown in Table 6.14. Storm V-65 is rare by all measures; in fact, it is the third largest storm of record on the basis of peak flow (and the second largest rainfall volume event). However, when storm V-52 is ranked by peak flow and when storm P-52 is ranked by volume, they are seen to be not especially rare, with return periods of 7.4 and 6.8 months, respectively. Return periods for rainfall volumes shown in Table 6.14 were

TABLE 6.13

Characteristics of Historical Design Storms

STORM NUMBER*	DATE	RUNOFF DURATION (hr)	RUNOFF VOLUME (in.)	RAINFALL VOLUME (in.)	PEAK FLOW (cfs)	TIME SINCE LAST EVENT (hr)	RAINFALL LAST EVENT (in.)
V-65	7/16/64	73	2.69	10.16	1670	74	0.27
V-52	7/26/75	97	2.39	9.30	685	31	0.07
P-65	7/21/69	58	1.61	6.11	1253	32	1.35
P-52	9/3/65	11	1.19	4.35	1224	67	0.11

* V means ranked by volume, P means ranked by peak, and numbers following are return periods in months.

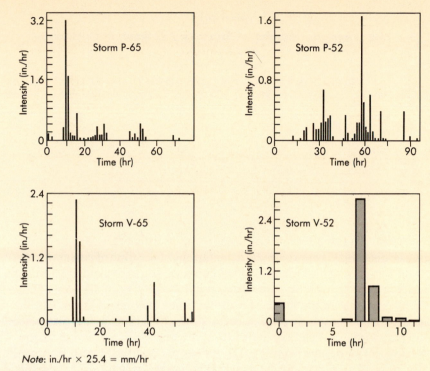

Note: in./hr × 25.4 = mm/hr

FIGURE 6.23

Hyetographs of historic storms from frequency analysis. P means based on peak flow, V means based on runoff volume, numbers refer to return period (months). (From Huber et al., 1986.)

obtained from the SYNOP results (Environmental Protection Agency, 1976; Hydroscience, 1979) shown earlier in Table 6.6. Thus, although the return periods were computed slightly differently, Table 6.14 further illustrates that return periods of the same event ranked by different parameters are rarely the same.

The four storms defined above were run again through the model using hourly rainfall inputs but a 15-min time step, instead of the 1-hr time step used in the continuous simulation. The results are described later.

Synthetic Design Storms

Using the method illustrated in Example 6.4, a 5-yr, 24-hr synthetic design storm was constructed according to the SCS type II distribution (Soil Conservation Service, 1964). The hourly hyetograph is shown in Fig. 6.8; this synthetic storm is certainly unlike any of the four storms shown in Fig. 6.23. The choice of a 24-hr duration for the storms was made for

TABLE 6.14

Return Periods (Months) of Storms by Volume, Peak Flow, and Rainfall

STORM NUMBER	DATE	RETURN PERIOD BY VOLUME	RETURN PERIOD BY PEAK FLOW	RETURN PERIOD BY RAINFALL VOLUME
V-65	7/16/64	65	87	156
V-52	7/26/75	52	7.4	78
P-65	7/21/69	14	65	12
P-52	9/3/65	6.8	52	7.8

two reasons: (1) it is commonly used in engineering practice in the Tallahassee area and (2) approximate calculations of the time of concentration of the basin using the kinematic wave equation (Eq. 6.6) yielded estimates ranging from 13 to 60 hr, depending on the choice of rainfall excess. Thus, 24 hr is at least in the range of possible times of concentration. But the idea of having to arbitrarily select a storm duration in the first place illustrates one of the major difficulties in using synthetic design storms. Various 24-hr depths from the IDF curves are compared with historical storms in Table 6.6. It may be seen that although the depths are comparable, the actual durations of the historical storms are certainly not 24 hr. Alternative design storm models are discussed by Arnell (1982) and illustrated for Tallahassee by Huber et al. (1986).

Other Design Techniques

The USGS (Franklin and Losey, 1984) has performed a flood frequency analysis on 15 basins in the Tallahassee area using the combination of continuous simulation and regression discussed in Section 6.4. The resulting regression equation for most of the Tallahassee area is

$$Q_p(T) = C_p A^a IA^b, \tag{6.27}$$

where

Q_p = predicted flood peak for return period T,

A = drainage area (mi^2),

IA = impervious area as percentage of total area A,

and C_p, a, and b are coefficients that are functions of return period, as given in Table 6.15. Peaks calculated by the log Pearson 3 frequency analysis are also listed in Table 6.15, along with predicted peaks using Eq. (6.27). Megginnis Arm has an area of 3.44 mi^2 and impervious percentage of 28.3.

TABLE 6.15

Coefficients for Flood Peak Prediction in Tallahassee, Using Eq. (6.27)

RETURN PERIOD (yr)	C_p	a	b	PREDICTED Q_p (cfs)*	Q_p BY LP3 (cfs)*
2	10.7	0.766	1.07	986	1040
5	24.5	0.770	0.943	1480	1570
10	39.1	0.776	0.867	1850	1960
25	63.2	0.787	0.791	2350	2530
50	88.0	0.797	0.736	2760	3020
100	118.0	0.808	0.687	3180	3560
200	218.0	0.834	0.589	4380	5150

* For Megginnis Arm, rounded to three significant figures.

Coefficients of determination (R^2) for the multiple regression ranged from 0.97 to 0.99 for the indicated equations. The regression predictions are mainly for use in ungaged catchments. Since they are clearly low for Megginnis Arm, the LP3 values will be used in subsequent comparisons with SWMM predictions.

The rational method could also be applied. The results for this method are also shown below, using a runoff coefficient of 0.27 (from the SWMM modeling), the IDF data of Fig. 6.6, and three t_c values based on a range of assumptions about rainfall intensity, imperviousness, and wave travel time in channels.

Results

The main objective of the above analysis is a comparison of peak flows developed using alternative techniques. For this purpose, the four historical and five synthetic storms were run on a single-event basis using the calibrated version of SWMM described earlier. For all nine runs, a time step of 15 min was used. Peak flows calculated by the model in this way are somewhat higher than for the identical storms during the continuous simulation because SWMM calculates average flows over the hourly time step during continuous simulation (to avoid large continuity errors), whereas instantaneous flows at the end of a time step are calculated during single-event simulation. Results for all storms are shown in Table 6.16, along with the other methods that have been applied to this basin.

Note that when SWMM is run on a single-event basis, and including antecedent conditions, the P-65 storm produces a lower peak than does the P-52 storm, in contrast to the results for the same storms during the continuous simulation (Table 6.13). This is primarily a result of the short time increment rainfall and shorter time step (15 min) used in the single-

TABLE 6.16

Comparison of 5-yr Peak Flows Using Several Methods

SWMM SIMULATIONS		ALTERNATIVE METHODS	
Storm number	Peak flow (cfs)	Method	Peak flow (cfs)
V-65	1996	USGS (LP3)	1570
V-52	938	USGS (Eq. 6.27)	1480
P-65	1384		
P-52	1748	Rational method	
		$t_c = 24$ hr	187
SCS type II	1926	$t_c = 13$ hr	289
		$t_c = 50$ min	1764

event simulations. This is the only such change in rankings that occurred during the simulations for all return periods.

The analysis for the seven highest peaks generated by historic storms was conducted analogously to that for the 5-yr peaks. Results for SWMM historical, SWMM-SCS type II, and USGS predictions are shown in Fig. 6.24. Although other studies (e.g., Marsalek, 1978) have indicated that synthetic design storms will produce higher peaks than historical storms, it may be seen in Fig. 6.24 that this is only partially true for the Tallahassee catchment. Somewhat surprisingly, the highest predicted value is that due

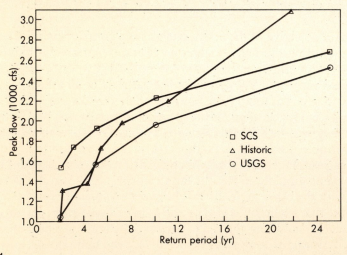

FIGURE 6.24

Predicted peak flows vs. return period for Megginnis Arm catchment. (From Huber et al., 1986.)

to the 22-yr historical storm. Although the SCS design storm is usually considered to be conservative (in the sense of erring on the high side of peak flow predictions), this case study illustrates that the SCS technique cannot always be assumed to be conservative. While peak flow predictions at lower return periods are indeed highest using the SCS storm, the historical storm of return period 22 yr (260 months) gives a higher peak than does the 25-yr SCS synthetic storm. This fact stands out even considering the very broad error bounds on such return period estimates (Chapter 3).

Assuming that the historical storms more accurately represent the true peak flows for the various return periods, Fig. 6.24 also shows that the SCS method is conservative and therefore uneconomical at lower return periods. That is, if a 5-yr design flow were sought, for example, it should be less expensive to design for the value of 1384–1748 cfs predicted using the historical storms than for the value of 1926 cfs predicted using the SCS design storm.

The USGS estimates (using the log Pearson type 3 frequency analysis of simulated flows; Table 6.16) are also shown in Fig. 6.24. They appear somewhat low but are in reasonable agreement with the SWMM simulations for low return periods. For the rational method to predict a peak flow of 1764 cfs would require a duration on the IDF curve of approximately 50 min, much less than the actual (but unknown!) time of concentration. Hence it does not appear to be applicable. Furthermore, the return period for the method is assigned on the basis of conditional frequency analysis of rainfall, as opposed to the storm event analysis of runoff peaks or volumes of the other methods.

Ideally, the measured runoff records should be analyzed to isolate storm events. But the gage for the Megginnis Arm catchment has been in operation only since 1973, making it difficult to identify storms for return periods greater than about 10 yr. Another difficulty is the changing land use within the catchment, with rapid urbanization altering the nature of the rainfall-runoff response. Ordinary frequency analysis (Chapter 3) thus does not apply, and the analysis is directed back to a properly calibrated model.

Case Study Summary

The SWMM model was calibrated and verified on the Megginnis Arm catchment in Tallahassee. Based on a continuous simulation, historical storm events were selected for prediction of peak flows at desired return periods. A comparison was made with the peaks predicted using the SCS type II distribution. Advantages of the peaks predicted using the historical storms reflect the advantages of continuous simulation (Linsley and Crawford, 1974) and include the following:

1. Simulation using historical storms implies a frequency analysis of peak flows (or runoff volumes), not a conditional frequency analysis

of rainfall depths. That is, the frequency analysis is on the parameter of interest.

2. The frequency analysis of the continuous time series of flows includes all effects of antecedent conditions, whereas analysis of the synthetic design storms does not.

3. The historical storms avoid the vexing questions of storm duration, shape, and hyetograph discretization. The duration is especially critical since peak flows may arise out of a storm that lasts several days (e.g., storms V-65, V-52, P-65) or out of a short, intense one (e.g., storm P-52). There is simply no basis for establishing a standard duration, such as 24 hr, for all design work, as seems to be the unfortunate tendency in many urban areas.

4. Historical storms can also be used for analysis of volumes for design of basins for detention or retention. The volume of synthetic storms is arbitrarily linked to the assumed rainfall duration. A given volume can result from infinitely many combinations of intensity and duration, with a corresponding range of return periods.

5. If historical storms rather than "unreal" synthetic storms are used for design, local citizens can be confident that the design will withstand a real storm that they may remember for its flooding. Thus the engineers or agency can make a statement such as "Our design will avoid the flooding that resulted from the storm of September 3, 1965."

SUMMARY

A wealth of analytical options exists for application to quantity problems in urban hydrology. These range from the simple, desktop calculations of the rational method (properly applied) to sophisticated computer models for the microcomputer or mainframe. However, only the latter option permits the use of continuous simulation, which avoids knotty problems of antecedent conditions and assignment of frequency to hydrographs that are inherent in the use of IDF curves and synthetic design storms. Regardless of whether an operational model is used for analysis of urban runoff, the same hydrologic tasks as in nonurban areas must be performed. These tasks include collection and analysis of rainfall data, computation of losses, conversion of rainfall excess to runoff by any of several methods, and flow routing down the drainage system. Although many of the analytical methods that are applied to nonurban areas may also be applied to computation of urban runoff, urban areas present special problems associated with high imperviousness, fast response times, and complex hydraulic phenomena in the sewer system. New methods and models are continually revealed; the reader should maintain contact with the hydrologic literature to stay abreast of new developments in a field subject to rapid change.

PROBLEMS

6.1. Land use and population density data are given for several cities in the table on the following page. Estimate the percent imperviousness for each city using two methods: (1) imperviousness as a function of population density and (2) weighted imperviousness as a function of land use. Assume that undeveloped land is the same as "open" in Table 6.2, and "other" is 50% institutional and 50% open. (Data sources: Heaney et al., 1977; Sullivan et al., 1978.)

6.2. A small, 2-ha, mostly impervious urban catchment has an average slope of 1.5% and the following average Horton infiltration parameters: $f_0 = 4$ mm/hr, $f_c = 1$ mm/hr, $k = 2.2$ hr^{-1}. (Infiltration occurs through cracks in the paving.) Consider the rainfall hyetograph in Fig. P6.2.

a) Determine the depression storage using Fig. 6.3.

b) Determine the time of beginning of runoff.

c) Calculate and sketch the hyetograph of rainfall excess. Use the same time intervals as for the rainfall hyetograph, and average the first nonzero rainfall excess over the whole time interval.

d) Determine the runoff coefficient.

e) Compute the volume of runoff, in m^3.

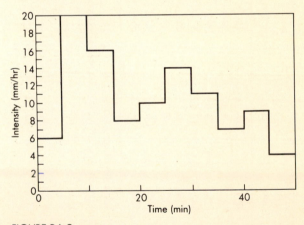

FIGURE P6.2

6.3. Using the runoff volume found in Problem 6.2, determine the depth of storage in a retention basin that has vertical walls and a base area of 500 m^2.

6.4. For the hyetograph of hourly rainfall values shown in Fig. P6.4, determine the number of events corresponding to minimum inter-event times (MITs) of 0, 1, 2, 3, 4, and 5 hr. What MIT is needed to have the entire 40-hr sequence treated as one event?

CITY	URBANIZED AREA (1000-ac)	POPULATION DENSITY (persons/ac)	LAND USE (PERCENT)						
			Residential	Commercial	Industrial	Other*	Undeveloped†	Total	
Boston, Mass.	425	6.24	38.2	5.6	9.7	11.9	34.6	100	
Trenton, N.J.	42	6.59	39.3	5.8	10.0	12.3	32.6	100	
Tallahassee, Fla.	19	4.06	29.1	4.3	7.4	9.1	50.1	100	
Houston, Tex.	345	4.86	33.0	4.8	8.3	10.2	43.7	100	
Chicago, Ill.	626	9.13	46.0	6.8	11.7	14.3	21.2	100	
Denver, Colo.	188	5.58	35.8	5.3	9.1	11.2	38.6	100	
San Francisco, Calif.	436	6.86	40.2	5.9	10.2	12.5	31.2	100	
Windsor, Ont.	26	7.63	38.0	6.0	10.0	18.7	27.3	100	

* "Other" = recreational, schools and colleges, cemeteries.

† High "undeveloped" results from definition of urbanized area, which includes population densities as low as 1 person/ac.

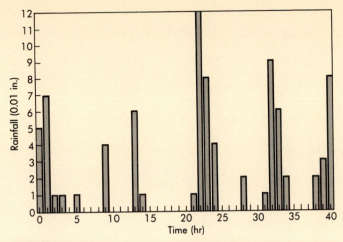

FIGURE P6.4

6.5. The EPA SYNOP program (Environmental Protection Agency, 1976; Hydroscience, 1979) has been run for hourly rainfall data for Houston for the period 1948–1979. A minimum interevent time of 16 hr was used to separate independent storm events, giving the following results:

RANK	DATE	RAINFALL VOLUME (in.)	DURATION (hr)
1	6/11/73	11.55	52
2	6/24/60	11.33	63
3	10/11/70	7.15	17
4	10/14/57	6.78	40
5	11/12/61	6.59	22
6	7/14/49	6.33	53
7	6/3/62	5.73	25
8	6/20/63	5.69	11
9	4/14/66	5.49	17
10	9/9/71	5.45	29
11	9/4/73	5.28	41
12	7/9/61	5.22	77
13	4/14/73	4.81	47
14	10/22/70	4.80	21
15	10/6/49	4.60	46
16	7/29/54	4.55	24
17	9/19/67	4.43	47
18	8/24/67	4.38	27
19	3/20/72	4.22	7
20	10/31/74	4.10	34

(continued)

RANK	DATE	RAINFALL VOLUME (in.)	DURATION (hr)
21	5/12/72	3.98	4
22	5/21/70	3.90	10
23	6/17/68	3.86	8
24	8/2/71	3.84	46
25	5/15/70	3.74	28
26	7/7/73	3.70	11
27	12/10/63	3.60	94

	VOLUME (in.)	DURATION (hr)
Mean	5.37	33.4
Unbiased Standard Deviation	2.01	22.2

a) Perform a Gumbel frequency analysis (Chapter 3) to determine the storm volume magnitudes corresponding to return periods of 2, 5, 10, 25, 50 and 100 yr. Use the method of moments fit.

b) Compare the volumes of part (a) with the 24-hr volumes indicated on the Houston IDF curve shown in Fig. 1.8. The good agreement between the Gumbel-fitted storm event depths and the 24-hr IDF depths indicates that the IDF curves were probably derived from these very same data, extracting the largest 24-hr depth out of a much longer storm event (see the durations listed in the preceding table).

6.6. Using the IDF curves for Houston shown in Fig. 1.8 and the SCS hyetograph distribution given in Table 6.7, prepare a 25-yr SCS type II design storm for Houston. Plot the hyetograph of hourly values.

6.7. Consider a circular pipe of diameter 1 m and a trapezoidal channel of maximum depth 1 m. The trapezoidal channel has a bottom width of 1 m and side slopes (vertical/horizontal) of 0.25.

a) For a Manning roughness of 0.020 and slope of 0.008 for each channel, calculate the water velocity under uniform flow at depths of 0.25, 0.50 and 0.75 m. (*Note:* Appendix E contains a program for uniform flow computations.)

b) Calculate the wave speeds (in downstream direction) in each channel for depths of 0.25, 0.50, and 0.75 m.

c) For each wave speed and each channel, calculate the travel time over a length of 300 m.

6.8. A planned 5.43-ac subdivision is sketched in Fig. P6.8. The soils are generally sandy, and the only runoff will occur from the directly connected (i.e., hydraulically effective) impervious street and driveway surfaces shown in the figure. (Only the 20 × 30 ft portion of each driveway that drains to the street is shown in the figure.) The street is

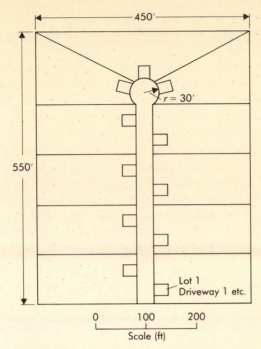

FIGURE P6.8

30 ft wide and the cul-de-sac has a radius of 30 ft. For storm drainage, the plan is to let the stormwater run along the street gutters in lieu of installing a pipe. For purposes of this problem, the entire drainage system can be treated as overland flow. The street slopes from an elevation of 165 ft at the center of the cul-de-sac to 160 ft at the entrance to the subdivision and has a Manning roughness of 0.016.

Drainage regulations specify a 5-yr return period design. Local IDF curves can be approximated functionally as

$$i = \frac{a}{b + t_r},$$

where

i = rainfall intensity (in./hr),

t_r = duration (min),

a, b = constants for different return periods.

For this hypothetical location, $a = 160$ and $b = 18$ for a return period of 5 yr.

a) Estimate the directly connected impervious area (ac).
b) Estimate the maximum drainage length along the directly connected impervious area (ft).

c) Determine the kinematic wave parameters α and m.
d) Estimate the 5-yr peak flow at the outlet. Assume that the impervious surface experiences no losses.

6.9. The subdivision of Problem 6.8 drains to the upstream end of a circular pipe that has an n value of 0.013 and a slope of 0.005 ft/ft. Determine the size of pipe needed to carry the flow from the 5-yr storm. (*Note:* Standard diameters in the United States start at 12 in. and increase in 6-in. increments to 60 in., then continue to increase in 12-in. increments.)

6.10. A 14.7-ac multifamily residential catchment in Miami has a total impervious area of 10.4 ac, but it has a hydraulically effective impervious area of only 6.48 ac. The pervious portions of the basin consist of lawns over a Perrine marl, with a very slow infiltration rate. Rainfall and runoff data monitored by the USGS are reported below for 16 storm events. (Data from Hardee et al., 1979.)

EVENT	RAINFALL (in.)	RUNOFF (in.)
1	2.85	1.983
2	1.17	0.657
3	2.08	1.426
4	1.86	1.176
5	1.67	0.668
6	0.53	0.217
7	0.84	0.541
8	1.50	0.900
9	0.70	0.308
10	0.73	0.277
11	2.02	0.712
12	1.56	0.423
13	0.74	0.330
14	0.75	0.238
15	0.61	0.266
16	1.01	0.444

a) Determine a linear relationship between runoff and rainfall using linear regression analysis (least squares). Test the significance of the regression and plot the data points and the fitted line.
b) What is the value of depression storage for this basin?

6.11. Using the data of Problem 6.10, determine the average runoff coefficient by
a) computing the runoff coefficient for each storm and finding the average;
b) dividing the total runoff for all storms by the total rainfall for all storms;

c) finding the slope of a runoff vs. rainfall regression line that is "forced" through the origin.

Discuss the computed runoff coefficients in relation to the values of imperviousness and hydraulically effective imperviousness.

6.12. A detention pond has the shape of an inverted truncated pyramid, shown in Fig. P6.12(a). It has a rectangular bottom of dimension 120×80 ft, a maximum depth of 5 ft, and uniform side slopes of 3 : 1 (horizontal : vertical). Hence the dimensions at 5-ft depth are also rectangular, with length 150 ft and width 110 ft. The outlet from the basin behaves as an orifice, with a diameter of 1 ft and a discharge coefficient of 0.9. The opening of the orifice (a pipe draining from the center of the basin) is effectively at a depth of zero (i.e., at the bottom of the pond). (The pond floor would typically slope toward the outlet, but this slope will be ignored in this problem. In addition, the orifice will be assumed to follow its theoretical behavior even at very small depths.)

a) What is the total volume of the pond, in ft^3 and ac-ft?

b) Develop the depth vs. surface area and depth vs. volume curves for the pond. Use 1-ft intervals. Tabulate and plot.

(a)

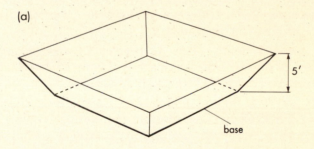

base

5′

(b)

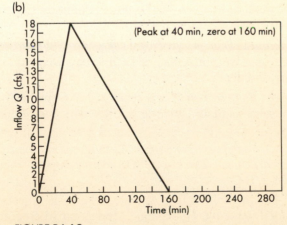

FIGURE P6.12

c) A triangular inflow hydrograph is shown in Fig. P6.12(b). Route it through the detention pond. The storage indication method (Puls method) of Section 4.3 is recommended, with a time step of 10 min. Plot the outflow hydrograph on the same graph as the inflow hydrograph. (*Note:* Appendix E contains a computer program that will perform storage indication routing.)

6.13. Use the USGS regression equation (Eq. 6.27) to compute the 5-yr peak flow for a 4-ac Tallahassee catchment that is 95% impervious. Compare your answer with the value computed in Example 6.6. The reason for the large discrepancy is that the regression equations were developed for much larger catchments (the smallest being 0.21 mi^2) and probably do not apply to a 4-ac catchment.

REFERENCES

Ackers, P., 1957, "A Theoretical Consideration of Side Weirs as Stormwater Overflows," *Proceedings, Institute of Civil Engineers,* London, vol. 120, p. 255, February.

Adams, B. J., and C. D. D. Howard, 1985, *The Pathology of Design Storms,* Publication 85-03, University of Toronto, Dept. of Civil Engineering, Toronto, Ontario, January.

Alley, W. M., and P. E. Smith, 1982a, *Distributed Routing Rainfall-Runoff Model: Version II,* USGS Open File Report 82-344, Gulf Coast Hydroscience Center, NSTL Station, Mississippi.

Alley, W. M., and P. E. Smith, 1982b, *Multi-Event Urban Runoff Quality Model,* USGS Open File Report 82-764, Reston, Virginia.

American Society of Civil Engineers and Water Pollution Control Federation, 1969, *Design and Construction of Sanitary and Storm Sewers,* WPCF Manual of Practice No. 9, Water Pollution Control Federation, Washington, D.C.

Arnell, V., 1982, *Rainfall Data for the Design of Sewer Pipe Systems,* Report Series A:8, Chalmers University of Technology, Dept. of Hydraulics, Göteborg, Sweden.

Aron, G., 1987, *Penn State Runoff Model for IBM-PC,* Dept. of Civil Engineering, Pennsylvania State University, University Park, Pennsylvania, January.

Bensen, M. A., 1962, *Factors Influencing the Occurrence of Floods in a Humid Region of Diverse Terrain,* USGS Water Supply Paper 1580-B, Washington, D.C.

Bontje, J. B., I. K. Ballantyne, and B. J. Adams, 1984, "User Interfacing Techniques for Interactive Hydrologic Models on Microcomputers," *Proceedings Conference on Stormwater and Water Quality*

Management Modeling, Burlington, Ontario, Report R128, Computational Hydraulics, McMaster University, Hamilton, Ontario, pp. 75–89, September.

Brandstetter, A. B., 1977, *Assessment of Mathematical Models for Storm and Combined Sewer Management,* EPA-600/2-76-175a (NTIS PB-259597), Environmental Protection Agency, Cincinnati, Ohio, August.

Brater, E. F., and J. D. Sherrill, 1975, *Rainfall-Runoff Relations on Urban and Rural Areas,* EPA-670/2-75-046 (NTIS PB-242830), Environmental Protection Agency, Cincinnati, Ohio, May.

Bridges, W. C., 1982, *Technique for Estimating Magnitude and Frequency of Floods on Natural-Flow Streams,* USGS Water Resources Investigations 82-4012, Tallahassee, Florida.

Capece, J. C., K. L. Campbell, and L. B. Baldwin, 1984, "Estimating Runoff Peak Rates and Volumes from Flat, High-Water-Table Watersheds," Paper No. 84-2020, American Society of Agricultural Engineers, St. Joseph, Michigan, June.

Carlisle, V. W., R. E. Caldwell, F. Sodek, III, L. C. Hammond, F. G. Calhoun, M. A. Granger, and H. L. Breland, 1978. *Characterization Data for Selected Florida Soils,* Soil Science Dept., University of Florida, Gainesville, Florida, June.

Chiang, C. Y., and P. B. Bedient, 1986, "PIBS Model for Surcharged Pipe Flow," *Journal Hydraulic Engineering, ASCE,* vol. 112, no. 3, pp. 181–192, March.

Chow, V. T., 1959, *Open Channel Hydraulics,* McGraw-Hill, New York.

Clark, C. O., 1945, "Storage and the Unit Hydrograph," *Transactions, ASCE,* vol. 110, pp. 1416–1446.

Crawford, N. H., and R. K. Linsely, 1966, *Digital Simulation in Hydrology: Stanford Watershed Model IV,* Technical Report No. 39, Civil Engineering Dept., Stanford University, Palo Alto, California, July.

Cunge, J. A., F. M. Holly, Jr., and A. Verwey, 1980, *Practical Aspects of Computational River Hydraulics,* Pitman Publishing, Boston.

Daugherty, R. L., J. G. Franzini, and E. J. Finnemore, 1985, *Fluid Mechanics with Engineering Applications,* McGraw-Hill, New York.

Davis, C. V., 1952, *Handbook of Applied Hydraulics,* 2nd ed., McGraw-Hill, New York.

Dawdy, D. R., R. W. Lichty, and J. M. Bergmann, 1972, *A Rainfall-Runoff Simulation Model for Estimation of Flood Peaks for Small Drainage Basins,* USGS Professional Paper 506-B, Washington, D.C.

DeGroot, W. (editor), 1982, *Stormwater Detention Facilities,* Proceedings of the Conference, ASCE, New York, August.

Delleur, J. W., and S. A. Dendrou, 1980, "Modeling the Runoff Process in Urban Areas," *CRC Critical Reviews in Environmental Control,* pp. 1–64, July.

Diskin, M. H., 1970, "Definition and Uses of the Linear Regression Model," *Water Resources Research,* vol. 6, no. 6, pp. 1668–1673, December.

Dooge, J. C. I., 1973, *Linear Theory of Hydrologic Systems,* Technical Bulletin No. 1468, Agricultural Research Service, U.S. Dept. of Agriculture, Washington, D.C.

Eagleson, P. S., 1962, "Unit Hydrograph Characteristics for Sewered Areas," *Journal Hydraulics Division, Proc. ASCE,* vol. 88, no. HY2, pp. 1–25, March.

Eagleson, P. S., 1970, *Dynamic Hydrology,* McGraw-Hill, New York.

Environmental Protection Agency, 1976, *Areawide Assessment Procedures Manual,* Three Volumes, EPA-600/9-76-014, Environmental Protection Agency, Cincinnati, Ohio, July.

Environmental Protection Agency, 1977, *Sewer System Evaluation, Rehabilitation, and New Construction: A Manual of Practice,* EPA-600/2-77-017d (NTIS PB-279248), Environmental Protection Agency, Cincinnati, Ohio.

Environmental Protection Agency, 1983, *EPA Environmental Data Base and Model Directory,* Information Clearinghouse (PM 211A), Environmental Protection Agency, Washington, D.C., July.

Espey, W. H., Jr., D. G. Altman, and C. B. Graves, Jr., 1977, *Nomographs for Ten-Minute Unit Hydrographs for Small Urban Watersheds,* Technical Memorandum No. 32 (NTIS PB-282158), ASCE Urban Water Resources Research Program, ASCE, New York (also, Addendum 3 in *Urban Runoff Control Planning,* EPA-600/9-78-035, EPA, Washington, D. C.), December.

Esry, D. H., and J. E. Bowman, 1984, *Final Construction Report, Lake Jackson Clean Lakes Restoration Project,* Northwest Florida Water Management District, Havana, Florida.

Farnsworth, R. K., and E. S. Thompson, 1982, *Mean Monthly, Seasonal, and Annual Pan Evaporation for the United States,* NOAA Technical Report NWS 34, Office of Hydrology, National Weather Service, Washington, D.C., December.

Field, R., 1984, "The USEPA Office of Research and Development's View of Combined Sewer Overflow Control," *Proceedings of the Third International Conference on Urban Storm Drainage,* Chalmers University, Göteborg, Sweden, vol. 4, pp. 1333–1356, June.

Field, R., and E. J. Struzeski, Jr., 1972, "Management and Control of Combined Sewer Overflows," *Journal Water Pollution Control Federation,* vol. 44, no. 7, pp. 1393–1415, July.

Franklin, M. A., 1984, *Magnitude and Frequency of Flood Volumes for Urban Watersheds in Leon County, Florida,* USGS Water Resources Investigations Report 84-4233, Tallahassee, Florida.

Franklin, M. A., and G. T. Losey, 1984, *Magnitude and Frequency of Floods from Urban Streams in Leon County, Florida,* USGS Water Resources Investigations Report 84-4004, Tallahassee, Florida.

Haan, C. T., 1977, *Statistical Methods in Hydrology,* Iowa State University Press, Ames, Iowa.

Hager, W. H., 1987, "Lateral Outflow over Side Weirs," *Journal of Hydraulic Engineering, ASCE,* vol. 113, no. 4, pp. 491–504, April.

Hardee, J., R. A. Miller, and H. C. Mattraw, Jr., 1979, "Stormwater Runoff Data for a Multifamily Residential Area, Dade County, Florida," USGS Open File Report 79-1295, Tallahassee, Florida.

Harley, B. M., R. E. Perkins, and P. S. Eagleson, 1970, *A Modular Distributed Model of Catchment Dynamics,* Parsons Laboratory Report No. 133, M.I.T., Cambridge, Massachusetts, December.

Harremöes, P. (editor), 1983, *Rainfall as the Basis for Urban Runoff Design and Analysis,* Proceedings of the Seminar, Pergamon Press, New York, August.

Heaney, J. P., W. C. Huber, M. A. Medina, Jr., M. P. Murphy, S. J. Nix, and S. M. Hasan, 1977, *Nationwide Evaluation of Combined Sewer Overflows and Urban Stormwater Discharges, Vol. II: Cost Assessment and Impacts,* EPA-600/2-77-064b (NTIS PB-242290), Environmental Protection Agency, Cincinnati, Ohio, March.

Heaney, J. P., and S. J. Nix, 1977, *Storm Water Management Model: Level I—Comparative Evaluation of Storage-Treatment and Other Management Practices,* EPA-600/2-77-083 (NTIS PB-265671), Environmental Protection Agency, Cincinnati, Ohio, April.

Henderson, F. M., 1966, *Open Channel Flow,* Macmillan, New York.

Hershfield, D. M., 1961, *Rainfall Frequency Atlas of the United States for Durations from 30 Minutes to 24 Hours and Return Periods from 1 to 100 Years,* Technical Paper No. 40, Weather Bureau, U.S. Dept. of Commerce, Washington, D.C., May.

Hicks, W. I., 1944, "A Method of Computing Urban Runoff," *Transactions, ASCE,* vol. 109, pp. 1217–1253.

Hoff-Clausen, N. E., K. Havnø, and A. Key, 1981, "System 11 Sewer: A Storm Sewer Model," *Urban Stormwater Hydraulics and Hydrology,* Proceedings of the Second International Conference on Urban

Storm Drainage, B. C. Yen (editor), Water Resources Publications, Littleton, Colorado, pp. 137–145, June.

Hornbeck, R. W., 1982, *Numerical Methods,* Prentice-Hall, Englewood Cliffs, New Jersey.

Huber, W. C., 1985, "Deterministic Modeling of Urban Runoff Quality," in *Urban Runoff Pollution,* Proceedings of the NATO Advanced Research Workshop on Urban Runoff Pollution, H. C. Torno, J. Marsalek, and M. Desbordes (editors), Springer-Verlag, New York, Series G: Ecological Sciences, vol. 10., pp. 167–242.

Huber, W. C., 1987, Discussion of "Estimating Urban Time of Concentration," by R. H. McCuen et al., *Journal of Hydraulic Engineering, ASCE,* vol. 113, no. 1, pp. 122–124, January.

Huber, W. C., B. A. Cunningham, and K. A. Cavender, 1986, "Use of Continuous SWMM for Selection of Historic Design Events in Tallahassee," *Proceedings of Stormwater and Water Quality Model Users Group Meeting,* Orlando, Florida, EPA/600/9-86/023, Environmental Protection Agency, Athens, Georgia, pp. 295–321, March.

Huber, W. C., and J. P. Heaney, 1982, "Analyzing Residuals Discharge and Generation from Urban and Non-Urban Land Surfaces," in *Analyzing Natural Systems, Analysis for Regional Residuals: Environmental Quality Management,* D. J. Basta and B. T. Bower (editors), Resources for the Future, Washington, D.C., the Johns Hopkins University Press, Baltimore, Maryland, Chapter 3, pp. 121–243 (also EPA-600/3-83-046, NTIS PB83-223321).

Huber, W. C., J. P. Heaney, and B. A. Cunningham, 1985, *Storm Water Management Model (SWMM) Bibliography,* EPA/600/3-85/077 (NTIS PB86-136041/AS), Environmental Protection Agency, Athens, Georgia, September.

Huber, W. C., J. P. Heaney, S. J. Nix, R. E. Dickinson, and D. J. Polmann, 1981, *Storm Water Management Model User's Manual, Version III,* EPA-600/2-84-109a (NTIS PB84-198423), Environmental Protection Agency, Athens, Georgia, November.

Hydrologic Engineering Center, 1977, *Storage, Treatment, Overflow, Runoff Model, STORM, User's Manual,* Generalized Computer Program 723-S8-L7520, U.S. Army Corps of Engineers, Davis, California, August.

Hydroscience, Inc., 1979, *A Statistical Method for Assessment of Urban Stormwater Loads-Impacts-Controls,* EPA-440/3-79-023, Environmental Protection Agency, Washington, D.C., January.

James, L. D., and R. R. Lee, 1971, *Economics of Water Resources Planning,* McGraw-Hill, New York.

James, W., and M. A. Robinson, 1984, "PCSWMM3: Version 3 of the Executive, Runoff and Extended Transport Blocks, Adapted for the IBM-PC," *Proceedings of the Stormwater and Water Quality Modelling Conference,* Burlington, Ontario, McMaster University, Dept. of Civil Engineering, Hamilton, Ontario, pp. 39–52, September.

James, W., and R. Scheckenberger, 1984, "RAINPAK: A Program Package for Analysis of Storm Dynamics in Computing Rainfall Dynamics," *Proceedings of Stormwater and Water Quality Model Users Group Meeting,* Detroit, Michigan, EPA-600/9-85-003 (NTIS PB85-168003/AS), Environmental Protection Agency, Athens, Georgia, April.

James, W., and Z. Shtifter, 1981, "Implications of Storm Dynamics on Design Storm Inputs," *Proceedings of Stormwater and Water Quality Management Modeling and SWMM Users Group Meeting,* Niagara Falls, Ontario, McMaster University, Dept. of Civil Engineering, Hamilton, Ontario, pp. 55–78, September.

Johanson, R. C., J. C. Imhoff, and H. H. Davis, Jr., 1980, *User's Manual for Hydrological Simulation Program: Fortran (HSPF),* EPA-600/9-80-015, Environmental Protection Agency, Athens, Georgia, April.

Kibler, D. F. (editor), 1982, *Urban Stormwater Hydrology,* American Geophysical Union, Water Resources Monograph 7, Washington, D.C.

Kibler, D. F., and G. Aron, 1980, "Observations on Kinematic Response in Urban Runoff Models," *Water Resources Bulletin,* vol. 16, no. 3, pp. 444–452, June.

Kibler, D. F., D. C. Froelich, and G. Aron, 1981, "Analyzing Urbanization Impacts on Pennsylvania Flood Peaks," *Water Resources Bulletin,* vol. 17, no. 3, pp. 270–274, April.

Kidd, C. H. R., 1978, *Rainfall-Runoff Processes over Urban Surfaces,* Proceedings of the International Workshop held at the Institute of Hydrology, Wallingford, Oxon, United Kingdom, April.

Kohlhaas, C. A. (editor), 1982, *Compilation of Water Resources Computer Program Abstracts,* U.S. Committee on Irrigation, Drainage and Flood Control, Denver, Colorado.

Lager, J. A., W. G. Smith, W. G. Lynard, R. G. Finn, and E. J. Finnemore, 1977, *Urban Stormwater Management and Technology: Update and User's Guide,* EPA-600/8-77-014 (NTIS PB-275264), Environmental Protection Agency, Cincinnati, Ohio, September.

Lakatos, D. F., 1976, *Analysis of the Timing of Subwatershed Response to Storms with Use of Computer Simulations Models,* M.S. Thesis, Dept. of Civil Engineering, Penn State University, University Park, Pennsylvania.

Leopold, L. B., 1968, *Hydrology for Urban Land Planning: A Guide-*

book on the Hydrologic Effects of Urban Land Use, USGS Circular 554, Washington, D.C.

Linsley, R. K., and N. H. Crawford, 1974, "Continuous Simulation Models in Urban Hydrology," *Geophysical Research Letters,* vol. 1, no. 1, pp. 59–62, May.

Maalel, K., and W. C. Huber, 1984, "SWMM Calibration Using Continuous and Multiple Event Simulation," *Proc. Third International Conference on Urban Storm Drainage,* Chalmers University, Göteborg, Sweden, vol. 2, pp. 595–604, June.

Marsalek, J., 1978, *Research on the Design Storm Concept,* ASCE Urban Water Resources Research Program Tech. Memo. No. 33, NTIS PB-291936, ASCE, New York (also Addendum 2 to EPA-600/9-78-035, Environmental Protection Agency, Washington, D.C.), September.

McCorquodale, J. A., and M. A. Hamam, 1983, "Modeling Surcharged Flow in Sewers," *Proc. 1983 International Symposium on Urban Hydrology, Hydraulics and Sediment Control,* Report UKY BU131, University of Kentucky, Lexington, Kentucky, pp. 331–338, July.

McCuen, R. H., 1982, *A Guide to Hydrologic Analysis Using SCS Methods,* Prentice-Hall, Englewood Cliffs, New Jersey.

McCuen, R. H., S. L. Wong, and W. J. Rawls, 1984, "Estimating Urban Time of Concentration," *Journal of Hydraulic Engineering, ASCE,* vol. 110, no. 7, pp. 887–904, July.

McPherson, M. B., 1978, *Urban Runoff Control Planning,* EPA-600/9-78-035, Environmental Protection Agency, Washington, D.C., October.

Medina, M. A., W. C. Huber, and J. P. Heaney, 1981, "Modeling Stormwater Storage/Treatment Transients: Theory," *Journal Environmental Engineering Division, Proc. ASCE,* vol. 107, no. EE4, pp. 781–797, August.

Metcalf and Eddy, Inc., 1914, *American Sewerage Practice, Design of Sewers,* vol. 1, 1st ed., McGraw-Hill, New York.

Metcalf and Eddy, Inc., University of Florida, and Water Resources Engineers, Inc., 1971, *Storm Water Management Model, Volume I: Final Report,* EPA Report 11024DOC07/71 (NTIS PB-203289), Environmental Protection Agency, Washington, D.C., July.

Meyer, A. F., 1928, *Elements of Hydrology,* 2nd ed., John Wiley, New York.

Najarian, T. O., T. T. Griffin, and V. K. Gunawardana, 1986, "Development Impacts on Water Quality: A Case Study," *Journal Water*

Resources Planning and Management, ASCE, vol. 112, no. 1, pp. 20–35, January.

Overton, D. E., and M. E. Meadows, 1976, *Stormwater Modeling,* Academic Press, New York.

Patry, G., and M. B. McPherson (editors), 1979, *The Design Storm Concept,* Report EP80-R-8, GREMU-79/02, Civil Engineering Dept., Ecole Polytechnique de Montréal, Montreal, Quebec, December.

Poertner, H. G. (editor), 1981, *Urban Stormwater Management,* Special Report No. 49, American Public Works Association, Chicago.

Restrepo-Posada, P. J., and P. S. Eagleson, 1982, "Identification of Independent Rainstorms," *Journal of Hydrology,* vol. 55, pp. 303–319.

Ring, S. L., 1983, "Analyzing Storm Water Flow on Urban Streets: The Weakest Link," *Proc. 1983 International Symposium on Urban Hydrology, Hydraulics and Sediment Control,* Report UKY BU131, University of Kentucky, Lexington, Kentucky, pp. 351–358, July.

Roesner, L. A., 1982, "Urban Runoff Processes," Chapter 5 in *Urban Stormwater Hydrology,* D. F. Kibler (editor), American Geophysical Union, Water Resources Monograph 7, Washington, D. C.

Roesner, L. A., H. M. Nichandros, R. P. Shubinski, A. D. Feldman, J. W. Abbott, and A. O. Friedland, 1974, *A Model for Evaluating Runoff-Quality in Metropolitan Master Planning,* ASCE Urban Water Resources Research Program Technical Memorandum No. 23, NTIS PB-234312, American Society of Civil Engineers, New York, April.

Roesner, L. A., R. P. Shubinski, and J. A. Aldrich, 1981, *Storm Water Management Model User's Manual Version III: Addendum I, EX-TRAN,* EPA-600/2-84-109b (NTIS PB84-198431), Environmental Protection Agency, Cincinnati, Ohio, November.

Sauer, V. B., W. O. Thomas, Jr., V. A. Stricker, and K. B. Wilson, 1983, *Flood Characteristics of Urban Watersheds in the United States,* USGS Water Supply Paper 2207, U.S. Government Printing Office, Washington, D.C.

Sevuk, A. S., B. C. Yen, and G. E. Peterson, 1973, *Illinois Storm Sewer System Simulation Model: User's Manual,* Research Report No. 73, UILU-WRC-73-0073, Water Resources Research Center, University of Illinois, Urbana, Illinois, October.

Sharp, J. J., and P. G. Sawden, 1984, *BASIC Hydrology,* Butterworths, Boston.

SOGREAH, 1977, *Mathematical Model of Flow in an Urban Drain-*

age Programme, SOGREAH Brochure 06-77-33-05-A, SOGREAH, Grenoble, France.

Soil Conservation Service, 1971, *SCS National Engineering Handbook, Section 4, Hydrology,* Soil Conservation Service, U.S. Dept. Agriculture, U.S. Government Printing Office, Washington, D.C.

Soil Conservation Service, 1986, *Urban Hydrology for Small Watersheds,* Technical Release 55, 2nd ed., U.S. Dept. Agriculture, NTIS PB87-101580 (microcomputer version 1.11, NTIS PB87-101598), Springfield, Virginia, June.

Song, C. C. S., J. A. Cardle, and K. S. Leung, 1983, "Transient Mixed-Flow Models for Storm Sewers," *Journal of Hydraulic Engineering, ASCE,* vol. 109, no. 11, pp. 1487–1504, November.

Stall, J. B., and M. L. Terstriep, 1972, *Storm Sewer Design: An Evaluation of the RRL Method,* EPA-R2-72-068, Environmental Protection Agency, Washington, D.C., October.

Stankowski, S. J., 1974, *Magnitude and Frequency of Floods in New Jersey with Effects of Urbanization,* Special Report 38, USGS, Water Resources Division, Trenton, New Jersey.

Stubchaer, J. M., 1975, "The Santa Barbara Urban Hydrograph Method," *Proc. National Symposium on Urban Hydrology and Sediment Control,* Report UKY BU109, University of Kentucky, Lexington, Kentucky, pp. 131–141, July.

Sullivan, R. H., W. D. Hurst, T. M. Kipp, J. P. Heaney, W. C. Huber, and S. J. Nix, 1978, *Evaluation of the Magnitude and Significance of Pollution from Urban Stormwater Runoff in Ontario,* Research Report No. 81, Canada-Ontario Agreement, Environment Canada, Ottawa, Ontario.

Surkan, A. J., 1974, "Simulation of Storm Velocity Effects on Flow from Distributed Channel Networks," *Water Resources Research,* vol. 10, no. 6, pp. 1149–1160, December.

Tavares, L. V., 1975, "Continuous Hydrological Time Series Discretization," *Journal Hydraulics Division, Proc. ASCE,* vol. 101, no. HY1, pp. 49–63, January.

Terstriep, M. L., and J. B. Stall, 1969, "Urban Runoff by Road Research Laboratory Method," *Journal Hydraulics Division, Proc. ASCE,* vol. 95, no. HY6, pp. 1809–1834, November.

Terstriep, M. L., and J. B. Stall, 1974, *The Illinois Urban Drainage Area Simulator, ILLUDAS,* Bulletin 58, Illinois State Water Survey, Urbana, Illinois.

Tholin, A. L., and C. J. Keifer, 1960, "Hydrology of Urban Runoff," *Transactions, ASCE,* vol. 125, pp. 1308–1379.

Viessman, W., Jr., J. W. Knapp, G. L. Lewis, and T. E. Harbaugh, 1977, *Introduction to Hydrology*, 2nd ed., Harper and Row, New York.

Watkins, L. H., 1962, *The Design of Urban Sewer Systems*, Road Research Technical Paper No. 55, Dept. of Scientific and Industrial Research, Her Majesty's Stationery Office, London.

Weldon, K. E., 1985, "FDOT Rainfall Intensity-Duration-Frequency Curve Generation," in *Stormwater Management, An Update*, M. P. Wanielista and Y. A. Yousef (editors), Environmental Systems Engineering Institute, University of Central Florida, Orlando, Florida, July, pp. 11–31.

Whipple, W., N. S. Grigg, T. Grizzard, C. W. Randall, R. P. Shubinski, and L. S. Tucker, 1983, *Stormwater Management in Urbanizing Areas*, Prentice-Hall, Englewood Cliffs, New Jersey.

Wood, D. J., and G. C. Heitzman, 1983, *Hydraulic Analysis of Surcharged Storm Sewer Systems*, Research Report No. 137, Water Resources Research Institute, University of Kentucky, Lexington, Kentucky.

Wylie, E. B., and V. L. Streeter, 1978, *Fluid Transients*, McGraw-Hill, New York.

Yen, B. C., 1986, "Hydraulics of Sewers," in *Advances in Hydroscience*, vol. 14, B. C. Yen (editor), Academic Press, New York, pp. 1–122.

Zoch, R. T., 1934, "On the Relation Between Rainfall and Streamflow, Part I," *Monthly Weather Review*, vol. 62, pp. 315–322.

7

Floodplain Hydraulics

7.1

UNIFORM FLOW

This chapter on **open channel flow** is limited to coverage of steady flow problems, such as **uniform flow** and **nonuniform flow.** Unsteady channel flow hydraulics is beyond the scope of this text but is available in more advanced fluid mechanics references. Channel and floodplain computations of **water surface profiles** are critical to defining levels of flood inundation and are normally treated as steady flow problems.

Uniform open channel flow is the hydraulic condition in which the water depth and the channel cross section do not change over some reach of the channel. These criteria require that the energy line, water surface, and channel bottom are all parallel. In other words, the total energy change over the channel reach is exactly equal to the energy losses of boundary friction and turbulence.

Uniform flow is eventually established over any long channel reach that has constant flow and an unchanging channel cross section. Strict uniform flow is rare in natural streams because of the constantly changing channel conditions. But it is often assumed for natural streams for engineering calculations, with the understanding that the results are general approximations of the actual hydraulic conditions. Uniform flow assumptions should not be made in cases where a well-defined and consistent channel does not exist and where rapidly varying flow is caused by changes in the channel. Man-made channels are usually very consistent, and uniform flow calculations are much more exact for them.

Two uniform flow equations are frequently used for open channel flow problems: the Chezy formula and Manning's equation, which relate the following variables:

A = cross sectional area of channel,

V = channel velocity,

P = the wetted perimeter of the channel,

R = hydraulic radius, or cross-sectional area A divided by the wetted perimeter P,

S = energy slope, which is equal to the slope of the channel bed under the uniform flow assumptions,

C or n = roughness coefficient, related to the friction loss associated with water flowing over the bottom and sides of the channel.

The Chezy formula was developed in 1775; it relates channel velocity to roughness, hydraulic radius, and channel slope:

$$V = C\sqrt{RS}. \tag{7.1}$$

The C factor can be related to Darcy's friction factor f, used in pipe flow, with the relationship

$$C = \sqrt{8g/f}, \tag{7.2}$$

where g is the gravitational constant.

The Chezy equation is based on two main assumptions: the frictional force, proportional to the square of the velocity, and the uniform flow assumption that the gravity force is balanced by the frictional resistance of the stream. The Chezy equation has been used for pressure pipe flow as well as open channel flow. The Chezy C coefficient for open channel conditions can be calculated using relationships from Chow (1959).

For most present applications, however, Manning's equation is used instead of the Chezy formula for open channel computations. Manning's equation was presented in 1890; it describes roughness with the Manning's roughness coefficient n:

$$V = \frac{1}{n} R^{2/3} \sqrt{S}. \tag{7.3}$$

The roughness coefficients were originally developed in the metric system (m and s), and a conversion coefficient must be used to apply Manning's equation with U.S. customary units:

$$V = \frac{1.49}{n} R^{2/3} \sqrt{S}. \tag{7.4}$$

The conversion coefficient 1.49 is the cube root of 3.28, which converts m^3 to ft^3. A dimensional analysis of n shows units of $TL^{-1/3}$ and illustrates a theoretical problem with Manning's equation. The empirical nature of the relationship does not restrict its usefulness, however, and it is widely used to solve numerous types of open channel flow problems.

The selection of the roughness coefficient n is usually based on "best engineering judgment" or on values prescribed by municipal design ordinances. Several tables are available in the general literature for the selection of Manning's roughness coefficient for a particular open channel (see Table 7.1 or Chow, 1959).

7.2

UNIFORM FLOW COMPUTATIONS

Uniform flow problems usually involve the application of Manning's equation to compute the **normal depth** y_n, the only water depth at which flow is uniform. The selection of Manning's n requires more judgment and experience on the part of the engineer or hydrologist than any other parameters in the equation. One usually solves for normal depth as a function of given bed slope S_0, flow rate, and channel geometry. Likewise, one can solve for the design width of a channel when the normal depth is specified. Various types of slide rules, nomographs, and tables are available for the solution of open channel flow problems for channels of varying shape and dimension (King and Brater, 1976).

A cross section can be characterized by its shape, normal depth, cross-sectional area and hydraulic radius, defined as the ratio of area to wetted perimeter. Figure 7.1 indicates the properties of various geometries of open channel flow sections. Depending on the shape of the sections, Manning's equation can be used to solve for normal depth or width given the other parameters. Examples 7.1 and 7.2 indicate the solution of uniform flow problems for a rectangular and trapezoidal open channel, respectively.

FIGURE 7.1

Geometric properties of common open channel shapes.

SHAPE	SECTION	FLOW AREA A	WETTED PERIMETER P	HYDRAULIC RADIUS R
Trapezoidal	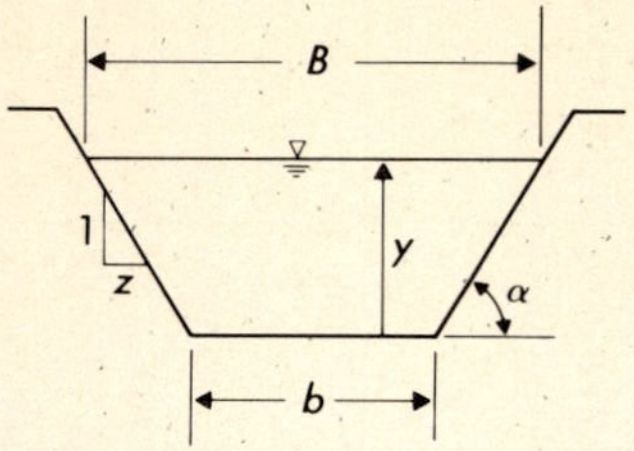	$y(b + y \cot \alpha)$	$b + \dfrac{2y}{\sin \alpha}$	$\dfrac{y(b + y \cot \alpha)}{b + \dfrac{2y}{\sin \alpha}}$
Triangular	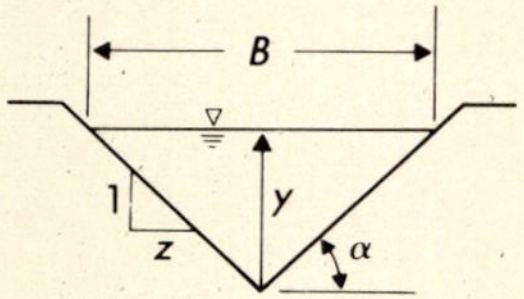	$y^2 \cot \alpha$	$\dfrac{2y}{\sin \alpha}$	$\dfrac{y \cos \alpha}{2}$
Rectangular	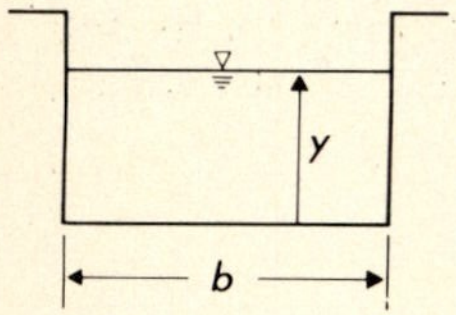	by	$b + 2y$	$\dfrac{by}{b + 2y}$

SHAPE	SECTION	FLOW AREA A	WETTED PERIMETER P	HYDRAULIC RADIUS R
Wide flat		by	b	y
Circular		$(\alpha - \sin \alpha)\dfrac{D^2}{8}$	$\dfrac{\alpha D}{2}$	$\dfrac{D}{4}\left(1 - \dfrac{\sin \alpha}{\alpha}\right)$

TABLE 7.1

Values of Roughness Coefficient n in Manning's formula*

NATURE OF SURFACE	n	
	MIN	MAX
Neat cement surface	0.010	0.013
Wood-stave pipe	0.010	0.013
Plank flumes, planed	0.010	0.014
Vitrified sewer pipe	0.010	0.017
Metal flumes, smooth	0.011	0.015
Concrete, precast	0.011	0.013
Cement mortar surfaces	0.011	0.015
Plank flumes, unplaned	0.011	0.015
Common clay drainage tile	0.011	0.017
Concrete, monolithic	0.012	0.016
Brick with cement mortar	0.012	0.017
Cast iron	0.013	0.017
Cement rubble surfaces	0.017	0.030
Riveted steel	0.017	0.020
Canals and ditches, smooth earth	0.017	0.025
Metal flumes, corrugated	0.022	0.030
Canals		
Dredged in earth, smooth	0.025	0.033
In rock cuts, smooth	0.025	0.035
Rough beds and weeds on sides	0.025	0.040
Rock cuts, jagged and irregular	0.035	0.045
Natural streams		
Smoothest	0.025	0.033
Roughest	0.045	0.060
Very weedy	0.075	0.150

EXAMPLE 7.1

UNIFORM FLOW IN A RECTANGULAR CHANNEL

A rectangular open channel is to be designed to carry a flow of 10 m³/s. The channel is to be built using concrete (Manning's $n = 0.010$) on a slope of 0.005 m/m. Assuming normal flow in the channel, determine the dimensions y_n and b (see Fig. E7.1) if $b = 2y_n$.

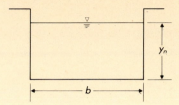

FIGURE E7.1

SOLUTION

Given

$$Q = 10 \text{ m}^3/\text{s},$$
$$n = 0.010,$$
$$S_0 = 0.005 \text{ m/m},$$
$$b = 2y_n.$$

Equation (7.3) gives

$$V = \frac{1}{n} R^{2/3} \sqrt{S_0}$$

for metric units. It is known that $Q = VA$, so

$$Q = \frac{1}{n} AR^{2/3} \sqrt{S_0}.$$

Knowing that

$$R = A/P$$

and referring to the figure, we have

$$A = y_n b = 2y_n^2,$$
$$P = 2y_n + b = 4y_n,$$

so

$$R = 2y_n^2/4y_n = 0.5y_n.$$

Therefore,

$$Q = \frac{1}{n} AR^{2/3} \sqrt{S_0},$$

$$10 = \frac{1}{0.01} (2y_n^2)(0.5y_n)^{2/3} \sqrt{0.005},$$

$$10 = 8.909 y_n^{8/3},$$
$$y_n = (1.1225)^{3/8},$$

$$\boxed{y_n = 1.04 \text{ m.}}$$

Then

$$\boxed{b = 2.08 \text{ m.}}$$

EXAMPLE 7.2

UNIFORM FLOW IN A TRAPEZOIDAL CHANNEL

A trapezoidal channel with side slopes of 2 to 1 is designed to carry a normal flow of 200 cfs. The channel is grass-lined with a Manning's n of 0.025 and has a bottom slope of 0.0006 ft/ft. Determine the normal depth, bottom width, and top width (see Fig. E7.2) assuming normal flow and the bottom width (BW) as 1.5 times the normal depth.

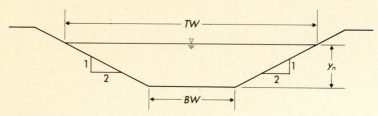

FIGURE E7.2

SOLUTION

Given

$$Q = 200 \text{ cfs,}$$
$$n = 0.025,$$
$$S_0 = 0.0006 \text{ ft/ft,}$$
$$BW = 1.5 \, y_n.$$

Equation (7.4) gives

$$V = \frac{1.49}{n} R^{2/3} \sqrt{S_0}$$

for U.S. customary units. Since $Q = VA$,

$$Q = \frac{1.49}{n} AR^{2/3} \sqrt{S_0}.$$

Referring to the figure, we have

$$P = BW + 2(y_n^2 + 2y_n^2)^{1/2}$$
$$= 1.5y_n + 2(\sqrt{5}y_n)$$
$$= y_n(1.5 + 2\sqrt{5})$$

and

$$A = BWy_n + 2(1/2)(2y_n)(y_n)$$
$$= 1.5y_n^2 + 2y_n^2$$
$$= 3.5y_n^2.$$

Then

$$R = A/P$$
$$= \frac{(3.5y_n^2)}{(y_n)(1.5 + 2\sqrt{5})}$$
$$= 0.586y_n,$$

so

$$Q = \frac{1.49}{n} AR^{2/3}\sqrt{S_0},$$

$$200 = \frac{1.49}{0.025}(3.5y_n^2)(0.586y_n)^{2/3}\sqrt{0.0006},$$

$$200 = 3.578y_n^{8/3},$$

$$y_n = (55.89)^{3/8},$$

$$\boxed{y_n = 4.5 \text{ ft.}}$$

Then

$$\boxed{BW = 6.8 \text{ ft}}$$

and

$$\boxed{TW = 24.8 \text{ ft.}}$$

For a given slope, flow rate, and roughness, an **optimum channel cross section** can be found that requires a minimum flow area. The optimum cross section is the one for which the hydraulic radius R is a maximum and the wetted perimeter is a minimum, since $R = A/P$. Figure 7.2 indicates the properties of optimum open channel sections based on minimizing the wetted perimeter for each shape. Example 7.3 indicates the procedure for finding the optimum cross section for a trapezoidal channel.

FIGURE 7.2

Properties of optimum open channel sections.

SHAPE	SECTION	OPTIMUM GEOMETRY	NORMAL DEPTH y_n	CROSS-SECTIONAL AREA A
Trapezoidal	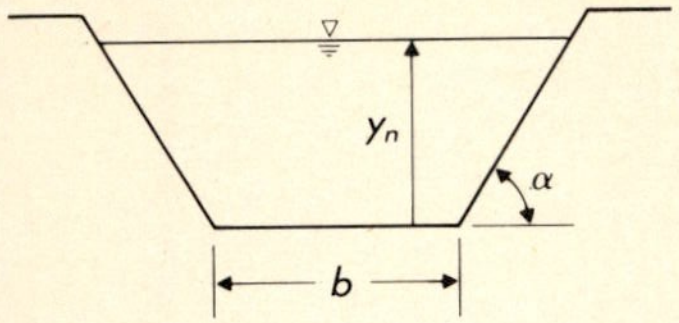	$\alpha = 60°$ $b = \dfrac{2}{\sqrt{3}}\, y_n$	$0.968\left[\dfrac{Qn}{S_b^{1/2}}\right]^{3/8}$	$1.622\left[\dfrac{Qn}{S_b^{1/2}}\right]^{3/4}$
Rectangular	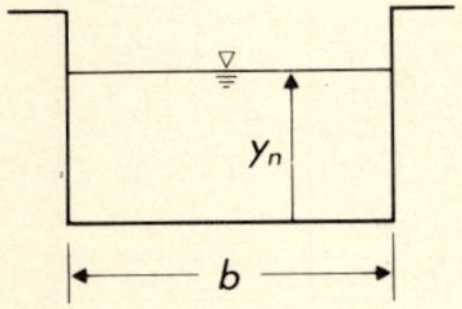	$b = 2y_n$	$0.917\left[\dfrac{Qn}{S_b^{1/2}}\right]^{3/8}$	$1.682\left[\dfrac{Qn}{S_b^{1/2}}\right]^{3/4}$
Triangular	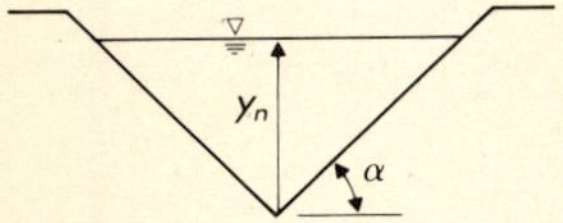	$\alpha = 45°$	$1.297\left[\dfrac{Qn}{S_b^{1/2}}\right]^{3/8}$	$1.682\left[\dfrac{Qn}{S_b^{1/2}}\right]^{3/4}$

SHAPE	SECTION	OPTIMUM GEOMETRY	NORMAL DEPTH y_n	CROSS-SECTIONAL AREA A
Wide flat		None	$1.00\left[\dfrac{(Q/b)n}{S_b^{1/2}}\right]^{3/8}$	—
Circular		$D = 2y_n$	$1.00\left[\dfrac{Qn}{S_b^{1/2}}\right]^{3/8}$	$1.583\left[\dfrac{Qn}{S_b^{1/2}}\right]^{3/4}$

EXAMPLE 7.3

DETERMINATION OF AN OPTIMUM CROSS SECTION

Given the trapezoidal open channel shown in Fig. E7.3, determine the optimum side slope angle θ and the optimum ratio of side length to bottom width L/b, where θ and L/b are defined as shown. Assume normal flow depth in the channel.

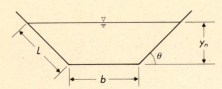

FIGURE E7.3

SOLUTION

As discussed previously, the optimum cross section is the one for which R is maximized. To maximize R, P must be minimized. From the figure, it can be seen that

$$A = by_n + (L \cos \theta)(L \sin \theta)$$

and that

$$L = y_n/\sin \theta.$$

Therefore,

$$A = by_n + (y_n/\sin \theta)^2(\cos \theta)(\sin \theta)$$

or

$$A = by_n + (y_n^2)(\cos \theta/\sin \theta).$$

Solving for b yields

$$b = \frac{A - y_n^2(\cos \theta/\sin \theta)}{y_n}.$$

The wetted perimeter P is given by

$$P = b + 2L$$

or, substituting for b,

$$P = \frac{A}{y_n} - y_n \frac{\cos \theta}{\sin \theta} + 2 \frac{y_n}{\sin \theta}.$$

To minimize P, differentiate with respect to y_n and set $dP/dy_n = 0$:

$$\frac{dP}{dy_n} = -\frac{A}{y_n^2} - \frac{\cos\theta}{\sin\theta} + \frac{2}{\sin\theta} = 0.$$

Solving this equation for y_n^2 gives

$$y_n^2 = \frac{A\sin\theta}{2 - \cos\theta}.$$

To find the optimum value for θ, we differentiate y_n^2 with respect to θ and $dy_n/d\theta$ is set to 0:

$$2y_n\frac{dy_n}{d\theta} = \frac{A\cos\theta}{2 - \cos\theta} - \frac{A\sin^2\theta}{(2 - \cos\theta)^2}$$

or

$$0 = A\frac{\cos\theta(2 - \cos\theta) - \sin^2\theta}{(2 - \cos\theta)^2}.$$

Dividing through by A and simplifying gives

$$\cos\theta = 1/2$$

or

$$\boxed{\theta = 60°.}$$

From the equation for y_n^2, we get

$$A = \frac{(2 - \cos\theta)y_n^2}{\sin\theta}$$

or

$$A = \sqrt{3}\,y_n^2.$$

But we also have

$$A = by_n + y_n^2(\cos\theta/\sin\theta)$$

These two equations for A are now set equal and solved for b:

$$\sqrt{3}y_n^2 = by_n + y_n^2(\cos\theta/\sin\theta),$$

$$\sqrt{3}y_n = \frac{b + y_n(1/2)}{\sqrt{3}/2},$$

$$b = y_n(2/\sqrt{3}),$$

or

$$b = \frac{y_n}{\sin\theta}.$$

Referring again to the figure,

$$L = \frac{y_n}{\sin \theta}.$$

Thus $L = b$ for the optimum channel, and the optimum ratio is

$$\boxed{L/b = 1.}$$

7.3
SPECIFIC ENERGY AND CRITICAL FLOW

Specific energy is a special case of **total energy** that can be defined for any location along an open channel. The total energy can be defined as the sum of the pressure head, the elevation head, and the velocity head for any cross section. The energy equation becomes

$$H = \frac{p}{\gamma} + z + \frac{V^2}{2g}, \tag{7.5}$$

where $p/\gamma = y$ for a free surface and $\gamma = \rho g$. Specific energy E at a single section is referenced to the channel bed and therefore is a sum of depth y and **velocity head** $V^2/2g$:

$$E = y + \frac{V^2}{2g}, \tag{7.6}$$

where the depth y is measured normal to the channel bed. For uniform flow at a section, specific energy can be written in terms of flow rate $\dot{Q}$ by substituting $V = Q/A$ in Eq. (7.6):

$$E = y + \frac{Q^2}{2gA^2}. \tag{7.7}$$

For simplicity, the following discussion is limited to a wide rectangular channel where $V = Q/A = q/y$, where q is flow per unit width in the open channel Q/b. Thus E can be written as a function of y only:

$$E = y + \frac{Q^2}{2gb^2y^2}, \tag{7.8}$$

where b is channel width.

Figure 7.3 shows the variation of depth as a function of E for a given flow rate. It can be seen that for a given flow rate and specific energy there are two possible values of depth y, called alternate depths. The curve for constant q gives a curve of depth values and the corresponding values of E. As q increases, the curves are shifted to the right. For each curve in Fig. 7.3,

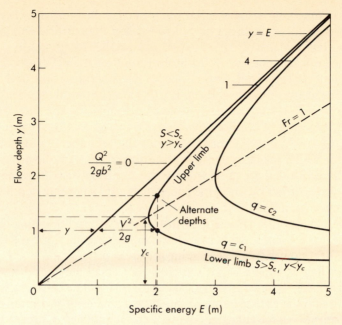

FIGURE 7.3

Specific energy diagram.

there is a value of depth y_c that gives a minimum E, which may be determined by differentiating Eq. (7.8) and setting the result to 0:

$$\frac{dE}{dy} = 1 - (q^2/gy^3). \qquad (7.9)$$

Solving for y, we obtain the **critical depth** y_c, defined as

$$y_c = (q^2/g)^{1/3}. \qquad (7.10)$$

Summarizing the results for a rectangular channel, **critical flow** can be characterized by the following relationships:

$$E_{min} = (3/2)y_c = (V_c^2/2g) + y_c,$$

$$\frac{V_c^2}{2g} = \frac{y_c}{2} \quad \text{or} \quad \frac{V_c}{\sqrt{gy_c}} = 1,$$

$$y_c = (q^2/g)^{1/3}. \qquad (7.11)$$

The two arms of the curve in Fig. 7.3 provide additional information for classifying open channel flows. On the upper limb of the curve, the flow is said to be tranquil, while on the lower limb it is called rapid flow. The velocity and flow rate occurring at the critical depth are termed V_c and q_c, the critical velocity and critical flow, respectively. The velocity of the upper limb flow is slower than critical and is called **subcritical** velocity, and the velocity of the lower limb flow is faster than critical and is called **supercritical** velocity.

The critical condition occurs when **Froude number** (Fr) is equal to 1, where $Fr = V/\sqrt{gy}$. Thus, Fr < 1 occurs for subcritical flow and Fr > 1 occurs for supercritical flow. From Eq. (7.11) a condition of subcritical or supercritical flow may be easily tested by comparing the velocity head $V^2/2g$ to the value of $y/2$. For any value of E there exists a critical depth for which the flow is a maximum. For any value of q there exists a critical depth for which specific energy is a minimum. For any flow condition other than critical, there exists an alternate depth at which the same rate of discharge is carried by the same specific energy. Alternate depth can be found by solving Eq. (7.8).

The condition of critical flow just defined is used to define channel slope. If, for a given roughness and shape, the channel slope is such that the uniform flow is subcritical, slope is said to be **mild** and $y > y_c$. If the uniform flow is supercritical, the slope is termed **steep** and $y < y_c$. A **critical** slope S_c is the slope that will just sustain a given rate of discharge in a uniform flow at critical depth. When flow is near critical, a small change in E results in a large change in depth, and an undulating stream surface will result. Because of this phenomenon, shown in Fig. 7.3, it is undesirable to design channels with slopes near the critical condition.

For nonrectangular channels, the specific energy equation becomes

$$E = y + \frac{Q^2}{2gA^2},\tag{7.12}$$

where $A = F(y)$. Differentiating with respect to y and realizing that $dA = B\,dy$, where B is the width of the water surface, we obtain

$$\frac{dE}{dy} = 1 - \frac{Q^2}{2g}\left(\frac{2}{A^3}\frac{dA}{dy}\right), \quad \text{or}$$

$$\frac{Q^2}{g} = \left(\frac{A^3}{B}\right)_{y=y_c}.\tag{7.13}$$

If the channel is rectangular, $A = By$, and the above reduces to Eq. (7.10). Example 7.4 indicates the computation of critical flow conditions for an open channel based on the above equations.

EXAMPLE 7.4

CRITICAL FLOW COMPUTATION

Water is flowing uniformly in an open triangular channel at a rate of 14 m³/s. The channel has 1:1 side slopes and a roughness coefficient of

$n = 0.0012$ (see Fig. E7.4). Determine whether the flow is subcritical or supercritical if the bottom slope is 0.006 m/m.

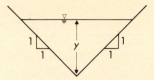

FIGURE E7.4

SOLUTION

Critical depth is found using Eq. (7.13), which states

$$Q^2/g = (A^3/B),$$

when $y = y_c$. Referring to the figure, it can be seen that

$$A = y^2,$$
$$P = 2\sqrt{2}y,$$
$$R = y/(2\sqrt{2}),$$
$$B = 2y.$$

Thus for $y = y_c$,

$$A = y_c^2 \quad \text{and} \quad B = 2y_c.$$

Therefore

$$(A^3/B) = (Q^2/g),$$
$$(y_c^6/2y_c) = (14 \text{ m}^2/9.81$$
$$y_c^5 = 19.98 \text{ m}^5,$$
$$y_c = 1.82 \text{ m}.$$

Since the flow is assumed to be uniform, the depth can be found using Manning's equation (Eq. 7.3):

$$Q = (1/n)AR^{2/3}\sqrt{S_0},$$
$$14 = (1/0.0012)(y^2)(y/2\sqrt{2})^{2/3}(\sqrt{0.006}),$$
$$y^{8/3} = 0.433,$$
$$y = 0.731 \text{ m}.$$

Comparing the uniform flow depth for these conditions to the critical flow depth shows that $y < y_c$. Therefore the flow in the channel is *supercritical*.

7.4

OCCURRENCE OF CRITICAL DEPTH

When flow changes from subcritical to supercritical or vice versa, the depth must pass through critical depth. The condition of critical depth implies a unique relationship between y and V or Q. This condition can occur only at a **control section.** In the transition from supercritical to subcritical flow, a **hydraulic jump** will occur, as described in Section 7.8. By measuring the depth at a control section, we can compute a value of Q for a channel based on critical flow equations.

Critical depth occurs when flow passes over a weir or a free outfall with subcritical flow in the channel prior to the control section. Critical depth may also occur in a channel if the bottom is suddenly elevated or if the sidewalls are moved in by contraction. In fact, flumes are designed so as to force flow to pass through critical depth by adjusting the bottom and sides of a channel. In this way, one simply measures the depth in the flume to estimate Q.

In mountainous streams, a sudden change in slope from mild to steep will force the flow condition to pass through critical and will set up the possibility for standing waves or whitewater. One should never design a channel on a slope that is near critical because of the unpredictable water

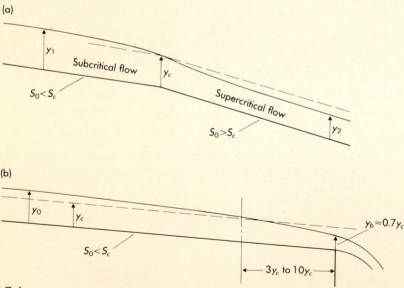

FIGURE 7.4

Occurrence of critical depth. (a) Change in flow from subcritical to supercritical at a break in slope. (b) Free outfall. Mild slope.

surface. Figure 7.4 indicates two possible conditions under which critical depth can occur in an open channel.

7.5

NONUNIFORM FLOW OR GRADUALLY VARIED FLOW

The previous discussion of uniform flow indicates that a channel of constant shape and slope is a requirement for the condition of uniform flow. However, with a natural stream, the shape, size, and slope of the cross section typically varies along the stream length. Such variations produce the condition of nonuniform or **gradually varied flow** in most problems of interest to the engineering hydrologist. Equations from uniform flow can be applied to the nonuniform case by dividing the stream into lengths or reaches within which uniform conditions apply.

In an open channel or natural stream the effect of a grade or slope tends to produce a flow with a continually increasing velocity along the flow path. Gravity is opposed by frictional resistance, which increases with velocity, and eventually the two will be in balance for the case of uniform flow. When the two forces are not in balance, the flow is nonuniform. Nonuniform flow may be called gradually varied flow if changing conditions occur over a long distance. **Rapidly varied flow** occurs when there is an abrupt change or transition confined to a short distance.

Gradually varied flow can occur at the entrance and exit of a channel, at changes in cross-sectional shape or size, at bends, and at structures such as dams, bridges, or weirs. Of particular interest in watershed analysis is the case of the natural stream and associated bridge crossings. This represents one of the most complex applications of nonuniform flow theory, and computer models have been written to handle all of the necessary computations. An example of a very popular flood profile model is the U.S. Army Corps of Engineers HEC-2 model (Hydrologic Engineering Center, 1982), which is described in detail later in this chapter.

7.6

GRADUALLY VARIED FLOW EQUATIONS

When flow in an open channel or stream encounters a change in bed slope or a change in cross-sectional shape, the flow depth may change gradually. Such a flow condition where depth and velocity may change along the channel must be analyzed numerically. The energy equation is applied to a differential control volume, and the resulting equation relates change in depth to distance along the flow path. A solution is possible if one assumes that head loss at each section is the same as that for normal flow with the

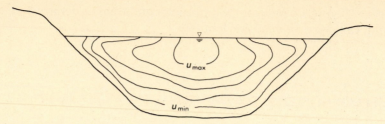

FIGURE 7.5

Typical velocity distribution in an open channel.

same velocity and depth of the section. Thus a nonuniform flow problem is assumed to be approximated by a series of uniform flow stream segments.

The total energy of a given channel section can be written

$$H = z + y + \frac{\alpha V^2}{2g},$$ (7.14)

where $z + y$ is the potential energy head above a datum and the kinetic energy head is represented by the velocity head term. The value of α ranges from 1.05 to 1.40 for most channels and is an indication of the velocity distribution across the cross section. In many cases, the value is assumed to be 1.0 (see Fig. 7.5).

The energy equation for steady flow between two sections, 1 and 2, a distance L apart becomes (Fig. 7.6)

$$z_1 + y_1 + \frac{\alpha_1 V_1^2}{2g} = z_2 + y_2 + \frac{\alpha_2 V_2^2}{2g} + h_L,$$ (7.15)

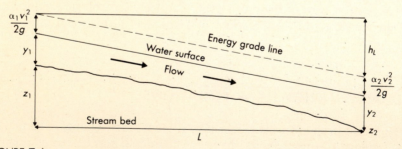

FIGURE 7.6

Nonuniform flow energy equation. For steady flow between two stations, 1 and 2, a distance L apart,

$$z_1 + y_1 + \frac{\alpha_1 v_1^2}{2g} = z_2 + y_2 + \frac{\alpha_2 v_2^2}{2g} + h_L.$$

where h_L is the head loss from section 1 to section 2. If we assume that $\alpha = 1$, $z_1 - z_2 = S_0 L$, and $h_L = SL$, the energy equation becomes

$$y_1 + \frac{V_1^2}{2g} = y_2 + \frac{V_2^2}{2g} + (S - S_0)L. \tag{7.16}$$

The energy slope is determined by assuming that the rate of head loss at a section is the same as that for flow at normal depth with the same velocity of the section. Thus, using Manning's equation and solving for S, we have

$$S = \left(\frac{nV_m}{1.49 R_m^{2/3}}\right)^2, \tag{7.17}$$

where the subscript m refers to a mean value for the reach. If we differentiate Eq. (7.14) with respect to x, the distance along the channel, the rate of energy dissipation is found to be

$$\frac{dH}{dx} = \frac{dz}{dx} + \frac{dy}{dx} + \frac{\alpha}{2g} \frac{d(V^2)}{dx}. \tag{7.18}$$

Equation (7.18) describes the variation of water surface profile for gradually varying flows. S_0 and S terms can be substituted. The sign of the slope of the water surface profile depends on whether flow is subcritical or supercritical and on the relative magnitudes of S and S_0. We observe $V = q/y$ for Eq. (7.18), and the last term can be written as ($\alpha = 1$)

$$\frac{1}{2g} \frac{d}{dx}(V^2) = \frac{1}{2g} \frac{d}{dx}\left(\frac{q^2}{y^2}\right) = -\left(\frac{q^2}{g}\right)\left(\frac{1}{y^3}\right)\frac{dy}{dx}. \tag{7.19}$$

Thus

$$-S = -S_0 + \frac{dy}{dx}\left(1 - \frac{q^2}{gy^3}\right). \tag{7.20}$$

If we include the definition of the Froude number (Fr), then the water surface profile for a rectangular section can be written as

$$\frac{dy}{dx} = \frac{S_0 - S}{1 - (V^2/gy)} = \frac{S_0 - S}{1 - \text{Fr}^2}. \tag{7.21}$$

With S_0 and n known and the depths and velocities at both ends of the reach given, the length L of the reach can be computed as follows:

$$L = \frac{[y_1 + (V_1^2/2g)] - [y_2 + (V_2^2/2g)]}{S - S_0}. \tag{7.22}$$

Example 7.5 illustrates the water surface profile computation using Eq. (7.22) and is referred to as the standard step method.

EXAMPLE 7.5

WATER SURFACE PROFILE DETERMINATION

A trapezoidal channel with the dimensions shown in Fig. E7.5(a) is laid on a slope of 0.001 ft/ft. The Manning's n value for this channel is 0.025 and the rate of flow through the channel is 1000 cfs. Calculate and plot the water surface profile from the point where the channel ends (assume a free outfall) to the point where $y \geq 0.9y_n$.

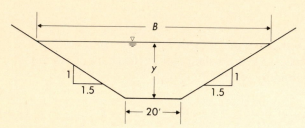

FIGURE E7.5(a)

SOLUTION

At the point of free outfall, flow will pass through critical depth. Equation (7.13) is used to find the value of critical depth:

$$Q^2/g = (A^3/B).$$

Referring to the figure, we have

$$A = 20y_c + 2(1/2)(y_c)(1.5y_c)$$
$$= y_c(20 + 1.5y_c),$$
$$B = 20 + 2(1.5y_c)$$
$$= 20 + 3y_c.$$

Then

$$\frac{(1000 \text{ cfs})^2}{(32.2 \text{ ft/s}^2)} = \frac{y_c^3(20 + 1.5y_c)^3}{(20 + 3y_c)},$$

or

$$\frac{y_c^3(20 + 1.5y_c)^3}{(20 + 3y_c)} = 31{,}056 \text{ ft}^5.$$

Solving by trial and error yields

$$\boxed{y_c = 3.853 \text{ ft.}}$$

The value of normal depth is found using Manning's equation (Eq. 7.3):

$$Q = (1.49/n)AR^{2/3}\sqrt{S_0}.$$

Referring again to the figure, we have

$$P = 20 + 2\sqrt{3.25}y_n$$

$$R = \frac{y_n(20 + 1.5y_n)}{20 + 3.61y_n}.$$

So

$$1000 \text{ cfs} = \frac{1.49}{0.025}[y_n(20 + 1.5y_n)]\left[\frac{y_n(20 + 1.5y^n)}{(20 + 3.61y_n)}\right]^{2/3}\sqrt{.001},$$

or

$$\frac{[y_n(20 + 1.5y_n)]^{5/3}}{(20 + 3.61y_n)^{2/3}} = 530.58.$$

Solving by trial and error yields

$$\boxed{y_n = 6.55 \text{ ft}}$$

and

$$0.9y_n = 5.90 \text{ ft}.$$

Thus the range of depth for which the profile is desired is 3.85 ft to 5.90 ft. For chosen values of y_1 and y_2, V_1 and V_2 can be found by $Q = V/A$, and S can be calculated using Eq. (7.17). Thus for each pair of chosen y_1 and y_2, a reach length $L = \Delta x$ can be determined from Eq. (7.22). Values of y, A, P, R, V, V_m, R_m, S, $y + V^2/2g$, Δx, and x are given in the table on page 444.

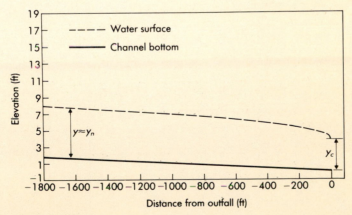

FIGURE E7.5(b)

y (ft)	A (ft²)	P (ft)	R (ft)	V (ft/s)	V_m (ft/s)	R_m (ft)	S	$y + \dfrac{V^2}{2g}$ (ft)	Δx (ft)	$x = \Sigma\Delta x$ (ft)
3.85	99.23	33.88	2.93	10.08				5.43		0
4.10	107.22	34.78	3.08	9.33	9.71	3.01	0.0061	5.45	−3.92	−3.9
4.40	117.04	35.86	3.26	8.54	8.94	3.17	0.0048	5.53	−21.05	−25.0
4.70	127.14	36.95	3.44	7.87	8.21	3.35	0.0038	5.66	−46.43	−71.4
5.00	137.50	38.03	3.61	7.27	7.57	3.53	0.0030	5.82	−80.00	−151.4
5.30	148.14	39.11	3.79	6.75	7.01	3.70	0.0024	6.01	−135.70	−287.1
5.60	159.04	40.19	3.95	6.29	6.52	3.87	0.0020	6.21	−200.00	−487.1
5.90	170.22	41.27	4.12	5.87	6.08	4.04	0.0016	6.44	−383.30	−870.4
6.20	181.70	42.35	4.29	5.50	5.69	4.21	0.0013	6.67	−766.70	−1637.1
6.50	193.40	43.44	4.45	5.17	5.34	4.37	0.0011	6.92	−2500.00	−4137.1

The negative sign for x values indicates that the profile is a backwater profile. As seen in column 1, y values are chosen at a regular interval. It is possible to use smaller intervals for y near the free outfall since this is the steepest part of the profile and y is changing quickly over small distances of x. The water surface profile is plotted in Fig. E7.5(b) on page 443 with respect to the channel bottom, using the free outfall as the datum.

7.7

CLASSIFICATION OF WATER SURFACE PROFILES

The classification of water surface profiles for nonuniform flow can be studied most easily for rectangular channels. Then Eq. (7.21) becomes

$$\frac{dy}{dx} = \frac{S_0 - S}{1 - \text{Fr}^2} = \frac{S_0(1 - S/S_0)}{1 - \text{Fr}^2},$$

where

$$S = \frac{n^2 V^2}{y^{4/3}} = \frac{n^2 Q^2}{b^2 y^{10/3}}$$

in metric units. From Manning's equation for a wide rectangular channel of width b, Q is related to normal depth:

$$Q = \frac{R^{2/3}\sqrt{S_0} A}{n} = \frac{y_n^{2/3}\sqrt{S_0} b y_n}{n}. \tag{7.23}$$

Solving for S_0, we have

$$S_0 = \frac{n^2 Q^2}{b^2 y_n^{10/3}} \tag{7.24}$$

and

$$S/S_0 = (y_n/y)^{10/3}. \tag{7.25}$$

With the Froude number expressed in terms of critical depth,

$$\text{Fr} = (y_c/y)^{3/2}, \tag{7.26}$$

the slope of the water surface for this case becomes

$$\frac{dy}{dx} = S_0 \left[\frac{1 - (y_n/y)^{10/3}}{1 - (y_c/y)^3} \right]. \tag{7.27}$$

The sign of dy/dx, the water slope, depends on the relations among depths y, y_n, and y_c in Eq. (7.27) (see Table 7.2).

TABLE 7.2

Surface Profiles for Gradually Varied Flow

SURFACE PROFILES	CURVE	DEPTH	FLOW	SURFACE SLOPE
Mild slope, $S_b < S_c$	$M1$	$y > y_n > y_c$	Subcritical	Positive
	$M2$	$y_n > y > y_c$	Subcritical	Negative
	$M3$	$y_n > y_c > y$	Supercritical	Positive
Steep slope, $S_b > S_c$	$S1$	$y > y_c > y_n$	Subcritical	Positive
	$S2$	$y_c > y > y_n$	Supercritical	Negative
	$S3$	$y_c > y_n > y$	Supercritical	Positive
Critical slope, $S_b = S_c$	$C1$	$y > y_c = y_n$	Subcritical	Positive
	$C3$	$y < y_c = y_n$	Supercritical	Positive
Horizontal slope, $S_b = 0$	$H2$	$y > y_c$	Subcritical	Negative
	$H3$	$y < y_c$	Supercritical	Positive
Adverse slope, $S_b < 0$	$A2$	$y > y_c$	Subcritical	Negative
	$A3$	$y < y_c$	Supercritical	Positive

From earlier consideration of critical flow, in Section 7.3, the following definitions are presented:

If $y_n > y_c$,   then $S_0 < S_c$ and the slope is mild (subcritical).

If $y_n = y_c$,   then $S_0 = S_c$ and the slope is critical.

If $y_n < y_c$,   then $S_0 > S_c$ and the slope is steep (supercritical).

The three conditions are referred to as mild (M), critical (C), and steep (S) slopes. Other slopes may be **horizontal** (H) or **adverse** (A). The derivation presented here can also be shown to apply to a general cross-sectional channel.

For a given bed slope, the shape of the water surface profile depends on actual depth y relative to y_n and y_c. If the water surface lies above both normal and critical depth lines, it is type 1; if it lies between these lines it is type 2; and if it is below both lines, it is type 3. There are a total of twelve possibilities, as shown in Table 7.2. For most cases of interest in the analysis of floodplains or floodways, the $M1$ profile is the most important and most observed case.

Mild Slope Cases

For the $M1$ profile, called the **backwater curve**, $y > y_n$ and $y_n > y_c$, and both the numerator and denominator are positive in Eq. (7.27). The water surface slope dy/dx is positive, and as y increases, the water slope approaches S_0 and the free surface approaches the horizontal. As depth increases, velocity must decrease gradually to maintain constant Q. The $M1$ profile typically occurs where dams, bridges, or other control structures tend to create a backwater effect along a stream (see Fig. 7.7a). For a uniform, prismatic channel, Eq. (7.21) can be solved using the **standard step method** depicted in Example 7.5. Lengths upstream of the dam are calculated incrementally as depths are gradually decreased toward y_n, since the depth y will asymptotically approach normal depth at some point upstream. Accuracy of the resulting calculation depends on the number of increments selected between the starting water surface at the dam and the value of y_n.

For a natural stream, cross-sectional areas and slopes may change at various points along the stream. Since Manning's n values may vary both with location in the cross section and with location along the stream, more sophisticated computer programs must be used for calculation of the $M1$ profile. Sections 7.9–7.15 describe in detail the HEC-2 model, one of the most widely used water surface profile models. The model is primarily used to define floodplains and floodways for natural streams.

For the $M2$ profile, $y_n > y > y_c$, and flow is subcritical, with a negative numerator and positive denominator in Eq. (7.27). Thus, dy/dx is negative, and depth decreases in the direction of flow, with a strong

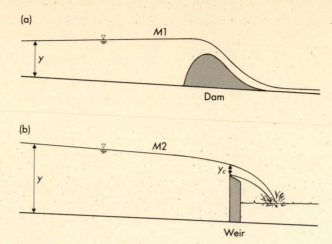

FIGURE 7.7

(a) **Example of an M1 profile. (b) Example of an M2 profile.**

curvature when $y \rightarrow y_c$, where the equation predicts $dy/dx \rightarrow \infty$. Thus the equation does not strictly hold at $y = y_c$ and Eq. (7.11) applies. The $M2$ curve, called the drawdown curve, can occur upstream from a section where the channel slope changes from mild to critical or supercritical, as in flow over a spillway crest or weir (see Fig. 7.7b). The $M2$ profile is created due to a control condition *downstream,* since flow is subcritical.

The $M3$ profile is for supercritical flow where $y_n > y_c > y$; both numerator and denominator are negative, so dy/dx is positive, and depth increases in the direction of flow. The $M3$ curve occurs downstream of a spillway or sluice gate, and as critical depth is approached from below, a sudden transition called the hydraulic jump occurs from supercritical to subcritical conditions (see Figs. 7.8 and 7.9).

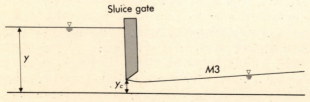

FIGURE 7.8

Example of an M3 profile.

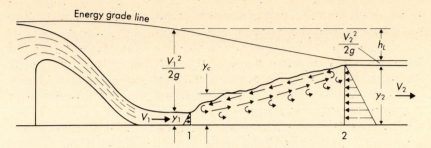

FIGURE 7.9
Hydraulic jump on horizontal bed following spillway.

Steep Slope Cases

The steep curves that occur on steep slopes ($S_0 > S_c$), where $y_c > y_n$, can be analyzed in the same way as the M curves, with due regard for downstream control for subcritical flow ($S1$) and upstream control for supercritical flow ($S2$ and $S3$). For example, a dam on a steep slope will produce an $S1$ curve, which is preceded by a hydraulic jump. A sluice gate on a steep channel will produce an $S3$ curve, which will smoothly approach the uniform depth line ($y = y_n$). Table 7.2 illustrates the steep slope cases.

Horizontal and Adverse Slope Cases

The $H2$ and $H3$ profiles correspond to the $M2$ and $M3$ curves for $S_0 = 0$. $H1$ cannot exist since for $S_0 = 0$, normal depth is infinite. The profile slope can be determined from Eq. (7.21), but this profile can exist only for a short section contained in more complex channel reaches, since flow cannot continue indefinitely on a horizontal bed.

The adverse slope, $S_0 < 0$, occurs where flows sometimes move against gravity; as in the case of horizontal slopes, the adverse slope can occur only for a short section in a more complex channel system. In some natural channel systems, there are occasional reaches where these conditions can and do occur.

7.8
HYDRAULIC JUMP

The hydraulic jump is covered in detail in most fluid mechanics textbooks (Daugherty and Franzini, 1985; Fox and McDonald, 1985). The transition from supercritical flow to subcritical flow produces a marked discontinuity in the surface, characterized by a steep upward slope of the profile and a significant loss of energy through turbulence. The hydraulic jump can best be explained by reference to the $M3$ curve of Table 7.2 and Fig. 7.8, downstream of a sluice gate. The flow decelerates because the mild slope is not great enough to maintain constant supercritical flow, and the specific energy E decreases as the depth increases. When downstream conditions require a change to subcritical flow, the need for the change cannot be telegraphed upstream. Theory calls for a vertical slope of the water surface. Therefore, a hydraulic jump occurs and flow changes from supercritical to subcritical conditions.

When flow at a section is supercritical and downstream conditions require a change to subcritical flow, the abrupt change in depth involves a significant energy loss through turbulent mixing. A jump will form when slope changes from steep to mild, as on the apron at the base of a spillway. In fact, stilling basins below spillways are designed to dissipate the damaging energy of supercritical velocities, and downstream erosion is held in check.

An equation can be derived for horizontal or very mild slopes to relate depth before and after the jump has occurred. Assuming hydrostatic pressure and neglecting gravity and friction forces in this case, Newton's equation for momentum change becomes (Fig. 7.9)

$$\Sigma F_x = \gamma h_1 A_1 - \gamma h_2 A_2 = (\gamma Q/g)(V_2 - V_1),$$
$$Q^2/Ag + Ah = \text{Constant}. \tag{7.28}$$

For a rectangular channel of unit width,

$$\frac{V_1^2 y_1}{g} + \frac{y_1^2}{2} = \frac{V_2^2 y_2}{g} + \frac{y_2^2}{2}.$$

The continuity equation implies, for section 1 and 2, that

$$V_1 y_1 = V_2 y_2. \tag{7.29}$$

Solving Eqs. (7.28) and (7.29) simultaneously after eliminating V_2 and rearranging:

$$y_2^2 - y_1^2 = \left(\frac{2V_1^2 y_1}{g}\right)\left(\frac{(y_2 - y_1)}{y_2}\right). \tag{7.30}$$

Dividing by $y_2 - y_1$, multiplying by y_2/y_1^2 and solving for y_2/y_1, using the quadratic formula, we obtain

$$y_2/y_1 = \frac{\sqrt{1 + 8\text{Fr}_1^2} - 1}{2}, \tag{7.31}$$

where Fr_1 is the upstream Froude number. Thus the ratio y_2/y_1 is dependent only on upstream Froude number Fr_1, and y_1 and y_2 are called conjugate depths. An increase in depth requires $\text{Fr}_1 > 1$, or supercritical conditions upstream. The head loss and location of a hydraulic jump can also be calculated but this is beyond the scope of the present coverage. Example 7.6 illustrates the use of Eq. (7.31) to evaluate depths for the hydraulic jump.

<hr />

EXAMPLE 7.6

CALCULATION OF THE DEPTH OF A HYDRAULIC JUMP

A sluice gate is constructed across an open channel. When the sluice gate is open, water flows under it, creating a hydraulic jump, as shown in Fig. E7.6. Determine the depth just downstream of the jump (point b) if the depth of flow at point a is 0.0563 m and the velocity at point a is 5.33 m/s.

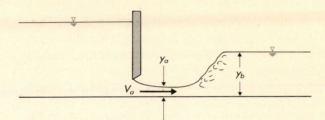

FIGURE E7.6

SOLUTION

Equation (7.31) gives

$$\frac{y_b}{y_a} = \frac{(\sqrt{1 + 8Fr_a{}^2}) - 1}{2}.$$

The upstream Froude number is found from

$$Fr_a = \frac{V_a}{\sqrt{gy_a}}$$

$$= \frac{5.33 \text{ m/s}}{\sqrt{(9.81 \text{ m/s}^2)(0.0563 \text{ m})}}$$

$$Fr_a = 7.17.$$

Then

$$y_b = [(\sqrt{1 + (8)(7.17)^2}) - 1][(0.0563 \text{ m})/2]$$

$$\boxed{y_b = 0.544 \text{ m.}}$$

7.9

INTRODUCTION TO THE HEC-2 MODEL

The HEC-2 model was developed in the 1970s by the Hydrologic Engineering Center (1982) of the U.S. Army Corps of Engineers. The program is designed to calculate water surface profiles for steady, gradually varied flow in natural or man-made channels. The computational procedure is based on solution of the one-dimensional energy equation (Eq. 7.16.) using the standard step method. The program can be applied to floodplain management and flood insurance studies to evaluate **floodway encroachments** and to delineate **flood hazard zones.** The model can also be used to evaluate effects on water surface profiles of channel improvements and levees as well as the presence of bridges or other structures in the floodplain.

The main objective of the HEC-2 program is to simply compute water surface elevations at all locations of interest for given flow values. Data requirements include flow regime, starting elevation, discharge, loss coefficients, cross-sectional geometry, and reach lengths.

Profile computations begin at a cross-section with known or assumed starting conditions and proceed upstream for subcritical flow or downstream for supercritical flow. Subcritical profiles are constrained to critical depth or above, and supercritical profiles are constrained to critical depth or below. The program will not allow profile computations to cross critical depth in most cases because the governing equations do not apply for $y = y_c$, as discussed in Section 7.4.

The HEC-2 program is often used in association with the HEC-1 program for determination of flood flows and flood elevations in a particular watershed. Peak flows at various locations along the main stream or channel are computed by HEC-1 for a given design rainfall (Chapter 5). These peak flows are then used in HEC-2 to calculate the steady-state, nonuniform water surface profile along the stream. For example, the 100-yr rainfall could be used in HEC-1 to calculate 100-yr flows to be used in HEC-2 to predict the 100-yr floodplain.

HEC-2 is a very sophisticated computer program designed to handle a number of hydraulic computations in a single run. The basic program and the basic input data requirements are relatively easy to learn and are covered in detail in the following sections. Special features and other options are described in the user's manual (Hydrologic Engineering Center, 1982). HEC-2 has recently become available for personal computers.

7.10

THEORETICAL BASIS FOR HEC-2

Equations presented in Section 7.6 are solved using the standard step method to compute an unknown water surface elevation at a particular cross section. In HEC-2 the following two equations are solved for subcritical flow by an iterative procedure for upstream (subscript 2) and downstream (subscript 1) sections:

$$WS_2 + \frac{\alpha_2 V_2^2}{2g} = WS_1 + \frac{\alpha_1 V_1^2}{2g} + h_e, \tag{7.32}$$

$$h_e = L\bar{S}_f + C\left(\frac{\alpha_2 V_2^2}{2g} - \frac{\alpha_1 V_1^2}{2g}\right), \tag{7.33}$$

where

WS_1, WS_2 = water surface elevations at ends of reach,

V_1, V_2 = mean velocities (total discharge/total flow area) at ends of reach,

α_1, α_2 = velocity coefficients for flow at ends of reach,

g = gravitational constant,

h_e = energy head loss,

L = discharge-weighted reach length,

S_f = representative friction slope for reach,

C = expansion or contraction loss coefficient.

The discharge-weighted reach length L is computed by weighting lengths in the **left overbank,** channel, and **right overbank** with their respective flows at the end of the reach. A representative friction slope in HEC-2 is usually expressed as follows, although alternative equations can be used:

$$\bar{S}_f = \left(\frac{Q_1 + Q_2}{K_1 + K_2}\right)^2, \tag{7.34}$$

where K_1 and K_2 represent the conveyance at the beginning and end of a reach. **Conveyance** is defined from Manning's equation as

$$K = \frac{1.49}{n} AR^{2/3}. \tag{7.35}$$

The total conveyance for a cross section is obtained by summing the conveyance from the left and right overbanks and the channel. The velocity coefficient α is obtained with the equation

$$\alpha = \left(\frac{A_T^2}{K_T^3}\right)\left(\frac{K_{LOB}^3}{A_{LOB}^2} + \frac{K_{CH}^3}{A_{CH}^2} + \frac{K_{ROB}^3}{A_{ROB}^2}\right), \tag{7.36}$$

where the subscript T is for cross-sectional total, LOB is for left overbank, CH is for channel, and ROB is for right overbank.

The computational procedure for the iterative solution of Eqs. (7.32) and (7.33) is as follows:

1. Assume a water surface elevation at the downstream cross section and proceed upstream (proceed downstream for a supercritical profile).

2. Based on assumed elevation, determine the corresponding total conveyance and velocity head.

3. With values from step 2, compute friction slope S_f and solve Eq. (7.33) for head loss h_e.

4. With values from steps 2 and 3, solve Eq. (7.32) for WS_2.

5. Compare the computed value of WS_2 with the values assumed in step 1 and repeat steps 1 – 5 until values agree to within 0.01 ft (0.01 m).

The first iterative trial is based on the friction slope from the previous two cross sections. The second trial is an average of the computed and assumed elevations from the first trial. Once a balanced water surface elevation has been obtained for a cross section, checks are made to be sure that the elevation is on the correct side of the critical water surface elevation. If otherwise, critical depth is assumed and a message to that effect is provided. The occurrence of critical depth in the program is usually the result of a problem with reach lengths or flow areas unless a critical flow condition actually occurs.

The following assumptions are implicit in the equations and procedures used in the program: (1) flow is steady, (2) flow is gradually varied, (3) flow is one-dimensional, and (4) river channels have small slopes (less than 1 : 10). If any of these assumptions are violated, the results from the HEC-2 program may be in error.

7.11
BASIC DATA REQUIREMENTS

An overview of the data required of the HEC-2 model is shown in Fig. 7.10. The model begins its computation with a water surface elevation and a beginning discharge at the most downstream cross section. Discharges are specified at every cross section as one moves in an upstream direction. Thus lateral inflows can be accommodated. Cross-sectional geometry is specified in terms of ground surface profiles oriented perpendicular to the stream and the measured distances between them in an upstream direction. Spacing of cross sections is dependent on a number of variables, described below. Within each cross section, the left overbank, the right overbank, and the channel locations must be specified (Fig. 7.11).

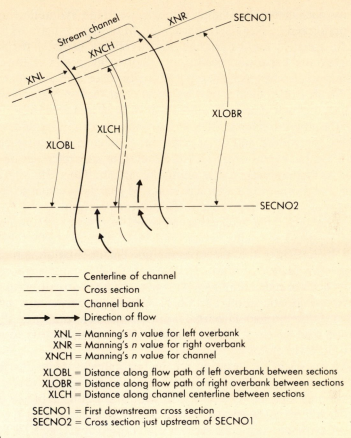

————·——— Centerline of channel
————————— Cross section
————————— Channel bank
————➤————➤ Direction of flow

XNL = Manning's *n* value for left overbank
XNR = Manning's *n* value for right overbank
XNCH = Manning's *n* value for channel

XLOBL = Distance along flow path of left overbank between sections
XLOBR = Distance along flow path of right overbank between sections
XLCH = Distance along channel centerline between sections

SECNO1 = First downstream cross section
SECNO2 = Cross section just upstream of SECNO1

FIGURE 7.10

HEC-2 data requirements.

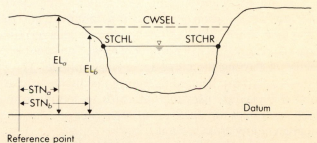

EL$_a$, EL$_b$ = Ground elevation above datum for points *a* and *b*, respectively
STN$_a$, STN$_b$ = Distance from reference point for points *a* and *b*, respectively
STCHL = Left channel bank station (when looking downstream)
STCHR = Right channel bank station (when looking downstream)

Note: Up to 100 points (EL, STN) may be defined for each section.

FIGURE 7.11

Typical cross section.

Loss coefficients are of great importance to the hydraulic computations performed by HEC-2; in particular, Manning's roughness factor n must be specified for the channel and overbank areas of the stream. Contraction or expansion coefficients due to changes in channel cross section must also be specified, especially around bridges and other structures. Figure 7.11 indicates how a typical cross section is specified.

Once the basic data requirements have been met for HEC-2, execution of the program will provide water surface elevations, velocities, areas, volumes, top widths, and other pertinent geometric data for each cross section. Output can be organized into tables for easy reference or can be plotted as longitudinal profiles along the stream. Thus HEC-2 uses hydraulic and geometric data for a stream and peak flow values to arrive at estimates of the resulting water surface at each section.

A number of different card images (lines on a computer terminal) can be used to specify the many options and data requirements for the HEC-2 model. In particular, the categories include documentation, job control, change, cross-section, and bridge data cards. Data cards are formatted with ten fields of eight columns each, where the first two card columns are used for card identification (i.e., T1, J1, GR). All numbers must be right-justified within the field if decimal points are not indicated, and all blank fields are read as zeros. The program uses selected integers to specify certain program options.

The water surface elevation for the beginning cross section may be specified as follows: (1) as critical depth, (2) as a known elevation, or (3) by the slope area method. The critical depth assumption should be used at locations where critical conditions are known to exist, such as at a waterfall, weir, or rapids. The slope area method is used by setting STRT on the J1 card equal to the estimated energy slope and WSEL to the initial water surface elevation. The depth is adjusted by the program until the computed flow is within 1% of the starting flow, which is then used as the starting water surface elevation for the computation. In the following discussion, the term "card" signifies one line image on a terminal.

Discharge can be specified in a number of ways. The variable Q on the J1 card specifies a starting discharge for a single profile. The variable QNEW on the X2 card can be used to change the discharge at any cross section. Another procedure uses the QT cards to specify from one to nineteen discharge values for single or multiple runs and can be used to change discharges at any cross section.

Several types of loss coefficients are utilized by the program, such as Manning's n for friction loss, contraction and expansion losses, and bridge loss coefficients for weir or pressure flow. Manning's n values are typically indicated on the NC card for the channel and overbank areas of the stream. Any of these values may be permanently changed at any cross section using another NC card. The NH or NV card can be used to describe a more detailed variation of n for a cross section. Typical values of Manning's n for various channels were given in Table 7.1.

Contraction or expansion of flow due to changes in the channel cross section is a common cause of energy losses within a reach. These losses can be computed by specifying contraction and expansion coefficients as CCHV and CEHV, respectively, on the NC card. For small changes in river cross section, coefficients are typically 0.1 and 0.3 for contraction and expansion. For abrupt changes such as at bridges, the coefficients may be as high as 0.6 and 1.0, respectively. The coefficients can be changed at any cross section by inserting a new NC card.

Cross-sectional geometry for natural streams is specified in terms of ground surface profiles (cross sections) and the measured distances between them (reach lengths). Generally cross sections should cover the entire floodplain and should be perpendicular to the main flow line. Small ponds or inlets should not be included in the geometry. Cross sections are required at representative locations along a stream length, and where abrupt changes occur, several cross sections should be used, as in the case of a bridge. Cross-section spacing is a function of stream size, slope, and the number of abrupt changes.

Each cross section is identified and described by X1 and GR cards. Each data point is given a station number corresponding to the horizontal distance from a zero point on the left. The elevation EL and corresponding station STA of each data point are input on the GR cards, with a maximum of 100 points allowed. It is important that the cross section be oriented looking downstream so that the lowest station numbers are on the left. The left and right overbank stations are indicated as variables STCHL and STCHR on the X1 card, and endpoints that are too low will automatically be extended vertically and noted in the output. A number of program options are available to easily add or modify cross-sectional data; for example, a cross section can be easily repeated or adjusted vertically or horizontally using X1 card variables. Channel improvement or encroachment options are slightly more difficult to implement and are described later.

The last basic data requirement involves the definition of reach lengths for the left overbank (XLOBL), right overbank (XLOBR) and the channel (XLCH). Channel reach lengths are typically measured along the curvature of the stream (thalweg), and overbank reach lengths are measured along the center of mass of overbank flow. These parameters are indicated as variables on the X1 card. Section 7.13 describes data input features in more detail.

7.12

OPTIONAL HEC-2 CAPABILITIES

The HEC-2 model has been designed with maximum flexibility so that the user has a number of optional capabilities. Thirteen special options are available, but only those most commonly used will be discussed here.

More details can be found in the HEC-2 user's manual (Hydrologic Engineering Center, 1982). In this section the only options described are multiple profile analysis, bridge losses, encroachment, channel improvement, and storage outflow data generation. Options not described here include critical depth, effective flow area, friction loss equations, interpolated cross sections, tributary stream profiles, Manning's n calculation, split-flow option, and ice-covered streams.

HEC-2 can compute up to fourteen profiles in a single run by using NPROF on the J2 card. The first profile has NPROF equal to 1, and the remaining have NPROF equal to 2, 3, 4, 5, and so on. This option allows the user to evaluate flows of different magnitudes (return periods) in a single run. The multiple-profile option also allows for the generation of stream routing data for the modified Puls method in HEC-1. For this purpose, storage-outflow data in a tabular form are automatically created by HEC-2.

Energy losses through culverts and bridges are computed in two ways. First, losses due to expansion and contraction upstream and downstream of the structure are computed. Second, the **normal bridge** or **special bridge method** is used to compute the loss through the structure itself. The normal bridge method is particularly suited for bridges without piers, bridges under high submergence, and low flow through culverts. The special bridge method should be used for bridges with piers where low flow controls, for pressure flow, and when flow passes through critical depth within the structure. The special bridge method allows for both pressure flow and weir flow or any combination of the two. Special bridge methods are described in Section 7.14.

A number of methods are available in HEC-2 for specifying encroachments for floodway studies. Encroachment can be defined as the filling in of a portion of the floodplain, reducing the flow capacity of the floodplain and increasing the water surface elevation. Method 1 and method 4 are the most popular options for floodway determination. Stations and elevations of the left or right encroachment can be specified for individual cross sections using method 1. Method 4 requires an equal loss of conveyance on each side of the channel, and each modified cross section will have the same discharge as the natural cross section.

Cross-sectional data may be modified automatically by the CHIMP option to analyze improvements made to natural stream sections. Trapezoidal excavation is assumed, and geometric data are provided on the CI card that specify the location, elevation, roughness, side slopes, and bottom width.

Another useful feature of the model is the insertion of additional cross sections between those specified by input. The variable HVINS on the J1 card is the maximum allowable change in velocity head between adjacent cross sections and is specified by the user. If it is exceeded, up to three interpolated cross sections will be automatically inserted between two existing cross sections in the model.

7.13

INPUT AND OUTPUT FEATURES

Data Organization and Documentation

Table 7.3 illustrates the organization of data for a typical profile run. The minimum data set would require cards T1, T2, T3, J1, NC, X1, GR, EJ, three blanks, and ER. Multiple-profile data sets are constructed by inputting successive sets of T1, T2, T3, J1, and J2 cards following the EJ cards.

Documentation cards allow the identification of stream name, location, frequency, or data sources. Comment cards (C) and title cards (T1, T2, and T3) are used for data explanation, title information, and summary tables.

Job Control and Change Cards

The job control cards specify the level of printout, select various computation procedures, terminate execution of the program, and generally control the processing of data in HEC-2. The J1 card specifies starting conditions of flow, elevation, or energy slope. It also controls the printing of the data input list and other related options. The J2 card is required for each profile, except the first, of a multiple-profile run. This card controls the reading of data, plotting, modification of Manning's n, calculation of critical depth, and channelization options. J2 also handles flow distribution and summary tables. The J3 card is an optional card and allows the user to select from a list of variables for the summary printout.

The J4 through J6 cards are optional and handle the punching of routing data, output control, and equation selection for friction loss, respectively. The EJ card serves to terminate the reading of all data cards and is required. The ER card, when preceded by three blank cards, terminates the execution of the program.

Change cards provide options to initialize and change values associated with Manning's n, discharge, cross section by encroachment, and channel improvement. Once initial values are changed, they remain changed for all subsequent cross sections until another change card occurs. The NC card is required to initialize n values and loss coefficients prior to the first cross section. Subsequent NC cards may be used to change values at any cross section. NH or NV cards are optional and are used to specify up to twenty values of Manning's n in the horizontal or vertical direction for a cross section. The QT card allows a table of nineteen discharge values for multiple-profile runs, as specified on the J1 card. The ET card allows a table of up to nine encroachment specifications that correspond to the QT card. Finally, the CI card allows the user to simulate channel improvement by excavation. Invert elevations, side slopes, n values, and bottom widths may be specified.

TABLE 7.3

Typical HEC-2 Input Data

CARD	MAJOR VARIABLES TO BE INPUT						
T1–T3	Title cards for output						
J1	INQ	STRT	HVINS	Q		WSEL	FQ
J2	NPROF	IPLOT	PRFVS	FN			
J3	Numbers for summary output or detailed output						
J4–J5	Optional						
J6	IHLEQ	ICOPY					
NC	XNL	XNR	XNCH	CCHV	CEHV		
QT	Table of flow values for multiple profiles						
X1	SECNO	NUMST	STCHL	STCHR	XLOBL	XLOBR	XLCH
X3	IEARA = 10 to contain flow between levees until topped						
GR	EL(1)	STN(1)	EL(2),	STN(2) etc. (up to 100 points from left to right looking downstream)			
NH	NUMNH VAL(N) STN(N) (to change Manning's *n* from left to right)						

Special bridge cards

X2	IBRID	ELLC	ELTRD	(fields 3, 4, 5)			
SB	XK	XKOR	COFQ	RDLEN BWC BWP BAREA SS ELCHU ELCHD (to describe the bridge geometry)			
BT	NRD	RDST(1)	RDEL(1)	XLCEL(1)			
		RDST(2)	RDEL(2)	XLCEL(2) etc. up to (100 points)			
		(for normal bridge, the BT and GR STNs must coincide)					

EJ	
ER	End of run

(Parameters are defined in Appendix D.4.)

Cross Section Cards

The cross section cards contain the basic data that describe the cross-sectional geometry of a stream. X1 and GR are required and provide the basic representation of a reach of stream. X2–X5 provide a series of options related to bridges, effective flow areas, and high-water elevations. The X1 card indicates the number of GR data points to be read on the following GR cards. It locates the cross section by indicating the distance to the immediate downstream cross section, locates bank stations, and raises or lowers elevations on the GR cards. The X1 card can also be used to request a plot of the cross-section data. The X2 card is optional but is required for each application of the special bridge routine. The X3 card allows a specification of ineffective flow areas to be removed from the GR data. The X4 card allows additional points to be added to the GR cards. The X5 card is used to input water surface elevations at a cross section. And finally, the GR card represents a profile of a stream taken perpendicular to the direction of flow with up to 100 pairs of elevation station data to describe the ground profile.

Split-Flow Cards

Split-flow cards are used to specify input data for cases where the flow splits into two separate channels, as in the case of a braided river system. Refer to the HEC-2 user's manual (Hydrologic Engineering Center, 1982) for more details.

7.14
NORMAL AND SPECIAL BRIDGE CONSIDERATIONS

The HEC-2 program offers two methods for computing head losses through bridge structures. The normal bridge method is based on Manning's equation and uses the standard step method to determine bridge losses. The special bridge method utilizes a series of hydraulic equations to compute losses through the bridge structure. Depending on the physical configuration of the bridge and the expected flow condition, the user selects one of the two bridge methods.

The Normal Bridge Method

The normal bridge method computes losses through a bridge in the same way that losses in a normal stream reach are computed, based on Manning's equation and the standard step method. The HEC-2 program accounts for the presence of the bridge by subtracting that portion of the bridge below the water surface from the total flow area and by increasing the wetted perimeter where water is in contact with the bridge. The bridge deck is described by defining the top of roadway (ELTRD) and the low chord (ELLC) on the X2 card. Also, a table of roadway stations and elevations for the top of road and low chord can be specified on the BT cards. Thus horizontal or sloping bridge profiles can be handled in HEC-2.

The Special Bridge Method

The special bridge method uses specific hydraulic equations for a number of possible flow conditions, including **low flow, pressure flow, weir flow,** or any possible combination. In this method, the opening under the bridge is approximated by a trapezoid with given bottom elevation, bottom width, and side slopes. The presence of piers is accounted for by specifying a total width of flow obstruction due to the piers.

Low flow conditions occur when flow passes underneath a bridge without touching the low chord, and none of the flow passes over the top of the road surface. For a bridge without piers in a low flow condition, HEC-2 automatically selects the normal bridge method for calculation because it is accurate for such conditions.

For low flow under bridges with piers, HEC-2 considers three conditions, labeled Class A, Class B, and Class C. Class A low flow assumes that the flow is subcritical, and the Yarnell equation is used to compute the change in water surface elevation:

$$H3 = 2K(K + 10w - 0.6)(a + 15a)^4(V3^2/2g), \tag{7.37}$$

where

$H3$ = change in water surface elevation through bridge,

K = pier shape coefficient,

w = ratio of velocity head to depth downstream of bridge,

$$a = \frac{\text{obstructed area}}{\text{total unobstructed area}},$$

$V3$ = velocity downstream from the bridge.

The value of $H3$ is added to the downstream water surface elevation after the computation.

Class B low flow occurs when the water surface profile passes through critical depth underneath the bridge. HEC-2 uses a momentum balance for cross sections adjacent to and under the bridge, as described in more detail in the HEC-2 user's manual (Hydrologic Engineering Center, 1982). Class C low flow occurs when the flow condition is supercritical, and the same procedure used for Class B is applied in this case.

Pressure flow occurs when the bridge deck becomes submerged such that the low chord is in contact with water and a head buildup occurs on the upstream side of the bridge. Pressure flow through a bridge is similar to orifice flow in fluid mechanics and can be described by the equation

$$Q = A(2gH/K)^{0.5}, \tag{7.38}$$

where

H = total energy difference upstream and downstream,

K = loss coefficient,

A = cross-sectional area of the bridge opening,

Q = total orifice flow.

HEC-2 defines H as the distance from the energy grade line to the centroid of the orifice area.

Weir flow occurs when water begins to flow over the bridge and elevated roadway approaches, and HEC-2 uses the standard weir equation for this flow condition:

$$Q = CLH^{3/2}, \tag{7.39}$$

where

C = discharge coefficient,

L = effective length of weir,

H = total energy difference upstream of the bridge and top of the roadway,

Q = flow over the weir.

The coefficient of discharge C varies according to the configuration of the bridge and the roadway and is reduced in the HEC-2 model whenever the weir or bridge is submerged by high tailwater.

Combination flows are handled by HEC-2 using an iterative procedure to balance energy elevations and computed flow rate.

Selection of a Bridge Method in HEC-2

The normal bridge method is most applicable when friction losses are the major consideration in head loss through a bridge. In particular, when a bridge has no piers, for culverts under low flow conditions, and when the top of road is expected to have a high level of submergence, the normal bridge method may be applied.

The special bridge method should be applied (1) when the bridge has piers, (2) when pressure flow or weir flow is expected to occur, (3) when flow is expected to pass through critical depth at the bridge, or (4) for combinations of low flow or pressure flow or weir flow. In situations where it is not clear what types of flow are expected at a bridge or for a wide range of flow rates, the special bridge method should be used.

Cross-Section Layout for Bridge Modeling

Figure 7.12 illustrates the cross section requirements for both the normal and special bridge methods. The normal bridge method requires a total of six cross sections: (1) one cross section downstream of the bridge where flow is not affected by the bridge, (2) one just downstream of the bridge, (3) one at the downstream face of the bridge, (4) one at the upstream face of the bridge, (5) one just upstream of the bridge, and (6) one cross section located far enough upstream to be out of the backwater influence.

The special bridge method requires only four cross sections, which correspond to the first, third, fourth, and sixth sections for the normal bridge method (Fig. 7.12).

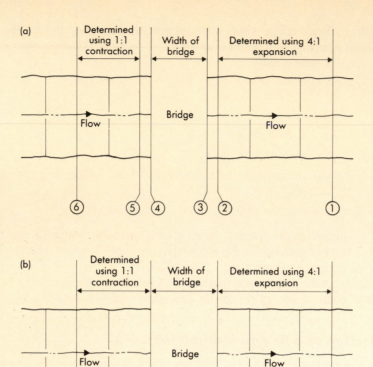

FIGURE 7.12

Cross-section layout for bridge modeling. (a) Normal bridge method. (b) Special bridge method.

The spacing of cross sections downstream of the bridge should be based on a 4:1 expansion of flow, which corresponds to 1 ft laterally for every 4 ft traveled in the flow direction. This phenomenon is shown in Fig. 7.13. The location of the final bridge cross section in either the normal bridge or special bridge method should be based on a 1:1 contraction of flow. To avoid complications where reach lengths become too long upstream or downstream of bridges, the expansion and contraction criteria should be used to limit the lateral extent of any intermediate (inserted) cross sections.

Effective Area Option

One of the limitations of HEC-2 in bridge modeling is the inability of the program to differentiate between effective and ineffective flow areas. Ef-

fective flow area is that portion of the total cross-sectional area where flow velocity is normal to the cross section in the downstream direction. Any other areas that do not convey flow in the downstream direction are called **ineffective flow areas.** Figure 7.14 illustrates some examples of ineffective flow areas created by the presence of bridges.

The ineffective flow area may be eliminated from consideration by using GR and BT cards to define the top of road and low chord profiles of a bridge. The effective area option in HEC-2 may be used to "block out" portions of the channel cross section that are not effective in conveying flow downstream. For example, all flow can be restricted to a channel cross section until the computed water surface elevation exceeds the elevation of one or both channel banks coded on the GR card. Encroachment stations for a cross section can be considered using this same method. All parameters necessary for implementing the effective area method are contained on the X3 card.

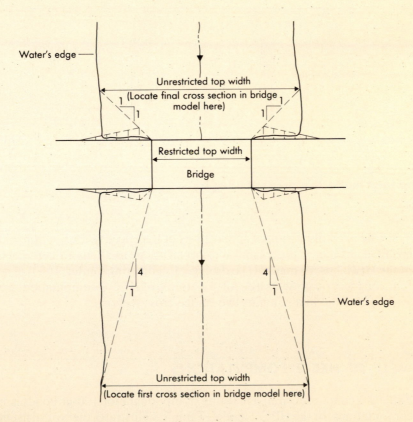

FIGURE 7.13

Expansion and contraction of flow at bridges.

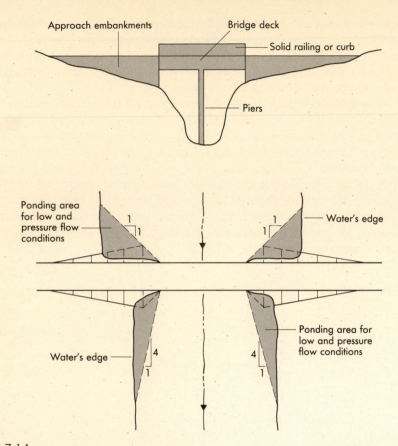

FIGURE 7.14

Examples of ineffective areas at bridges.

A detailed discussion of applications of the effective area option and coefficients for both the normal and special bridge methods are contained in the HEC-2 user's manual (Hydrologic Engineering Center, 1982). The following section presents a detailed example of HEC-2 computations for a small channel that contains two bridge cross sections.

7.15

EXAMPLE OF HEC-2 COMPUTATIONS

Table 7.4 contains a complete listing of input data for 100-yr floodplain computations for the HEC-2 model. Table 7.3 illustrates the typical list of card types and variables required by HEC-2. For example, the T1, T2, and T3 cards provide for a title. The J1 card provides for a starting energy slope

TABLE 7.4
HEC-2 Input Data

T1	GARNERS BAYOU		P130-00-00						
T2	EXISTING CHANNEL		100-YEAR FLOWS						
T3	EXAMPLE OF HEC2		GARNERS BAYOU						
J1		2			.00146			48.0	
J2	-1								
J3	150	100							
J6	1								
NC	.12	.12	.05	.1	.3				
QT	1	8400							
X1	158	26	20914	21004	0	0	0		
X3	10								
GR	55	16300	50	19300	51	20000	50.9	20100	50.7	20200
GR	50.3	20300	50.3	20400	50	20500	51.6	20600	53.6	20630
GR	56.3	20700	53.7	20700	54.4	20725	50.8	20914	43.1	20926
GR	32.7	20940	28.5	20950	32.3	20960	40.3	20972	49.5	21004
GR	53.4	21100	55.9	21150	54.4	21200	53.9	21250	54.7	22254
GR	56.0	24550							
NH	4	.12	25905	.05	25995	.12	27100	.07	28700
X1	2481	12	25905	25995	2100	1800	2323		
GR	55.5	22699	55.0	22699.9	55.0	22700	50.0	25900	50.8	25905
GR	32.7	25931	31.9	25941	32.3	25951	49.5	25995	50.0	26000
GR	55.0	27100	59.0	28700					
NH	4	.12	25598	.05	25663	.12	27050	.07	30000
X1	4382	13	25598	25663	2000	1500	1901		
GR	55.8	22299.8	55.0	22299.9	55.0	22300	51.0	25550	50.3	25598
GR	34.7	25622	34.5	25633	34.6	25644	50.5	25663	51	25750
GR	55.0	27050	60.0	29550	61.5	30000			
NH	4	.12	25038	.05	25106	.12	26400	.07	29600
X1	5222	20	25038	25106	845	645	840		
GR	60.6	21999.9	60.6	22000	57.1	22287	57.1	22487	56.4	22600
GR	56.6	22800	54.9	23400	52.8	24000	51.8	25000	51.8	25038
GR	34.2	25065	33.7	25073	34.2	25082	51.3	25106	51.4	25200
GR	52.3	25600	52.9	26400	55.8	27600	60.0	28800	60.5	29600
NH	3	.11	27113	.05	27199	.11	31200		
X1	6825	24	27113	27199	2713	813	1603		
X3	10								
GR	59.8	22799.7	59.8	22799.8	59.8	22799.9	59.8	22800	57.4	24000
GR	55.5	25200	53.9	26000	53.8	26400	54.5	26800	54.9	27113
GR	52.2	27114	51.9	27127	42.0	27141	38.2	27146	37.0	27153
GR	37.7	27163	41.7	27170	44.1	27178	51.0	27195	55.0	27199
GR	53.4	27600	54.3	28400	59.3	30000	61.1	31200		
NC	0	0	0	.3	.5				
NH	6	.12	25200	.07	26400	.12	27113	.05	27199	.12
NH	28400	.07	31200						
X1	6925	0	0	0	100	100	100		
X3	10						55.0	55.0	
NH	6	.12	25200	.07	26400	.12	27113	.05	27199	.12
NH	28400	.07	31200						
SB	1.25	1.5	3	0	14.0	1	713	2	37.0	37.0
X1	6963	0	0	0	38	38	38		
X2			1	53.6	54.5				
X3	10						55.0	55.0	
BT	17	22800	60.81	0	22800	60.82	0	22800	60.83	0
BT	22800	60.8	0	24000	58.4	0	25200	56.5	0	26000

(continued)

TABLE 7.4 (continued)
HEC-2 Input Data

BT	54.9	0	26400	54.8	0	26800	55.5	0	27113	55.9
BT	0	27114	56.6	0	27195	56.6	0	27199	56.0	0
BT	27600	54.4	0	28400	55.3	0	30000	60.3	0	31200
BT	62.1	0								
NH	6	.12	25200	.09	26800	.12	27071	.35	27239	.12
NH	28400	.08	31200							
X1	7063	18	27071	27239	100	100	100			
X3	10									
GR	59.8	22799.7	59.8	22799.8	59.8	22799.9	59.8	22800	57.4	24000
GR	55.5	25200	53.9	26000	53.8	26400	54.5	26800	54.0	27071
GR	36.0	27125	36.0	27155	36.0	27185	54.0	27239	54.0	27600
GR	54.3	28400	59.3	30000	61.1	31200				
NC	0	0	0	.1	.3					
NH	7	.12	26000	.09	27281	.12	27484	.04	27663	.12
NH	29400	.08	31000	.07	31600					
QT	1	8100								
X1	8192	37	27484	27663	862	762	1129			
X3	10									
GR	59.1	22799.6	59.1	22799.7	59.1	22799.8	59.1	22799.9	59.1	22800
GR	58.5	23000	59.8	23200	60.0	23400	58.8	23600	58.1	24600
GR	57.5	25000	56.3	25400	57.3	25600	57.1	25800	56.2	26000
GR	56.4	26200	54.9	27281	53.1	27400	53.0	27484	38.7	27538
GR	37.6	27562	39.5	27607	52.5	27663	52.6	27707	57.1	27749
GR	57.3	27800	56.5	27887	53.9	27900	53.9	28000	54.6	28200
GR	56.2	28500	57.5	29000	59.0	29400	60.0	30000	59.8	30200
GR	60.8	31000	61.5	31600						
NH	5	.12	25200	.07	26200	.12	26300	.04	26448	.12
NH	30600									
QT	1	5600								
X1	9670	16	26300	26448	2500	1300	1478			
X3	10									
GR	60.0	20000	55.0	25200	53.8	26200	54.3	26300	52.5	26322
GR	40.8	26367	39.1	26389	39.6	26400	53.7	26448	53.4	26506
GR	56.3	26535	57.0	26600	54.8	26758	54.5	26787	60.0	28725
GR	65.0	30600								
NH	4	.08	24800	.12	24932	.04	25056	.12	29300	
X1	11835	13	24932	25056	3000	1400	2165			
X3	10									
GR	63.0	21400	63.0	21700	60.0	24100	55.4	24800	56.5	24900
GR	54.9	24932	41.3	24966	40.4	24986	40.9	25005	56.8	25056
GR	55.7	25113	60.0	26900	65.0	29050				
NH	5	.09	24500	.12	24828	.04	24974	.12	26900	.08
NH	29300									
X1	12785	31	24828	24974	1100	800	950			
X3	10									
GR	65.0	20000	63.2	21500	63.2	21900	62.6	22408	62.0	22700
GR	60.3	23122	59.0	23597	57.4	23900	56.3	24828	55.1	24855
GR	41.3	24890	40.8	24900	41.1	24931	55.5	24974	55.8	25026
GR	58.9	25048	59.3	25100	58.9	25121	56.0	25141	56.4	25500
GR	56.2	25700	59.3	26300	61.7	26500	61.1	26900	63.4	27500
GR	63.3	27700	64.1	28100	64.1	28500	64.6	28700	64.6	29100
GR	64.9	29300								
NH	5	.09	24600	.12	25294	.04	25394	.12	26700	.09
NH	28550									

TABLE 7.4 (*continued*)

HEC-2 Input Data

QT	1	5000							
X1	14950	16	25294	25394	2300	1300	2165		
GR	70	20000	65.0	22550	60.0	24600	57.6	25150	57.9	25237
GR	58.9	25246	58.8	25282	56.8	25294	45.5	25329	45.1	25344
GR	45.5	25359	58.4	25394	58.5	25522	60.0	25750	65.0	26700
GR	65.0	28550							
NH	5	.07	23000	.12	23647	.04	23737	.12	26600	.08
NH	27100								
X1	16481	31	23647	23737	1600	1000	1531		
GR	69.0	20000	67.7	20500	67.4	20800	66.7	21000	66.7	21100
GR	62.1	21700	61.3	21900	62.1	22000	61.2	22100	58.3	22900
GR	59.0	23000	58.8	23200	58.7	23400	58.4	23548	57.9	23555
GR	58.6	23570	59.2	23591	57.4	23647	47.5	23677	47.0	23692
GR	47.5	23707	57.5	23737	58.9	23800	58.6	24000	59.2	24200
GR	63.7	25000	64.1	25400	65.0	26200	65.0	26600	65.0	27000
GR	65.6	27100							
NH	5	.07	23600	.12	24470	.04	24566	.12	24950	.09
NH	27850								
QT	1	4700							
X1	17590	11	24470	24566	900	1100	1109		
X4	2	61.0	23600	66.5	26400				
GR	70.0	20000	65.0	21950	59.0	24300	60.3	24470	49.3	24504
GR	48.3	24525	48.8	24545	58.1	24566	58.9	24623	65.0	24950
GR	68.0	27850							
NH	5	.08	23100	.12	24060	.04	24156	.12	25025	.10
NH	27750								
X1	19649	12	24060	24156	1950	1750	2059		
X4	1	66.0	22000						
GR	70.0	20000	67.0	20800	65.0	23100	60.6	24000	58.4	24060
GR	50.3	24088	49.5	24106	50.4	24125	61.5	24156	60.3	24210
GR	70.0	25025	70.0	27750					
NH	6	.07	22200	.10	25000	.12	25155	.04	25258	.12
NH	26400	.09	27800						
X1	21233	21	25155	25258	75	1550	1584		
X4	2	63.5	23500	62.5	24200				
GR	70.0	20000	65.0	22200	62.0	25000	62.0	25155	51.9	25187
GR	49.8	25208	52.0	25229	61.1	25258	62.7	25400	64.4	25600
GR	66.2	25800	67.0	26000	67.7	26200	68.3	26400	68.5	26600
GR	68.7	26800	68.6	27000	69.0	27200	69.1	27400	69.1	27600
GR	69.1	27800							
NH	5	.07	24200	.12	25940	.04	26040	.12	27800	.09
NH	29200								
X1	22300	32	25940	26040	700	1800	1067		
GR	70.0	20000	70.0	21500	65.6	24200	65.1	24600	65.5	24800
GR	65.2	25000	66.1	25215	64.2	25600	64.2	25700	63.9	25800
GR	63.7	25940	53.1	25970	53.1	25990	53.1	26010	63.4	26040
GR	63.8	26100	64.4	26200	64.6	26400	65.5	26600	66.1	26800
GR	66.7	27000	66.9	27200	66.9	27300	67.6	27400	67.2	27500
GR	67.9	27800	68.4	28200	67.3	28400	67.9	28600	67.5	28800
GR	68.1	29165	69.2	29200					
NC	0	0	0	.3	.5				
NH	5	.07	24200	.12	25954	.04	26055	.12	27800	.09
NH	29200								
X1	22500	31	25954	26055	200	200	200		

TABLE 7.4 (continued)
HEC-2 Input Data

X3	10							67.0	67.0	
GR	70.0	20000	70.0	21500	65.6	24200	65.1	24600	65.5	24800
GR	65.2	25000	66.1	25215	64.2	25600	64.2	25700	63.9	25800
GR	63.7	25954	53.1	25979	53.1	26029	63.4	26055	63.8	26100
GR	64.4	26200	64.6	26400	65.5	26600	66.1	26800	66.7	27000
GR	66.9	27200	66.9	27300	67.6	27400	67.2	27500	67.9	27800
GR	68.4	28200	67.3	28400	67.9	28600	67.5	28800	68.1	29165
GR	69.2	29200								
NH	5	.07	24200	.12	25954	.04	26055	.12	27800	.09
NH	29200									
SB	1.05	2.56	2.63	0	50.0	.01	1081	2.35	53.1	53.1
X1	22525	0	0	0	25	25	25			
X2	0	0	1	66.2	69.0					
X3	10							69.0	69.0	
BT	21.0	20000	70.0	70.0	21500	70.0	70.0	24200	65.5	65.5
BT	24600	65.1	65.1	24800	65.5	65.5	25000	65.2	65.2	25215
BT	66.1	66.1	25600	66.4	66.4	25700	68.2	68.2	25954	69.0
BT	66.2	26055	69.0	66.2	26200	67.5	64.0	26400	67.7	64.6
BT	26600	67.7	66.7	26800	67.7	67.7	27000	68.0	66.7	27200
BT	66.8	66.8	27300	66.9	66.9	28200	68.4	68.4	28600	69.7
BT	69.7	29200	69.2	69.2						
NH	5	.07	24600	.12	25935	.04	26045	.12	27500	.09
NH	29200									
X1	22551	32	25935	26045	26	26	26			
X3	10									
GR	70.0	20000	70.0	21500	65.6	24200	65.1	24600	65.5	24800
GR	65.2	25000	66.1	25215	64.2	25600	64.2	25700	63.9	25800
GR	63.7	25935	53.1	25965	53.1	25990	53.1	26015	63.4	26045
GR	63.8	26100	64.4	26200	64.6	26400	65.5	26600	66.1	26800
GR	66.7	27000	66.9	27200	66.9	27300	67.6	27400	67.2	27500
GR	67.9	27800	68.4	28200	67.3	28400	67.9	28600	67.5	28800
GR	68.1	29165	69.2	29200						
EJ										
ER										

for the slope area method (STRT) and an initial guess for the starting water surface elevation (WSEL). The J2 card arranges for output and in this case indicates that only a single profile will be run and a summary printout will be provided. The J3 card is an optional card for summary printout, and in this case the standard summary and special bridge summary are requested. Finally, the J6 card allows the program to select the most appropriate friction equation based on flow conditions (IHLEQ).

The next group of cards defines the various cross-sectional geometry and Manning coefficients. The NC card specifies Manning coefficients for the left and right overbanks and the channel. Contraction and expansion coefficients are also included. The QT card specifies the flow rate, 8400 cfs in this case, at the first cross section. The NC card can be used to change coefficients at any cross section.

An X1 card is required for each cross section and allows up to 100 stations on the following GR cards. Station locations of the left bank (STCHL) and the right bank (STCHR) are indicated in fields 3 and 4. The distance from the next downstream cross section of the left overbank, right overbank, and channel are indicated in fields 5, 6, and 7. These last three values are zero for the first cross section. Next, the GR cards specify an elevation and station number for the cross section oriented left to right looking downstream.

For example, section number 158 (first field of the X1 card) has twenty-six stations on the GR cards, and the channel is contained between stations 20914 and 21004, as shown on the X1 card. The X3 card with a 10 in the first field implies that flow is contained in the channel unless the water surface elevation exceeds elevations of the bank stations.

The NH card placed immediately before section 2481 indicates that four new Manning's coefficients will be inserted at this point. This section contains twelve stations on the GR cards. The same format is repeated for cross sections 4382, 5222, and 6825 with only minor differences.

Section number 6925 is located 100 ft upstream of 6825 and is a duplicate of 6825, as indicated by the zero in the second field of the X1 card. The X3 card is used to define the effective flow area through the special bridge at section 6963. The X2, SB, and BT cards for section 6963 are used to define the special bridge method variables. The BT card contains a number of points that describe the bridge roadway and low chord elevations and stations.

Routine cross sections occur from section 7063 until another special bridge is encountered at cross section 22525. The last cross section is followed by an EJ card, which indicates the end of the job. The EJ card is followed by three blank lines and an ER card. The ER card indicates the end of the run; if this had been a multiple-profile run, the job control data for each additional profile would have been entered between the ER and EJ cards.

After the HEC-2 program is executed, a detailed output for each profile requested appears, part of which is shown in Table 7.5. All of the variables listed at the top of that table are calculated for each cross section, and additional messages are provided as warnings or where flow conditions may approach critical flow. Note the special bridge output for section 6963. In addition, a plotted profile may be requested, as shown in Fig. 7.15, where the various letters in the plot are defined as shown.

Summary printouts are shown in Table 7.6, which is referred to as Table 150 in HEC-2. These tables provide the basic information required to define water surface elevation (CWSEL), channel velocity (VCH), area, flow rate (Q), top width, and a host of other variables that could be requested on the J3 card. See the user's manual (Hydrologic Engineering Center, 1982). A more complete listing of input and output parameters is provided in Appendix D.4.

TABLE 7.5

Detailed Output from HEC-2

8/06/87   14.5659

SECNO	DEPTH	CWSEL	CRIWS	WSELK	EG	HV	HL	OLOSS	BANK ELEV
Q	QLOB	QCH	QROB	ALOB	ACH	AROB	VOL	TWA	LEFT/RIGHT
TIME	VLOB	VCH	VROB	XNL	XNCH	XNR	WTN	ELMIN	SSTA
SLOPE	XLOBL	XLCH	XLOBR	ITRIAL	IDC	ICONT	CORAR	TOPWID	ENDST

*PROF 1

IHLEQ = 1. THEREFORE FRICTION LOSS (HL) IS CALCULATED AS A FUNCTION OF
PROFILE TYPE, WHICH CAN VARY FROM REACH TO REACH. SEE DOCUMENTATION FOR
DETAILS.

CCHV=   0.100 CEHV=   0.300
*SECNO 158.000

3265 DIVIDED FLOW

158.00	23.62	52.12	0.0	48.00	52.50	0.38	0.0	0.0	50.80
8400.	2034.	6317.	48.	3462.	1115.	85.	0.	0.	49.50
0.0	0.59	5.67	0.57	0.120	0.050	0.120	0.0	28.50	18025.68
0.001479	0.	0.	0.	0	0	8	0.0	2806.27	21068.59

1490 NH CARD USED
*SECNO 2481.000

2481.00	22.19	54.09	0.0	0.0	54.23	0.14	1.70	0.02	50.80
8400.	2431.	5121.	848.	5369.	1339.	1861.	319.	152.	49.50
0.25	0.45	3.82	0.46	0.120	0.050	0.120	0.0	31.90	23282.98
0.000514	2100.	2323.	1800.	3	0	0	0.0	3616.62	26899.60

1490 NH CARD USED
*SECNO 4382.000

4382.00	20.59	55.09	0.0	0.0	55.18	0.09	0.95	0.00	50.30
8400.	3225.	3670.	1504.	6995.	992.	3092.	738.	331.	50.50
0.52	0.46	3.70	0.49	0.120	0.050	0.120	0.0	34.50	22299.89
0.000500	2000.	1901.	1500.	2	0	0	0.0	4793.54	27093.43

1490 NH CARD USED
*SECNO 5222.000

5222.00	21.77	55.47	0.0	0.0	55.54	0.07	0.35	0.00	51.80
8400.	2138.	3302.	2959.	4353.	1008.	5516.	932.	411.	51.30
0.66	0.49	3.28	0.54	0.120	0.050	0.108	0.0	33.70	23196.14
0.000405	845.	840.	645.	1	0	0	0.0	4270.44	27466.58

1490 NH CARD USED

TABLE 7.5 (continued)
Detailed Output from HEC-2

8/06/87   14.5659

SECNO	DEPTH	CWSEL	CRIWS	WSELK	EG	HV	HL	OLOSS	BANK ELEV
Q	QLOB	QCH	QROB	ALOB	ACH	AROB	VOL	TWA	LEFT/RIGHT
TIME	VLOB	VCH	VROB	XNL	XNCH	XNR	WTN	ELMIN	SSTA
SLOPE	XLOBL	XLCH	XLOBR	ITRIAL	IDC	ICONT	CORAR	TOPWID	ENDST

*SECNO 6825.000

6825.00	19.41	56.41	0.0	0.0	56.52	0.11	0.96	0.01	54.90
8400.	2358.	3786.	2256.	4086.	985.	3642.	1316.	588.	55.00
0.88	0.58	3.84	0.62	0.110	0.050	0.110	0.0	37.00	24626.92
0.000759	2713.	1603.	813.	2	0	0	0.0	4447.43	29074.36

CCHV=   0.300 CEHV=   0.500
1490 NH CARD USED
*SECNO 6925.000

6925.00	19.52	56.52	0.0	0.0	56.59	0.07	0.07	0.01	54.90
8400.	2947.	3332.	2122.	4365.	994.	3851.	1337.	598.	55.00
0.90	0.68	3.35	0.55	0.084	0.050	0.110	0.0	37.00	24557.09
0.000569	100.	100.	100.	2	0	0	0.0	4552.64	29109.73

1490 NH CARD USED

SPECIAL BRIDGE

SB XK	XKOR	COFQ	RDLEN	BWC	BWP	BAREA	SS	ELCHU	ELCHD
1.25	1.50	3.00	0.0	14.00	1.00	713.00	2.00	37.00	37.00

*SECNO 6963.000
PRESSURE AND WEIR FLOW

EGPRS	EGLWC	H3	QWEIR	QPR	BAREA	TRAPEZOID AREA	ELLC	ELTRD
59.75	56.59	0.00	7056.	1349.	713.	767.	53.60	54.50

6963.00	19.53	56.53	0.0	0.0	56.60	0.07	0.01	0.0	54.90
8400.	2956.	3316.	2127.	4395.	995.	3873.	1345.	37.00	55.00
0.91	0.67	3.33	0.55	0.084	0.050	0.110	0.0	37.00	24549.70
0.000562	38.	38.	38.	2	0	7	0.0	4563.79	29113.48

1490 NH CARD USED
*SECNO 7063.000

7063.00	20.68	56.68	0.0	0.0	56.69	0.01	0.07	0.02	54.00
8400.	3853.	1886.	2661.	4832.	2503.	3902.	1369.	613.	54.00
0.94	0.80	0.75	0.68	0.094	0.350	0.111	0.0	36.00	24453.27
0.000900	100.	100.	100.	2	0	0	0.0	4709.07	29162.34

PROFILE FOR STREAM EXAMPLE OF HEC2   GARNER

PLOTTED POINTS (BY PRIORITY)=E-ENERGY,W-WATER SURFACE,I-INVERT,C-CRITICAL W.S.,L-LEFT BANK,R-RIGHT BANK,M-LOWER END STA

ELEVATION
SECNO   CUMDIS   25.   30.   35.   40.   45.   50.   55.   60.   65.   70.

158.00   0.   C
                  200.   C
                  400.   C
                  600.   C
                  800.   C
                 1000.   C
                 1200.   C
                 1400.   C
                 1600.   C
                 1800.   C
                 2000.   C
                 2200.   C
2481.00   2400.   C
                 2600.   C
                 2800.   C
                 3000.   C
                 3200.   C
                 3400.   C
                 3600.   C
                 3800.   C
                 4000.   C
                 4200.   C
4382.00   4400.   C
                 4600.   C
                 4800.   C
5222.00   5000.   C
                 5200.   C
                 5400.   C
                 5600.   C
                 5800.   C
                 6000.   C
                 6200.   C
                 6400.   C
                 6600.   C
6825.00   6800.   C
6925.00   7000.   C
6963.00   7200.   C
7063.00   7400.   C
                 7600.   C
                 7800.   C
                 8000.   C
8192.00   8200.   C
                 8400.   C
                 8600.   C
                 8800.   C
                 9000.   C
                 9200.   C
9670.00   9400.   C
                 9600.   C
                 9800.   C
                10000.   C
                10200.   C
                10400.   C

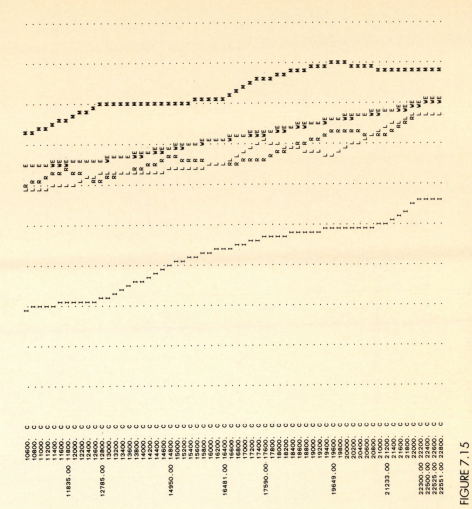

FIGURE 7.15

Portion of water surface profile plotted by HEC-2.

TABLE 7.6

Summary Printout Tables from HEC-2

8/06/87   14.5659

THIS RUN EXECUTED   8/06/87   14.5661

```
***********************************************
  HEC2 RELEASE DATED NOV 76 UPDATED MAY  1984
  ERROR CORR -  01,02,03,04,05,06
  MODIFICATION -  50,51,52,53,54,55
***********************************************
```

NOTE- ASTERISK (*) AT LEFT OF CROSS-SECTION NUMBER INDICATES MESSAGE IN SUMMARY OF ERRORS LIST

EXAMPLE OF HEC2   GARNER

SUMMARY PRINTOUT TABLE 100

SECNO	EGLWC	ELLC	EGPRS	ELTRD	QPR	QWEIR	CLASS	H3	DEPTH	CWSEL	VCH	EG
6963.000	56.59	53.60	59.75	54.50	1348.94	7055.75	30.00	0.00	19.53	56.53	3.33	56.60
22525.000	65.30	66.20	0.0	69.00	4700.00	0.0	1.00	0.00	11.80	64.90	5.08	65.30

EXAMPLE OF HEC2   GARNER

SUMMARY PRINTOUT TABLE 150

SECNO	XLCH	ELTRD	ELLC	ELMIN	Q	CWSEL	CRIWS	EG	10K*S	VCH	AREA	.01K
158.000	0.0	0.0	0.0	28.50	8400.00	52.12	0.0	52.50	14.79	5.67	4661.82	2184.46
2481.000	2323.00	0.0	0.0	31.90	8400.00	54.09	0.0	54.23	5.14	3.82	8568.99	3706.42
4382.000	1901.00	0.0	0.0	34.50	8400.00	55.09	0.0	55.18	5.00	3.70	11078.95	3757.65
5222.000	840.00	0.0	0.0	33.70	8400.00	55.47	0.0	55.54	4.05	3.28	10876.77	4173.06
6825.000	1603.00	0.0	0.0	37.00	8400.00	56.41	0.0	56.52	7.59	3.84	8712.70	3048.78
6925.000	100.00	0.0	0.0	37.00	8400.00	56.52	0.0	56.59	5.69	3.35	9210.23	3520.05
6963.000	38.00	54.50	53.60	37.00	8400.00	56.53	0.0	56.60	5.62	3.33	9263.63	3542.16
7063.000	100.00	0.0	0.0	36.00	8400.00	56.68	0.0	56.69	9.00	0.75	11237.19	2800.52
8192.000	1129.00	0.0	0.0	37.60	8100.00	56.91	0.0	57.01	1.76	2.81	6471.46	6099.49
9670.000	1478.00	0.0	0.0	39.10	5600.00	57.23	0.0	57.27	1.28	2.06	9282.82	4950.95
11835.000	2165.00	0.0	0.0	40.40	5600.00	57.70	0.0	57.88	3.71	3.57	2988.03	2906.72
12785.000	950.00	0.0	0.0	40.80	5600.00	58.08	0.0	58.21	3.27	3.15	4249.71	3098.26
14950.000	2165.00	0.0	0.0	45.10	5000.00	59.26	0.0	59.63	9.34	4.99	1585.95	1635.77
16481.000	1531.00	0.0	0.0	47.00	5000.00	60.55	0.0	60.72	5.73	3.98	3971.89	2089.01
17590.000	1109.00	0.0	0.0	48.30	4700.00	61.22	0.0	61.49	8.04	4.55	2319.52	1658.08
19649.000	2059.00	0.0	0.0	49.50	4700.00	62.80	0.0	63.09	7.62	4.55	1986.51	1702.54
21233.000	1584.00	0.0	0.0	49.80	4700.00	63.81	0.0	63.98	5.43	3.79	3450.95	2016.65
22300.000	1067.00	0.0	0.0	53.10	4700.00	64.60	0.0	65.05	13.47	5.49	1183.06	1280.54
22500.000	200.00	0.0	0.0	53.10	4700.00	64.90	0.0	65.30	10.26	5.08	925.88	1467.08
22525.000	25.00	69.00	66.20	53.10	4700.00	64.90	0.0	65.30	10.26	5.08	925.87	1467.08
22551.000	26.00	0.0	0.0	53.10	4700.00	65.07	0.0	65.36	7.81	4.44	1758.43	1682.24

(continued)

TABLE 7.6 (continued)
Summary Printout Tables from HEC-2

8/06/87   14.5659

EXAMPLE OF HEC2   GARNER

SUMMARY PRINTOUT TABLE   150

SECNO	Q	CWSEL	DIFWSP	DIFWSX	DIFKWS	TOPWID	XLCH
158.000	8400.00	52.12	0.0	0.0	4.12	2806.27	0.0
2481.000	8400.00	54.09	0.0	1.96	0.0	3616.62	2323.00
4382.000	8400.00	55.09	0.0	1.00	0.0	4793.54	1901.00
5222.000	8400.00	55.47	0.0	0.38	0.0	4270.44	840.00
6825.000	8400.00	56.41	0.0	0.94	0.0	4447.43	1603.00
6925.000	8400.00	56.52	0.0	0.11	0.0	4552.64	100.00
6963.000	8400.00	56.53	0.0	0.01	0.0	4563.79	38.00
7063.000	8400.00	56.68	0.0	0.15	0.0	4709.07	100.00
8192.000	8100.00	56.91	0.0	0.22	0.0	3154.68	1129.00
9670.000	5600.00	57.23	0.0	0.32	0.0	4864.04	1478.00
11835.000	5600.00	57.70	0.0	0.47	0.0	1491.88	2165.00
12785.000	5600.00	58.08	0.0	0.38	0.0	2208.15	950.00
14950.000	5000.00	59.26	0.0	1.18	0.0	864.74	2165.00
16481.000	5000.00	60.55	0.0	1.29	0.0	2161.13	1531.00
17590.000	4700.00	61.22	0.0	0.67	0.0	1235.73	1109.00
19649.000	4700.00	62.80	0.0	1.59	0.0	871.06	2059.00
21233.000	4700.00	63.81	0.0	1.00	0.0	2297.85	1584.00
22300.000	4700.00	64.60	0.0	0.79	0.0	871.92	1067.00
22500.000	4700.00	64.90	0.0	0.31	0.0	101.00	200.00
22525.000	4700.00	64.90	0.0	0.00	0.0	101.00	25.00
22551.000	4700.00	65.07	0.0	0.17	0.0	1081.40	26.00

SUMMARY

This chapter presents a review of uniform and nonuniform open channel flow. Uniform flow computations with Manning's equation are derived and examples are presented for various channel shapes. Critical flow conditions are defined in which the Froude number is used to characterize channel conditions. The gradually varied flow equations are derived based on energy loss through a channel. These equations are used to classify water surface profiles for backwater computations, which are presented in the examples.

The HEC-2 model is an extremely flexible tool that can be used to calculate water surface profiles for natural or man-made streams. It contains all of the necessary hydraulic coefficients and equations for most flow conditions that would be encountered and can handle a number of special options for normal bridges, special bridges, multiple profiles, effective area, encroachment, and split flow. The detailed example presented in the chapter is designed to represent typical input and output, and more detailed applications require familiarity with the HEC-2 user's manual.

PROBLEMS

7.1. The Colorado River System Aqueduct has the cross section shown in Fig. P7.1. When the water in the aqueduct is 10.2 ft deep, flow is measured as 1600 cfs. If $n = 0.014$, what is S_0 in (a) ft/ft and (b) ft/mi?

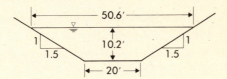

FIGURE P7.1

7.2. A rectangular open channel that is 2 m wide has water flowing at a depth of 0.45 m. Using $n = 0.014$, find the rate of flow in the channel if S_0 is (a) 0.002 m/m; (b) 0.006 m/m, and (c) 0.012 m/m.

7.3. A channel has the irregular shape shown in Fig. P7.3, with a bottom slope of 0.0016 ft/ft. The indicated Manning's n values apply to the corresponding areas only. Assuming that $Q = Q_1 + Q_2 + Q_3$, find the rate of flow in the channel if $y_1 = 2$ ft, $y_2 = 10$ ft, and $y_3 = 3$ ft.

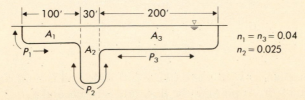

FIGURE P7.3

7.4. Water is flowing 2 m deep in a rectangular channel that is 2.5 m wide. The average velocity is 5.8 m/s and $C = 100$. What is the slope of the channel? (Use Chezy's formula.)

7.5. A triangular channel with side slopes at 45° to the horizontal has water flowing through it at a velocity of 10 ft/s. Find Chezy's roughness coefficient C if the bed slope is 0.03 ft/ft and the depth is 4 ft.

7.6. Water is flowing at a rate of 900 cfs in a trapezoidal open channel. Given that $S_0 = 0.001$, $n = 0.015$, bottom width $b = 20$ ft, and the side slopes are $1 : 1.5$, what is the normal depth y_n?

7.7. Find the normal depth y_n for the triangular channel shown in Fig. P7.7 if $S_0 = 0.0005$ m/m, $Q = 40$ m³/s, and $n = 0.030$.

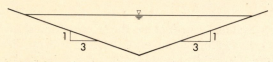

FIGURE P7.7

7.8. Determine the critical depth and the critical velocity for the Colorado River System Aqueduct (Problem 7.1) if $Q = 1500$ cfs.

7.9. Find the critical depth and critical velocity for the triangular channel of Problem 7.7 if Q is (a) 10 m³/s and (b) 50 cfs.

7.10. Determine the local change in water surface elevation caused by a 0.2-ft-high obstruction in the bottom of a 10-ft-wide rectangular channel on a slope of 0.0005 ft/ft. The rate of flow is 20 cfs and the unobstructed flow depth is 0.9 ft. See Fig. P7.10. Assume no head loss.

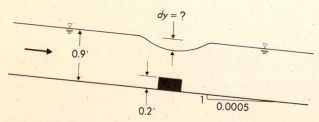

FIGURE P7.10

7.11. A rectangular channel with $n = 0.012$ is 5 ft wide and is built on a slope of 0.0006 ft/ft. At point a, the flow rate is 60 cfs and $y_a = 3$ ft. Using one reach, find the distance to point b where $y_b = 2.5$ ft and determine whether this point is upstream or downstream of point a.

7.12. If a channel with the same cross-sectional and flow properties as the channel of Problem 7.11 is laid on a slope of 0.01 ft/ft, determine whether the flow is supercritical or subcritical. Find the depth of flow at a point 1000 ft downstream from the point where $y = 1.5$ ft. (A trial-and-error solution may be necessary.)

7.13. Classify the water surface profiles according to Table 7.2 of (a) Problem 7.11 and (b) Problem 7.12.

7.14. Classify the bed slopes (mild, critical, steep) of the channels of the following problems: (a) Problem 7.1; (b) Problem 7.6; and (c) Problem 7.7.

7.15. A stream bed has a rectangular cross section 5 m wide and a slope of 0.0002 m/m. The rate of flow in the stream is 8.75 m³/s. A dam is built across the stream, causing the water surface to rise to 2.5 m just upstream of the dam. See Fig. P7.15. Using the step method illustrated in Example 7.5, determine the water surface profile upstream of the dam to a point where $y = y_n \pm 0.1$ m. How far upstream does this point occur?

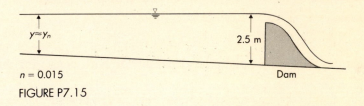

$y \approx y_n$ 2.5 m

$n = 0.015$ Dam

FIGURE P7.15

7.16. An unfinished concrete channel ($n = 0.020$) changes from a mild slope to a steep slope. The channel is 20 m wide throughout, and the rate of flow is 180 m³/s. If the slope of the mild portion of the channel is 0.0006 m/m, determine the distance upstream from the slope change to the point where $y = 3.0$ m. (*Hint:* Use the same step method as in Example 7.5, with y intervals of 0.1 m.)

7.17. A rectangular channel 1.4 m wide on a slope of 0.0026 m/m has water flowing through it at a rate of 0.5 m³/s and a depth of 0.6 m. A cross section of the channel is constricted to a width of 0.9 m. What is the change in water surface elevation at this point?

Problems 7.18–7.20 refer to the watershed shown in Figs. P7.18(a) and (b). Cypress Creek has a rectangular cross section with a bottom width of 200 ft, $n = 0.03$, and $S_0 = 0.001$ ft/ft. East Creek has the same characteristics except that the bottom width is 100 ft. The 100-yr storm hydrographs are shown for both creeks. The assumption is made that the flow in the creeks remains constant above point C and that the 100-yr hydrograph for Cypress Creek includes the inflow from East Creek at point C.

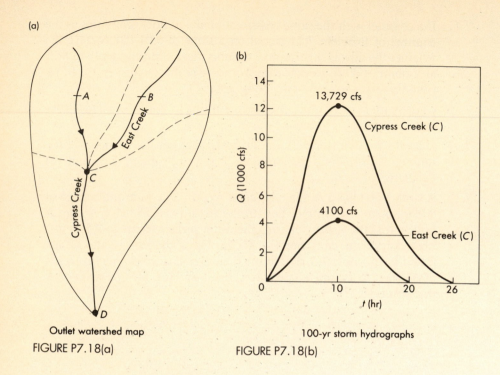

(a)

(b)

Outlet watershed map

FIGURE P7.18(a)

100-yr storm hydrographs

FIGURE P7.18(b)

7.18. Determine the normal and critical depths for both creeks for the 100-yr peak flow.

7.19. Assume that East Creek meets Cypress Creek at a right angle. Using a starting elevation at point C consistent with the 100-yr flow in Cypress Creek, develop the 100-yr water surface profile for East Creek. Use three points between the starting elevation and the elevation $y = 1.1y_n$.

7.20. A developer proposes improvements to the East Creek subwatershed that will increase the peak flow of the 100-yr storm by 1000 cfs. The developer contends that there will be no change in the 100-yr storm event on the Cypress Creek subarea above point C. Determine the 100-yr water surface profile for East Creek. Use the same number of intervals as in Problem 7.19.

REFERENCES

Chow, V. T., 1959, *Open Channel Hydraulics*, McGraw-Hill Book Company, New York.

Chow, V. T. (editor), 1964, *Handbook of Applied Hydrology*, McGraw-Hill Book Company, New York.

Daugherty, R. L., J. B. Franzini, and E. J. Finnemore, 1985, *Fluid Mechanics with Engineering Applications,* 8th ed., McGraw-Hill Book Company, New York.

Fox, R. W., and A. T. McDonald, 1985, *Introduction to Fluid Mechanics,* John Wiley and Sons, New York.

Hydrologic Engineering Center, 1982, *HEC-2 Water Surface Profiles, User's Manual,* U.S. Army Corps of Engineers, Davis, California.

King, H. W., and E. F. Brater, 1976, *Handbook of Hydraulics,* 6th ed., McGraw-Hill Book Company, New York.

8

Ground Water Hydrology

8.1
INTRODUCTION

Our study of hydrology up to this point has concentrated on various aspects of surface water processes, including rainfall, runoff, infiltration, evaporation, channel flow, and storage in lakes and reservoirs. An engineering hydrologist also must be able to address processes of ground water hydrology and flow. This chapter presents a concise treatment of ground water topics, including properties of ground water, ground water flow, well hydraulics, and ground water modeling techniques. This represents a minimum coverage for the practicing hydrologist or engineering student. Several excellent textbooks in the ground water area are available for more detailed treatment (see DeWiest, 1965; Bear, 1979; Freeze and Cherry, 1979; Todd, 1980; Fetter, 1980).

Ground water hydrology is of great importance because of the use of **aquifer** systems for water supply and because of the threat of contamination from waste sites at or below the ground surface. Properties of the **porous media** and subsurface geology govern both the rate and direction of ground water flow in any aquifer system. The injection or accidental spill of waste into an aquifer or the pumping of the aquifer for water supply may alter the natural hydraulic flow patterns. The hydrologist

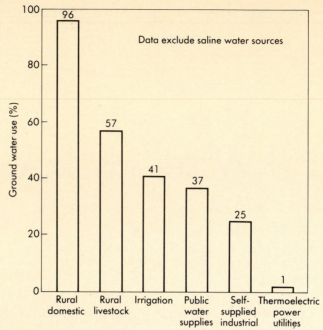

FIGURE 8.1

Percentage of ground water use to total water use for various types of use in the United States, 1975. (After Murray and Reeves, 1977.)

must have a working knowledge of methods that have been developed to predict rates of flow and directions of movement in ground water systems.

Ground water is an important source of water supply for municipalities, agriculture, and industry. Figure 8.1 indicates the percentage of various types of ground water use in the United States, and Fig. 8.2 depicts ground water use relative to total water use for each state in the continental United States. As the figures show, western and midwestern areas of the United States are generally much more dependent on ground water. Techniques for the design of water supply systems using ground water aquifers are an important part of engineering hydrology.

The U.S. Geological Survey (USGS) has primary responsibility for the collection of ground water data and evaluation of these data in terms of impacts on water supply, water quality, water depletion, and potential contamination. Results published by the USGS provide information on ground water levels and water quality data throughout the United States.

FIGURE 8.2   ➡

Ground water usage. (Reprinted with permission from Geraghty et al., 1973, courtesy of Water Information Center, Inc., Plainview, New York.)

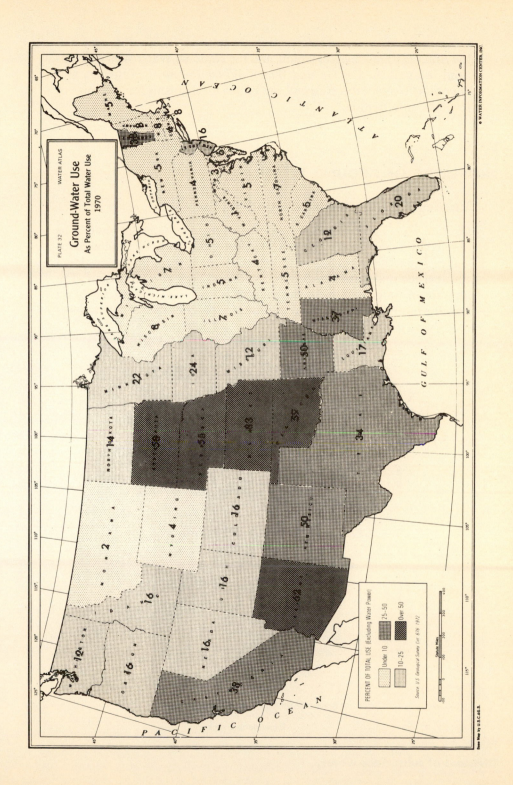

PLATE 32

WATER ATLAS

Ground-Water Use
As Percent of Total Water Use
1970

PERCENT OF TOTAL USE (Excluding Water Power)

Under 10
10-25
25-50
Over 50

Source: U.S. Geological Survey Circ. 676, 1972

Statute Miles

Base Map by U.S.C.&G.S.

© WATER INFORMATION CENTER, INC.

ATLANTIC OCEAN

GULF OF MEXICO

PACIFIC OCEAN

Other primary sources of information are state water resources agencies, the American Geophysical Union, and the National Water Well Association. Journals such as *Ground Water, Water Resources Research,* and *Journal of Geophysical Research* are major mechanisms for exchange of technical information.

8.2

PROPERTIES OF GROUND WATER

Vertical Distribution of Ground Water

Ground water can be characterized according to vertical distribution as defined by Todd (1980). Figure 8.3 indicates the divisions of subsurface water. The **soil water zone,** which extends from the ground surface down through the major **root zone,** varies with soil type and vegetation. The amount of water present in the soil water zone depends primarily on recent exposure to rainfall and infiltration. **Hygroscopic water** remains

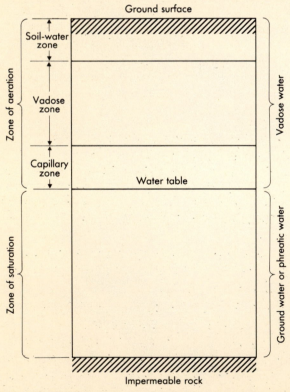

FIGURE 8.3

Divisions of subsurface water.

adsorbed to the surface of soil grains, while **gravitational water** drains through the soil under the influence of gravity. **Capillary water** is held by surface tension forces just above the **water table,** which is defined as the level to which water will rise in a well drilled into the **saturated zone.**

The **vadose zone** extends from the lower edge of the soil water zone to the upper limit of the capillary zone (see Fig. 8.3). Thickness may vary from zero for high water table conditions to more than 100 m in arid regions of the country, such as Arizona or New Mexico. Vadose zone water is held in place by hygroscopic and capillary forces, and infiltrating water passes downward toward the water table as gravitational flow, subject to retardation by these forces.

The capillary zone, or fringe, extends from the water table up to the limit of **capillary rise,** which varies inversely with the pore size of the soil and directly with the surface tension. Capillary rise can range from 2.5 cm for fine gravel to more than 200 cm for silt (Lohman, 1972). Just above the water table almost all pores contain capillary water, and the water content decreases with height depending on the type of soil. A typical soil moisture curve is shown in Figure 8.4.

In the zone of saturation, which occurs beneath the water table, the **porosity** is a direct measure of the water contained per unit volume, expressed as the ratio of the volume of voids to the total volume. Only a portion of the water can be removed from the saturated zone by drainage or by pumping from a well. **Specific yield** is defined as the volume of water released from an unconfined aquifer per unit surface area per unit head decline in the water table. Fine-grained materials yield little water whereas coarse-grained materials provide significant water and thus serve as aquifers. In general, specific yields for unconsolidated formations fall in the range of 7 to 25%.

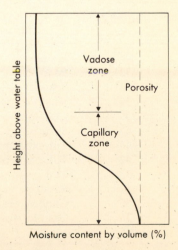

FIGURE 8.4
Typical soil moisture relationship.

A vertical hole dug into the earth penetrating an aquifer is referred to as a **well.** Wells are used for pumping, recharge, disposal, and water level observation. Often the portion of the well hole that is open to the aquifer is screened to prevent aquifer material from entering the well. For more detail on well construction methods, see Section 8.7.

Aquifers

An **aquifer** can be defined as a formation that contains sufficient saturated permeable material to yield significant quantities of water to wells and springs. Aquifers are generally areally extensive and may be overlain or underlain by a confining bed, defined as a relatively impermeable material. An **aquiclude** is a relatively impermeable confining unit, such as clay, and an **aquitard** is a poorly permeable stratum, such as a sandy clay unit, that may leak water to adjacent sand aquifers.

Aquifers can be characterized by the porosity of a rock or soil, expressed as the ratio of the volume of voids V_v to the total volume V. Porosity may also be expressed by

$$n = \frac{V_v}{V} = 1 - \frac{\rho_b}{\rho_m}, \tag{8.1}$$

where ρ_m is the **density** of the grains and ρ_b is the **bulk density,** defined as the oven-dried mass of the sample divided by its original volume. Table 8.1 shows a range of porosities for a number of aquifer materials.

Unconsolidated geologic materials are normally classified according

TABLE 8.1

Representative Values of Porosity*

MATERIAL	POROSITY (%)	MATERIAL	POROSITY (%)
Gravel, coarse	28†	Loess	49
Gravel, medium	32†	Peat	92
Gravel, fine	34†	Schist	38
Sand, coarse	39	Siltstone	35
Sand, medium	39	Claystone	43
Sand, fine	43	Shale	6
Silt	46	Till, predominantly silt	34
Clay	42	Till, predominantly sand	31
Sandstone, fine-grained	33	Tuff	41
Sandstone, medium-grained	37	Basalt	17
Limestone	30	Gabbro, weathered	43
Dolomite	26	Granite, weathered	45
Dune sand	45		

* After Morris and Johnson, 1967.

† These values are for repacked samples; all others are undisturbed.

TABLE 8.2

Soil Classification Based on Particle Size*

MATERIAL	PARTICLE SIZE (mm)
Clay	<0.004
Silt	0.004–0.062
Very fine sand	0.062–0.125
Fine sand	0.125–0.25
Medium sand	0.25–0.5
Coarse sand	0.5–1.0
Very coarse sand	1.0–2.0
Very fine gravel	2.0–4.0
Fine gravel	4.0–8.0
Medium gravel	8.0–16.0
Coarse gravel	16.0–32.0
Very coarse gravel	32.0–64.0

* After Morris and Johnson, 1967.

to their size and distribution. Soil classification based on particle size is shown in Table 8.2. Particle sizes are measured by mechanically sieving grain sizes larger the 0.05 mm and measuring rates of settlement for smaller particles in suspension. A typical particle size distribution graph is shown in Fig. 8.5. The **uniformity coefficient** indicates the relative uniformity of the material. A uniform material such as dune sand has a low coefficient, while a well-graded material such as **alluvium** has a high coefficient (Fig. 8.5).

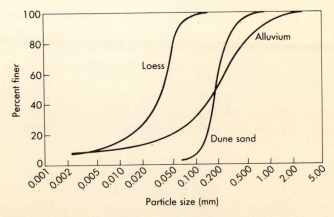

FIGURE 8.5

Particle size distribution graph for three geologic samples.

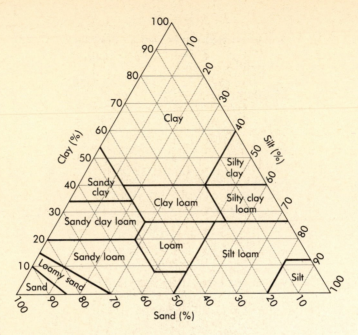

FIGURE 8.6

Triangle representation of soil textures for describing various combinations of sand, silt, and clay.

The texture of a soil is defined by the relative proportions of sand, silt and clay present in the particle size analysis and can be expressed most easily on a triangle of soil textures (Fig. 8.6). For example, a soil with 30% clay, 60% silt, and 10% sand is referred to as a silty clay loam.

Most aquifers can be considered as underground storage reservoirs that receive **recharge** from rainfall or from an artificial source. Water flows out of an aquifer due to gravity or to pumping from wells. Aquifers may be classified as **unconfined** or as **confined,** depending on the existence of a water table. A leaky confined aquifer represents a stratum that allows water to flow through the confining zone.

Figure 8.7 shows a vertical cross section illustrating unconfined and confined aquifers. An unconfined aquifer is one in which a water table exists and rises and falls with changes in volume of water stored. A **perched water table** is an example where an unconfined water body sits on top of a clay lens, separated from the main aquifer.

Confined aquifers, called **artesian aquifers,** occur where ground water is confined by a relatively impermeable stratum, or confining unit, and water is under pressure greater than atmospheric. If a well penetrates such an aquifer, the water level will rise above the bottom of the confining unit.

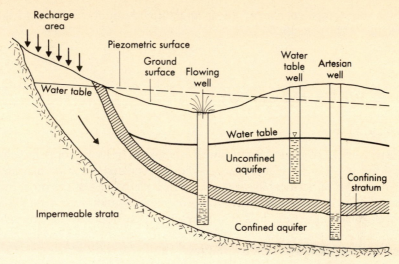

FIGURE 8.7

Schematic cross section illustrating unconfined and confined aquifers.

If the water level rises above the land surface, a **flowing well** or spring results and is referred to as an **artesian well** or spring.

A recharge area supplies water to a confined aquifer, and such an aquifer can convey water from the recharge area to locations of natural or artificial discharge. The **piezometric surface** (or **potentiometric surface**) of a confined aquifer is the hydrostatic pressure level of water in the aquifer, defined by the water level that occurs in a lined penetrating well. It should be noted that a confined aquifer can become unconfined when the piezometric surface falls below the bottom of the upper confining bed.

Contour maps and profiles can be prepared of the water table for an unconfined aquifer or the piezometric surface for a confined aquifer. These **equipotential lines** will be described in more detail in Section 8.3. Once determined from a series of wells in an aquifer, orthogonal lines can be drawn to indicate the general direction of ground water flow, in the direction of decreasing head.

A parameter of some importance relates to the water-yielding capacity of an aquifer. The **storage coefficient** S is defined as the volume of water that an aquifer releases from or takes into storage per unit surface area per unit change in piezometric head. For a confined aquifer, values of S fall in the range of 0.00005 to 0.005, indicating that large pressure changes produce small changes in the storage volume. For unconfined aquifers, a change in storage volume is expressed simply by the product of the volume of aquifer lying between the water table at the beginning and end of a period of time and the average specific yield of the formation.

Thus, the storage coefficient for an unconfined aquifer is equal to the specific yield (typically 7–25%).

8.3

GROUND WATER MOVEMENT

Darcy's Law

The movement of ground water is well established by hydraulic principles reported in 1856 by Henri Darcy, who investigated the flow of water through beds of permeable sand. Darcy discovered one of the most important laws in hydrology — that the flow rate through porous media is proportional to the head loss and inversely proportional to the length of the flow path. Darcy's law serves as the basis for present-day knowledge of ground water flow and well hydraulics.

Figure 8.8 depicts the experimental setup for determining head loss through a sand column, with **piezometers** located a distance L apart. Total energy for this system can be expressed by the Bernoulli equation

$$\frac{p_1}{\gamma} + \frac{v_1^2}{2g} + z_1 = \frac{p_2}{\gamma} + \frac{v_2^2}{2g} + z_2 + h_1, \tag{8.2}$$

where

p = pressure,

γ = specific weight of water,

v = velocity,

z = elevation,

h_1 = head loss.

Because velocities are very small in porous media, velocity heads may be neglected, allowing head loss to be expressed

$$h_1 = \left(\frac{p_1}{\gamma} + z_1\right) - \left(\frac{p_2}{\gamma} + z_2\right). \tag{8.3}$$

It follows that the head loss is independent of the inclination of the column. Darcy related flow rate to head loss and length of column through a proportionality constant referred to as K, the **hydraulic conductivity**, a measure of the **permeability** of the porous media. Darcy's law can be stated

$$V = -\frac{Q}{A} = -K\frac{dh}{dL}. \tag{8.4}$$

The negative sign indicates that flow of water is in the direction of decreasing head.

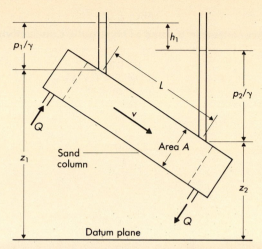

FIGURE 8.8
Head loss through a sand column.

The Darcy velocity that results from Eq. (8.4) is an average discharge velocity through the entire cross section of the column. The actual flow is limited to the pore space only, so that the seepage velocity V_s is equal to the Darcy velocity divided by porosity:

$$V_s = \frac{Q}{nA}. \tag{8.5}$$

Thus, actual velocities are usually much higher (by a factor of 3) than the Darcy velocities.

It should be pointed out that Darcy's law applies to laminar flow in porous media, and experiments indicate that Darcy's law is valid for Reynolds numbers less than 1 and perhaps as high as 10. This represents an upper limit to the validity of Darcy's law, which turns out to be applicable in most ground water systems. Deviations can occur near pumped wells and in fractured aquifer systems with large openings.

Hydraulic Conductivity

The hydraulic conductivity of a soil or rock depends on a variety of physical factors and is an indication of an aquifer's ability to transmit water. Thus, sand aquifers have K values many orders of magnitude larger than clay units. Table 8.3 indicates representative values of hydraulic conductivity for a variety of materials. As can be seen, K can vary over many orders of magnitude in an aquifer that may contain different types of material. Thus, velocities and flow rates can also vary over the same range, as expressed by Darcy's law.

TABLE 8.3

Representative Values of Hydraulic Conductivity*

MATERIAL	HYDRAULIC CONDUCTIVITY (m/day)	TYPE OF MEASUREMENT†
Gravel, coarse	150	R
Gravel, medium	270	R
Gravel, fine	450	R
Sand, coarse	45	R
Sand, medium	12	R
Sand, fine	2.5	R
Silt	0.08	H
Clay	0.0002	H
Sandstone, fine-grained	0.2	V
Sandstone, medium-grained	3.1	V
Limestone	0.94	V
Dolomite	0.001	V
Dune sand	20	V
Loess	0.08	V
Peat	5.7	V
Schist	0.2	V
Slate	0.00008	V
Till, predominantly sand	0.49	R
Till, predominantly gravel	30	R
Tuff	0.2	V
Basalt	0.01	V
Gabbro, weathered	0.2	V
Granite, weathered	1.4	V

* After Morris and Johnson, 1967.

† H is horizontal hydraulic conductivity, R is a repacked sample, and V is vertical hydraulic conductivity.

Transmissivity is a term often used in ground water hydraulics as applied to confined aquifers. It is defined as the product of K and the **saturated thickness** of the aquifer b. Hydraulic conductivity is usually expressed in m/day (ft/day) and transmissivity T is expressed in m²/day (ft²/day). An older unit for T that is still reported in some applications is gal/day/ft. Conversion factors for these units are contained in Appendix B.

The **intrinsic permeability** of a rock or soil is a property of the medium only, independent of fluid properties. Intrinsic permeability k can be expressed

$$k = \frac{K\mu}{\rho g} \tag{8.6}$$

where

$\mu =$ dynamic viscosity,

$\rho =$ fluid density,

$g =$ gravitational constant.

Intrinsic permeability k has units of m^2 or darcy, equal to $0.987 \, (\mu m)^2$; k is often used in the petroleum industry, whereas K is used in ground water hydrology for categorizing aquifer systems.

Determination of Hydraulic Conductivity

Hydraulic conductivity in saturated zones can be determined by a number of techniques in the laboratory as well as in the field. **Constant head and falling head permeameters** are used in the laboratory for measuring K and are described in more detail below. In the field, **pump tests, slug tests,** and **tracer tests** are available for determination of K. These tests are described in more detail in Sections 8.5 and 8.6 under the general heading of well hydraulics.

A permeameter (Fig. 8.9) is used in the laboratory to measure K by maintaining flow through a small column of material and measuring flow rate and head loss. For a constant head permeameter, Darcy's law can be directly applied to find K, where V is volume flowing in time t through a sample of area A, length L, and with constant head h:

$$K = \frac{VL}{Ath}. \tag{8.7}$$

The falling head permeameter test consists of measuring the rate of fall of the water level in the tube or column and noting

$$Q = \pi r^2 \frac{dh}{dt}. \tag{8.8}$$

Darcy's law can be written for the sample as

$$Q = \pi r_c^2 K \frac{dh}{dl}. \tag{8.9}$$

After equating and integrating,

$$K = \frac{r^2 L}{r_c^2 t} \ln \left(\frac{h_1}{h_2} \right), \tag{8.10}$$

where L, r, and r_c are as shown in Fig. 8.9 and t is the time interval for water to fall from h_1 to h_2.

Auger hole tests involve the measurement of a change in water level after rapid removal or addition of a volume of water to a bore hole. The value of K is applicable in the immediate vicinity of the hole and is generally useful for shallow water table conditions. The slug test for shal-

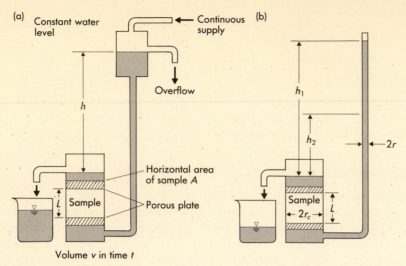

FIGURE 8.9

Permeameters for measuring hydraulic conductivity of geologic samples. (a) Constant head. (b) Falling head.

low wells operates in much the same fashion, with a measurement of decline or recovery of the water level in the well through time.

The pump test involves the constant removal of water from a single well and observations of water level declines at several adjacent wells. In this way, an integrated K value for a portion of the aquifer is obtained. Field methods generally yield significantly different values of K than corresponding laboratory tests performed on cores removed from the aquifer. Thus, field tests are preferable for the accurate determination of aquifer parameters.

Anisotropic Aquifers

Most real geologic systems tend to have variations in one or more directions due to the processes of deposition and layering that can occur. In the typical field situation in alluvial deposits, we find the hydraulic conductivity in the vertical direction K_z to be less than the value in the horizontal direction K_x. For the case of a two-layered aquifer of different K in each layer and different thicknesses, we can apply Darcy's law to horizontal flow to show

$$K_x = \frac{K_1 z_1 + K_2 z_2}{z_1 + z_2} \tag{8.11}$$

or, in general,

$$K_x = \frac{\Sigma K_i z_i}{\Sigma z_i}, \tag{8.12}$$

where

$K_i = K$ in layer i,

$z_i =$ thickness of layer i.

For the case of vertical flow through two layers, q_z is the same flow per unit horizontal area in each layer:

$$dh_1 + dh_2 = \left(\frac{z_1}{K_1} + \frac{z_2}{K_2} \right) q_z \qquad (8.13)$$

but

$$dh_1 + dh_2 = \left(\frac{z_1 + z_2}{K_z} \right) q_z, \qquad (8.14)$$

where K_z is the hydraulic conductivity for the entire system. Equating Eqs. (8.13) and (8.14), we have

$$K_z = \frac{z_1 + z_2}{(z_1/K_1) + (z_2/K_2)} \qquad (8.15)$$

or, in general,

$$K_z = \frac{\Sigma z_i}{\Sigma z_i/K_i}. \qquad (8.16)$$

Ratios of K_x/K_z usually fall in the range of 2 to 10 for alluvium, with values up to 100 where clay layers exist. In actual application to layered systems, it is usually necessary to apply ground water flow models that can properly handle complex geologic strata through numerical simulation. Selected modeling techniques are described in Section 8.8.

Flow Nets

Darcy's law was originally derived in one dimension, but because many ground water problems are really two- or three-dimensional, methods are available for the determination of flow rate and direction. A specified set of **streamlines** and **equipotential lines** can be constructed for a given set of boundary conditions to form a **flow net** (Fig. 8.10) in two dimensions.

Equipotential lines are prepared based on observed water levels in wells penetrating an **isotropic** aquifer. Flow lines are then drawn orthogonally to indicate the direction of flow. For the flow net of Fig. 8.10, the hydraulic gradient i is given by

$$i = \frac{dh}{ds}, \qquad (8.17)$$

and constant flow q per unit thickness between two adjacent flow lines is

$$q = K \frac{dh}{ds} dm. \qquad (8.18)$$

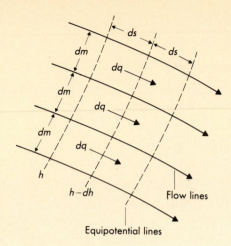

FIGURE 8.10
Typical flow net.

If we assume $ds = dm$ for a square net, then for n squares between two flow lines over which total head is divided ($h = H/n$) and for m divided flow channels,

$$Q = mq = \frac{KmH}{n},\tag{8.19}$$

where

K = hydraulic conductivity of the aquifer,
m = number of flow channels,

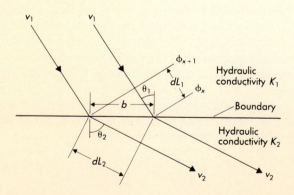

FIGURE 8.11

Refraction of flow lines across a boundary between media of different hydraulic conductivities.

(a)

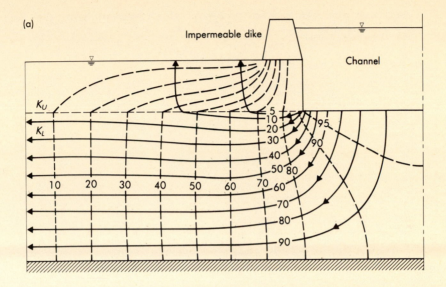

(b)

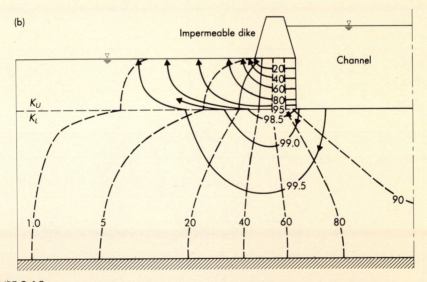

FIGURE 8.12

Flow nets for seepage from one side of a channel through two different anisotropic two-layer systems. (a) $K_u/K_L = 1/50$. (b) $K_u/K_L = 50$. The anisotropy ratio for all layers is $K_x/K_z = 10$. (After Todd and Bear, 1961.)

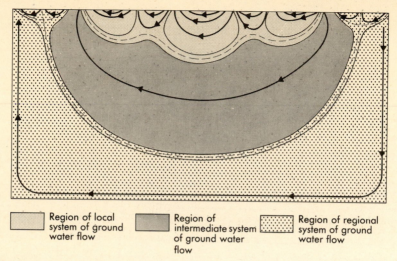

Region of local
system of ground
water flow

Region of
intermediate system
of ground water
flow

Region of regional
system of ground
water flow

FIGURE 8.13

Local, intermediate, and regional systems of ground water flow.

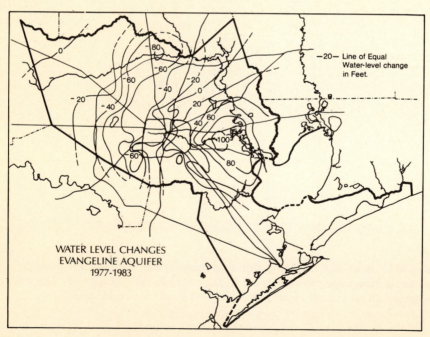

−20− Line of Equal
Water-level change
in Feet.

WATER LEVEL CHANGES
EVANGELINE AQUIFER
1977-1983

FIGURE 8.14

Drawdown in Harris and Galveston Counties, Texas.

n = number of squares over the direction of flow,

H = total head loss in direction of flow.

Flow nets are useful graphical methods to display streamlines and equipotential lines. Since no flow can cross an impermeable boundary, streamlines must parallel it. Also, streamlines are usually horizontal through high K material and vertical through low K material because of refraction of lines across a boundary between different K media. It can be shown (Fig. 8.11) that

$$\frac{K_1}{K_2} = \frac{\tan \theta_1}{\tan \theta_2}. \tag{8.20}$$

Flow nets for seepage through a layered system may not be orthogonal and are shown in Fig. 8.12. In Fig. 8.12(a), $K_u/K_L = 1/50$ and in part (b) $K_u/K_L = 50$. Such bending of streamlines becomes important in determining regional flow patterns. Figure 8.13 indicates possible local and regional patterns for a homogeneous isotropic aquifer that has a water table with a sinusoidal distribution.

Flow nets can be used to evaluate the effects of pumping on ground water levels and directions of flow. Figure 8.14 depicts the contour map resulting from heavy pumping near Houston, Texas, over a period of years. Equipotential lines and streamlines are discussed later in Section 8.4.

8.4

GENERAL FLOW EQUATIONS

The governing flow equations for ground water are derived in most of the standard texts in the field (DeWiest, 1965; Bear, 1979; Freeze and Cherry, 1979; Todd, 1980). The equation of continuity from fluid mechanics is combined with Darcy's law in three dimensions to yield a partial differential equation of flow in porous media, as shown in the next section. Both steady-state and transient flow equations can be derived. Mathematical solutions for specific boundary conditions are well known for the governing ground water flow equation. For complex boundaries and heterogeneous systems, numerical computer solutions must be used (Section 8.8).

Steady-State Saturated Flow

Consider a unit volume of porous media (Fig. 8.15) called an elemental control volume. The law of conservation of mass requires that

Mass in − Mass out = Change in storage per time.

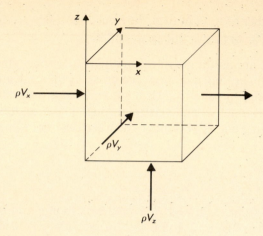

FIGURE 8.15

Elemental control volume.

For steady-state conditions, the right-hand side is zero, and the equation of continuity becomes (Fig. 8.15)

$$-\frac{\partial}{\partial x}(\rho V_x) - \frac{\partial}{\partial y}(\rho V_y) - \frac{\partial}{\partial z}(\rho V_z) = 0. \tag{8.21}$$

The units of ρV are mass/area/time as required. For an incompressible fluid $\rho(x, y, z) = $ Constant, and ρ can be divided out of Eq. (8.21). Substitution of Darcy's law for V_x, V_y, and V_z yields

$$\frac{\partial}{\partial x}\left(K_x \frac{\partial h}{\partial x}\right) + \frac{\partial}{\partial y}\left(K_y \frac{\partial h}{\partial y}\right) + \frac{\partial}{\partial z}\left(K_z \frac{\partial h}{\partial z}\right) = 0. \tag{8.22}$$

For an isotropic, homogeneous medium, $K_x = K_y = K_z = K$ and can be divided out of the equation to yield

$$\frac{\partial^2 h}{\partial x^2} + \frac{\partial^2 h}{\partial y^2} + \frac{\partial^2 h}{\partial z^2} = 0. \tag{8.23}$$

Equation (8.23) is called Laplace's equation and is one of the best-understood partial differential equations. The solution is $h = h(x, y, z)$, the hydraulic head at any point in the flow domain. In two dimensions, the solution is equivalent to the graphical flow nets described in Section 8.3. If there were no variation of h with z, then the equation would reduce to two terms on the left-hand side of Eq. (8.23).

Transient Saturated Flow

The transient equation of continuity for a confined aquifer becomes

$$-\frac{\partial}{\partial x}(\rho V_x) - \frac{\partial}{\partial y}(\rho V_y) - \frac{\partial}{\partial z}(\rho V_z) = \frac{\partial}{\partial t}(\rho n) = n\frac{\partial \rho}{\partial t} + \rho\frac{\partial n}{\partial t}. \tag{8.24}$$

The first term on the right-hand side of Eq. (8.24) is the mass rate of water produced by an expansion of water under a change in ρ. The second term is the mass rate of water produced by compaction of the porous media (change in n). The first term relates to the compressibility of the fluid β and the second term to the aquifer compressibility α. Freeze and Cherry (1979) indicate that a change in h will produce a change in ρ and n, and the volume of water produced for a unit head decline is S_s, the specific storage.

Theoretically, we can show that

$$S_s = \rho g(\alpha + n\beta), \tag{8.25}$$

and the mass rate of water produced is $S_s(\partial h/\partial t)$. Equation (8.24) becomes, after substituting Eq. (8.25) and Darcy's law

$$\frac{\partial}{\partial x}\left(K_x \frac{\partial h}{\partial x}\right) + \frac{\partial}{\partial y}\left(K_y \frac{\partial h}{\partial y}\right) + \frac{\partial}{\partial z}\left(K_z \frac{\partial h}{\partial z}\right) = S_s \frac{\partial h}{\partial t}. \tag{8.26}$$

For homogeneous and isotropic media,

$$\frac{\partial^2 h}{\partial x^2} + \frac{\partial^2 h}{\partial y^2} + \frac{\partial^2 h}{\partial z^2} = \frac{S_s}{K}\frac{\partial h}{\partial t}. \tag{8.27}$$

For the special case of a horizontal confined aquifer of thickness b,

$$S = S_s b, \quad \text{where } S \text{ is the storativity or storage coefficient}$$

$$T = Kb,$$

$$\nabla^2 h = \frac{S}{T}\frac{\partial h}{\partial t} \quad \text{in two dimensions.} \tag{8.28}$$

Solution of Eqs. (8.28) requires knowledge of S and T to produce $h(x, y)$ over the flow domain. The classical development of Eq. (8.28) was first advanced by Jacob (1940) along with considerations of storage concepts. More advanced treatments consider the problems of a fixed elemental control volume in a deforming media (Cooper, 1966), but these are unnecessary considerations for most practical problems.

Dupuit Assumption

For the case of unconfined ground water flow, Dupuit developed a theory that allows for a simple solution based on several important assumptions:

1. The water table or free surface is only slightly inclined.
2. Streamlines may be considered horizontal and equipotential lines vertical.
3. Slopes of the free surface and hydraulic gradient are equal.

Figure 8.16 shows the graphical example of Dupuit's assumptions for essentially one-dimensional flow. The free surface from $x = 0$ to $x = L$

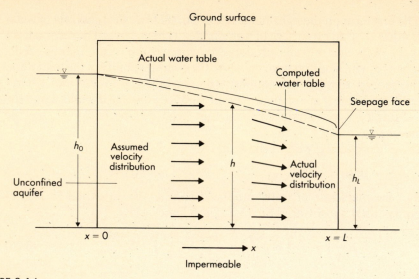

FIGURE 8.16

Steady flow in an unconfined aquifer between two water bodies with vertical boundaries.

can be derived by considering Darcy's law and the governing one-dimensional equation. Example 8.1 shows the derivation of the Dupuit equations.

EXAMPLE 8.1

DUPUIT EQUATION

Derive the equation for one-dimensional flow in an unconfined aquifer using the Dupuit assumptions (Fig. 8.16).

SOLUTION

Darcy's law gives the one-dimensional flow per unit width as

$$q = -Kh \frac{dh}{dx},$$

where h and x are as defined in Fig. 8.16. At steady state, the rate of change of q with distance is zero, or

$$\frac{d}{dx}\left[-Kh \frac{dh}{dx}\right] = 0,$$

$$-\frac{K}{2}\frac{d^2 h^2}{dx^2} = 0,$$

or

$$\frac{d^2 h^2}{dx^2} = 0.$$

Integration yields

$$h^2 = ax + b,$$

where a and b are constants. Setting the boundary condition $h = h_0$ at $x = 0$,

$$b = h_0^2.$$

Differentiation of $h^2 = ax + b$ gives

$$a = 2h \frac{dh}{dx}.$$

From Darcy's law,

$$h \frac{dh}{dx} = -\frac{q}{K},$$

so, by substitution,

$$h^2 = h_0^2 - \frac{2qx}{K}.$$

Setting $h = h_L$ at $x = L$ and neglecting flow across the seepage face yields

$$h_L^2 = h_0^2 - \frac{2qL}{K}.$$

Rearrangement gives

$$\boxed{q = \frac{K}{2L}(h_0^2 - h_L^2) \qquad \text{Dupuit equation.}}$$

Then the general equation for the shape of the parabola is

$$\boxed{h^2 = h_0^2 - \frac{x}{L}(h_0^2 - h_L^2) \qquad \text{Dupuit parabola.}}$$

The derivation of the Dupuit equations in Example 8.1 does not consider recharge to the aquifer. For the case of a system with recharge, the **Dupuit parabola** will take the mounded shape shown in Fig. 8.17. The point where $h = h_{max}$ is known as the **water divide**. At the water divide,

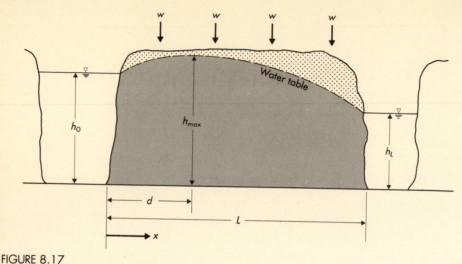

FIGURE 8.17

Dupuit parabola with recharge.

$q = 0$ since the gradient is zero. Example 8.2 derives the Dupuit equation for recharge and illustrates the use of the water divide concept.

EXAMPLE 8.2

DUPUIT EQUATION WITH RECHARGE

a) Derive the general Dupuit equation with the effect of recharge.

b) Two rivers located 1000 m apart fully penetrate an aquifer (see Fig. E8.2). The aquifer has a K value of 0.5 m/day. The region receives an average rainfall of 15 cm/yr and evaporation is about 10 cm/yr. Assume that the water elevation in River 1 is 20 m and the water elevation in River 2 is 18 m. Using the equation derived in part (a), determine the location and height of the water divide.

c) What is the daily discharge per m width into each river?

SOLUTION

a) Designating recharge intensity as W, it can be seen that

$$\frac{dq}{dx} = W.$$

From Darcy's law for one-dimensional flow, the flow per unit width is

$$q = -Kh\frac{dh}{dx}.$$

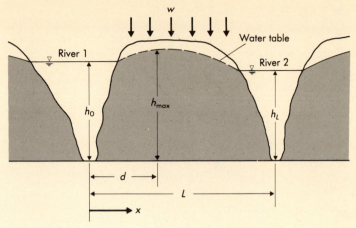

FIGURE E8.2

Substituting the second equation into the first yields

$$-\frac{K}{2}\frac{d^2h^2}{dx^2} = W,$$

or

$$\frac{d^2h^2}{dx^2} = -\frac{2W}{K}.$$

Integration gives

$$h^2 = -\frac{Wx^2}{K} + ax + b,$$

where a and b are constants. The boundary condition $h = h_0$ at $x = 0$ gives

$$b = h_0^2,$$

and the boundary condition $h = h_L$ at $x = L$ gives

$$a = \frac{(h_L^2 - h_0^2)}{L} + \frac{WL}{K}.$$

Substitution of a and b into the previous equation for h^2 yields

$$\boxed{h^2 = h_0^2 + \frac{(h_L^2 - h_0^2)}{L}x + \frac{Wx}{K}(L - x)} \qquad \text{Dupuit parabola.}$$

This equation will give the shape of the Dupuit parabola shown in Fig. 8.17. If $W = 0$, this equation will reduce to the parabolic equation found in Example 8.1. Differentiation of the parabolic equation gives

$$2h\frac{dh}{dx} = \frac{(h_L{}^2 - h_0{}^2)}{L} + \frac{W}{K}(L - 2x).$$

But Darcy's law gives

$$h\frac{dh}{dx} = -\frac{q}{K},$$

so

$$-2\frac{q}{K} = \frac{1}{L}(h_L{}^2 - h_0{}^2) + \frac{W}{K}(L - 2x).$$

Simplifying,

$$\boxed{q = \frac{K}{2L}(h_0{}^2 - h_L{}^2) + W\left(x - \frac{L}{2}\right).}$$

b) Given

$L = 1000$ m,

$K = 0.5$ m/day,

$h_0 = 20$ m,

$h_L = 18$ m,

$W = 5$ cm/yr $= 1.369 \times 10^{-4}$ m/day.

At $x = d$, $q = 0$ (see Fig. E8.2),

$$0 = \frac{K}{2L}(h_0{}^2 - h_L{}^2) + W\left(d - \frac{L}{2}\right),$$

$$d = \frac{L}{2} - \frac{K}{2WL}(h_0{}^2 - h_L{}^2)$$

$$= \frac{1000 \text{ m}}{2} - \frac{(0.5 \text{ m/day})(20^2 \text{ m}^2 - 18^2 \text{ m}^2)}{(2)(1.369 \times 10^{-4} \text{ m/day})(1000 \text{ m})}$$

$$= 500 \text{ m} - 138.8 \text{ m},$$

$$\boxed{d = 361.2 \text{ m}.}$$

At $x = d$, $h = h_{max}$,

$$h_{max}^2 = h_0{}^2 + \frac{(h_L{}^2 - h_0{}^2)}{L}d + \frac{Wd}{K}(L - d)$$

$$= (20 \text{ m})^2 + \frac{(18^2 \text{ m}^2 - 20^2 \text{ m}^2)}{1000 \text{ m}}(361.2 \text{ m})$$

$$+ \frac{(1.369 \times 10^{-4} \text{ m/day})(361.2 \text{ m})}{0.5 \text{ m/day}}(1000 \text{ m} - 361.2 \text{ m})$$

$$= 400 \text{ m}^2 - 27.5 \text{ m}^2 + 63.2 \text{ m}^2$$
$$= 435.7 \text{ m}^2$$

$$\boxed{h_{\max} = 20.9 \text{ m.}}$$

Thus, the water divide is located 361.2 m from the edge of River 1 and is 20.9 m high.

c) For discharge into River 1, set $x = 0$ m:

$$q = \frac{K}{2L}(h_0{}^2 - h_L{}^2) + W\left(0 - \frac{L}{2}\right)$$

$$= \frac{0.5 \text{ m/day}}{(2)(1000 \text{ m})}(20^2 \text{ m}^2 - 18^2 \text{ m}^2)$$
$$+ (1.369 \times 10^{-4} \text{ m/day})(- 1000 \text{ m}/2),$$

$$q = -0.0495 \text{ m}^2/\text{day}.$$

The negative sign indicates that flow is in the opposite direction from the x-direction. Therefore,

$$\boxed{q = 0.0495 \text{ m}^2/\text{day into River 1.}}$$

For discharge into River 2, set $x = L = 1000$ m:

$$q = \frac{K}{2L}(h_0{}^2 - h_L{}^2) + W\left(1000 \text{ m} - \frac{L}{2}\right)$$

$$= \frac{0.5 \text{ m/day}}{(2)(1000 \text{ m})}(20^2 \text{ m}^2 - 18^2 \text{ m}^2)$$
$$+ (1.369 \times 10^{-4} \text{ m/day})(1000 \text{ m} - 1000 \text{ m}/2),$$

$$\boxed{q = 0.08745 \text{ m}^2/\text{day into River 2.}}$$

Streamlines and Equipotential Lines

The equation of continuity for steady, incompressible, isotropic flow in two dimensions becomes

$$\frac{\partial u}{\partial x} + \frac{\partial v}{\partial y} = 0.$$

The governing steady-state flow equation is

$$\nabla^2 h = \frac{\partial^2 h}{\partial x^2} + \frac{\partial^2 h}{\partial y^2} = 0. \tag{8.29}$$

The **velocity potential** ϕ is a scalar function and can be written

$$\phi(x, y) = -K(z + p/\gamma) + c, \tag{8.30}$$

where K and c are assumed constant. From Darcy's law in two dimensions,

$$u = \frac{\partial \phi}{\partial x}, \qquad v = \frac{\partial \phi}{\partial y}. \tag{8.31}$$

Using Eq. (8.29) in two dimensions, we have

$$\nabla^2 \phi = \frac{\partial^2 \phi}{\partial x^2} + \frac{\partial^2 \phi}{\partial y^2} = 0, \tag{8.32}$$

where $\phi(x, y) = $ Constant represents a family of equipotential curves on a two-dimensional surface. It can be shown that the **stream function** $\psi(x, y) = $ Constant is orthogonal to $\phi(x, y) = $ Constant and that both satisfy the equation of continuity and Laplace's equation. The stream function $\psi(x, y)$ is defined by

$$u = \frac{\partial \psi}{\partial y}, \qquad v = -\frac{\partial \psi}{\partial x}. \tag{8.33}$$

Combining Eqs. (8.31) and (8.33), the Cauchy-Riemann equations become

$$\frac{\partial \phi}{\partial x} = \frac{\partial \psi}{\partial y}, \qquad \frac{\partial \phi}{\partial y} = -\frac{\partial \psi}{\partial x}. \tag{8.34}$$

It can be shown that ψ also satisfies Laplace's equation

$$\nabla^2 \psi = \frac{\partial^2 \psi}{\partial x^2} + \frac{\partial^2 \psi}{\partial y^2} = 0. \tag{8.35}$$

EXAMPLE 8.3

STREAMLINES AND EQUIPOTENTIAL LINES

Prove that ψ and ϕ are orthogonal for isotropic flow, given Fig. E8.3, where V is a velocity vector tangent to ψ_2. Show that flow between two streamlines is constant.

SOLUTION

We can write for the slope of the streamline

$$\frac{v}{u} = \frac{dy}{dx} = \tan \alpha$$

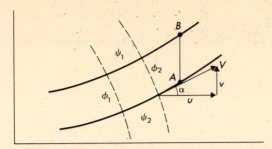

FIGURE E8.3

and

$$vdx - udy = 0.$$

Since

$$u = \frac{d\psi}{dy}, \qquad v = -\frac{d\psi}{dx},$$

then

$$\frac{\partial \psi}{\partial x} dx + \frac{\partial \psi}{\partial y} dy = 0,$$

or $d\psi(x, y) = 0$. The total differential equals zero, and $\psi(x, y) =$ Constant, as required by the stream function. The ϕ_1 and ϕ_2 lines represent equipotential lines and can be represented by the total differential

$$d\phi = \frac{\partial \phi}{\partial x} dx + \frac{\partial \phi}{\partial y} dy = 0.$$

Substituting for $\partial \phi / \partial x$ and $\partial \phi / \partial y$ from Darcy's law produces

$$udx + vdy = 0$$

and

$$\frac{dy}{dx} = -\frac{u}{v}, \qquad \text{the slope of the equipotential line.}$$

Thus, since the two slopes are negative inverses, equipotential lines are normal to streamlines. The system of orthogonal lines forms a flow net. Consider the flow crossing a vertical section AB between streamlines ψ_1 and ψ_2. The discharge across the section is designated Q, and it is apparent from fluid mechanics that

$$Q = \int_{\psi_2}^{\psi_1} udy,$$

or

$$Q = \int_{\psi_2}^{\psi_1} d\psi$$

or

$$Q = \psi_1 - \psi_2 .$$

Thus the flow between streamlines is constant and the spacing between streamlines reveals the relative magnitude of flow velocities between them.

Once either streamlines or equipotential lines are determined in a domain, the other can be evaluated from the Cauchy-Riemann equations (Eq. 8.34). Thus,

$$\psi = \int \left(\frac{\partial \phi}{\partial x} \, dy - \frac{\partial \phi}{\partial y} \, dx \right)$$

and

$$\phi = \int \left(\frac{\partial \psi}{\partial y} \, dx - \frac{\partial \psi}{\partial x} \, dy \right) .$$

8.5
STEADY-STATE WELL HYDRAULICS

The case of steady flow to a well implies that the variation of head occurs only in space and not in time. The governing equations presented in Section 8.4 can be solved for pumping wells in unconfined or confined aquifers under steady or unsteady conditions. Boundary conditions must be kept relatively simple and aquifers must be assumed to be homogeneous and isotropic in each layer. More complex geometries can be handled by numerical simulation models in two or three dimensions (Section 8.8).

Steady One-Dimensional Flow

For the case of ground water flow in the x-direction in a confined aquifer, the governing equation becomes

$$\frac{d^2h}{dx^2} = 0 \tag{8.36}$$

and has the solution

$$h = \frac{-vx}{K}, \tag{8.37}$$

where $h = 0$ and $x = 0$ and $dh/dx = -v/K$, according to Darcy's law. This states that head varies linearly with flow in the x-direction.

The case of steady one-dimensional flow in an unconfined aquifer was presented in Section 8.4 using Dupuit's assumptions. The resulting variation of head with x is called the Dupuit parabola and represents the approximate shape of the water table for relatively flat slopes. In the presence of steep slopes near wells, the Dupuit approximation may be in error, and more sophisticated computer methods should be used.

Steady Radial Flow to a Well — Confined

The **drawdown curve** or **cone of depression** varies with distance from a pumping well in a confined aquifer (Fig. 8.18). The flow is assumed two-dimensional for a completely penetrating well in a homogeneous, isotropic aquifer of unlimited extent. For horizontal flow, the above assumptions apply and Q at any r equals, from Darcy's law,

$$Q = -2\pi r b K \frac{dh}{dr} \tag{8.38}$$

for steady radial flow to a well. Integrating after separation of variables, with $h = h_w$ at $r = r_w$ at the well, yields

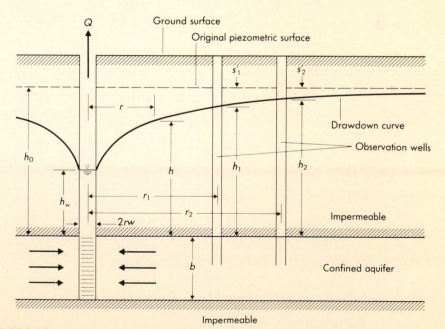

FIGURE 8.18

Radial flow to a well penetrating an extensive confined aquifer.

$$Q = 2\pi Kb \frac{h - h_w}{\ln(r/r_w)}. \tag{8.39}$$

Equation (8.39) shows that h increases indefinitely with increasing r, yet the maximum head is h_0 for Fig. 8.18. Near the well the relationship holds and can be rearranged to yield an estimate for transmissivity T

$$T = Kb = \frac{Q}{2\pi(h_2 - h_1)} \ln \frac{r_2}{r_1} \tag{8.40}$$

by observing heads h_1 and h_2 at two adjacent observation wells located at r_1 and r_2, respectively, from the pumping well. In practice, it is often necessary to use unsteady-state analyses because of the long times required to reach steady state.

EXAMPLE 8.4

DETERMINATION OF K AND T IN A CONFINED AQUIFER

A well is constructed to pump water from a confined aquifer. Two observation wells, OW1 and OW2, are constructed at distances of 100 m and 1000 m, respectively. Water is pumped from the pumping well at a rate of 0.2 m³/min. At steady state, drawdown s' is observed as 2m in OW2 and 8 m in OW1. Determine the hydraulic conductivity K and transmissivity T if the aquifer is 20 m thick.

SOLUTION

Given

$$Q = 0.2 \text{ m}^3/\text{min},$$
$$r_2 = 1000 \text{ m},$$
$$r_1 = 100 \text{ m},$$
$$s_2' = 2\text{m},$$
$$s_1' = 8 \text{ m},$$
$$b = 20 \text{ m}.$$

Equation (8.40) gives

$$T = Kb = \frac{Q}{2\pi(h_2 - h_1)} \ln \left(\frac{r_2}{r_1}\right).$$

Knowing that $s_1' = h_0 - h_1$ and $s_2' = h_0 - h_2$, we have

$$T = Kb = \frac{Q}{2\pi(s_1' - s_2')} \left[\ln \left(\frac{r_2}{r_1}\right)\right]$$

$$= \frac{0.2 \text{ m}^3/\text{min}}{(2\pi)(8\text{m} - 2\text{m})} \ln \left(\frac{1000 \text{ m}}{100 \text{ m}}\right),$$

$$\boxed{T = 0.0122 \text{ m}^2/\text{min} = 2.04 \text{ cm}^2/\text{s}.}$$

Then

$$K = T/b$$
$$= (2.04 \text{ cm}^2/\text{s})/(20 \text{ m})(100 \text{ cm}/1 \text{ m}),$$

$$\boxed{K = 1.02 \times 10^{-3} \text{ cm/s.}}$$

Steady Radial Flow to a Well — Unconfined

Applying Darcy's law for radial flow in an unconfined, homogeneous, isotropic, and horizontal aquifer and using Dupuit's assumptions (Fig. 8.19),

$$Q = -2\pi r K h \frac{dh}{dr}. \tag{8.41}$$

Integrating, as before,

$$Q = \pi K \frac{h_2^2 - h_1^2}{\ln(r_2/r_1)}. \tag{8.42}$$

Solving for K,

$$K = \frac{Q}{\pi(h_2^2 - h_1^2)} \ln \frac{r_2}{r_1}, \tag{8.43}$$

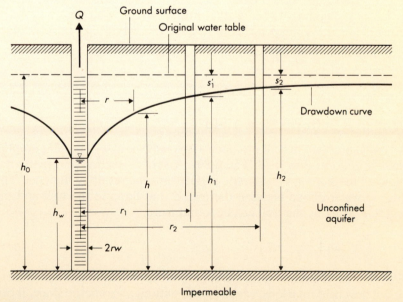

FIGURE 8.19
Radial flow to a well penetrating an unconfined aquifer.

where heads h_1 and h_2 are observed at adjacent wells located distances r_1 and r_2 from the pumping well, respectively.

EXAMPLE 8.5

DETERMINATION OF K IN AN UNCONFINED AQUIFER

A fully penetrating well discharges 75 gpm from an unconfined aquifer. The original water table was recorded as 35 ft. After a long time period the water table was recorded as 20 ft MSL in an observation well located 75 ft away and 34 ft MSL at an observation well located 2000 ft away. Determine the hydraulic conductivity of this aquifer in ft/s.

SOLUTION

Given

$$Q = 75 \text{ gpm,}$$
$$r_2 = 2000 \text{ ft,}$$
$$r_1 = 75 \text{ ft,}$$
$$h_2 = 34 \text{ ft,}$$
$$h_1 = 20 \text{ ft.}$$

Equation (8.43) gives

$$K = \frac{Q}{\pi(h_2{}^2 - h_1{}^2)} \ln\left(\frac{r_2}{r_1}\right)$$
$$= \frac{(75 \text{ gpm})(0.134 \text{ ft}^3/\text{gal})(1 \text{ min}/60 \text{ s})}{(\pi)(34^2 \text{ ft}^2 - 20^2 \text{ ft}^2)} \ln\left(\frac{2000 \text{ ft}}{75 \text{ ft}}\right),$$

$$\boxed{K = 2.32 \times 10^{-4} \text{ ft/s.}}$$

Well in a Uniform Flow Field

A typical problem in well mechanics involves the case of a well pumping from a uniform flow field (Fig. 8.20). A vertical section and plan view indicate the sloping piezometric surface and the resulting flow net. The ground water divide between the region that flows to the well and the region flowing by the well can be found from

$$-\frac{y}{x} = \tan\left(\frac{2\pi Kbi}{Q} y\right). \tag{8.44}$$

Equation (8.44) results from the superposition of radial and one-dimensional flow field solutions, where i is the natural piezometric slope. It can be shown that

$$y_L = \pm \frac{Q}{2Kbi}$$ (8.45)

as $x \to \infty$, and the stagnation point (no flow) occurs at

$$x_S = -\frac{Q}{2\pi Kbi}, \qquad y = 0.$$ (8.46)

Equations (8.45) and (8.46) may be applied to unconfined aquifers for cases of relatively small drawdowns, where b is replaced by h_0, the average saturated aquifer thickness. An important application of the well in a uniform flow field involves the evaluation of pollution sources and impacts on downgradient well fields and the potential for pumping and capturing a plume as it migrates downgradient.

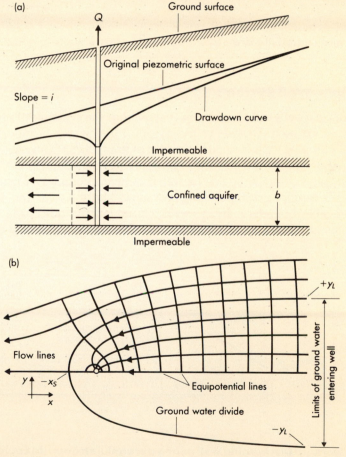

FIGURE 8.20

Flow to a well penetrating a confined aquifer having a sloping-plane piezometric surface. (a) Vertical section. (b) Plan view.

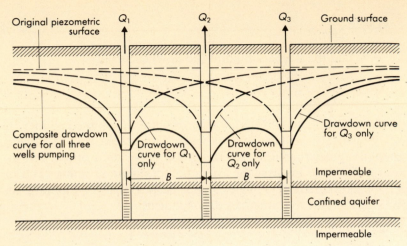

FIGURE 8.21

Individual and composite drawdown curves for three wells in a line.

Multiple-Well Systems

For multiple wells with drawdowns that overlap, the principle of superposition can be used. Drawdown at any point in the area of influence of several pumping wells is equal to the sum of drawdowns from each well in confined or unconfined aquifers (Fig. 8.21).

The same principle applies for well flow near a boundary. Figure 8.22 shows the case for a well pumping near a fixed head stream and near an impermeable boundary. Image wells placed on the other side of the boundary at a distance a can be used to represent the equivalent hydraulic

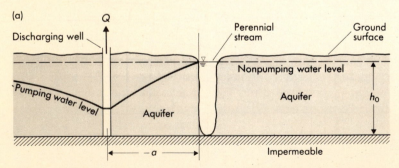

FIGURE 8.22

Sectional views. (a) Discharging well near a perennial stream. (b) Equivalent hydraulic system in an aquifer of infinite areal extent. (c) Discharging well near an impermeable boundary. (d) Equivalent hydraulic system in an aquifer of infinite areal extent. (After Ferris et al., 1962.)

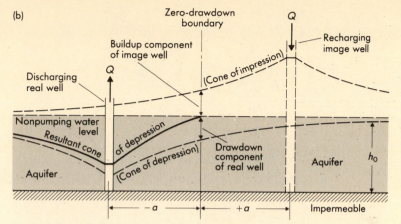

Aquifer thickness h_0 should be very large compared with resultant drawdown near real well.

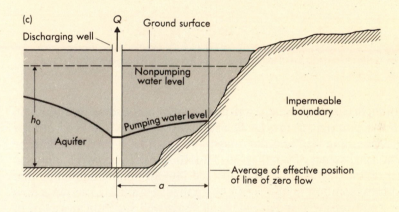

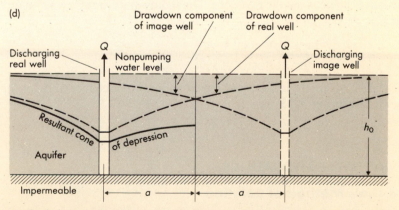

Aquifer thickness h_0 should be very large compared with resultant drawdown near real well.

FIGURE 8.22 (Continued)

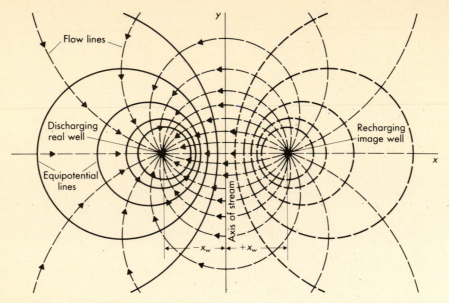

FIGURE 8.23

Flow net for a discharging real well and a recharging image well. (After Ferris et al., 1962.)

condition. In case A, the image well is recharging at the same rate Q, and in case B it is pumping at rate Q. The summation of drawdowns from the original pumping well and the image well provides a correct boundary condition at distance r from the well. Thus the use of image wells allows an aquifer of finite extent to be transformed into an infinite aquifer so that closed-form solution methods can be applied. Figure 8.23 shows a flow net for a pumping well and a recharging image well and indicates a line of constant head between the two wells. The steady-state drawdown s' at any point (x, y) is given by

$$s' = \frac{Q}{4\pi T} \ln \frac{(x + x_w)^2 + (y + y_w)^2}{(x - x_w)^2 + (y - y_w)^2},\tag{8.47}$$

where $\pm x_w, y_w$ are the locations of the recharge and discharge wells for the case shown in Fig. 8.23.

8.6

UNSTEADY WELL HYDRAULICS

The Theis Method of Solution

As a well penetrating a confined aquifer of infinite extent is pumped at a constant rate, a drawdown occurs radially extending from the well. The rate of decline of head times the storage coefficient summed over the area of influence equals the discharge. The rate of decline decreases continuously as the area of influence expands.

The governing ground water flow equation (Eq. 8.27) in plane polar coordinates is

$$\frac{\partial^2 h}{\partial r^2} + \frac{1}{r}\frac{\partial h}{\partial r} = \frac{S}{T}\frac{\partial h}{\partial t}, \tag{8.48}$$

where

h = head,
r = radial distance,
S = storage coefficient,
T = transmissivity.

Theis (1935) obtained a solution for Eq. (8.48) by assuming that the well is a mathematical sink of constant strength and by using boundary conditions $h = h_0$ for $t = 0$ and $h \to h_0$ as $r \to \infty$ for $t \geq 0$:

$$s' = \frac{Q}{4\pi T}\int_u^\infty \frac{e^{-u}du}{u} = \frac{Q}{4\pi T}W(u), \tag{8.49}$$

where s' is drawdown, Q is discharge at the well, and

$$u = \frac{r^2 S}{4Tt}. \tag{8.50}$$

Equation (8.49) is known as the nonequilibrium, or Theis, equation. The integral is written as $W(u)$ and is known as the exponential integral, or well function, which can be expanded as a series:

$$W(u) = -0.5772 - \ln(u) + u - \frac{u^2}{2 \cdot 2!} + \frac{u^3}{3 \cdot 3!} - \frac{u^4}{4 \cdot 4!} + \cdots. \tag{8.51}$$

The equation can be used to obtain aquifer constants S and T by means of pumping tests at fully penetrating wells. It is widely used because a value of S can be determined, only one observation well and a relatively short pumping period are required, and large portions of the flow field can be sampled with one test.

The assumptions inherent in the Theis equation should be included since they are often overlooked:

1. The aquifer is homogeneous, isotropic, uniformly thick, and of infinite areal extent.
2. Prior to pumping, the piezometric surface is horizontal.
3. The fully penetrating well is pumped at a constant rate.
4. Flow is horizontal within the aquifer.
5. Storage within the well can be neglected.

6. Water removed from storage responds instantaneously with a declining head.

These assumptions are seldom completely satisfied for a field problem, but the method still provides one of the most useful and accurate techniques for aquifer characterization. The complete Theis solution requires the graphical solution of two equations with four unknowns:

$$s' = \frac{Q}{4\pi T}\, W(u), \tag{8.52}$$

$$\frac{r^2}{t} = \left(\frac{4T}{S}\right) u. \tag{8.53}$$

The relation between $W(u)$ and u must be the same as that between s' and r^2/t because all other terms are constants in the equations. Theis suggested a solution based on graphical superposition. Example 8.6 indicates how a plot of $W(u)$ vs. u, called a **type curve,** is superimposed over observed time-drawdown data while keeping the coordinate axes parallel. The two plots are adjusted until a position is found by trial, such that most of the observed data fall on a segment of the type curve. Any convenient point is selected, and values of $W(u)$, u, s', and r^2/t are used in Eqs. (8.52) and (8.53) to determine S and T (see Fig. 8.24).

It is also possible to use Theis's solution for the case where several wells are sampled for drawdown simultaneously near a pumped well.

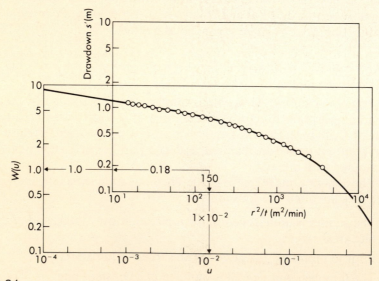

FIGURE 8.24

Theis method of superposition for solution of the nonequilibrium equation.

Distance-drawdown data are then fitted to the type curve similar to the method just outlined.

EXAMPLE 8.6

DETERMINATION OF *T* AND *S* BY THE THEIS METHOD

A fully penetrating well in a 25-m-thick confined aquifer is pumped at a rate of 0.2 m³/s for 1000 min. Drawdown is recorded vs. time at an observation well located 100 m away. Compute the transmissivity and storativity using the Theis method.

TIME (min)	s' (m)	TIME (min)	s' (m)
1	0.11	60	1.02
2	0.20	70	1.05
3	0.28	80	1.08
4	0.34	90	1.11
6	0.44	100	1.15
8	0.50	200	1.35
10	0.54	400	1.55
20	0.71	600	1.61
30	0.82	800	1.75
40	0.85	1000	1.80
50	0.92		

SOLUTION

A plot of s' vs. r^2/t is made on log-log paper. This is superimposed on a plot of $W(u)$ versus u, which is also on log-log paper. A point is chosen at some convenient point on the matched curve, and values for s', r^2/t, $W(u)$, and u are read (see Fig. E8.6 and the accompanying table of values). From the plot,

$$r^2/t = 180 \text{ m}^2/\text{min},$$
$$s' = 1.0 \text{ m},$$
$$u = 0.01,$$
$$W(u) = 4.0.$$

Equation (8.52) gives

$$s' = \frac{Q}{4\pi T} W(u),$$
$$T = QW(u)/4\pi s',$$
$$T = \frac{(0.2 \text{ m}^3/\text{s})(4.0)}{(4\pi)(1.0 \text{ m})},$$

$$\boxed{T = 6.37 \times 10^{-2} \text{ m}^2/\text{s}.}$$

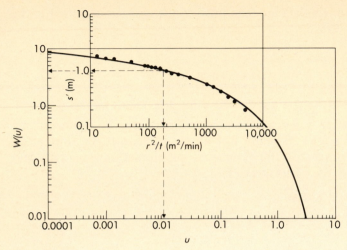

FIGURE E8.6

Values of the Function W(u) for Various Values of u

u	W(u)	u	W(u)	u	W(u)	u	W(u)	u	W(u)
1×10^{-10}	22.45	7×10^{-8}	15.90	4×10^{-5}	9.55	1×10^{-2}	4.04		
2	21.76	8	15.76	5	9.33	2	3.35		
3	21.35	9	15.65	6	9.14	3	2.96		
4	21.06	1×10^{-7}	15.54	7	8.99	4	2.68		
5	20.84	2	14.85	8	8.86	5	2.47		
6	20.66	3	14.44	9	8.74	6	2.30		
7	20.50	4	14.15	1×10^{-4}	8.63	7	2.15		
8	20.37	5	13.93	2	7.94	8	2.03		
9	20.25	6	13.75	3	7.53	9	1.92		
1×10^{-9}	20.15	7	13.60	4	7.25	1×10^{-1}	1.823		
2	19.45	8	13.46	5	7.02	2	1.223		
3	19.05	9	13.34	6	6.84	3	0.906		
4	18.76	1×10^{-6}	13.24	7	6.69	4	0.702		
5	18.54	2	12.55	8	6.55	5	0.560		
6	18.35	3	12.14	9	6.44	6	0.454		
7	18.20	4	11.85	1×10^{-3}	6.33	7	0.374		
8	18.07	5	11.63	2	5.64	8	0.311		
9	17.95	6	11.45	3	5.23	9	0.260		
1×10^{-8}	17.84	7	11.29	4	4.95	1×10^{0}	0.219		
2	17.15	8	11.16	5	4.73	2	0.049		
3	16.74	9	11.04	6	4.54	3	0.013		
4	16.46	1×10^{-5}	10.94	7	4.39	4	0.004		
5	16.23	2	10.24	8	4.26	5	0.001		
6	16.05	3	9.84	9	4.14				

Equation (8.53) gives

$$r^2/t = \frac{4Tu}{S},$$

$$S = \frac{4Tu}{r^2/t}$$

$$= (4)(6.37 \times 10^{-2} \text{ m}^2/\text{s})(0.01)/(180 \text{ m}^2/\text{min})(1 \text{ min}/60 \text{ s}),$$

$$\boxed{S = 8.49 \times 10^{-4}.}$$

Cooper-Jacob Method of Solution

Cooper and Jacob (1946) noted that for small values of r and large values of t, the parameter u in Eq. (8.49) becomes very small so that the infinite series can be approximated by

$$s' = \frac{Q}{4\pi T}\left[-0.5772 - \ln\left(\frac{r^2 S}{4Tt}\right)\right]. \tag{8.54}$$

Further rearrangement and conversion to decimal logarithms yields

$$s' = \frac{2.30Q}{4\pi T}\log\left(\frac{2.25Tt}{r^2 S}\right). \tag{8.55}$$

Thus a plot of drawdown s' vs. logarithm of t forms a straight line as shown in Fig. 8.25. A projection of the line to $s' = 0$, where $t = t_0$, yields

$$0 = \frac{2.3Q}{4\pi T}\log\left(\frac{2.25Tt_0}{r^2 S}\right), \tag{8.56}$$

and it follows that, since $\log(1) = 0$,

$$S = \frac{2.25Tt_0}{r^2}. \tag{8.57}$$

Finally, by replacing s' by $\Delta s'$, where $\Delta s'$ is the drawdown difference of data per log cycle of t, Eq. (8.55) becomes

$$T = \frac{2.3Q}{4\pi\,\Delta s'}. \tag{8.58}$$

The Cooper-Jacob method first solves for T with Eq. (8.58) and then for S with Eq. (8.57) and is applicable for small values of u (less than 0.01). Calculations with the Theis method were presented in Example 8.6, and the Cooper-Jacob method is used in Example 8.7.

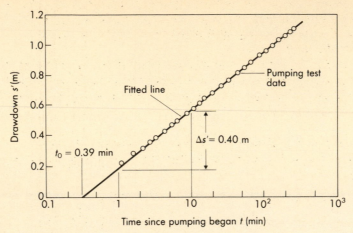

FIGURE 8.25

Cooper-Jacob method for solution of the nonequilibrium equation.

EXAMPLE 8.7

DETERMINATION OF *T* AND *S* BY THE COOPER-JACOB METHOD

Using the data given in Example 8.6, determine the transmissivity and storativity of the 25-m-thick confined aquifer using the Cooper-Jacob method.

SOLUTION

Values of s' and t are plotted on semilog paper with the t-axis logarithmic (see Fig. E8.7). A line is fitted through the later time periods and is

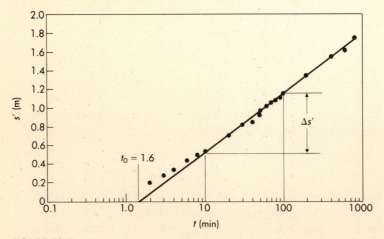

FIGURE E8.7

projected back to a point where $s' = 0$. This point determines t_0. $\Delta s'$ is measured over 1 log cycle of t.

From the plot,

$$t_0 = 1.6 \text{ min,}$$

$$\Delta s' = 0.65 \text{ m.}$$

Equation (8.58) gives

$$T = \frac{2.3Q}{4\pi \, \Delta s'}$$

$$= \frac{(2.3)(0.2 \text{ m}^3/\text{s})}{(4\pi)(0.65 \text{ m})},$$

$$\boxed{T = 5.63 \times 10^{-2} \text{ m}^2/\text{s.}}$$

Equation (8.57) gives

$$S = \frac{2.25 T t_0}{r^2}$$

$$= \frac{(2.25)(5.63 \times 10^{-2} \text{ m}^2/\text{s})(1.6 \text{ min})(60 \text{ s}/1 \text{ min})}{100^2 \text{ m}^2},$$

$$\boxed{S = 1.22 \times 10^{-3}.}$$

Slug Tests

Slug tests involve the use of a single well for the determination of aquifer formation constants. Rather than pumping the well for a period of time, as described above, a volume of water is suddenly removed or added to the well casing and observations of recovery or drawdown are noted through time. By careful evaluation of the drawdown curve and knowledge of the well screen geometry, it is possible to derive K or T for an aquifer.

Typical procedure for a slug test requires use of a rod of slightly smaller diameter than the well casing or a pump to evacuate the well casing. The simplest slug test method in a piezometer was published by Hvorslev (1951), who used the recovery of water level over time to calculate hydraulic conductivity of the porous media. Hvorslev's method relates the flow $q(t)$ at the piezometer at any time to the hydraulic conductivity and the unrecovered head distance, $H_0 - h$ in Fig. 8.26, by

$$q(t) = \pi r^2 \frac{dh}{dt} = FK(H_0 - h), \tag{8.59}$$

where F is a factor that depends on the shape and dimensions of the piezometer intake. If $q = q_0$ at $t = 0$, then $q(t)$ will decrease toward zero as

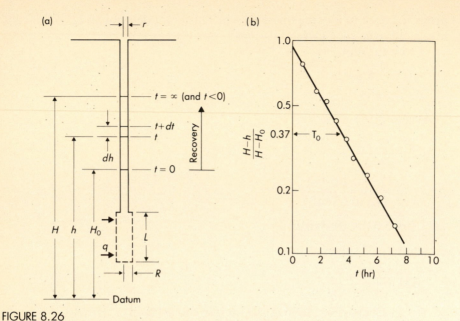

FIGURE 8.26

Hvorslev piezometer test. (a) Geometry. (b) Method of analysis.

time increases. Hvorslev defined the basic time lag

$$T_0 = \frac{\pi r^2}{FK}$$

and solved Eq. (8.59) with initial conditions $h = H_0$ at $t = 0$. Thus

$$\frac{H - h}{H - H_0} = e^{-t/T_0}. \tag{8.60}$$

By plotting recovery $(H - h)/(H - H_0)$ vs. time on semilog graph paper, we find that $t = T_0$, where recovery equals 0.37 (Fig. 8.27). For piezometer intake length divided by radius (L/R) greater than 8, Hvorslev has evaluated the shape factor F and obtained an equation for K:

$$K = \frac{r^2 \ln(L/R)}{2LT_0}. \tag{8.61}$$

Several other slug test methods have been developed by Cooper et al. (1967) and Papadopoulos et al. (1973) for confined aquifers. These methods are similar to Theis's in that a curve-matching procedure is used to obtain S and T for a given aquifer. A family of type curves $H(t)/H_0$ vs. Tt/r_c^2 was published for five values of the variable α, defined as $(r_s^2/r_c^2)S$ in Fig. 8.27. Papadopoulos et al. (1973) added five additional values of α.

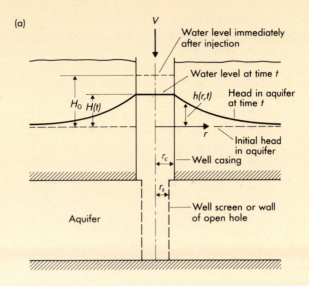

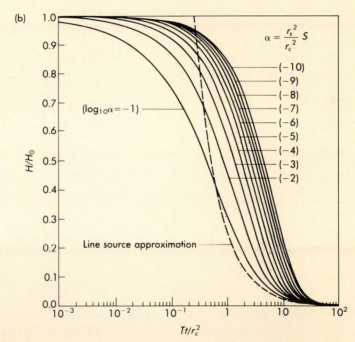

FIGURE 8.27

Illustration of a slug test. (a) Well used for slug test. (b) Type curves. (After Papadopoulos et al., 1973.)

The solution method is graphical and requires a semilogarithmic plot of measured $H(t)/H_0$ vs. t, where H_0 is the assumed initial excess head. The data are then curve-matched to the plotted type curves by horizontal translation until the best match is achieved (Fig. 8.27), and a value of α is selected for a particular curve. The vertical time axis t, which overlays the vertical axis for $Tt/r_c^2 = 1.0$ is selected, and a value of T can then be found from $T = 1.0r_c^2/t_1$. The value of S can be found from the definition of α. The use of the method is representative of the formation only in the immediate vicinity of the test hole and should be used with caution. Example 8.8 indicates how the type curves for slug tests are applied to field data.

EXAMPLE 8.8

DETERMINATION OF *T, K,* AND *S* FROM A SLUG TEST

A screened, cased well penetrates a confined aquifer that is 10 m thick. The casing radius is 8.5 cm and the radius of the screen is 5.7 cm. A slug of water is injected that raises the water level by 0.39 m. The change in water level vs. time was measured and is listed in the following table. Using the type curves of Fig. 8.27, determine T, K, and S for the aquifer.

TIME (s)	H (m)
3	0.34
6	0.30
12	0.23
20	0.18
30	0.13
40	0.09
90	0.03
110	0.02

SOLUTION

A plot of H/H_0 vs. t points is made with t on a logarithmic axis (see Fig. E8.8). This plot is superimposed on the type curves of Fig. 8.27 and shifted horizontally until the data fit one of the curves. The curve that best fits the given data is the curve of $\log(\alpha) = 10^{-4}$. It can be seen that for $Tt/r_c^2 = 1.0$, $t = 10$ s. If $Tt/r_c^2 = 1.0$, then

$$T = 1.0r_c^2/t$$
$$= (1.0)(8.5 \text{ cm})^2/(10 \text{ s}),$$

$$\boxed{T = 7.23 \text{ cm}^2/\text{s}.}$$

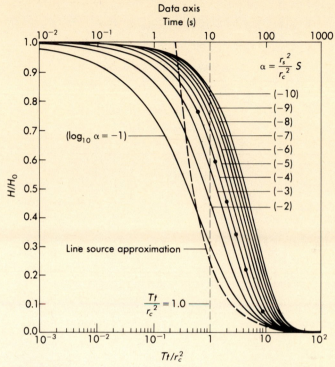

FIGURE E8.8

Using this value of T,

$$K = T/b,$$
$$= (7.23 \text{ cm}^2/\text{s})/(10 \text{ m})(100 \text{ cm}/1 \text{ m}),$$

$$\boxed{K = 7.23 \times 10^{-3} \text{ cm/s.}}$$

By definition,

$$\alpha = \left(\frac{r_s^2}{r_c^2}\right) S,$$

so

$$S = \frac{(10^{-4})(8.5 \text{ cm})^2}{(5.7 \text{ cm})^2},$$

$$\boxed{S = 2.22 \times 10^{-4}.}$$

Radial Flow in a Leaky Aquifer

Leaky aquifers represent a unique and complex problem in well mechanics. When a leaky aquifer is pumped, as shown in Fig. 8.28, water is
withdrawn both from the lower aquifer and from the saturated portion of

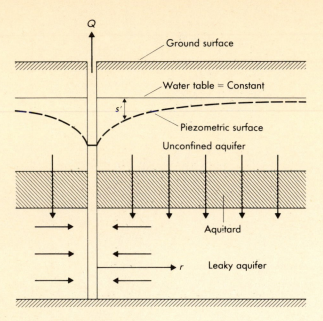

FIGURE 8.28

Well pumping from a leaky aquifer.

the overlying aquitard. By creating a lowered piezometric surface below
the water table, ground water can migrate vertically downward and then
move horizontally to the well. While steady-state conditions in a leaky
system are possible, a more general nonequilibrium analysis for unsteady
flow is more applicable and more often occurs in the field. When pumping
starts from a well in a leaky aquifer, drawdown of the piezometric surface
can be given by

$$ s' = \frac{Q}{4\pi T}\, W\left(u, \frac{r}{B}\right), \tag{8.62} $$

where the quantity r/B is given by

$$ \frac{r}{B} = \frac{r}{\sqrt{T/(K'/b')}}, $$

where T is transmissivity of the aquifer, K' is vertical hydraulic conductivity of the aquitard, and b' is thickness of the aquitard. Values of the
function $W(u, r/B)$ have been tabulated by Hantush (1956) and have been
used by Walton (1960) to prepare a family of type curves, shown in Fig.
8.29. Equation (8.61) reduces to the Theis equation for $K' = 0$ and $r/B =
0$. The method of solution for the leaky aquifer works in the same way as
the Theis solution with a superposition of drawdown data on top of the
leaky type curves. A curve of best fit is selected, and values of W, $1/u$, s',

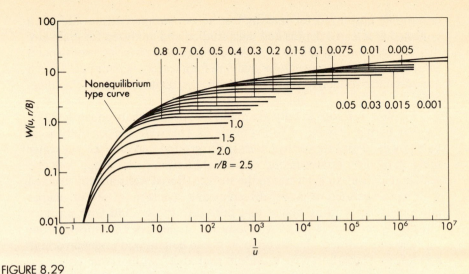

FIGURE 8.29

Type curves for analysis of pumping test data to evaluate storage coefficient and transmissivity of leaky aquifers. (After Walton, 1960, Illinois State Water Survey.)

and t are found, which allows T and S to be determined. Finally, based on the value of r/B, it is possible to calculate K' and b'.

In general, leaky aquifers are much more difficult to deal with than confined or unconfined systems. But the method just described does provide a useful tool for evaluating leaky systems analytically. For more complex geologies and systems with lenses, a three-dimensional computer simulation must be employed to properly represent ground water flow. These types of models are described in detail in Section 8.8.

8.7
WATER WELLS

Shallow Well Construction

A water well is a vertical hole dug into the earth, usually designed for bringing ground water to the surface. Wells can be used for pumping, artificial recharge, waste disposal, water level observation, and water quality monitoring. A variety of methods exists for constructing wells, depending on the flow rate, depth to ground water, geologic condition, casing material, and economic factors.

Prior to drilling a well in a new area, a test hole is normally drilled and a record, or log, is kept of various geologic formations and the depths at which they are encountered. Cable tool, rotary, and jetting methods are commonly employed for making test holes. Sample cuttings are often collected at selected depths and later studied and analyzed for grain size distribution. Borings can be removed using a Shelby tube or a split-spoon sampler, and these cores can be analyzed for a variety of parameters, such as hydraulic conductivity, porosity, grain size, texture, and soil classification.

Shallow wells are usually less than 15 m deep and are constructed by digging, boring, driving, or jetting methods. **Dug wells** are generally excavated by hand and are vertical holes in the ground that intersect the water table. A modern domestic dug well should have a rock curb, a concrete seal, and a pump to lift water to the surface. **Bored wells** are constructed with hand-operated or power-driven augers, available in several shapes and sizes, with cutting blades that bore into the ground with a rotary motion. When the blades fill with loose earth, the auger is removed from the hole and emptied. Depths up to 30 m can be achieved under favorable conditions. Hand-bored wells seldom exceed 20 cm in diameter.

A continuous-flight power auger has a spiral extending from the bottom of the hole to the surface, and sections of the auger may be added as depth increases. Cuttings are carried to the surface in the drilling process, and depths up to 50 m can be achieved with truck-mounted equipment. Hollow-stem augers are often used to construct small-diameter wells, which are associated with waste sites for monitoring or pumping. Augers work best in formations that do not cave, such as where some clay or silt is present in the aquifer.

Driven wells consist of a series of pipe lengths driven vertically downward by repeated impacts into the ground. Water enters the wells through a drive point at the lower end, which consists of a screened, cylindrical section, protected by a steel cone. Driven wells usually have diameters of less than 10 cm, with depths generally below 15 m. Yields from driven wells are usually fairly small, but a battery of such wells connected to a single pipe and pump are effective for lowering water table elevations. Such a system is known as a well-point system and is used for dewatering excavations for subsurface construction. Figure 8.30 illustrates how a well-point system reduces ground water levels.

Jetted wells are constructed with a high-velocity stream of water directed vertically downward, while the casing that is lowered into the hole conducts the water and cuttings to the surface. Small-diameter holes, up to 10 cm, with depths up to 15 m, can be installed in unconsolidated formations. Because of the speed of installation, jetted wells are useful for observation wells and well-point systems.

Deep Well Construction

Most **deep wells** of high capacity are constructed by drilling with a cable tool or with one of several rotary methods. Construction techniques vary depending on the type of geologic formation, depth of the well, and

(a)

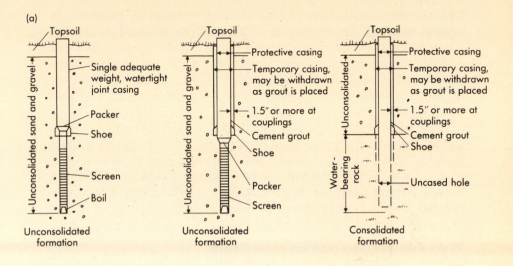

(b)

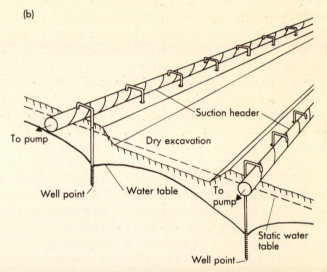

FIGURE 8.30

(a) Typical well designs for consolidated and unconsolidated formations. (b) A well-point system to dewater an excavation site.

required flow rate. Cable tool drilling is accomplished by regular lifting and dropping of a string of tools, with a sharp bit on the lower end to break rock by impact. The method is most useful for consolidated rock materials to depths of 600 m.

A rapid method for drilling in unconsolidated formations is the rotary method, which consists of drilling with a hollow, rotating bit, with drilling mud or water used to increase efficiency. No casing is required with drilling mud because the mud forms a clay lining on the wall of the well.

Drilling mud consists of a suspension of water, bentonite clay, and various organic additives. Proper mixture is essential for trouble-free drilling. This method is used extensively for oil wells. Advantages are the rapid drilling rate and the convenience for electric logging. Air rotary methods use compressed air in place of drilling mud and are convenient for consolidated formations. Drilling depths can exceed 150 m under favorable conditions.

The reverse-circulation rotary method is popular for unconsolidated formations. Water is pumped up through the drill pipe, using a large-capacity pump, and after cuttings settle out in a large pit, the water is recycled into the hole. Velocities must be kept low, and hole diameters must be greater than 40 cm. This method can drill to depths of 125 m, and the wells usually must be gravel-packed.

Well Completion and Development

Once a well has been drilled, it must be completed in such a way that it remains an efficient producer of water. This may involve installation of casing, cementing, placement of well screens, and gravel packing. **Well casings** serve as a lining to maintain an open hole up to the ground surface. They provide structural support against caving materials and seal out surface water. Casing materials include wrought iron, steel, and PVC pipe.

Wells are cemented in the annular space around the casing to prevent entrance of water of unsatisfactory quality. Cement grout, which may contain additives, is usually introduced at the bottom of the space to ensure proper sealing. **Grout seals** are common for layered aquifers, where it is desirable to pump from one layer while not disturbing another. **Well screens,** typically used for unconsolidated formations, are perforated sections of pipe of variable length that allow ground water to flow into the well. Their main purpose is to prevent aquifer material, such as sand or gravel, from entering the well and to minimize hydraulic resistance to flow. Screens are made of a variety of metals and metal alloys, PVC, and wood. Where a well screen is to be surrounded by an artificial **gravel pack,** the size of the screen openings is governed by the size of the gravel.

A gravel pack that envelops a well screen is designed to stabilize the aquifer and provide an annular zone of high-permeability material. In layered systems, gravel packs can provide conduits of flow from one layer to another and should be separated by grout plugs. Details on gravel pack material and well screen sizes are available from the American Water Works Association (1967).

Well development follows completion and is designed to increase the hydraulic efficiency by removing the finer material from the formation surrounding the screen. The importance of well development is often overlooked but is required to produce full-potential yields. Development procedures include pumping, surging, use of compressed air, hydraulic jetting, chemical addition, and use of explosives. The methods will not be described in detail here, but are discussed in Campbell and Lehr (1973).

Whenever ground water is pumped and is to be used for human consumption, proper sanitary precautions must be taken to protect water quality. Surface pollution can enter wells through the annular space or through the top of the well itself. Cement grout outside the casing and a watertight cover of concrete should be installed for protection. Samples of well water should be evaluated for quality after development and after periods of excessive flooding if contamination is suspected. Wells are often associated with septic tank systems in rural areas, and most communities have requirements on the distance necessary to separate a water well from the nearest septic tank as a measure of sanitary protection. Ground water contamination in wells is discussed in detail in Freeze and Cherry (1979).

8.8
GROUND WATER MODELING TECHNIQUES

Types of Ground Water Models

Ground water models have become significant tools of analysis in the past decade, with increasing use of digital computers. Prickett (1975) identified three broad categories of ground water flow models: sand tank models, analog models, and mathematical models, including analytical and numerical types. A sand model consists of a tank filled with an unconsolidated porous material through which water is induced to flow. These models are seriously limited, however, because of the scale problem from laboratory to field and are applicable only to problems with well-specified boundary conditions.

Analog models are used to study the movement of ground water, using theories of viscous flow or theories for the flow of electricity. Viscous fluid models are known as Hele-Shaw models, and they use a fluid such as oil flowing between two closely spaced parallel plates. The viscous model then represents a vertical or horizontal cross section through an aquifer. These models have been used to study various seepage and drainage problems associated with artificial recharge, saltwater intrusion, and irregular boundaries.

Darcy's law for ground water and Ohm's law for electricity are similar. Changes in voltage in an electrical analog model are similar to changes in ground water head. The major drawback of the electrical analog model is that a specific model must be designed for each aquifer system to be studied. The electrical analog model has been largely replaced by mathematical models that use numerical techniques. However, the electrical analogs were used to model a number of aquifer systems in the 1960s and 1970s prior to the advent of the high-speed digital computer.

Mathematical models designed for digital computers have been greatly expanded and improved in recent years. One great advantage of a computer model is its flexibility — it can be used to model a large number of different problems with the same program or set of instructions. Some computer codes are extremely versatile, treating both saturated and unsaturated flow in two or three dimensions. Heterogeneous and anisotropic aquifer systems can be modeled using finite-difference and finite-element numerical techniques, as described in the next section.

Finite-Difference Methods

The two-dimensional transient flow of ground water in a confined aquifer is governed by the partial differential equation

$$\frac{\partial}{\partial x}\left(T_x \frac{\partial h}{\partial x}\right) + \frac{\partial}{\partial y}\left(T_y \frac{\partial h}{\partial y}\right) = S\frac{\partial h}{\partial t} + Q, \qquad (8.63)$$

where

T_x = transmissivity in x-dimension,

T_y = transmissivity in y-dimension,

h = head,

S = storativity,

Q = ground water recharge or withdrawal,

x, y = rectilinear coordinates.

The equation is valid for a continuous aquifer for which there are values of T, S, and h everywhere. Finite-difference methods utilize a grid system of points separated by a small distance and replace the partial differential flow equations by finite-difference equations. Figure 8.31 indicates the grid network in space and time that is used by the computer to solve the finite-difference equations.

The grid network is specified by an integer pair (i, j) to represent locations in a two-dimensional array. The ordered pair $(1, 1)$ is located in the upper left-hand corner in Fig. 8.31. Values of i increase in the positive x-direction and values of j increase in a negative y-direction. The value of head at point (i, j) is indicated $(h_{i,j})$. In the finite-difference approximation, derivatives are replaced by differences taken between nodal points. For example, as in Sections 4.6 and 4.7,

$$\frac{dh}{dx} = \frac{h_{(i+1,j)} - h_{(i,j)}}{\Delta x} \qquad (8.64)$$

and

$$\frac{\partial^2 h}{\partial x^2} = \frac{h_{(i-1,j)} - 2h_{(i,j)} + h_{(i+1,j)}}{(\Delta x)^2}.$$ (8.65)

Similarly, let $Q_{(i,j)}$ be the net withdrawal at node (i, j) and Q_n be the net leakage or recharge into the aquifer. Define $T_{(i,j,1)}$ and $T_{(i,j,2)}$ according to Fig. 8.32, and $h_{(i,j)}$ is the head at node (i, j) and $h\phi_{(i,j)}$ is the head calculated at the previous time increment. The finite-difference form of Eq. (8.63) according to Prickett and Lonnquist (1971) is

$$h_{(i,j)}[T_{(i-1,j,2)} + T_{(i,j,2)} + T_{(i,j,1)} + T_{(i,j-1,1)}] - T_{(i-1,j,2)}h_{(i-1,j)}$$

$$- T_{(i,j,2)}h_{(i+1,j)} - T_{(i,j,1)}h_{(i,j+1)} - T_{(i,j-1,1)}h_{(i,j-1)} + h_{(i,j)}\frac{S\,\Delta x^2}{\Delta t}$$

$$= \left(\frac{S\,\Delta x^2}{\Delta t}\right)h\phi_{(i,j)} - Q_{(i,j)} + Q_n.$$ (8.66)

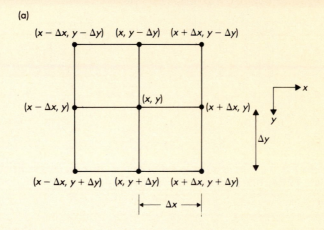

(a)

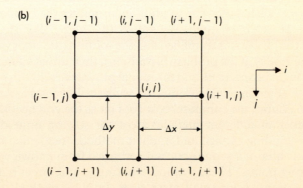

(b)

FIGURE 8.31

(a) Finite-difference grid. (b) Notation used by computer programs for the grid.

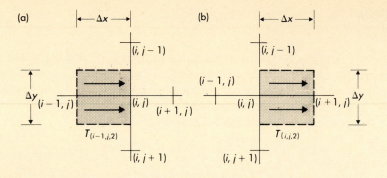

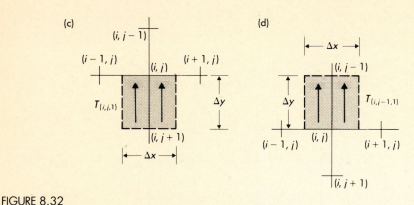

FIGURE 8.32

Vector volumes for aquifer transmissivity for Eq. (8.66). (After Prickett and Lonnquist, 1971, Illinois State Water Survey.)

A difference equation is written for each node in the grid system, at each time step, and a large number of equations must be solved simultaneously for h at each node. The solution incorporates the boundary conditions and proceeds in increasing values of i and j. For steady flow conditions, the problem is simplified and the solution provides values of h at all interior nodes. If the problem is transient, then initial values of head at all nodes must be specified, and the model provides change in head with time. As with any numerical scheme, the accuracy increases as Δx and Δy decrease. Of course, the computer cost goes up as the grid becomes finer.

A number of different methods exist for solving finite-difference equations, including explicit and implicit methods. In explicit methods, an unknown value of h at one node is determined from known values of h at neighboring points. These methods are simple and direct but can become very expensive for small mesh sizes. Implicit methods are more complicated mathematically, in that a set of simultaneous equations is written at each time step, but they have the advantage of using larger time steps. Generally, implicit methods are preferred to explicit for solving ground water flow problems.

Solution methods for the steady-state problem can be handled with an iterative procedure. The governing flow equation for isotropic, homogeneous conditions, without recharge or pumping, can be simplified to

$$\frac{\partial^2 h}{\partial x^2} + \frac{\partial^2 h}{\partial y^2} = 0, \tag{8.67}$$

and either the Gauss-Seidel or the successive overrelaxation (SOR) method can be used to obtain a solution when Eq. (8.65) is used to represent each term of Eq. (8.67). A Gauss-Seidel computer program is shown in Example 8.9, which solves the resulting equation for $h_{(i,j)}$ at iteration $m + 1$:

$$h_{(i,j)}^{m+1} = \frac{h_{(i-1,j)}^{m+1} + h_{(i,j-1)}^{m+1} + h_{(i+1,j)}^{m} + h_{(i,j+1)}^{m}}{4}, \tag{8.68}$$

where m is the number of iterations (see Fig. 8.31 for grid definition).

If we consider Eq. (8.67) with recharge or pumping included, the resulting equation is called Poisson's equation, which can be written

$$\frac{\partial^2 h}{\partial x^2} + \frac{\partial^2 h}{\partial y^2} = -\frac{R(x, y)}{T}, \tag{8.69}$$

where $R(x, y)$ is the rate of pumping. Examples of recharge from precipitation, drawdown from a well, and seepage through a dam under steady-state conditions can be handled by solving this equation using finite differences. The Gauss-Seidel iteration method must be changed to consider the constant right-hand term, or

$$h_{(i,j)} = \frac{1}{4}\left(h_{(i-1,j)} + h_{(i+1,j)} + h_{(i,j-1)} + h_{i,j+1} + \frac{\Delta x^2 R}{T}\right). \tag{8.70}$$

Example 8.10 uses the computer program from Appendix E to solve Eq. (8.69).

EXAMPLE 8.9

GAUSS-SEIDEL ITERATIVE SOLUTION OF LAPLACE'S EQUATION

The FORTRAN program listed below has been written to solve Laplace's equation (Eq. 8.67) using the Gauss-Seidel iterative method. A well is located at the lower left-hand corner of a 10×10 grid, as shown in Fig. E8.9. Given the head values along the boundaries, use the program to solve for the values of head at the interior points of the grid.

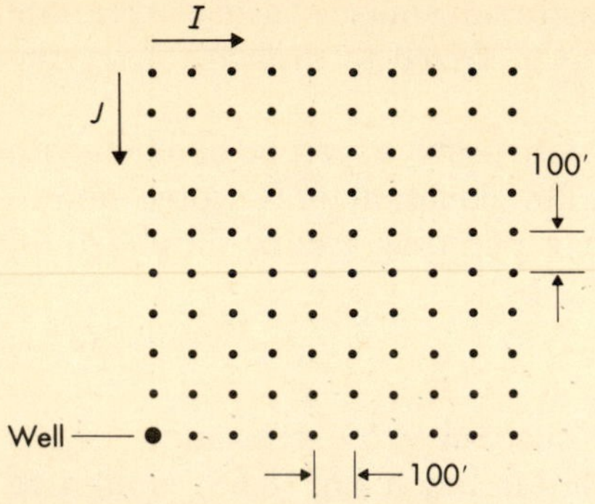

FIGURE E8.9

20.00	20.00	20.00	20.00	20.00	20.00	20.00	20.00	20.00	20.00
19.55									20.00
19.07									20.00
18.55									20.00
17.95									20.00
17.22									20.00
16.29									20.00
14.95									20.00
12.68									20.00
7.68	12.68	14.95	16.29	17.22	17.92	18.55	19.07	19.55	20.00

```
      PROGRAM GAUSS
C
C.... This program solves Laplace's equation (Eqn. 8.67) using
C.... a Gauss-Seidel iterative procedure.  The program will
C.... solve for the unknown interior values of head in a grid,
C.... with I used to denote grid columns and J used to denote
C.... grid rows.
C
      REAL H(25,25),MAX,OLDH,ERROR,TOL
      INTEGER N,UNIT,NI,NJ
      CHARACTER*1 ANS
      CHARACTER*4 UNITS
C
      WRITE(*,*) '        GAUSS-SEIDEL SOLUTION OF EQN. 8.67'
      WRITE(*,*)
      WRITE(*,*)
      WRITE(*,*) '(E)nglish or (M)etric units?'
      READ(*,'(A)') ANS
      UNITS = '(ft)'
      IF ((ANS .EQ. 'M') .OR. (ANS .EQ. '(m)')) UNITS = '(m)'
C
```

```
C.... Determine size of grid
      WRITE(*,*) 'Enter the number of columns'
      READ(*,*) NI
      WRITE(*,*) 'Enter the number of rows'
      READ(*,*) NJ
C
C.... Enter the boundary conditions
      DO 20 J = 1,NJ
        DO 10 I = 1,NI
          IF ((J .EQ. 1) .OR. (J .EQ. NJ)) THEN
            WRITE(*,1000) UNITS,I,J
            READ(*,*) H(I,J)
          ELSEIF ((I .EQ. 1) .OR. (I .EQ. NI)) THEN
            WRITE(*,1000) UNITS,I,J
            READ(*,*) H(I,J)
          ENDIF
   10     CONTINUE
   20   CONTINUE
C
C.... Enter initial guess for interior nodes
      WRITE(*,*) 'Enter the initial guess for the head values',
     &' at the interior nodes.'
      DO 40 J = 2,NJ-1
        DO 30 I = 2,NI-1
          WRITE(*,1000) UNITS,I,J
          READ(*,*) H(I,J)
   30     CONTINUE
   40   CONTINUE
C
C.... Enter error tolerance
      WRITE(*,*) 'Enter the allowable error ',UNITS
      READ(*,*) TOL
C
C.... Sweep interior points with a 5-point operator, keeping track of
C.... the number of iterations needed.
      N = 0
   50 MAX = 0.
      N = N + 1
      DO 70 J = 2,NJ-1
        DO 60 I = 2,NI-1
          OLDH = H(I,J)
          H(I,J) = (H(I-1,J) + H(I+1,J) + H(I,J-1) + H(I,J+1))/4.
          ERROR = ABS(H(I,J) - OLDH)
          IF (ERROR .GT. MAX) MAX = ERROR
   60     CONTINUE
   70   CONTINUE
      IF (MAX .GT. TOL) GOTO 50
```

```
C
C.... Print results on screen
      WRITE(*,*)
      WRITE(*,*) 'The number of iterations needed was',N
      WRITE(*,*)
      WRITE(*,*) 'The array of head values is:'
      WRITE(*,*)
      DO 80 J = 1,NJ
         WRITE(*,1010) (H(I,J), I = 1,NI)
   80 CONTINUE
 1000 FORMAT(1X,'Enter head ',A4,' at node ',I3,',',I3)
 1010 FORMAT(1X,10F7.2)
      END
```

SOLUTION

The program is run and the size of the array is entered when prompted as 10 columns by 10 rows. The boundary head values are entered and then an initial guess at the values at the each interior point is entered. Using an error tolerance of 0.001, the resulting head array is as follows:

20.00	20.00	20.00	20.00	20.00	20.00	20.00	20.00	20.00	20.00
19.55	19.55	19.58	19.62	19.68	19.74	19.80	19.87	19.93	20.00
19.07	19.09	19.14	19.23	19.34	19.47	19.60	19.73	19.87	20.00
18.55	18.58	18.67	18.82	19.00	19.20	19.40	19.60	19.80	20.00
17.95	18.00	18.16	18.38	18.65	18.92	19.20	19.47	19.74	20.00
17.22	17.32	17.57	17.91	18.28	18.65	19.00	19.35	19.68	20.00
16.29	16.48	16.90	17.41	17.91	18.38	18.82	19.23	19.62	20.00
14.95	15.41	16.15	16.90	17.57	18.16	18.68	19.14	19.58	20.00
12.68	14.04	15.41	16.48	17.32	18.00	18.58	19.09	19.55	20.00
7.68	12.68	14.95	16.29	17.22	17.95	18.55	19.07	19.55	20.00

EXAMPLE 8.10

DRAWDOWN TO A WELL IN A CONFINED AQUIFER, STEADY STATE

Appendix E contains a program that solves Poisson's equation (Eq. 8.69) for drawdown to a well in a confined aquifer under steady-state conditions. A well is located in a 10×10 grid as shown in Fig. E8.10 and pumps water at a rate of 10,000 gpd. The transmissivity of the aquifer is 500 ft^2/day and the initial head is 20 ft. The zone of influence of the pumping extends 1000 ft radially beyond the well. Using a value of 1.8 for ω, determine drawdown near the well. Use an error tolerance of 0.001 and limit the number of iterations to 200.

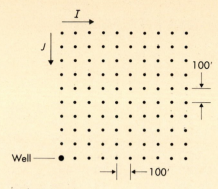

FIGURE E8.10

SOLUTION

Data are entered into the program in the following manner:

```
        DRAWDOWN NEAR A WELL IN A CONFINED AQUIFER

    Compute (H)ead or (D)rawdown?
   D
    Enter the number of columns in the grid.
   10
    Enter the number of rows in the grid.
   10
    Enter the distance between nodes (ft,m).
   100
    Enter the radial limit (ft,m) of the zone of influence.
   1000


    Enter the transmissivity (sq ft/day,sq m/day).
   500
    Enter the pumping rate (gal/day,cu m/day).
   10000


    Enter the relaxation factor omega (0<omega<2).
   1.8
    Enter the error tolerance (ft, m)
   0.001
    Enter the limit of iterations
   200


    Enter the initial head (ft,m) before pumping started.
   20
```

The results are as follows:

```
The number of iterations made was              65

The drawdown (ft,m) array is:
    .00    .00    .00    .00    .00    .00    .00    .00    .00    .00
    .45    .44    .41    .35    .28    .18    .00    .00    .00    .00
    .93    .90    .83    .73    .59    .42    .22    .00    .00    .00
   1.45   1.41   1.30   1.13    .93    .70    .46    .22    .00    .00
   2.05   1.99   1.82   1.58   1.30   1.00    .70    .42    .18    .00
   2.78   2.67   2.41   2.06   1.68   1.30    .93    .59    .28    .00
   3.71   3.51   3.08   2.57   2.06   1.58   1.13    .73    .35    .00
   5.05   4.59   3.84   3.08   2.41   1.82   1.30    .83    .41    .00
   7.32   5.96   4.59   3.51   2.67   1.99   1.41    .90    .44    .00
  12.32   7.32   5.05   3.71   2.78   2.05   1.45    .93    .45    .00
```

An option of the program allows one to compute head instead of draw-down. The following values are obtained:

```
The head (ft,m) array is:
 20.00 20.00 20.00 20.00 20.00 20.00 20.00 20.00 20.00 20.00
 19.55 19.56 19.59 19.65 19.72 19.82 20.00 20.00 20.00 20.00
 19.07 19.10 19.17 19.27 19.41 19.58 19.78 20.00 20.00 20.00
 18.55 18.59 18.70 18.87 19.07 19.30 19.54 19.78 20.00 20.00
 17.95 18.01 18.18 18.42 18.70 19.00 19.30 19.58 19.82 20.00
 17.22 17.33 17.59 17.94 18.32 18.70 19.07 19.41 19.72 20.00
 16.29 16.49 16.92 17.43 17.94 18.42 18.87 19.27 19.65 20.00
 14.95 15.41 16.16 16.92 17.59 18.18 18.70 19.17 19.59 20.00
 12.68 14.04 15.41 16.49 17.33 18.01 18.59 19.10 19.56 20.00
  7.68 12.68 14.95 16.29 17.22 17.95 18.55 19.07 19.55 20.00
```

Note that there is no drawdown on the boundaries or outside the zone of influence. Compare results with those of Example 8.9.

Finite-Difference Methods for Transient Flow

The **transient flow** problem of Eq. (8.63) can be solved by either explicit or implicit methods. In the following discussion, an implicit method will be employed to solve the classical Theis well drawdown problem in an x-y coordinate system. The second partial derivative can be represented by weighting terms with α, where $0 < \alpha < 1$ and n denotes the time step:

$$\frac{\partial^2 h}{\partial x^2} = \alpha \left[\frac{h_{(i+1,j)}^{n+1} - 2h_{(i,j)}^{n+1} + h_{(i-1,j)}^{n+1}}{\Delta x^2} \right]$$

$$+ (1 - \alpha) \left[\frac{h_{(i+1,j)}^{n} - 2h_{(i,j)}^{n} + h_{(i-1,j)}^{n}}{\Delta x^2} \right]. \tag{8.71}$$

A similar expression can be written for $\partial^2 h / \partial y^2$. We can use the notation $\bar{h}$ as

$$\bar{h}_{(i,j)}^{n} = \frac{1}{4} [h_{(i-1,j)} + h_{(i+1,j)} + h_{(i,j-1)} + h_{(i,j+1)}]^{n}. \tag{8.72}$$

Finally, the governing transient flow equation can be written (Wang and Anderson, 1982)

$$\alpha [(\bar{h}_{(i,j)}^{n+1} - h_{(i,j)}^{n+1})] + (1 - \alpha) [(\bar{h}_{(i,j)}^{n} - h_{(i,j)}^{n}]$$

$$= \frac{a^2 S}{4T} \frac{h_{(i,j)}^{n+1} - h_{(i,j)}^{n}}{\Delta t} - \frac{a^2 R_{(i,j)}^{n}}{4T}, \tag{8.73}$$

where $\Delta x = \Delta y = a$. For $\alpha = 0$, the finite-difference equation is explicit. If the terms involving time level n are known on the right-hand side, then a system of equations results for time level $n + 1$ on the left-hand side of Eq. (8.73). These can be solved at each time step $n + 1$ by an iterative method such as Gauss-Seidel or by matrix methods. The resulting iterative form is

$$h_{(i,j)}^{n+1} = \frac{\left[\alpha \bar{h}_{(i,j)}^{n+1} + \frac{a^2 S}{4T \Delta t} h_{(i,j)}^{n} + (1 - \alpha)[\bar{h}_{(i,j)}^{n} - h_{(i,j)}^{n}] + \frac{a^2 R_{(i,j)}^{n}}{4T} \right]}{\frac{a^2 S}{4T \Delta t} + \alpha}. \tag{8.74}$$

For steady-state problems, only one system of equations must be solved. For $\alpha = 1$, the numerical scheme is fully implicit. For $\alpha = 1/2$, the equation is balanced between time level n and time level $n + 1$ and is called the Crank-Nicolson method.

A program that solves Eq. (8.74) for the case of a pumping well at one corner of a square geometry is listed in Appendix E. The solution can be compared against the analytical Theis method at various points in time. It can be shown that the selection of Δx, Δy, Δt, and α will strongly influence the accuracy of the numerical results.

EXAMPLE 8.11

NUMERICAL SOLUTION FOR DRAWDOWN TO A WELL, TRANSIENT CONDITIONS

A FORTRAN program for solving the governing equation for drawdown to a well in a confined aquifer under transient conditions is listed in Appendix E. Assume that flow to the well in the aquifer described in

Example 8.10 is transient. The well is pumping at a rate of 2500 gpd. If the storativity of the aquifer is 0.005, determine the drawdown near the well after 1 week. Use a starting time period of 0.05 days and a factor of 1.4 to increase the time increment (12 time iterations will equal about 7 days). Use the Crank-Nicolson scheme with $\alpha = 0.5$, a tolerance of 0.001, and no more than 200 iterations in the drawdown computation for each time step.

SOLUTION

Input data are entered as follows:

```
     NUMERICAL SOLUTION OF DRAWDOWN IN A CONFINED
            AQUIFER UNDER TRANSIENT CONDITIONS


  Compute (H)ead or (D)rawdown?
D
  Enter the number of columns
10
  Enter the number of rows
10
  Enter the distance between nodes (ft,m).
100


  Enter the transmissivity (sq ft/day, sq m/day).
500
  Enter the storativity.
0.005
  Enter the pumping rate (gal/day, cu m/day).
2500


  Enter alpha.
0.5
  Enter the error tolerance.
0.001
  Enter the limit of iterations used to solve for drawdown.
100


  Enter the starting time increment (day).
0.05
  Enter the factor to multiply the time increment.
1.4
  Enter the number of time increments to be made.
12


  Enter the initial head (ft,m) before pumping starts.
20
```

The results are as follows:

```
Time =  6.96

The drawdown array is:
       .88     .88     .87     .85     .82     .80     .78     .76     .75     .74
       .90     .89     .88     .86     .83     .81     .78     .76     .75     .75
       .93     .93     .91     .89     .86     .83     .80     .78     .76     .76
      1.00     .99     .97     .93     .90     .86     .82     .80     .78     .78
      1.10    1.09    1.05    1.00     .95     .90     .86     .83     .81     .80
      1.24    1.22    1.16    1.09    1.02     .95     .90     .86     .83     .82
      1.45    1.40    1.30    1.19    1.09    1.00     .93     .89     .86     .85
      1.76    1.65    1.47    1.30    1.16    1.05     .97     .91     .88     .87
      2.32    1.98    1.65    1.40    1.22    1.09     .99     .93     .89     .88
      3.56    2.32    1.76    1.45    1.25    1.10    1.00     .94     .90     .89
```

For the explicit case, if Δt is too large the scheme becomes unstable and yields useless answers. Boundary conditions and choice of error tolerance in the iterative method also contribute to numerical errors. Results in Table 8.4 show the comparison of a numerical example with $Q = 1000$ m³/day, $T = 4.50$ m²/day, $S = 0.0005$, for various values of Δt and α. Values of $\Delta x = \Delta y = 100$ m were chosen.

Advanced Ground Water Models

Two of the most widely used and well-documented two-dimensional finite-difference models for ground water flow are by Prickett and Lonnquist (1971) and Trescott et al. (1976). Both models solve the following transient equation with heterogeneous and anisotropic conditions allowed (see Eq. 8.63):

$$\frac{\partial}{\partial x}\left(T_x \frac{\partial h}{\partial x}\right) + \frac{\partial}{\partial y}\left(T_y \frac{\partial h}{\partial y}\right) = S \frac{\partial h}{\partial t} - R(x, y, t). \tag{8.75}$$

The equation can be used for unconfined aquifers by allowing T_x and T_y to vary as saturated thickness changes. Steady-state conditions can be treated by using large time steps or setting $S = 0$.

The Prickett-Lonnquist (1971) model uses an alternating direction implicit (ADI) method, combined with the Gauss-Seidel iteration to solve the governing equation for transient ground water flow. The ADI method is beyond the scope of this chapter and is covered in detail in Wang and Anderson (1982) and Carnahan et al. (1969). The ADI method basically converts the governing finite-difference equation into a tridiagonal matrix, which can be conveniently solved using Gaussian elimination. The method alternates between columns and rows in an attempt to compen-

TABLE 8.4

Comparison of Drawdown at 100 m from a Well Using Numerical Solutions (Example 8.11) with Selected α and Theis Analytical Solutions

TIME (days)	Δt	$\alpha = 0$	$\alpha = 0.5$	$\alpha = 1$	Theis
			DRAWDOWN IN FEET FOR		
0.05	0.05	0.00	0.19 (4)*	0.33 (4)	0.011
0.12	0.07	0.68	1.03 (4)	1.28 (5)	0.56
0.24	0.11	2.66	2.98 (5)	3.20 (5)	3.00
0.41	0.17	6.57	6.41 (5)	6.37 (6)	6.90
0.66	0.25	12.35	11.40 (6)	10.85 (8)	11.82
1.04	0.38	18.95	17.54 (7)	16.43 (10)	17.56
1.61	0.57	25.90	24.26 (9)	22.76 (12)	23.79
2.46	0.85	32.60	31.21 (11)	29.52 (16)	30.30
3.74	1.28	42.29	38.25 (14)	36.48 (21)	37.06

* The numbers in parentheses indicate the number of iterations used.

sate for errors generated in either direction. Prickett and Lonnquist also use ADI iteration to speed convergence of the solution.

Input data are required for the definition of boundary conditions, values of T_x, T_y, S, and R for each node, and initial conditions for head at all nodes. No flow, constant head, or recharge boundaries can be used, and versions of the model are available to simulate water table (unconfined) conditions and leaky aquifers. No flow boundary implies that head is equivalent on either side of the boundary, and a constant head condition implies that head does not vary in time, as for a river or lake.

Boundary conditions are sometimes difficult to select for a particular problem. A common approach is to make the model grid area large enough so that heads in the interior portion of the grid are not affected by the boundary conditions. Physically realistic boundary values should be used whenever possible, and constant head conditions are most commonly used.

Usually only a few values of T and S will be available from pumping test data for the model area. This information must be combined with data on stratigraphy and recharge to create an estimate for the distribution of T and S over the grid. Variations that are not properly represented in the model can lead to significant errors.

Initial conditions for head values across the grid are required for transient simulations. The solution will converge faster if the initial conditions represent a reasonable first guess of the solution for transient or steady-state cases. The ADI and iterative methods depend to a large extent on the initial conditions and on the amount of recharge or withdrawal for the system.

The Prickett-Lonnquist code is relatively easy to use and can be easily changed to represent local conditions. The code consists of less than 200 lines of FORTRAN (Example 8.12). The model is extremely well documented and represents the basic aquifer model used by the Illinois State Water Survey. Versions of the model are also available for small computer systems.

EXAMPLE 8.12

PRICKETT-LONNQUIST CODE

A confined aquifer with a pumping well, which produces 2000 m³/day, can be suitably modeled by using a grid of 10 columns by 10 rows. The well is located at node 5,5. The average transmissivity and storage factor are 30,000 m²/day and 6000 m³/m, respectively. Assuming a time step of 0.25 day and an error tolerance of 0.01, determine the drawdown for this aquifer after 12 hr of pumping, using the Prickett-Lonnquist code. Assume that the values given for the transmissivity and storage factor are constant throughout the aquifer and that the initial head is zero at all points. Where is the greatest drawdown and what is its magnitude?

Prickett-Lonnquist Code

```
C      ILLINOIS STATE WATER SURVEY BASIC CONFINED AQUIFER SIMULATION
C      MODEL BY PRICKETT-LONQUIST
C
C      DEFINITION OF VARIABLES
C
C      HO(I,J)   HEADS AT START OF TIME INCREMENT
C      H(I,J)    HEADS AT END OF TIME INCREMENT   (M)
C      SF 1(I,J) STORAGE FACTOR FOR ARTESIAN CONDITIONS  (M**2)
C      Q(I,J)    CONSTANT WITHDRAWL RATES (M**3/DAY)
C      T(I,J,1)  AQUIFER TRANSMISSIVITY BETWEEN I,J & I,J+1 (M**2/DAY)
C      T(I,J,2)  AQUIFER TANSMISSIVITY BETWEEN I,J & I+1,J (M**2/DAY)
C      AA,BB,CC,DD COEFFICIENTS IN WATER BALANCE EQUATIONS
C      NR        NO. OF ROWS IN MODEL
C      NC        NO. OF COLUMNS IN MODEL
C      NSTEPS    NO. OF TIME INCREMENTS
C      DELTA     TIME INCREMENTS (DAYS)
C      HH,S1,QQ,TT DEFAULT VALUES
C      I         MODEL COLUMN NUMBER
C      J         MODEL ROW NUMBER
C
C
```

```
      DIMENSION H(50,50),HO(50,50),SF1(50,50),Q(50,50),T(50,50,2),
     #B(50),G(50),DL(50,50)
C
C     TURN OFF UNDERFLOW TRAP
C
      CALL ERRSET(208,256,-1,1)
C
C     DEFINE INPUT AND OUTPUT DEVICE NUMBERS
C
      IN=05
      OUT=06
C
C     READ PARAMETER CARD AND DEFAULT VALUE CARD
C
      READ(05,10)NSTEPS,DELTA,ERROR,NC,NR,TT,S1,HH,QQ
   10 FORMAT(I6,2F6.0/2I6,4F6.0)
C
C     FILL ARRAYS WITH DEFAULT VALUES
C
      DO 20 I=1,NC
      DO 20 J=1,NR
      T(I,J,1)=TT
      T(I,J,2)=TT
      SF1(I,J)=S1
      H(I,J)=HH
      HO(I,J)=HH
   20 Q(I,J)=QQ
C
C     READ NODE CARDS
C
   30 READ(05,40,END=50)I,J,T(I,J,1),T(I,J,2),SF1(I,J),H(I,J),Q(I,J)
   40 FORMAT(2I3,2F6.0,2F4.0,1F6.0)
      GO TO 30
C
C     START OF SIMULATION
C
   50 TIME=0
      DO 320 ISTEP=1,NSTEPS
      TIME=TIME+DELTA
C
C     PREDICT HEADS FOR NEXT TIME INCREMENT
C
      DO 70 I=1,NC
      DO 70 J=1,NR
      D=H(I,J)-HO(I,J)
```

```
        HO(I,J)=H(I,J)
        F=1.0
        IF(DL(I,J).EQ.0.0) GO TO 60
        IF(ISTEP.GT.2)F=D/DL(I,J)
        IF(F.GT.5)F=5.0
        IF(F.LT.0.0)F=0.0
     60 DL(I,J)=D
     70 H(I,J)=H(I,J)+D*F
C
C      REFINE ESTIMATES OF HEADS BY IADI METHOD
C
        ITER=0
     80 E=0.0
        ITER=ITER+1
C
C      COLUMN CALCULATIONS
C
        DO 190  II=1,NC
        I=II
        IF(MOD(ISTEP+ITER,2).EQ.1) I=NC-I+1
        DO 170 J=1,NR
C
C      CALCULATE B AND G ARRAYS
C
        BB=SF1(I,J)/DELTA
        DD=HO(I,J)*SF1(I,J)/DELTA-Q(I,J)
        AA=0.0
        CC=0.0
        IF(J-1)90,100,90
     90 AA=-T(I,J-1,1)
        BB=BB+T(I,J-1,1)
    100 IF(J-NR)110,120,110
    110 CC=-T(I,J,1)
        BB=BB+T(I,J,1)
    120 IF(I-1)130,140,130
    130 BB=BB+T(I-1,J,2)
        DD=DD+H(I-1,J)*T(I-1,J,2)
    140 IF(I-NC)150,160,150
    150 BB=BB+T(I,J,2)
        DD=DD+H(I+1,J)*T(I,J,2)
    160 W=BB-AA*B(J-1)
        B(J)=CC/W
    170 G(J)=(DD-AA*G(J-1))/W
C
C      RE-ESTIMATE HEADS
```

```
C
      E=E+ABS(H(I,NR)-G(NR))
      H(I,NR)=G(NR)
      N=NR-1
  180 HA=G(N)-B(N)*H(I,N+1)
      E=E+ABS(HA-H(I,N))
      H(I,N)=HA
      N=N-1
      IF(N)190,190,180
  190 CONTINUE
C
C     ROW CALCULATIONS
C
      DO 300 JJ=1,NR
      J=JJ
      IF(MOD(ISTEP+ITER,2).EQ.1)J=NR-J+1
      DO 280 I=1,NC
      BB=SF1(I,J)/DELTA
      DD=HO(I,J)*SF1(I,J)/DELTA-Q(I,J)
      AA=0.0
      CC=0.0
      IF(J-1)200,210,200
  200 BB=BB+T(I,J-1,1)
      DD=DD+H(I,J-1)*T(I,J-1,1)
  210 IF(J-NR)220,230,220
  220 DD=DD+H(I,J+1)*T(I,J,1)
      BB=BB+T(I,J,1)
  230 IF(I-1)240,250,240
  240 BB=BB+T(I-1,J,2)
      AA=-T(I-1,J,2)
  250 IF(I-NC)260,270,260
  260 BB=BB+T(I,J,2)
      CC=-T(I,J,2)
  270 W=BB-AA*B(I-1)
      B(I)=CC/W
  280 G(I)=(DD-AA*G(I-1))/W
C
C     RE-ESTIMATE HEADS
C
      E=E+ABS(H(NC,J)-G(NC))
      H(NC,J)=G(NC)
      N=NC-1
  290 HA=G(N)-B(N)*H(N+1,J)
      E=E+ABS(H(N,J)-HA)
      H(N,J)=HA
```

```
      N=N-1
      IF(N)300,300,290
  300 CONTINUE
      IF(E.GT.ERROR) GO TO 80
C
C     PRINT RESULTS
C
      WRITE(06,310)TIME,E,ITER
  310 FORMAT(6H2TIME=,F6.2///,E20.7,I5)
      DELTA=DELTA*1.2
      DO 320 J=1,NR
  320 WRITE(06,330)J,(H(I,J),I=1,NC)
  330 FORMAT(I5,5X,10F10.4/(12X,10F10.4))
      STOP
      END
```

SOLUTION

The Prickett-Lonnquist code uses three types of input data cards (or lines). The Parameter card is used to define the number of time steps NSTEPS that will be used, the length of the time step DELTA, and the allowable error tolerance ERROR. A Default card sets the number of columns NC and rows NR in the grid. The values of average transmissivity TT, storage factor SS, initial head HH, and flow QQ are also entered on the Default card. These values will be used at all nodes unless overridden by separate Node cards. The Node cards are used to enter specific data for individual nodes. The location of the node, I, J, is entered first. Next the transmissivities in the x-direction and y-direction $T1$ and $T2$ are entered. The storage factor for the node SF1, the initial head H, and pumping rate Q are also contained on the Node card. Any number of node cards may be used to model all pumping and injection wells, nodes where an initial head is different from the default value, and so on.

Great care must be taken to enter the variables on the proper line with the proper format. To do this, one needs to check the format statements pertaining to the read step for each type of card.

Input for this example is entered in the following manner:

```
  20   0.25   0.01
  10      1030000. 6000.    0.0    0.0
 5   530000.30000.6.E3 0.0   2.E3
```

The output from the program for this input data is symmetrical, and a portion of the results at time $t = 0.55$ day are listed below. As can be seen, the maximum drawdown occurs at node 5,5, which is the pumping well. For $t = 0.55$ day (approximately 12 hr) the drawdown is 0.0249 m.

```
TIME=   0.55


       0.9045407E-02     2
  1          -0.0003   -0.0004   -0.0006   -0.0009   -0.0010   -0.0008


  2          -0.0004   -0.0006   -0.0010   -0.0014   -0.0017   -0.0014


  3          -0.0006   -0.0010   -0.0017   -0.0028   -0.0038   -0.0028


  4          -0.0009   -0.0014   -0.0028   -0.0056   -0.0093   -0.0056


  5          -0.0010   -0.0017   -0.0038   -0.0093   -0.0249   -0.0093


  6          -0.0008   -0.0014   -0.0028   -0.0056   -0.0093   -0.0056


  7          -0.0006   -0.0009   -0.0017   -0.0028   -0.0038   -0.0028


  8          -0.0004   -0.0006   -0.0009   -0.0013   -0.0016   -0.0013


  9          -0.0003   -0.0003   -0.0005   -0.0006   -0.0007   -0.0006


 10          -0.0002   -0.0002   -0.0003   -0.0004   -0.0004   -0.0004
```

A second model that is widely used for two-dimensional ground water flow was written by Trescott et al. (1976). Three options are available for the numerical solution including iterative ADI, successive overrelaxation (SOR), and the strongly implicit procedure (SIP). It has been demonstrated that SIP converges faster than the other two methods for solving problems in heterogeneous or anisotropic media.

A completely revised and updated version of the Trescott et al. (1976) model has been released recently by the U.S. Geological Survey, with McDonald and Harbaugh (1984) as the major authors. The new code is a modular, three-dimensional finite-difference ground water flow model. The model was designed so that it could be readily modified, can be executed on a variety of computers, and is relatively efficient with respect to computer memory and execution time. The model's structure consists of a main program and a series of highly independent subroutines called modules. Each module deals with a specific feature of the hydrologic system, such as flow from rivers, flow from wells, or flow into drains. The numerical procedures are included in separate modules and can be selected by the user. The model consists of an abbreviated set of input instructions as part of the 500-page user's manual. The model documentation is extremely complete and very easy to use and represents the state of the art in ground water flow models.

The interested reader should consult the user's manuals for more details on the models described in this section.

SUMMARY

Ground water hydrology is reviewed in this chapter. Ground water flow according to Darcy's law is described and the general flow equations are presented for confined and unconfined aquifers. The theory of streamlines and equipotential lines is derived to explain flow net concepts. Equations for steady radial flow to a well are derived for single- and multiple-well systems. The concept of superposition can be used for complex flow fields.

Theis's method for unsteady well hydraulics is presented in theory and application; it forms the modern basis for aquifer pump tests. The simpler Cooper-Jacob method can be used to solve the unsteady flow equation under certain conditions. Finally, radial flow in a leaky aquifer is also presented in theory and example. Slug tests are used to determine aquifer properties using a single well, whereas pump tests require more than one well.

Ground water modeling techniques include both analytical and numerical solutions to the governing flow equations. Usually, finite-difference methods for steady-state or transient flow are used, especially where heterogeneous properties exist. Both simple and state-of-the-art flow models are reviewed with examples to provide a useful level of coverage for the engineer or hydrologist.

PROBLEMS

8.1. Compute the discharge velocity and seepage velocity for water flowing through a sand column with the following characteristics:

$$K = 10^{-4} \text{ cm/s},$$

$$\frac{dh}{dl} = 0.01,$$

$$\text{Area} = 75 \text{ cm}^2,$$

$$n = 0.20.$$

8.2. The average water table elevation has dropped 5 ft due to the removal of 100,000 ac-ft from an unconfined aquifer over an area of 75 mi². Determine the storage coefficient for the aquifer.

8.3. A confined aquifer is 50 m thick and 0.5 km wide. Two observation wells are located 1.4 km apart in the direction of flow. Head in well 1 is 50.0 m and in well 2 it is 42 m. Hydraulic conductivity K is 0.7 m/day.
   a) What is the total daily flow of water through the aquifer?
   b) What is the height z of the piezometric surface 0.5 km from well 1 and 0.9 km from well 2?

8.4. A well with a diameter of 18 in. penetrates an unconfined aquifer that is 100 ft thick. Two observation wells are located at 100 ft and 235 ft from the well, and the measured drawdowns are 22.2 ft and 21 ft, respectively. Flow is steady and the hydraulic conductivity is 1320 gpd/ft². What is the rate of discharge from the well?

8.5. Two piezometers are located 1000 ft apart with the bottom located at depths of 50 ft and 350 ft, respectively, in a 400-ft-thick unconfined aquifer. The depth to the water table is 50 ft in the deeper piezometer and 40 ft in the shallow one. Assume that hydraulic conductivity is 0.0002 ft/s.
   a) Use the Dupuit equation to calculate the height of the water table midway between the piezometers.
   b) Find the quantity of seepage through a 25-ft section.

8.6. In a fully penetrating well, the equilibrium drawdown is 30 ft with a constant pumping rate of 20 gpm. The aquifer is unconfined and the saturated thickness is 100 ft. What is the steady-state drawdown measured in the well when the constant discharge rate is 10 gpm?

8.7. A soil sample 6 in. in diameter and 1 ft long is placed in a falling head permeameter. The falling-head tube diameter is 1 in. and the initial head is 6 in. The head falls 1 in. over a 2-hr period. Calculate the hydraulic conductivity.

8.8. A constant head permeameter containing very fine-grained sand has a length of 12 cm and a cross-sectional area of 30 cm². With a head of 10 cm, a total of 100 ml of water is collected in 25 min. Find the hydraulic conductivity.

8.9. Bull Creek completely penetrates a confined aquifer 10 ft thick. A long drought decreases the flow in the stream by 16 cfs between two gaging stations located 4 mi apart. On the west side of the creek, the piezometric contours parallel the bank and slope toward the channel at 0.0004 ft/ft. The piezometric contours on the east side of the creek slope away from the channel toward a well field at a slope of 0.0006 ft/ft. Using Darcy's law and the continuity equation, compute the transmissivity of the aquifer along this section of Bull Creek.

8.10. Three geologic formations overlie one another with the characteristics listed below. A constant-velocity vertical flow field exists across the three formations. The hydraulic head is 33 ft at the top of the formations and 21 ft at the bottom. Calculate the hydraulic head at the two internal boundaries.

$$b_1 = 50 \text{ ft} \qquad K_1 = 0.0002 \text{ ft/s}$$
$$b_2 = 20 \text{ ft} \qquad K_2 = 0.000005 \text{ ft/s}$$
$$b_3 = 210 \text{ ft} \qquad K_3 = 0.001 \text{ ft/s}$$

8.11. Two wells are located 100 m apart in a confined aquifer with transmissivity $T = 2 \times 10^{-4}$ m^2/s and storativity $S = 7 \times 10^{-5}$. One well is pumped at a rate of 6.6 m^3/hr and the other at a rate of 10.0 m^3/hr. Plot drawdown as a function of distance along the line joining the wells at 1 hr after the pumping starts.

8.12. At a burn pit, a Hvorslev slug test was performed in a confined aquifer with a piezometer intake length of 20 ft and a radius of 1 in. The radius of the rod was 0.68 in. The following recovery data for the well were observed. Given that the static water level is 7.58 ft and $H_0 = 6.88$ ft, calculate the hydraulic conductivity.

TIME (s)	20	45	75	101	138	164	199
h (ft)	6.94	7.00	7.21	7.27	7.34	7.38	7.40

8.13. A well casing with a radius of 6 in. is installed through a confining layer into a formation with a thickness of 10 ft. A screen with a radius of 3 in. is installed in the casing. A slug of water is injected, raising the water level by 0.5 ft. Given the following recorded data for head decline, find the values of T, K, and S for this aquifer.

TIME (s)	5	10	30	50	80	120	200
HEAD (ft)	0.47	0.42	0.32	0.26	0.10	0.05	0.01

8.14. A small municipal well was pumped for 2 hr at a rate of 15.75 liters/s. An observation well was located 50 ft from the pumping well and the following data were recorded. Using the Theis method outlined in Example 8.6, compute T and S.

TIME (min)	1	2	3	5	7	9	12	15	20	40	60	90	120
DRAWDOWN (ft)	1.5	4.0	6.2	8.5	10.0	12.0	13.7	14.9	17.0	21.7	23.1	26.0	28.0

8.15. A well in a confined aquifer is pumped at a rate of 833 liters/min for a period of over 8 hr. Time-drawdown data for an observation well located 250 m away are given below. The aquifer is 5 m thick. Use the Cooper-Jacob method to find values of T, K, and S for this aquifer.

TIME (hr)	DRAWDOWN (m)
0.050	0.091
0.083	0.214
0.133	0.397
0.200	0.640
0.333	0.976
0.400	1.100
0.500	1.250
0.630	1.430
0.780	1.560
0.830	1.620
1.000	1.740
1.170	1.860
1.333	1.920
1.500	2.040
1.670	2.130
2.170	2.290
2.670	2.530
3.333	2.590
4.333	2.810
5.333	2.960
6.333	3.110
8.333	3.320

8.16. Drawdown was observed in a well located 100 ft from a pumping well that was pumped at a rate of 1.11 cfs for a 30-hr period. Use the Cooper-Jacob method to compute T and S for this aquifer.

TIME (hr)	1	2	3	4	5	6	8	10	12	18	24	30
DRAWDOWN (ft)	0.6	1.4	2.4	2.9	3.3	4.0	5.2	6.5	7.5	9.1	10.5	11.5

8.17. Repeat Problem 8.16 using the Theis method.

8.18. Repeat Example 8.10 with one half the pumping rate.

REFERENCES

American Water Works Association, 1967, *AWWA Standard for Deep Wells,* AWWA-A100-66, Denver, Colorado.

Bear, J., 1979, *Hydraulics of Groundwater,* McGraw-Hill Book Company, New York.

Campbell, M. D., and J. H. Lehr, 1973, *Water Well Technology,* McGraw-Hill Book Company, New York.

Carnahan, B., H. A. Luther, and J. O. Wilkes, 1969, *Applied Numerical Methods,* John Wiley and Sons, New York.

Cooper, H. H., Jr., 1966, "The Equation of Groundwater Flow in Fixed and Deforming Coordinates," *J. Geophys. Res.,* vol. 71, pp. 4785–4790.

Cooper, H. H., Jr., J. D. Bredehoeft, and I. S. Papadopoulos, 1967, "Response of a Finite-Diameter Well to an Instantaneous Charge of Water," *Water Resour. Res.,* vol. 3, pp. 263–269.

Cooper, H. H., Jr., and C. E. Jacob, 1946, "A Generalized Graphical Method for Evaluating Formation Constants and Summarizing Well Field History," *Trans. Am. Geophys. Union,* vol. 27, pp. 526–534.

DeWiest, R. J. M., 1965, *Geohydrology,* John Wiley and Sons, New York.

Ferris, J. G., D. B. Knowles, R. H. Browne, and R. W. Stallman, 1962, *Theory of Aquifer Tests,* USGS Water Supply Paper 1536-E.

Fetter, C. W., Jr., 1980, *Applied Hydrogeology,* Charles E. Merrill Publishing Company, Columbus, Ohio.

Freeze, R. A., and J. A. Cherry, 1979, *Groundwater,* Prentice-Hall, Englewood Cliffs, New Jersey.

Geraghty, J. J., D. W. Miller, F. Van der Leeden, F. L. Troise, 1973, *Water Atlas of the United States,* Water Information Center, Inc., Plainview, New York.

Hantush, M. S., 1956, "Analysis of Data from Pumping Tests in Leaky Aquifers," *J. Geophys. Res.,* vol. 69, pp. 4221–4235.

Hvorslev, M. J., 1951, *Time Lag and Soil Permeability in Groundwater Observations,* U.S. Army Corps of Engineers Waterways Exp. Sta. Bull. 36, Vicksburg, Mississippi.

Jacob, C. E., 1940, "On the Flow of Water in an Elastic Artesian Aquifer," *Trans. Am. Geophys. Union,* vol. 2, pp. 574–586.

Lohman, S. W., 1972, *Ground-Water Hydraulics,* U.S. Geological Survey Prof. Paper 708.

McDonald, M. G., and A. W. Harbaugh, 1984, *A Modular Three-Dimensional Finite-Difference Ground-Water Flow Model,* Open File Report 83-875, U.S. Department of the Interior, USGS National Center, Reston, Virginia.

Morris, D. A., and A. I. Johnson, 1967, *Summary of Hydrologic and Physical Properties of Rock and Soil Materials, as Analyzed by the Hydrologic Laboratory of the U.S. Geological Survey,* USGS Water Supply Paper 1839-D.

Murray, C. R., and E. B. Reeves, 1977, *Estimated Use of Water in the United States in 1975,* USGS Circ. 533.

Papadopoulos, I. S., J. D. Bredehoeft, and H. H. Cooper, Jr., 1973, "On the Analysis of Slug Test Data," *Water Resour. Res.,* vol. 9, no. 4, pp. 1087–1089.

Prickett, T. A., 1975, "Modeling Techniques for Groundwater Evaluation," in Chow, V. T. (editor), *Advances in Hydroscience,* vol. 10, pp. 1–143, Academic Press, New York.

Prickett, T. A., and C. G. Lonnquist, 1971, *Selected Digital Computer Techniques for Groundwater Resource Evaluation,* Illinois State Water Surv. Bull. 55, Urbana, Illinois.

Theis, C. V., 1935, "The Relation Between the Lowering of the Piezometric Surface and the Rate and Duration of Discharge of a Well Using Ground-Water Storage," *Trans. Am. Geophys. Union,* vol. 16, pp. 519–524.

Todd, D. K., 1980, *Groundwater Hydrology,* 2nd ed., John Wiley and Sons, New York.

Todd, D. K., and J. Bear, "Seepage Through Layered Anisotropic Porous Media," *J. Hydraulic Div.,* ASCE, vol. 87, no. HY3, pp. 31–57.

Trescott, P. E., G. F. Pinder, and S. P. Larson, 1976, "Finite-Difference Model for Aquifer Simulation in Two Dimensions with Results of Numerical Experiments," in *Techniques of Water-Resources Investigations,* Book 7, Chap. C1, U.S. Geological Survey, Washington, D.C.

Walton, W. C., 1960, *Leaky Artesian Aquifer Conditions in Illinois,* Illinois State Water Surv. Rept. Invest. 39, Urbana, Illinois.

Wang, H. F., and M. P. Anderson, 1982, *Introduction to Ground Water Modeling Finite Difference and Finite Element Methods,* W. H. Freeman and Company, San Francisco, California.

Appendixes

A

SYMBOLS AND NOTATION

SYMBOL	DEFINITION	SYMBOL	DEFINITION
A	cross-sectional area of channel	C_{s1}	skew coefficient
A	drainage area	C_t	Snyder's timing coefficient
A	loss rate	C_w	weir coefficient
A_T	total area	CDF	cumulative density function
ADI	alternating direction implicit method	CN	SCS runoff curve number
API	antecedent precipitation index	CV	coefficient of variation
$Ar(H)$	surface area of detention basin as a function of head	c	wave celerity
		D	duration of unit hydrograph
		D	Muskingum constant
B	channel top width	D	rainfall duration
B	time of fall (SCS unit hydro-graph)	D_c	circular channel diameter
		DHM	melt coefficient
BF	baseflow	DRH	direct runoff hydrograph
b	aquifer thickness	DRO	direct runoff
b	bottom width of channel	d	distance to water divide
b'	thickness of aquitard	d_p	depression storage
C	Chezy coefficient	E	evaporation
C	runoff coefficient	E	specific energy
C_0, C_1, C_2	Muskingum constants	$E(x)$	expectation, mean, first moment
C_p	Snyder's storage coefficient	ET	evapotranspiration

SYMBOL	DEFINITION	SYMBOL	DEFINITION
e	vapor pressure	K	routing parameter
e_a	vapor pressure of air	K'	vertical hydraulic conductivity of aquitard
e_s	saturation vapor pressure		
		$K(\theta)$	capillary conductivity
F	field capacity	K_1	diffusion equation constant
F	infiltration volume	K_s	hydraulic conductivity at saturation
$F(x)$	cumulative density function		
$F(z)$	normal CDF	K_x	horizontal hydraulic conductivity
F_f	frictional force		
F_G	gravitational force	K_z	vertical hydraulic conductivity
F_H	hydrostatic force	KS	Kolmogorov-Smirnov statistic at confidence level alpha
Fr	Froude number		
f	infiltration capacity or rate	k	intrinsic permeability
$f(x)$	probability density function	k	kinematic flow number
f_c	ultimate infiltration capacity		
f_o	initial infiltration capacity	L	channel length
		L	distance to wetting front
G	ground water flow	L	evaporation capacity
GW	subsurface seepage to ground water flow	L	lag time
		L	loss rate
g	skew	L_c	latent heat of condensation
		L_c	length along channel to point nearest centroid of area
H	heat storage		
H	relative humidity	L_{ca}	length to centroid of drainage area
H	total head		
$H3$	change in water surface elevation through bridge	L_e	latent heat of vaporization
		L_f	latent heat of freezing
h	head	L_m	latent heat of melting
h_l	head loss	L_s	infiltration storage
h_o	weir crest elevation		
		M	daily snowmelt
I	heat index	M	rate of heat storage
I	inflow	MIT	minimum interevent time
I	percent imperviousness	m	mass
I_a	initial abstraction		
IDF	intensity-duration-frequency	N	effective roughness
IUH	instantaneous unit hydrograph	N_m	monthly daylight adjustment factor
i	natural piezometric slope		
i	rainfall intensity	n	Manning's roughness coefficient
$i(t)$	instantaneous inflow	n	porosity
i_e	rainfall excess		
		O	outflow
J_o	solar constant		
		P	pressure
K	conveyance factor	P	precipitation
K	hydraulic conductivity	P	wetted perimeter

SYMBOL	DEFINITION	SYMBOL	DEFINITION
$P(X_i)$	probability of X_i	S	potential abstraction
P_e	effective runoff	S	storage
PDF	probability density function	S	storativity
PE	potential evaporation over a 10-day period	S_{av}	average capillary suction
		S_c	critical slope
PET	potential evapotranspiration	S_D	soil moisture storage
PMF	probability mass function	S_f	friction slope
p	probability	S_o	channel slope
		S_o	overland slope
Q	outflow	S_s	specific storage
Q_1	inflow	S_x	standard deviation of x
Q_2	outflow	SMELT	melt rate
Q_c	channel flow	s'	drawdown
Q_d	direct solar radiation		
Q_e	energy of evaporation	T	return period
Q_h	sensible heat transfer	T	temperature
Q_i	inflow	T	transmissivity
$Q_{(i,j)}$	recharge or pumping at node (i, j)	T	transpiration
		T_a	temperature of saturated air
Q_N	net radiation	T_B	duration of flood wave
Q_o	daily radiation at top of atmosphere	T_b	time base of hydrograph
		T_d	dewpoint temperature
Q_p	peak flow	T_R	time of rise
Q_T	peak flow rate for T-yr flood	t	time
Q_v	advected energy of inflow and outflow	t_c	time of concentration
		t_p	time to peak
QOBS	observed hydrograph ordinate at time i		
		UH	unit hydrograph
q	flow per unit thickness	u	wave velocity
q	infiltration rate	$u(t - \tau)$	weighting function
q_o	overland flow		
q_t	discharge at time t	V	average velocity
		V	total volume
R	Bowen ratio	V_s	seepage velocity
R	hydraulic radius	Var(x)	variance, second moment
R	recharge	v	velocity
R	surface runoff		
R_s	average albedo	W	recharge intensity
RF	rainfall	W	weight
RMS	root mean square	W	width of overland flow
RO	runoff	$W(u)$	well function
r	radius	WS	water surface elevation
r_c	radius of core sample	w	channel width
S	channel slope	X	event
S	energy gradient	X_i	ith possible outcome

SYMBOL	DEFINITION	SYMBOL	DEFINITION
x	horizontal distance	ϵ	Muskingum-Cunge weighting
x	Muskingum weighting factor		factor
x_m	median	Γ	gamma function
		γ	psychometric constant
y	average watershed slope	γ	Gumbel distribution constant
y	depth		(Euler's constant)
y_c	critical depth	γ	specific weight of water
y_c	depth of flow in collector	λ	exponential distribution
	channel		parameter
y_n	normal depth	μ	dynamic viscosity
y_o	average depth of overland flow	μ	mean value
		ν	kinematic viscosity
z	elevation	ϕ	infiltration index
z	horizontal dimension for side	ϕ	velocity potential
	slope of a channel	ψ	capillary suction
z	standard normal variate	ψ	stream function
z_i	thickness of layer i	ρ	fluid density
		ρ_b	bulk density
α	kinematic wave coefficient	ρ_m	density of moist air
α	attenuation factor	ρ_w	vapor density
α	compressibility of aquifer	σ	standard deviation
	material	σ^2	variance
α	conveyance factor	τ	incremental time unit
α	Gumbel distribution parameter	θ	volumetric moisture content
β	compressibility of fluid		

B.1
FLOW CONVERSION

UNIT	m³/s	m³/day	ℓ/s	ft³/s	ft³/day	ac-ft/day	gal/min	gal/day	mgd
1 m³/s	1	8.64×10^4	10^3	35.31	3.051×10^6	70.05	1.58×10^4	2.282×10^7	22.824
1 m³/day	1.157×10^{-5}	1	0.0116	4.09×10^{-4}	35.31	8.1×10^{-4}	0.1835	264.17	2.64×10^{-4}
1 liter/s	0.001	86.4	1	0.0353	3051.2	0.070	15.85	2.28×10^4	2.28×10^{-2}
1 ft³/s	0.0283	2446.6	28.32	1	8.64×10^4	1.984	448.8	6.46×10^5	0.646
1 ft³/day	3.28×10^{-7}	0.02832	3.28×10^{-4}	1.16×10^{-5}	1	2.3×10^{-5}	5.19×10^{-3}	7.48	7.48×10^{-6}
1 ac-ft/day	0.0143	1233.5	14.276	0.5042	43,560	1	226.28	3.259×10^5	0.3258
1 gal/min	6.3×10^{-5}	5.451	0.0631	2.23×10^{-3}	192.5	4.42×10^{-3}	1	1440	1.44×10^{-3}
1 gal/day	4.3×10^{-8}	3.79×10^{-3}	4.382×10^{-5}	1.55×10^{-6}	11,337	3.07×10^{-6}	6.94×10^{-4}	1	10^{-6}
1 million gal/day (mgd)	4.38×10^{-2}	3785	43.82	1.55	1.337×10^5	3.07	694	10^6	1

B.2
VOLUME CONVERSION

UNIT	ml	liters	m³	in³	ft³	gal	ac-ft	million gal
1 ml	1	0.001	10^{-6}	0.06102	3.53×10^{-5}	2.64×10^{4}	8.1×10^{-10}	2.64×10^{-10}
1 liter	10^{3}	1	0.001	61.02	0.0353	0.264	8.1×10^{-7}	2.64×10^{-7}
1 m³	10^{6}	1000	1	61,023	35.31	264.17	8.1×10^{-4}	2.64×10^{-4}
1 in³	16.39	1.64×10^{-2}	1.64×10^{-5}	1	5.79×10^{-4}	4.33×10^{-3}	1.218×10^{-8}	4.329×10^{-9}
1 ft³	28,317	28.317	0.02832	1728	1	7.48	2.296×10^{-5}	7.48×10^{-6}
1 U.S. gal	3785.4	3.785	3.78×10^{-3}	231	0.134	1	3.069×10^{-6}	10^{-6}
1 ac-ft	1.233×10^{9}	1.233×10^{6}	1233.5	75.27×10^{6}	43,560	3.26×10^{5}	1	0.3260
1 million gallons	3.785×10^{9}	3.785×10^{6}	3785	2.31×10^{8}	1.338×10^{5}	10^{6}	3.0684	1

B.3
AREA CONVERSION

UNIT	cm^2	m^2	km^2	ha	in^2	ft^2	yd^2	mi^2	ac
1 cm^2	1	0.0001	10^{-10}	10^{-8}	0.155	1.08×10^{-3}	1.2×10^{-4}	3.86×10^{-11}	2.47×10^{-8}
1 m^2	10^4	1	10^{-6}	10^{-4}	1550	10.76	1.196	3.86×10^{-7}	2.47×10^{-4}
1 km^2	10^{10}	10^6	1	100	1.55×10^9	1.076×10^7	1.196×10^6	0.3861	247.1
1 hectare (ha)	10^8	10^4	0.01	1	1.55×10^7	1.076×10^5	1.196×10^4	3.861×10^{-3}	2.471
1 in^2	6.452	6.45×10^{-4}	6.45×10^{-10}	6.45×10^{-8}	1	6.94×10^{-3}	7.7×10^{-4}	2.49×10^{-10}	1.574×10^7
1 ft^2	929	0.0929	9.29×10^{-8}	9.29×10^{-6}	144	1	0.111	3.587×10^{-8}	2.3×10^{-5}
1 yd^2	8361	0.8361	8.36×10^{-7}	8.36×10^{-5}	1296	9	1	3.23×10^{-7}	2.07×10^{-4}
1 mi^2	2.59×10^{10}	2.59×10^6	2.59	259	4.01×10^9	2.79×10^7	3.098×10^6	1	640
1 ac	4.04×10^7	4047	4.047×10^{-3}	0.4047	6.27×10^6	43,560	4840	1.562×10^{-3}	1

B.4
LENGTH CONVERSION

UNIT	mm	cm	m	km	in.	ft	yd	mi
1 mm	1	0.1	0.001	10^{-6}	0.0397	0.00328	0.00109	6.21×10^{-7}
1 cm	10	1	0.01	0.0001	0.3937	0.0328	0.0109	6.21×10^{-6}
1 m	1000	100	1	0.001	39.37	3.281	1.094	6.21×10^{-4}
1 km	10^{6}	10^{5}	1000	1	39,370	3281	1093.6	0.621
1 in.	25.4	2.54	0.0254	2.54×10^{-5}	1	0.0833	0.0278	1.58×10^{-5}
1 ft	304.8	30.48	0.3048	3.05×10^{-4}	12	1	0.333	1.89×10^{-4}
1 yd	914.4	91.44	0.9144	9.14×10^{-4}	36	3	1	5.68×10^{-4}
1 mi	1.61×10^{6}	1.01×10^{5}	1.61×10^{3}	1.6093	63,360	5280	1760	1

B.5
METRIC CONVERSION FACTORS (SI UNITS TO U.S. CUSTOMARY)*

MULTIPLY THE SI UNIT		BY	TO OBTAIN THE U.S. CUSTOMARY UNIT	
Name	Symbol		Symbol	Name
Energy				
kilojoule	kJ	0.9478	Btu	British thermal unit
joule	J	2.7778×10^{-7}	kW-h	kilowatt-hour
joule	J	0.7376	ft-lb$_f$	foot-pound (force)
joule	J	1.0000	W-s	watt-second
joule	J	0.2388	cal	calorie
kilojoule	kJ	2.7778×10^{-4}	kW-h	kilowatt-hour
kilojoule	kJ	0.2778	W-h	watt-hour
megajoule	MJ	0.3725	hp-h	horsepower-hour
Force				
newton	N	0.2248	lb	pound force
Mass				
gram	g	0.0353	oz	ounce
gram	g	0.0022	lb	pound
kilogram	kg	2.2046	lb	pound
Power				
kilowatt	kW	0.9478	Btu/s	British thermal units per second
kilowatt	kW	1.3410	hp	horsepower
watt	W	0.7376	ft-lb$_f$/s	foot-pounds (force) per second

(continued)

B.5 (continued)

MULTIPLY THE SI UNIT		BY	TO OBTAIN THE U.S. CUSTOMARY UNIT	
Name	Symbol		Symbol	Name
Pressure (force/area)				
pascal (newtons per square meter)	Pa (N/m²)	1.4504×10^{-4}	lb_f/ft^2	pounds (force) per square inch
pascal (newtons per square meter)	Pa (N/m²)	2.0885×10^{-2}	lb_f/ft^2	pounds (force) per square foot
pascal (newtons per square meter)	Pa (N/m²)	2.9613×10^{-4}	in.Hg	inches of mercury (60°F)
pascal (newtons per square meter)	Pa (N/m²)	4.0187×10^{-3}	in.H₂O	inches of water (60°F)
kilopascal (kilonewtons per square meter)	kPa (kN/m²)	0.1450	lb_f/in^2	pounds (force) per square inch
kilopascal (kilonewtons per square meter)	kPa (kN/m²)	0.0099	atm	atmosphere (standard)
Temperature				
degree Celsius (centigrade)	°C	$1.8°C + 32$	°F	degree Fahrenheit
degree Kelvin	K	$1.8K - 459.67$	°F	degree Fahrenheit
Velocity				
kilometers per second	km/s	2.2369	mi/hr	miles per hour
meters per second	m/s	3.2808	ft/s	feet per second

* Adapted with permission from G. Tchobanoglous and E. Schroeder, 1985; *Water Quality*, Addison-Wesley Publishing Co., Reading, Massachusetts.

C

PROPERTIES OF WATER

The principal physical properties of water are summarized in Table C.1. They are described briefly below.

Specific Weight

The specific weight of a fluid, γ, is the gravitational attractive force acting on a unit volume of a fluid. In SI units, the specific weight is expressed as kilonewtons per cubic meter (kN/m^3). At normal temperatures γ is 9.81 kN/m^3 (62.4 lb/ft^3).

Density

The density of a fluid, ρ, is its mass per volume. For water, ρ is 1000 kg/m^3 (1.94 $slug/ft^3$) at $4°C$. There is a slight decrease in density with increasing temperature. The relationship between γ, ρ, and the acceleration due to gravity g is $\gamma = \rho g$.

Modulus of Elasticity

For most practical purposes, liquids may be regarded as incompressible. The bulk modulus of elasticity E is given by

$$E = \frac{\Delta p}{\Delta V/V},$$

where Δp is the increase in pressure that, when applied to a volume V, results in a decrease in volume ΔV. In SI units the modulus of elasticity is expressed as kilonewtons per square meter (kN/m^2). For water, E is approximately 2.150 kN/m^2 (44.9 lb/ft^2) at normal temperatures and pressures.

Dynamic Viscosity

The viscosity of a fluid, μ, is a measure of its resistance to tangential or shear stress. In SI units the viscosity is expressed as newton-seconds per square meter (N-s/m^2).

Kinematic Viscosity

In many problems concerning fluid motion, the viscosity appears with the density in the form μ/ρ, and it is convenient to use a single term ν, known as the kinematic viscosity. The kinematic viscosity of a liquid, expressed as square meters per second (m^2/s) in SI units, diminishes with increasing temperature.

TABLE C.1

Physical Properties of Water

TEMPERATURE (°C)	SPECIFIC WEIGHT γ (kN/m³)	DENSITY ρ (kg/m³)	MODULUS OF ELASTICITY* $E/10^6$ (kN/m²)	DYNAMIC VISCOSITY $\mu \times 10^3$ (N-s/m²)	KINEMATIC VISCOSITY $\nu \times 10^6$ (m²/s)	SURFACE TENSION† σ (N/m)	VAPOR PRESSURE p_v (kN/m²)
0	9.805	999.8	1.98	1.781	1.785	0.0765	0.61
5	9.807	1000.0	2.05	1.518	1.519	0.0749	0.87
10	9.804	999.7	2.10	1.307	1.306	0.0742	1.23
15	9.798	999.1	2.15	1.139	1.139	0.0735	1.70
20	9.789	998.2	2.17	1.002	1.003	0.0728	2.34
25	9.777	997.0	2.22	0.890	0.893	0.0720	3.17
30	9.764	995.7	2.25	0.798	0.800	0.0712	4.24
40	9.730	992.2	2.28	0.653	0.658	0.0696	7.38
50	9.698	988.0	2.29	0.547	0.553	0.0679	12.33
60	9.642	983.2	2.28	0.466	0.474	0.0662	19.92
70	9.589	977.8	2.25	0.404	0.413	0.0644	31.16
80	9.530	971.8	2.20	0.354	0.364	0.0626	47.34
90	9.466	965.3	2.14	0.315	0.326	0.0608	70.10
100	9.399	958.4	2.07	0.282	0.294	0.0589	101.33

* At atmospheric pressure.

† In contact with air.

Adapted from Vennard and Street, 1975.

Surface Tension

Surface tension of a fluid, σ, is the physical property that enables a drop of water to be held in suspension at a tap, a glass to be filled with liquid slightly above the brim and yet not spill, and a needle to float on the surface of a liquid. The surface-tension force across any imaginary line at a free surface is proportional to the length of the line and acts in a direction perpendicular to it. In SI units, surface tension is expressed as newtons per meter (N/m). There is a slight decrease in surface tension with increasing temperature.

Vapor Pressure

Liquid molecules that possess sufficient kinetic energy are projected out of the main body of a liquid at its free surface and pass into the vapor. The pressure exerted by this vapor is known as the vapor pressure p_v. In SI units the vapor pressure is expressed as pascals (Pa) or kilonewtons per square meter (kN/m²).

REFERENCES

Vennard, J. K., and R. L. Street, 1975, *Elementary Fluid Mechanics,* 5th ed., John Wiley and Sons, New York.

Webber, N. B., 1971, *Fluid Mechanics for Civil Engineers,* SI ed., Chapman and Hall, London.

D.1

COMPUTATION OF NET TOTAL RADIATION

Total Radiation

Net total radiation absorbed by a water (or snow) surface Q_N consists of shortwave and longwave components as follows:

$$Q_N = Q_s - Q_{sr} + Q_a - Q_{ar} - Q_b, \qquad\qquad \text{(D.1.1)}$$

where

Q_s = incoming solar (shortwave) radiation (also called insolation) (cal/cm²-min = ly/min),

Q_{sr} = reflected shortwave radiation (ly/min),

Q_a = incoming atmospheric (longwave) radiation (ly/min),

Q_{ar} = reflected longwave radiation (ly/min),

Q_b = longwave radiation emitted from water surface (back radiation) (ly/min).

Traditional units for radiation flux are langleys per minute (ly/min), where a langley is 1 gram-calorie per square centimeter ($g\text{-}cal/cm^2$). However, SI units use watts per square meter for radiation flux: 1 ly/min = 697.33 W/m^2. Shortwave and longwave components are usually measured and predicted separately.

Shortwave Radiation

In the United States, pyranometer measurements of daily insolation are taken at a few weather stations in each state and published monthly by the National Climatic Data Center, National Oceanic and Atmospheric Administration, Asheville, North Carolina. Since these data are sparse, predictive methods are usually necessary. The best summary of methods related to hydrology has been prepared by the Tennessee Valley Authority (1972).

Shortwave radiation incident on a horizontal surface at the base of the atmosphere may be approximated by

$$Q_s = J_o/r^2 \cdot a^m \cdot \sin \alpha \cdot f_1(C), \tag{D.1.2}$$

where

J_o = solar constant, approximately 1.97 ly/min (Frohlich, 1977),

r = normalized earth-sun radius, a function of date,

a = atmospheric transmission coefficient,

m = optical air mass (relative path length of sun's rays through the atmosphere) (csc α for $\alpha \geq 10°$),

α = solar altitude (elevation of sun above horizon),

$f_1(C)$ = cloudiness attenuation function, where C is the fraction of sky covered.

The solar altitude is a function of latitude, earth's declination, and time of day and may be calculated exactly as a function of latitude, longitude, date, and time (Eagleson, 1970; List, 1966; Tennessee Valley Authority, 1972). However, for prediction of daily radiation totals, Eq. (D.1.2) must be integrated numerically over the variable duration of daylight. Absorption and scattering of radiation in the atmosphere are functions of atmospheric content of water vapor, ozone, and dust and ideally may be predicted from measurements of these quantities (Tennessee Valley Authority, 1972). However, it is convenient to aggregate the calculation by using an empirical atmospheric transmission coefficient, a.

Results of the numerical integration for clear skies are presented for typical values of a in Table D.1.1 (List, 1966); interpolation provides values for other dates and atmospheric transmission coefficient values. The values in Table D.1.1 provide only direct solar radiation Q_d. The values need to be increased to account for diffuse (scattered) radiation also reaching the earth's surface. This is accomplished as follows (List, 1966), using Table D.1.2:

1. Find the daily radiation at the top of the atmosphere Q_o from Table D.1.2.

2. Decrease Q_o by 9% to allow for water vapor absorption (7%) and for ozone absorption (2%). The remaining radiation is $0.91Q_o$.

3. From Table D.1.1 find the direct radiation Q_d reaching the earth's surface and subtract from $0.91Q_o$. The resulting difference, $S = 0.91Q_o - Q_d$, approximates the energy scattered out of the direct solar beam.

4. Half of the scattered radiation is returned to space. Thus, add $S/2$ to the value of direct solar radiation from Table D.1.1, yielding $Q_s = Q_d + S/2$. (For clear skies only.)

EXAMPLE D.1.1

Compute Q_s for clear skies on August 8 at a latitude of 40°N assuming atmospheric transmission coefficient 0.8.

SOLUTION

From Table D.1.1, $Q_d = 635$ ly/day. From Table D.1.2, $Q_o = 901$ ly/day. Thus, $S = 0.91 \cdot 901 - 635 = 185$ ly/day and $Q_s = 635 + 185/2 = 728$ ly/day.

Since the values in Tables D.1.1 and D.1.2 are based on a solar constant J_o of 1.94 ly/min, they could be increased 2% to account for the more recent estimate of 1.97 ly/min (Frohlich, 1977), but the correction is probably unwarranted in light of the much higher error in determining the value of the atmospheric transmission coefficient a. This coefficient may be estimated using methods described by List (1966) and Tennessee Valley Authority (1972) or measured directly. Typical values of a are in the range of 0.8 to 0.9.

Cloudiness measurements are made hourly at most weather stations and are reported as fraction of sky covered, in tenths. An empirical cloudiness attenuation factor (Tennessee Valley Authority, 1972) is

$$f_1(C) = 1 - 0.65 \cdot C^2, \tag{D.1.3}$$

with C as a fraction. This equation works best for long time periods, such as months, but will provide a fair measure of attenuation using C as the average fraction of sky covered during *daylight* hours. More sophisticated methods require knowledge of cloud type and elevation (Tarpley, 1979).

TABLE D.1.1

Total Daily Direct Solar Radiation Reaching the Ground with Various Atmospheric Transmission Coefficients

The solar constant J_o is assumed to be 1.94 cal cm^{-2} min^{-1}. Values apply to a horizontal surface.

APPROXIMATE DATE

Transmission coefficient $a = 0.6$ cal cm^{-2}

LATITUDE	Mar. 21	May 6	June 22	Aug. 8	Sept. 23	Nov. 8	Dec. 22	Feb. 4
90°		127	299	125				
80	6	158	309	156	5			
70	47	234	349	232	46			
60	120	312	406	308	118	10		10
50	202	376	450	372	199	58	19	58
40	282	426	477	421	278	130	75	131
30	350	453	481	449	345	213	152	215
20	404	459	465	454	398	293	237	296
10	436	444	428	439	430	366	323	370
0	447	407	372	404	440	422	397	427
−10	436	353	303	349	430	461	457	465
−20	404	282	222	279	398	475	497	480
−30	350	206	143	204	345	470	514	475
−40	282	125	70	124	278	441	509	445
−50	202	56	18	55	199	391	481	395
−60	120	10		10	118	323	434	327
−70	47				46	242	373	245
−80	6				5	164	330	166
−90						131	319	133

APPROXIMATE DATE

Transmission coefficient $a = 0.7$ cal cm^{-2}

LATITUDE	Mar. 21	May 6	June 22	Aug. 8	Sept. 23	Nov. 8	Dec. 22	Feb. 4
90°		217	440	215				
80	13	243	442	242	13			
70	80	324	467	321	79			
60	174	408	520	404	172	22	1	23
50	272	477	563	472	268	91	37	92
40	363	529	587	524	358	182	111	184
30	440	556	588	550	434	281	210	283
20	499	561	568	556	491	374	309	378
10	534	542	524	537	527	456	408	460
0	546	501	462	496	538	519	493	524
−10	534	439	382	436	527	563	560	568
−20	499	361	290	357	491	582	606	588
−30	440	271	196	268	434	576	628	582
−40	363	176	103	174	358	548	627	554
−50	272	88	34	87	268	495	601	500
−60	174	22	1	21	172	423	555	427
−70	80				79	337	499	340
−80	13				13	253	472	255
−90						226	469	228

Transmission coefficient a = 0.8

Angle								
90°	29	349	615	346	29	1		
80	128	365	608	361	126	44	5	1
70	242	434	605	429	240	136	64	44
60	356	520	650	515	350	247	164	137
50	456	591	686	585	449	360	277	249
40	539	641	708	635	532	463	393	363
30	601	668	706	662	593	555	503	468
20	639	669	678	663	630	626	598	560
10	652	649	630	643	643	673	672	631
0	639	602	560	597	630	695	725	680
−10	601	534	471	530	593	694	754	700
−20	539	446	369	442	532	665	756	700
−30	456	347	260	343	449	612	732	671
−40	356	239	153	236	350	539	694	619
−50	242	131	60	130	240	449	646	544
−60	128	42	4	42	126	378	649	454
−70	29	1		1	29	363	656	381
−80								366
−90								

Transmission coefficient a = 0.9

Angle								
90°	67	532	826	526	66	5		
80	199	528	813	523	196	80	16	5
70	333	571	774	566	328	200	107	81
60	455	650	799	643	449	328	229	201
50	562	718	828	711	554	453	362	331
40	651	766	841	759	641	566	491	458
30	715	791	831	783	705	664	610	571
20	755	789	799	781	744	740	713	670
10	768	763	744	756	757	792	794	748
0	755	714	668	707	744	819	853	799
−10	715	640	571	634	705	820	888	826
−20	651	545	460	539	641	794	898	828
−30	562	436	340	433	554	745	884	802
−40	455	316	214	313	449	674	854	752
−50	333	192	100	190	328	593	826	681
−60	199	77	15	76	196	547	867	598
−70	67	5		4	66	551	883	553
−80								556
−90								

From List, 1966.

TABLE D.1.2
Total Daily Solar Radiation at the Top of the Atmosphere

The solar constant J_o is assumed to be 1.94 cal cm⁻² min⁻¹. Values apply to a horizontal surface.

APPROXIMATE DATE

cal cm⁻²

LATI-TUDE	Mar. 21	Apr. 13	May 6	May 29	June 22	July 15	Aug. 8	Aug. 31	Sept. 23	Oct. 16	Nov. 8	Nov. 30	Dec. 22	Jan. 13	Feb. 4	Feb. 26
90°		423	772	999	1077	994	765	418								
80	155	423	760	984	1060	980	754	418	153	7						7
70	307	525	749	939	1012	934	742	519	303	129	24				24	131
60	447	635	809	934	979	929	801	629	442	273	146	72	49	73	146	276
50	575	732	867	958	989	954	859	725	568	414	286	204	176	205	289	419
40	686	807	910	972	991	967	901	798	677	545	429	348	317	350	434	553
30	775	865	929	967	975	960	921	856	765	663	564	492	466	494	568	670
20	841	894	923	935	935	930	916	884	831	760	685	627	605	630	691	769
10	882	897	893	881	873	877	886	887	871	835	789	748	733	752	795	845
0	895	873	837	804	790	800	830	863	885	886	870	851	843	855	878	896
−10	882	824	760	707	687	704	753	814	871	910	927	931	933	936	936	921
−20	841	750	660	593	567	590	654	741	831	907	959	988	999	993	968	918
−30	775	654	543	465	436	463	538	646	765	877	964	1020	1041	1025	973	888
−40	686	538	413	329	297	328	409	533	677	819	944	1027	1059	1032	953	828
−50	575	408	276	193	165	192	274	404	568	743	901	1014	1056	1018	909	752
−60	447	269	140	68	47	68	139	266	442	644	840	987	1046	992	847	652
−70	307	127	23				23	126	303	532	778	993	1081	998	785	539
−80	155	7						7	153	429	790	1041	1132	1046	796	434
−90										429	801	1056	1149	1062	809	434

From List, 1966.

TABLE D.1.3

**Average Daily
Shortwave Albedos
for Water**

MONTH	R_s (%)
Jan.	9
Feb.	7
Mar.	7
Apr.	6
May	6
Jun.	6
Jul.	6
Aug.	6
Sep.	7
Oct.	7
Nov.	9
Dec.	10

(*Source:* Anderson, 1954.
Values are for Lake Hefner,
near Oklahoma City, OK,
latitude 36°N)

Reflected Shortwave Radiation

Reflected shortwave radiation from a water surface depends on the solar altitude, but for daily radiation an average albedo (reflection coefficient) R_s of approximately 0.07 may be used; it ranges from about 0.06 during summer months to 0.10 in December (Table D.1.3). The albedo of a snowpack depends on the age of the snow. New snow has an albedo of approximately 0.83 (U.S. Army Corps of Engineers, 1956), decreasing to a range of from 0.42 to 0.70 with age (Fig. D.1.1). Finally, reflected shortwave radiation is simply

$$Q_{sr} = R_s \cdot Q_s. \tag{D.1.4}$$

Longwave Atmospheric Radiation

Unlike shortwave radiation, which occurs only during daylight hours, longwave radiation is continually emitted by all molecules. Measurements are made using a total radiometer, from which pyranometer measurements of shortwave radiation are subtracted to yield longwave radiation. Unfortunately, routine radiometer measurements are not performed at weather stations; however, longwave radiation is somewhat easier to predict. In general, longwave atmospheric radiation is computed

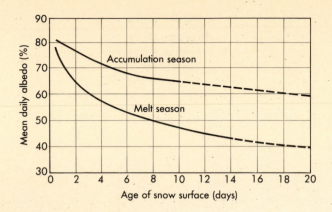

FIGURE D.1.1

Variation in snowpack albedo with time. (Adapted from U.S. Army Corps of Engineers, 1956, Fig. 4, Plate 5-2.)

using the Stefan-Boltzmann law,

$$Q_a = \epsilon_a \cdot \sigma \cdot T_a^4 \cdot f_2(C), \tag{D.1.5}$$

where

Q_a = atmospheric longwave radiation (ly/min),

ϵ_a = atmospheric emissivity,

T_a = air temperature (°K),

σ = Stefan-Boltzmann constant ($0.813 \cdot 10^{-10}$ ly/min-°K^4).

Air temperature is usually measured at the standard elevation of 2 m. The emissivity of water (or snow) is 0.97, and the emissivity of the atmosphere depends on its moisture content. In fact, under overcast conditions, the best estimate of Q_a is obtained if the cloud base temperature is known and used in Eq. (D.1.5) with an emissivity of 0.97 and $f_2(C) = 1$. For clear skies, several empirical formulas exist to relate ϵ_a to atmospheric vapor pressure e_a or air temperature T_a. From an evaluation of methods for clear skies and for U.S. conditions, Hatfield et al. (1983) found the Brunt equation to work well:

$$\epsilon_a = 0.51 + 0.066 \sqrt{e_a}, \tag{D.1.6}$$

where e_a is in mb and may be computed from

$$e_a = H \cdot e_s(T_a), \tag{D.1.7}$$

where

e_s = saturation vapor pressure (mb) at temperature T_a,

H = relative humidity (fraction).

Equation (1.6) in Chapter 1 provides a good estimate of e_s, or it may be obtained from Table C.1 or List (1966).

An empirical relationship for the increase of atmospheric radiation with cloud cover is again provided by Tennessee Valley Authority (1972):

$$f_2(C) = 1 + 0.17 \cdot C^2. \tag{D.1.8}$$

Reflected Longwave Radiation

For both water and snow, the longwave albedo is 0.03, which is equivalent to $(1 - \text{emissivity})$.

Back Radiation

The water surface temperature must be known to compute Q_b. Thus

$$Q_b = 0.97 \cdot \sigma \cdot T_s^4, \tag{D.1.9}$$

where T_s = water surface temperature (°K) and an emissivity of 0.97 has been used. This same equation may be used to compute back radiation from a snow surface.

EXAMPLE D.1.2

Compute net daily radiation into a water surface for the following conditions (daily averages):

Date: August 8

Latitude: 40°N

Cloudiness: 0.4

Atmospheric transmission coefficient: 0.8

Air temperature: 30°C

Relative humidity: 60%

Water surface temperature: 20°C

Clear sky insolation was computed in Example D.1.1 to be 728 ly/day. For a cloudiness of 0.4,

$$Q_s = 728 \cdot (1 - 0.65 \cdot 0.4^2) = 652 \text{ ly/day}.$$

Using a shortwave albedo of 0.07, reflected shortwave radiation is

$$Q_{sr} = 0.07 \cdot 652 = 46 \text{ ly/day}.$$

From Eq. (1.6), the saturation vapor pressure at an air temperature of 30°C is

$$e_s = 2.7489 \cdot 10^8 \cdot \exp\left(\frac{-4278.6}{30 + 242.79}\right) = 42.4 \text{ mb}.$$

Thus

$$e_a = 0.6 \cdot 42.4 = 25.4 \text{ mb}.$$

Using Brunt's formula for clear sky emissivity,

$$\epsilon_a = 0.51 + 0.066 \cdot \sqrt{25.4} = 0.843.$$

Longwave atmospheric radiation at an air temperature of $30 + 273 = 303°K$ is

$$Q_a = 0.843 \cdot 0.813 \cdot 10^{-10} \cdot 303^4$$
$$= 0.578 \text{ ly/min} = 832 \text{ ly/day}.$$

Reflected longwave radiation is

$$Q_{ar} = 0.03 \cdot 832 = 25 \text{ ly/day}.$$

Back radiation from the water surface at a temperature of $20 + 273 = 293°K$ is

$$Q_b = 0.97 \cdot 0.813 \cdot 10^{-10} \cdot 293^4$$
$$= 0.581 \text{ ly/min} = 837 \text{ ly/day}$$

Finally, the net total radiation absorbed in the water body is

$$Q_N = Q_s - Q_{sr} + Q_a - Q_{ar} - Q_b$$
$$= 652 - 46 + 832 - 25 - 837 = +576 \text{ ly/day}$$

Thus the water body absorbs a net total of 576 ly/day due to all radiation mechanisms. In water, the shortwave component is absorbed exponentially with depth, and the longwave component is absorbed at the surface (Eagleson, 1970; Tennessee Valley Authority, 1972). For snow, longwave radiation is absorbed at the surface, and at least 90% of the shortwave component is absorbed within 20 cm of the surface (U.S. Army Corps of Engineers, 1956).

REFERENCES

Anderson, E. R., 1954, "Energy-Budget Studies, Water-Loss Investigations: Lake Hefner Studies," USGS Professional Paper 269, Washington, D.C.

Eagleson, P. S., 1970, *Dynamic Hydrology*, McGraw-Hill Book Company, New York.

Frohlich, C., 1977, "Contemporary Measures of the Solar Constant," in *The Solar Output and Its Variations*, O. R. White (editor), Colorado Associated University Press, pp. 93–109.

Hatfield, J. L., R. J. Reginato, and S. B. Idso, 1983, "Comparison of Longwave Radiation Calculation Methods over the United States," *Water Resources Research,* vol. 19, no. 1, February, pp. 285–288.

List, R. J., 1966, *Smithsonian Meteorological Tables,* 6th ed., Smithsonian Institution, Washington, D.C.

Tarpley, J. D., 1979, "Estimating Incident Solar Radiation at the Surface from Geostationary Satellite Data," *J. Applied Meteorology,* vol. 18, no. 9, September, pp. 1172–1181.

Tennessee Valley Authority, 1972, "Heat and Mass Transfer Between a Water Surface and the Atmosphere," TVA Engineering Laboratory Report No. 14, Norris, Tennessee, April.

U.S. Army Corps of Engineers, 1956, "Snow Hydrology," North Pacific Division, U.S. Army Corps of Engineers, Portland, Oregon, NTIS PB-151660.

D.2

DERIVATION OF THE UNSATURATED FLOW EQUATION

A modified form of Darcy's law holds true for fluid flow through unsaturated media. In this form, hydraulic conductivity K becomes a function of the volumetric moisture content θ. Darcy's law for flow in the z-direction can be written as

$$v_z = -K(x, y, z, \theta) \frac{\partial h}{\partial z}. \tag{D.2.1}$$

Similar equations can be developed for both the x- and y-directions. The pressure head or capillary suction ψ is also a function of θ in unsaturated flow. Therefore, the hydraulic head h becomes

$$h = \psi(\theta) + z, \tag{D.2.2}$$

where z is the elevation head. The continuity equation can be written as

$$-\left[\frac{\partial(\rho v_x)}{\partial x} + \frac{\partial(\rho v_y)}{\partial y} + \frac{\partial(\rho v_z)}{\partial z}\right] = \frac{\partial}{\partial t}(\rho\theta). \tag{D.2.3}$$

Substituting the proper form of Darcy's law into the continuity equation yields the following equation:

$$\frac{\partial}{\partial x}\left[\rho K(x, y, z, \theta)\frac{\partial h}{\partial x}\right] + \frac{\partial}{\partial y}\left[\rho K(x, y, z, \theta)\frac{\partial h}{\partial y}\right]$$
$$+ \frac{\partial}{\partial z}\left[\rho K(x, y, z, \theta)\frac{\partial h}{\partial z}\right] = \rho\frac{\partial\theta}{\partial t} + \theta\frac{\partial\rho}{\partial t}. \tag{D.2.4}$$

For a homogeneous soil, ρ is constant and $\partial\rho/\partial t = 0$. K also reduces to become $K(\theta)$. Then the one-dimensional equation for unsaturated flow in the z-direction becomes

$$\frac{\partial}{\partial z}\left[K(\theta)\frac{\partial h}{\partial t}\right] = \frac{\partial \theta}{\partial t}. \tag{D.2.5}$$

Substituting Eq. (D.2.2) into Eq. (D.2.5) yields

$$\frac{\partial \theta}{\partial t} = \frac{\partial}{\partial z}\left[K(\theta)\frac{\partial \psi(\theta)}{\partial z}\right] + \frac{\partial K(\theta)}{\partial z}. \tag{D.2.6}$$

The elevation head z is measured from some datum, usually sea level. If the datum is taken as the ground surface and z is defined as the distance below the ground surface, Eq. (D.2.6) takes the following form

$$\frac{\partial \theta}{\partial t} = -\frac{\partial}{\partial z}\left[K(\theta)\frac{\partial \psi(\theta)}{\partial z}\right] - \frac{\partial K(\theta)}{\partial z}. \tag{D.2.7}$$

Equation (D.2.7) is the equation of flow through unsaturated media. It is called Richards equation and is used in theoretical infiltration methods.

NORMAL DISTRIBUTION TABLES

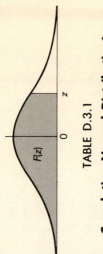

TABLE D.3.1

Cumulative Normal Distribution*

$$F(z) = \int_{-\infty}^{z} \frac{1}{\sqrt{2\pi}} e^{-z^2/2} \, dz$$

z	.00	.01	.02	.03	.04	.05	.06	.07	.08	.09
.0	.5000	.5040	.5080	.5120	.5160	.5199	.5239	.5279	.5319	.5359
.1	.5398	.5438	.5478	.5517	.5557	.5596	.5636	.5675	.5714	.5753
.2	.5793	.5832	.5871	.5910	.5948	.5987	.6026	.6064	.6103	.6141
.3	.6179	.6217	.6255	.6293	.6331	.6368	.6406	.6443	.6480	.6517
.4	.6554	.6591	.6628	.6664	.6700	.6736	.6772	.6808	.6844	.6879
.5	.6915	.6950	.6985	.7019	.7054	.7088	.7123	.7157	.7190	.7224
.6	.7257	.7291	.7324	.7357	.7389	.7422	.7454	.7486	.7517	.7549
.7	.7580	.7611	.7642	.7673	.7704	.7734	.7764	.7794	.7823	.7852
.8	.7881	.7910	.7939	.7967	.7995	.8023	.8051	.8078	.8106	.8133
.9	.8159	.8186	.8212	.8238	.8264	.8289	.8315	.8340	.8365	.8389
1.0	.8413	.8438	.8461	.8485	.8508	.8531	.8554	.8577	.8599	.8621
1.1	.8643	.8665	.8686	.8708	.8729	.8749	.8770	.8790	.8810	.8830
1.2	.8849	.8869	.8888	.8907	.8925	.8944	.8962	.8980	.8997	.9015
1.3	.9032	.9049	.9066	.9082	.9099	.9115	.9131	.9147	.9162	.9177

(continued)

TABLE D.3.1 (continued)

z	.00	.01	.02	.03	.04	.05	.06	.07	.08	.09
1.4	.9192	.9207	.9222	.9236	.9251	.9265	.9279	.9292	.9306	.9319
1.5	.9332	.9345	.9357	.9370	.9382	.9394	.9406	.9418	.9429	.9441
1.6	.9452	.9463	.9474	.9484	.9495	.9505	.9515	.9525	.9535	.9545
1.7	.9554	.9564	.9573	.9582	.9591	.9599	.9608	.9616	.9625	.9633
1.8	.9641	.9649	.9656	.9664	.9671	.9678	.9686	.9693	.9699	.9706
1.9	.9713	.9719	.9726	.9732	.9738	.9744	.9750	.9756	.9761	.9767
2.0	.9772	.9778	.9783	.9788	.9793	.9798	.9803	.9808	.9812	.9817
2.1	.9821	.9826	.9830	.9834	.9838	.9842	.9846	.9850	.9854	.9857
2.2	.9861	.9864	.9868	.9871	.9875	.9878	.9881	.9884	.9887	.9890
2.3	.9893	.9896	.9898	.9901	.9904	.9906	.9909	.9911	.9913	.9916
2.4	.9918	.9920	.9922	.9925	.9927	.9929	.9931	.9932	.9934	.9936
2.5	.9938	.9940	.9941	.9943	.9945	.9946	.9948	.9949	.9951	.9952
2.6	.9953	.9955	.9956	.9957	.9959	.9960	.9961	.9962	.9963	.9964
2.7	.9965	.9966	.9967	.9968	.9969	.9970	.9971	.9972	.9973	.9974
2.8	.9974	.9975	.9976	.9977	.9977	.9978	.9979	.9979	.9980	.9981
2.9	.9981	.9982	.9982	.9983	.9984	.9984	.9985	.9985	.9986	.9986
3.0	.9987	.9987	.9987	.9988	.9988	.9989	.9989	.9989	.9990	.9990
3.1	.9990	.9991	.9991	.9991	.9992	.9992	.9992	.9992	.9993	.9993
3.2	.9993	.9993	.9994	.9994	.9994	.9994	.9994	.9995	.9995	.9995
3.3	.9995	.9995	.9995	.9996	.9996	.9996	.9996	.9996	.9996	.9997
3.4	.9997	.9997	.9997	.9997	.9997	.9997	.9997	.9997	.9997	.9998

* For more extensive tables, see National Bureau of Standards, *Tables of Normal Probability Functions*, Washington, D.C., U.S. Government Printing Office, 1953 (Applied Mathematics Series 23). Note that they show $\int_{-z}^{z} f(z)\, dz$, not $\int_{-\infty}^{z} f(z)\, dz$.

Source: E. L. Crow, F. A. Davis, and M. W. Maxfield, 1960, *Statistics Manual*, Dover Publications, New York, Table 1, p. 229.

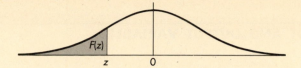

TABLE D.3.2

Percentiles of the Normal Distribution*

$$F(z) = \int_{-\infty}^{z} \frac{1}{\sqrt{2\pi}} e^{-z^2/2} \, dz$$

$F(z)$	z	$F(z)$	z
.0001	−3.719	.500	.000
.0005	−3.291	.550	.126
.001	−3.090	.600	.253
.005	−2.576	.650	.385
.010	−2.326	.700	.524
.025	−1.960	.750	.674
.050	−1.645	.800	.842
.100	−1.282	.850	1.036
.150	−1.036	.900	1.282
.200	−.842	.950	1.645
.250	−.674	.975	1.960
.300	−.524	.990	2.326
.350	−.385	.995	2.576
.400	−.253	.999	3.090
.450	−.126	.9995	3.291
.500	.000	.9999	3.719

* For a normally distributed variable x, we have $x = \mu + z\sigma$, where μ = mean of x and σ = standard deviation of x. For more extensive tables, see R. A. Fisher and F. Yates, *Statistical Tables,* 4th rev. ed., Edinburgh, Oliver & Boyd, Ltd., 1953, pp. 39, 60–62.

Source: E. L. Crow, F. A. Davis, and M. W. Maxfield, 1960, *Statistics Manual,* Dover Publications, New York, Table 2, p. 230.

D.4

HEC-2 INPUT AND OUTPUT VARIABLES

Input Data Variables

VARIABLE	LOCATION	DESCRIPTION
BAREA	SB.7	Net area of bridge opening used in special bridge methods.
BWC	SB.5	Bottom width of bridge opening used in special bridge methods.
BWP	SB.6	Total width of piers used in special bridge methods.
CCHV	NC.4	Contraction coefficient used in computing transition losses.
CEHV	NC.5	Expansion coefficient used in computing transition losses.
COFQ	SB.3	Coefficient of discharge for use in weir flow equation.
EL(N)	GR.1,3, . . .	Elevation of cross-section point at STN(N).
ELCHD	SB.10	Elevation of channel invert at downstream side of bridge.
ELCHU	SB.9	Elevation of channel invert at upstream side of bridge.
ELLC	X2.4	Elevation of low chord for bridge (normal bridge method) *or* maximum upstream low chord of bridge used to determine flow type (special bridge).
ELTRD	X2.5	Elevation of top of roadway (normal bridge) *or* minimum roadway elevation used to determine weir flow (special bridge).
IBRID	X2.3	Variable indicating whether special bridge method is being used.
IEARA	X3.1	Variable to determine how much of the cross-sectional area will be used in computation.
INQ	J1.2	Field number on QT, ET, or X5 card to be used.
IPLOT	J2.2,X1.10	Cross-section plot control.
NPROF	J2.1	Profile number.
NRD	BT.1	Number of points describing the bridge to be read from BT cards.

Input Data Variables *(continued)*

VARIABLE	LOCATION	DESCRIPTION
NUMNH	NH.1	Total number of Manning's *n* values to be entered.
NUMST	X1.2	Total number of stations on following GR cards.
Q	J1.8	Starting river flow.
RDEL(N)	BT.3,6,9	Top of roadway elevation at station RDST(N).
RDLEN	SB.4	Average length of roadway in feet (m) for use in the weir flow equation.
RDST(N)	BT.2,5,7	Roadway station corresponding to RDEL(N).
SECNO	X1.1	Cross section identification number, usually in river feet (m) from mouth.
SS	SB.8	Side slope of trapezoidal bridge opening.
STCHL	X1.3	Station of the left bank of the channel.
STCHR	X1.4	Station of the right bank of the channel.
STN(N)	NH.odd/GR.even	Station of ground surface.
STRT	J1.5	Variable that indicates method by which computations will be started.
VAL(N)	NH.even	Manning's *n* value between station/elevation N-1 and N.
WSEL	J1.9	Starting water surface elevation.
XK	SB.1	Pier shape coefficient K used in Yarnell's equation.
XKOR	SB.2	Total loss coefficient K between cross sections on either side of bridge for use in orifice flow equation.
XLCEL(N)	BT.4,7,10	Low chord elevation at station RDST(N).
XLCH	X1.7	Length of channel reach between current section and next downstream section.
XLOBL	X1.5	Length of left overbank reach between current section and next downstream section.
XLOBR	X1.6	Length of right overbank reach between current section and next downstream section.
XNCH	NC.3	Manning's *n* value for the channel.
XNL	NC.1	Manning's *n* value for the left overbank.
XNR	NC.2	Manning's *n* value for the right overbank.

(continued)

Output Data Variables

VARIABLE	DESCRIPTION
AREA	Cross-sectional area.
CLASS	Identification number for types of bridge flow.
CRIWS	Critical water surface elevation.
CWSEL	Computed water surface elevation.
DEPTH	Depth of flow.
DIFEG	Difference in energy elevation for each profile.
DIFKWS	Difference in water surface elevation between known and computed.
DIFWSP	Difference in water surface elevation for each profile.
DIFWSX	Difference in water surface elevation between sections.
EG	Energy gradient elevation for a cross section that is equal to the computed water surface elevation CWSEL plus the velocity head HV.
EGLWC	The energy grade line elevation computed assuming low flow.
EGPRS	The energy grade line elevation computed assuming pressure flow.
ELLC	Elevation of the bridge low chord. Equals ELLC entered on the X2 card if used; otherwise it equals maximum low chord in the BT table.
ELMIN	Minimum elevation in the cross section.
ELTRD	Elevation of the top of roadway. Equals ELTRD entered on the X2 card if used; otherwise it equals the minimum top of the road in the BT table.
H3	Drop in water surface elevation from upstream to downstream sides of the bridge computed using Yarnell's equation assuming class A low flow.
Q	Total flow in the cross section.
QPR	Total pressure or low flow at the bridge.
QWEIR	Total weir flow at the bridge.
SECNO	Identifying cross-section number. Equal to the number in the first field of the X1 card.
TOPWID	Width at the calculated water surface elevation.
VCH	Mean velocity in the channel.
XLCH	Distance in the channel between the previous cross section and the current cross section.
.01K	The total discharge (index Q) carried with $S^{1/2} = .01$ (equivalent to .01 times conveyance).

E.0

PROGRAM USE EXAMPLE

The following programs were written by Pamela J. Ross and John F. Haasbeek. Each program in Appendix E is executed by typing the program name indicated on the first line. For example, the convolution program is executed by typing CONVOL and pressing the return key. A prompt appears on the screen and the user responds with the correct option or input parameter. For example, most of the programs will ask whether a user wishes to use English or metric units. The correct response for English units is E or e.

The Kinematic Wave Routing, Uniform Flow Computation, and Critical Flow Computation programs all have several options to choose from. The letter or letters indicated in parentheses are entered to choose an option. If the return key is pressed without entering a letter or letters, the program will end that particular option.

The following is a list of prompts from and responses to the Unit Hydrograph Convolution program using the data listed in Example 2.5 in Chapter 2.

```
            UNIT HYDROGRAPH CONVOLUTION PROGRAM


 (E)nglish or (M)etric Units?
E
 The units for hydrograph intervals, starting and ending times and the time
  step must be hours or minutes.
 Enter a one (1) to use hours, or a two (2) to use minutes
1
 Enter the starting time of the rainfall (hrs)
0
 Enter the ending time of the rainfall (hrs)
5
 Enter the time interval of the rainfall (hrs)
1
 Enter the rainfall (in, cm) for time period              1
0.5
 Enter the rainfall (in, cm) for time period              2
1
 Enter the rainfall (in, cm) for time period              3
1.5
 Enter the rainfall (in, cm) for time period              4
0
 Enter the rainfall (in, cm) for time period              5
0.5
 Enter the number of unit hydrograph ordinates
using the same time interval as the rainfall data.
```

```
7
 Enter the UH ordinate (cfs, cms) for time period          1
0
 Enter the UH ordinate (cfs, cms) for time period          2
250
 Enter the UH ordinate (cfs, cms) for time period          3
500
 Enter the UH ordinate (cfs, cms) for time period          4
375
 Enter the UH ordinate (cfs, cms) for time period          5
250
 Enter the UH ordinate (cfs, cms) for time period          6
125
 Enter the UH ordinate (cfs, cms) for time period          7
0
```

TIME (hrs)	OUTFLOW (cfs)
.00	.00
1.00	125.00
2.00	500.00
3.00	1062.50
4.00	1250.00
5.00	1000.00
6.00	750.00
7.00	375.00
8.00	125.00
9.00	62.50
10.00	.00

E.1

UNIT HYDROGRAPH CONVOLUTION

```
      PROGRAM CONVOL
C
C.... This program will develop a storm hydrograph given rainfall
C.... data and a unit hydrograph using the Convolution Equation
C.... (Eqn. 2.3).  The rainfall data is entered as a net volume
C.... of rainfall (total - losses).  Either metric or English units
C.... may be specified.  The program may be applied to Problems
C.... 2.2, 2.12, and 2.15.
C
      INTEGER NQ,NUH,NRAIN,UNIT,NUM
      REAL RAIN(25),UH(25),Q(50),TIME,TO,T1,DT
      CHARACTER*1 ANS
      CHARACTER*5 UNITS(2)
      WRITE(*,1000)
```

```
      WRITE(*,*) '(E)nglish or (M)etric Units?'
      READ(*,'(A)') ANS
      UNITS(1) = '(hrs)'
      UNITS(2) = '(min)'
C
C.... Enter Timing Parameters
    5 WRITE(*,*) 'The units for hydrograph intervals, starting and ',
     &'ending times and the time step must be hours or minutes.'
      WRITE(*,*) 'Enter a one (1) to use hours, or a two (2) to use',
     &' minutes'
      READ(*,*) NUM
      IF ((NUM.LT.1).OR.(NUM.GT.2)) GOTO 5
      WRITE(*,*) 'Enter the starting time of the rainfall ',UNITS(NUM)
      READ(*,*) TO
      WRITE(*,*) 'Enter the ending time of the rainfall ',UNITS(NUM)
      READ(*,*) T1
      WRITE(*,*) 'Enter the time interval of the rainfall ',UNITS(NUM)
      READ(*,*) DT
      NRAIN = IFIX((T1-TO)/DT)
C
C.... Enter Rainfall Volumes and Unit Hydrograph Ordinates
      DO 10 N = 1,NRAIN
          WRITE(*,*) 'Enter the rainfall (in, cm) for time period ',N
          READ(*,*) RAIN(N)
   10 CONTINUE
      WRITE(*,*) 'Enter the number of unit hydrograph ordinates'
      WRITE(*,*) 'using the same time interval as the rainfall data.'
      READ(*,*) NUH
      DO 20 N = 1,NUH
          WRITE(*,*) 'Enter the UH ordinate (cfs, cms) for time period',N
          READ(*,*) UH(N)
   20 CONTINUE
C
C.... Compute Storm Hydrograph Ordinates
      NQ = NUH + NRAIN - 1
      DO 40 I = 1,NRAIN
        DO 30 J = 1,NUH
           N = I + J - 1
           Q(N) = Q(N) + (RAIN(I) * UH(J))
   30    CONTINUE
   40 CONTINUE
C
C.... Print Results on Screen
      ASSIGN 1002 TO UNIT
      IF ((ANS .EQ. 'M') .OR. (ANS .EQ. 'm')) ASSIGN 1003 TO UNIT
      WRITE(*,1001)
      WRITE(*,UNIT) UNITS(NUM)
      WRITE(*,1004)
      DO 50 N = 1,NQ
```

```
          TIME = TO + (N-1)*DT
          WRITE(*,1010) TIME,Q(N)
   50 CONTINUE
 1000 FORMAT(/,10X,'UNIT HYDROGRAPH CONVOLUTION PROGRAM'//)
 1001 FORMAT('            TIME          OUTFLOW')
 1002 FORMAT('          ',A5,'          (cfs)')
 1003 FORMAT('          ',A5,'          (cms)')
 1004 FORMAT('_____')
 1010 FORMAT(6X,F7.2,8X,F7.2)
      END
```

E.2

S-CURVE FROM UNIT HYDROGRAPH AND NEW UNIT HYDROGRAPH FROM S-CURVE

```
      PROGRAM SCURVE
C
C.... This program computes an S-curve from a unit hydrograph of
C.... given duration.  There is an option to determine a unit
C.... hydrograph for a different duration from the computed
C.... S-curve.  The program may be used with Problems 2.3, 2.7(b,c),
C.... 2.13, and 2.14.
C
      INTEGER NUH,NUHNEW,LAG,UNIT
      REAL RATIO,X,DT,D,TO,T1,DTNEW,DNEW,TEST
      LOGICAL OK
      CHARACTER*1 ANS
      CHARACTER*5 UNITS(2)
      REAL UH(25),S(25),UNEW(25),SS(25),T(25),TNEW(25)
      WRITE(*,*) '            S-CURVE PROGRAM'
      WRITE(*,*) ' with Option to Determine a New UH'
      WRITE(*,*)
      WRITE(*,*) '(E)nglish or (M)etric Units?'
      READ(*,'(A)') ANS
      ASSIGN 1001 TO UNIT
      IF ((ANS .EQ. 'M') .OR. (ANS .EQ. 'm')) ASSIGN 1002 TO UNIT
C
C.... Enter Timing Parameters
      UNITS(1) = '(hrs)'
      UNITS(2) = '(min)'
    5 WRITE(*,*) 'The units for hydrograph intervals, starting and ',
     &'ending times and the time step must be hours or minutes.'
      WRITE(*,*) 'Enter a one (1) to use hours, or a two (2) to use',
     &' minutes'
      READ(*,*) NUM
      IF ((NUM.LT.1).OR.(NUM.GT.2)) GOTO 5
   10 WRITE(*,*) 'Enter the rainfall duration of the UH ',UNITS(NUM)
      READ(*,*) D
```

```
          WRITE(*,*) 'Enter the starting time of the UH ',UNITS(NUM)
          READ(*,*) TO
          WRITE(*,*) 'Enter the ending time of the UH ',UNITS(NUM)
          READ(*,*) T1
          WRITE(*,*) 'Enter the time interval between ordinates ',UNITS(NUM)
          WRITE(*,*) '(Note: D must be evenly divisible by DT)'
          READ(*,*) DT
C
C.... Test Input Data
          TEST = MOD(D,DT)
          IF (TEST .NE. 0.0) THEN
              WRITE(*,*) 'D must be divisible by DT.  Try again!'
              GOTO 10
          ENDIF
C
C.... Enter Unit Hydrograph Ordinates
          NUH = IFIX((T1-TO)/DT) + 1
          DO 15 N = 1,NUH
              T(N) = TO + (N-1)*DT
   15 CONTINUE
          WRITE(*,*) 'Enter UH ordinate at prompt'
          DO 20 N = 1,NUH
              WRITE(*,UNIT) T(N),UNITS(NUM)
              READ(*,*) UH(N)
   20 CONTINUE
C
C.... Calculate S-Curve Ordinates
          LAG = IFIX(D/DT)
          DO 30 N = 1,LAG
              S(N) = UH(N)
   30 CONTINUE
          DO 35 N = LAG+1,NUH
              S(N) = S(N-LAG) + UH(N)
   35 CONTINUE
C
C.... Print Results on Screen
          ASSIGN 1006 TO UNIT
          IF ((ANS .EQ. 'M') .OR. (ANS .EQ. 'm')) ASSIGN 1007 TO UNIT
          WRITE(*,1005)
          WRITE(*,UNIT)
          WRITE(*,1008)
          DO 40 N = 1,NUH
              WRITE(*,1020) T(N),UH(N),S(N)
   40 CONTINUE
C
C.... Option to Determine UH of Different Duration
   50 WRITE(*,*)
          WRITE(*,*) 'Do you wish to find a UH of a different'
          WRITE(*,*) 'duration using this S-curve?  (Enter Y or N).'
```

```
      READ(*,'(A)') ANS
      IF ((ANS .EQ. 'N') .OR. (ANS .EQ. 'n')) THEN
         GOTO 2000
      ELSEIF ((ANS .EQ. 'Y') .OR. (ANS .EQ. 'y')) THEN
         GOTO 55
      ELSE
         GOTO 50
      ENDIF
C
C.... Enter New Timing Parameters
   55 WRITE(*,*) 'Enter duration of new UH ',UNITS(NUM)
      READ(*,*) DNEW
      WRITE(*,*) 'Enter time step of new UH ',UNITS(NUM)
      READ (*,*) DTNEW
C
C.... Calculate New Unit Hydrograph Ordinates
      NUHNEW = FLOAT((T1-TO)/DTNEW) + 1
      DO 60 N = 1,NUHNEW
         TNEW(N) = TO + (N-1)*DTNEW
   60 CONTINUE
      DO 70 N = 1,NUHNEW
         X = TNEW(N)
         SS(N) = XINTERP(NUH,S,T,X,OK)
   70 CONTINUE
      LAG = IFIX(DNEW/DTNEW)
      RATIO = D/DNEW
      DO 75 N = 1,LAG
         UNEW(N) = SS(N)*RATIO
   75 CONTINUE
      DO 80 N = LAG+1,NUHNEW
         UNEW(N) = (SS(N) - SS(N-LAG))*RATIO
   80 CONTINUE
C
C.... Print Results on Screen
      ASSIGN 1010 TO UNIT
      IF ((ANS .EQ. 'M') .OR. (ANS .EQ. 'm')) ASSIGN 1011 TO UNIT
      WRITE(*,1009)
      WRITE(*,UNIT)
      WRITE(*,1012)
      DO 90 N = 1,NUHNEW
         WRITE(*,1030) TNEW(N),UNEW(N)
   90 CONTINUE
 1001 FORMAT(1X,'At t = ',F7.1,1X,A5,'; UH ordinate (cfs) = ')
 1002 FORMAT(1X,'At t = ',F7.1,1X,A5,'; UH ordinate (cms) = ')
 1005 FORMAT(/,'         TIME        UH        S-CURVE')
 1006 FORMAT('         (hrs)      (cfs)       (cfs)')
 1007 FORMAT('         (hrs)      (cms)       (cms)')
 1008 FORMAT('_____ ')
 1009 FORMAT(/,'         TIME      UHNEW')
```

```
1010 FORMAT('          (hrs)      (cfs)')
1011 FORMAT('          (hrs)      (cms)')
1012 FORMAT('_____')
1020 FORMAT(7X,F7.1,5X,F6.2,5X,F6.2)
1030 FORMAT(7X,F7.1,5X,F6.2)
2000 END
C
C.... Interpolation Function
     FUNCTION XINTERP(NUM,A,B,X,OK)
     REAL A(100),B(100),X,FRACTION
     INTEGER NUM
     LOGICAL OK,LAST
     OK = .TRUE.
     LAST = .TRUE.
     IF (X .LT. B(1)) THEN
         OK = .FALSE.
         XINTERP = A(1)
         RETURN
     END IF
     IF (X .GT. B(NUM)) THEN
         OK = .FALSE.
         XINTERP = A(NUM)
         RETURN
     END IF
     DO 10 I=1,NUM-1
         IF (B(I+1) .GT. X) THEN
           LAST = .FALSE.
             GOTO 20
         END IF
  10 CONTINUE
  20 IF (X .EQ. B(I)) THEN
         XINTERP = A(I)
         RETURN
     END IF
     FRACTION = (X - B(I))/(B(I+1) - B(I))
     XINTERP = A(I) + (FRACTION * (A(I+1) - A(I)))
     RETURN
     END
```

E.3

MUSKINGUM ROUTING

```
     PROGRAM MUSK
C
C.... This program will route a given hydrograph by the Muskingum
C.... method (Eqn. 4.6).  The coefficient k must be entered in the
C.... same time units as the time interval of the hydrograph.  The
C.... program may be used with Problems 4.1, 4.2, and 4.14.
```

```
C
      CHARACTER*1 ANS
      CHARACTER*5 UNITS(2)
      REAL XX,DT,CK,CO,C1,C2,TEMP,T1,TO,D
      REAL T(50),INFLW(50),OUTFLW(50)
      INTEGER NI,NOUT,I,NUM,UNIT
      WRITE(*,1000)
      WRITE(*,*) '(E)nglish or (M)etric Units?'
      READ(*,'(A)') ANS
C
C.... Enter Timing Parameters
      UNITS(1) = '(hrs)'
      UNITS(2) = '(min)'
    5 WRITE(*,*) 'The units for hydrograph intervals, starting and ',
     &'ending times and the time step must be hours or minutes.'
      WRITE(*,*) 'Enter a one (1) to use hours, or a two (2) to use',
     &' minutes'
      READ(*,*) NUM
      IF ((NUM.LT.1).OR.(NUM.GT.2)) GOTO 5
      WRITE(*,*) 'Enter hydrograph starting time ',UNITS(NUM)
      READ(*,*) TO
      WRITE(*,*) 'Enter hydrograph ending time ',UNITS(NUM)
      READ(*,*) T1
      WRITE(*,*) 'Enter hydrograph time interval ',UNITS(NUM)
      READ(*,*) DT
C
C.... Enter Inflow Hydrograph and Muskingum Parameters
      WRITE(*,*) 'Enter inflow hydrograph ordinate at prompt'
      NI = IFIX((T1-TO)/DT) + 1
      DO 10 N = 1,NI
         T(N) = TO + (N-1)*DT
   10 CONTINUE
      DO 20 N = 1,NI
         WRITE(*,1005) T(N),UNITS(NUM)
         READ(*,*) INFLW(N)
   20 CONTINUE
      WRITE(*,1010)
      WRITE(*,1020) UNITS(NUM)
      READ(*,*) CK
      WRITE(*,1030)
      READ(*,*) XX
C
C.... Calculate the Muskingum Coefficients
      D = CK - (CK*XX) + (DT/2)
      CO = -((CK*XX) - (DT/2)) / D
      C1 = ((CK*XX) + (DT/2)) / D
      C2 = (CK - (CK*XX) - (DT/2)) / D
C
C.... Calculate the Outflow Hydrograph and Print Results on Screen
```

```
      WRITE(*,1040)
      READ(*,*) OUTFLW(1)
      WRITE(*,1050)
      READ(*,*) NOUT
      WRITE(*,*)
      WRITE(*,*) 'CO = ',CO
      WRITE(*,*) 'C1 = ',C1
      WRITE(*,*) 'C2 = ',C2
      WRITE(*,*)
      WRITE(*,1060)
      ASSIGN 1063 TO UNIT
      IF ((ANS .EQ. 'M') .OR. (ANS .EQ. 'm')) ASSIGN 1066 TO UNIT
      WRITE(*,UNIT) UNITS(NUM)
      WRITE(*,1070) TO,INFLW(1),OUTFLW(1)
      DO 30 I=2,NOUT
          IF (I.GT.NI) INFLW(I) = INFLW(NI)
          OUTFLW(I) = (CO*INFLW(I)) + (C1*INFLW(I-1)) + (C2*OUTFLW(I-1))
          TEMP = (I-1) * DT
          WRITE(*,1070) TEMP,INFLW(I),OUTFLW(I)
   30 CONTINUE
 1000 FORMAT(/,/,26X,'MUSKINGUM ROUTING METHOD',/)
 1005 FORMAT(1X,'Time =',F7.2,A6,'; inflow (cfs,cms) = ?')
 1010 FORMAT(1X,'The storage time constant must be in the same time',
     &' units',/,1X,'as the hydrograph time interval.')
 1020 FORMAT(/,1X,'Enter Muskingum method storage time constant',
     &' K ',A5)
 1030 FORMAT(/,1X,'Enter Muskingum method wedge storage weighting',
     &' factor x')
 1040 FORMAT(/,1X,'Enter initial outflow (cfs,cms)')
 1050 FORMAT(/,1X,'Enter number of outflow hydrograph ordinates',
     &' to print')
 1060 FORMAT(/,1X,'      TIME         INFLOW        OUTFLOW ')
 1063 FORMAT(1X,'      ',A5,'       (cfs)        (cfs)')
 1066 FORMAT(1X,'      ',A5,'       (cms)        (cms)')
 1070 FORMAT(1X,F10.2,5X,F10.2,5X,F10.2)
      END
```

E.4
STORAGE INDICATION ROUTING

```
      PROGRAM SIROUTE
C
C.... This program will route an inflow hydrograph through a
C.... reservoir using the Storage-Indication method (Eqn. 4.13).
C.... Values for the Q versus (2S/dt +Q) table must be entered.
C.... The more accurately that these values are input, the more
C.... accurate the results will be.  The program may be used
C.... with Problems 4.4, 4.6, 4.13 and 4.16.
```

```
C
      REAL SO,QO,RHS,TEMP,X,CONV,DT,TO,T1
      INTEGER NT,NSI,UNIT
      CHARACTER*1 ANS
      CHARACTER*5 UNITS(2)
      REAL INFLW(25),OUTFLW(25),Q(25),SI(25),T(25)
      LOGICAL OK
      WRITE(*,*) '        STORAGE-INDICATION ROUTING PROGRAM'
C
C.... Enter Timing Parameters
      WRITE(*,*)
      WRITE(*,*) '(E)nglish or (M)etric'
      READ(*,'(A1)') ANS
      CONV = 726.0
      IF ((ANS .EQ. 'M') .OR. (ANS .EQ. 'm')) CONV = 1.0/60.0
      UNITS(1) = '(hrs)'
      UNITS(2) = '(min)'
    5 WRITE(*,*) 'The units for hydrograph intervals, starting and ',
     &'ending times and the time step must be hours or minutes.'
      WRITE(*,*) 'Enter a one (1) to use hours, or a two (2) to use',
     &' minutes'
      READ(*,*) NUM
      IF ((NUM.LT.1).OR.(NUM.GT.2)) GOTO 5
      IF (NUM .EQ. 1) CONV = CONV/60.
      WRITE(*,*) 'Enter starting time ',UNITS(NUM)
      READ(*,*) TO
      WRITE(*,*) 'Enter ending time ',UNITS(NUM)
      READ(*,*) T1
      WRITE(*,*) 'Enter time step ',UNITS(NUM)
      READ(*,*) DT
      NT = IFIX((T1-TO)/DT) + 1
      DO 10 N = 1,NT
         T(N) = TO + (N-1)*DT
   10 CONTINUE
C
C.... Enter Inflow Hydrograph Ordinates and S-I Curve Ordinates
      DO 20 N=1,NT
         WRITE(*,1005) T(N),UNITS(NUM)
         READ(*,*) INFLW(N)
   20 CONTINUE
      WRITE(*,*) 'Enter number of points on SI curve'
      READ(*,*) NSI
      WRITE(*,*) 'Enter Q (cfs,cms) and corresponding value'
      WRITE(*,*) 'of 2S/dt + Q (cfs,cms) when prompted.'
      DO 30 N = 1,NSI
         WRITE(*,*) 'Values for point',N
         READ(*,*) Q(N),SI(N)
   30 CONTINUE
C
C.... Enter Initial Values of Storage and Outflow
```

```
      WRITE(*,*) 'Enter initial storage (Ac-ft,cu.m.)'
      READ(*,*) SO
      WRITE(*,*) 'Enter initial outflow (cfs,cms)'
      READ(*,*) QO
C
C.... Calculate Outflow
      TEMP = ((2*SO)/DT)*CONV - QO
      OUTFLW(1) = QO
      DO 40 N = 1,NT-1
          RHS = (INFLW(N) + INFLW(N+1)) + (TEMP)
          OUTFLW(N+1) = XINTERP(NSI,Q,SI,RHS,OK)
          TEMP = RHS - 2*OUTFLW(N+1)
   40 CONTINUE
C
C.... Print Results on Screen
      ASSIGN 1001 TO UNIT
      IF ((ANS .EQ. 'M') .OR. (ANS .EQ. 'm')) ASSIGN 1002 TO UNIT
      WRITE(*,1000)
      WRITE(*,UNIT) UNITS(NUM)
      WRITE(*,1003)
      DO 50 N = 1,NT
          WRITE(*,1010) T(N),INFLW(N),OUTFLW(N)
   50 CONTINUE
 1000 FORMAT(/,'          TIME        INFLOW      OUTFLOW          ')
 1001 FORMAT('          ',A5,'      (cfs)      (cfs)')
 1002 FORMAT('          ',A5,'      (cms)      (cms)')
 1003 FORMAT('_____')
 1005 FORMAT(1X,'Enter inflow (cfs, cms) at time t = ',F7.1,1X,A5)
 1010 FORMAT(7X,F7.2,5X,F6.1,5X,F7.1)
      END
C
C.... Interpolation Function
      FUNCTION XINTERP(NUM,A,B,X,OK)
      REAL A(100),B(100),X,FRACTION
      INTEGER NUM
      LOGICAL OK,LAST
      OK = .TRUE.
      LAST = .TRUE.
      IF (X .LT. B(1)) THEN
          OK = .FALSE.
          XINTERP = A(1)
          RETURN
      END IF
      IF (X .GT. B(NUM)) THEN
          OK = .FALSE.
          XINTERP = A(NUM)
          RETURN
      END IF
      DO 10 I=1,NUM-1
          IF (B(I+1) .GT. X) THEN
```

E.5
FOURTH-ORDER RUNGE-KUTTA ROUTING

```
      PROGRAM RK4
C
C.... This program uses the fourth order Runge-Kutta Methods
C.... (Eqns. 4.21 and 4.22) to route a given hydrograph.  The
C.... values for a table of depth versus outflow and for a table
C.... of depth versus area must be entered.  The more accurately
C.... each of these tables is entered, the more accurate the
C.... results.  The program may be used for Problem 4.7 or as
C.... a comparison to other methods of routing.
C
      REAL HEAD,TIME,INFLW,OUTFLW,AREA,HO,DH
      REAL TINO,TIN1,DT1,TOUTO,TOUT1,DT,XINTERP
      INTEGER NUM,NUM1,NUM2,NUM3
      CHARACTER*1 ANS
      CHARACTER*4 UHEAD
      CHARACTER*5 UNITS(2),UFLOW
      CHARACTER*7 UAREA
      REAL QIN(100),TIN(100),QOUT(100),HQOUT(100),
     &AR(100),HAR(100)
      REAL K1,K2,K3,K4,TTEMP,HTEMP,CONV
      LOGICAL OK
      WRITE(*,*) '    FOURTH ORDER RUNGE-KUTTA ROUTING METHODS'
      WRITE(*,*)
      WRITE(*,*) '(E)nglish or (M)etric Units?'
      READ(*,'(A)') ANS
      UFLOW = '(cfs)'
      UHEAD = '(ft)'
      UAREA = '(Acres)'
      IF ((ANS .EQ. 'M') .OR. (ANS .EQ. 'm')) THEN
          UFLOW = '(cms)'
          UHEAD = '(m)'
          UAREA = '(sq m)'
      ENDIF
C
C.... Enter Timing Parameters and Inflow Hydrograph Ordinates
      UNITS(1) = '(hrs)'
      UNITS(2) = '(min)'
    5 WRITE(*,*) 'The units for hydrograph intervals, starting and ',
     &'ending times and the time step must be hours or minutes.'
      WRITE(*,*) 'Enter a one (1) to use hours, or a two (2) to use',
     &' minutes'
      READ(*,*) NUM
      IF ((NUM.LT.1).OR.(NUM.GT.2)) GOTO 5
      CONV = 60.0
      IF (NUM.EQ.1) CONV = 3600.0
```

```
      IF ((ANS .EQ. 'E') .OR. (ANS .EQ. 'e')) CONV = 43560.0/CONV
      WRITE(*,*) 'Enter inflow hydrograph starting time ',UNITS(NUM)
      READ(*,*) TINO
      WRITE(*,*) 'Enter inflow hydrograph ending time ',UNITS(NUM)
      READ(*,*) TIN1
      WRITE(*,*) 'Enter inflow hydrograph time interval',UNITS(NUM)
      READ(*,*) DT1
      NUM1 = IFIX((TIN1-TINO)/DT1) + 1
      DO 10 I=1,NUM1
          TIN(I) = TINO + (I-1)*DT1
          WRITE(*,1015) UFLOW,TIN(I),UNITS(NUM)
          READ(*,*) QIN(I)
   10 CONTINUE
      WRITE(*,*)
C
C.... Enter the Values for the Head vs. Outflow Table
      WRITE(*,*) 'Enter number of entries in the head-outflow',
     &' table'
      READ(*,*) NUM2
      DO 20 I=1,NUM2
          WRITE(*,1016) UHEAD,I
          READ(*,*) HQOUT(I)
          WRITE(*,1017) UFLOW,HQOUT(I),UHEAD
          READ(*,*) QOUT(I)
   20 CONTINUE
      WRITE(*,*)
C
C.... Enter the Values for the Head vs. Area Table
      WRITE(*,*) 'Enter number of entries in the head-area',
     &' table'
      READ(*,*) NUM3
      DO 30 I=1,NUM3
          WRITE(*,1018) UHEAD,I
          READ(*,*) HAR(I)
          WRITE(*,1019) UAREA,HAR(I),UHEAD
          READ(*,*) AR(I)
   30 CONTINUE
C
C.... Enter Initial Values
      WRITE(*,*) 'Enter starting time of outflow hydrograph',UNITS(NUM)
      READ(*,*) TOUTO
      WRITE(*,*) 'Enter ending time of outflow hydrograph ',UNITS(NUM)
      READ(*,*) TOUT1
      WRITE(*,*) 'Enter time step ',UNITS(NUM)
      READ(*,*) DT
      WRITE(*,*) 'Enter initial head (ft,m)'
      READ(*,*) HO
C
C.... Calculate the Outflow and Print Results on Screen
```

```
C.... Classical Method
C
      TIME = TOUTO
      HEAD = HO
      DH = 0.0
      WRITE(*,1000)
      WRITE(*,1001) UNITS(NUM),UFLOW,UHEAD,UAREA,UHEAD,UFLOW
      WRITE(*,1003)
   40 INFLW = XINTERP(NUM1,QIN,TIN,TIME,OK)
      OUTFLW = XINTERP(NUM2,QOUT,HQOUT,HEAD,OK)
      AREA = XINTERP(NUM3,AR,HAR,HEAD,OK)
      WRITE(*,1010) TIME,INFLW,HEAD,AREA,DH,OUTFLW
      IF (TIME .GE. TOUT1) GOTO 50
      K1 = (INFLW-OUTFLW)/(AREA*CONV)
      TTEMP = TIME + (0.5 * DT)
      HTEMP = HEAD + (0.5 * K1 * DT)
      INFLW = XINTERP(NUM1,QIN,TIN,TTEMP,OK)
      OUTFLW = XINTERP(NUM2,QOUT,HQOUT,HTEMP,OK)
      AREA = XINTERP(NUM3,AR,HAR,HTEMP,OK)
      K2 = (INFLW-OUTFLW)/(AREA*CONV)
      TTEMP = TIME + (0.5 * DT)
      HTEMP = HEAD + (0.5 * K2 * DT)
      INFLW = XINTERP(NUM1,QIN,TIN,TTEMP,OK)
      OUTFLW = XINTERP(NUM2,QOUT,HQOUT,HTEMP,OK)
      AREA = XINTERP(NUM3,AR,HAR,HTEMP,OK)
      K3 = (INFLW-OUTFLW)/(AREA*CONV)
      TTEMP = TIME + DT
      HTEMP = HEAD + (K3 * DT)
      INFLW = XINTERP(NUM1,QIN,TIN,TTEMP,OK)
      OUTFLW = XINTERP(NUM2,QOUT,HQOUT,HTEMP,OK)
      AREA = XINTERP(NUM3,AR,HAR,HTEMP,OK)
      K4 = (INFLW-OUTFLW)/(AREA*CONV)
      DH = ((K1 + (2.0*K2) + (2.0*K3) + K4)/6.0) * DT
      HEAD = HEAD + DH
      TIME = TIME + DT
      GOTO 40
   50 CONTINUE
 1000 FORMAT(/,'      TIME      INFLOW      HEAD      AREA      dH
     &'OUTFLOW')
 1001 FORMAT(5X,A5,6X,A5,7X,A4,3X,A7,3X,A4,5X,A5)
 1003 FORMAT(1X,'_____
     &'_____')
 1010 FORMAT(5X,F5.1,5X,F7.2,5X,F5.2,4X,F5.1,2X,F5.2,3X,F8.2)
 1015 FORMAT(1X,'Enter the inflow ',A5,' for time = 'F7.2,1X,A5)
 1016 FORMAT(1X,'Enter the head ',A4,' for point number ',I2)
 1017 FORMAT(1X,'Enter the outflow ',A5,' at head = 'F7.2,1X,A4)
 1018 FORMAT(1X,'Enter the head ',A4,' for point number ',I2)
 1019 FORMAT(1X,'Enter the area ',A7,' at head = 'F7.2,1X,A4)
      END
```

```
C
C.... Interpolation Function
      FUNCTION XINTERP(NUM,A,B,X,OK)
      REAL A(100),B(100),X,FRACTION
      INTEGER NUM
      LOGICAL OK,LAST
      OK = .TRUE.
      LAST = .TRUE.
      IF (X .LT. B(1)) THEN
          OK = .FALSE.
          XINTERP = A(1)
          RETURN
      END IF
      IF (X .GT. B(NUM)) THEN
          OK = .FALSE.
          XINTERP = A(NUM)
          RETURN
      END IF
      DO 10 I=1,NUM-1
          IF (B(I+1) .GT. X) THEN
              LAST = .FALSE.
              GOTO 20
          END IF
   10 CONTINUE
   20 IF (X .EQ. B(I)) THEN
          XINTERP = A(I)
          RETURN
      END IF
      FRACTION = (X - B(I))/(B(I+1) - B(I))
      XINTERP = A(I) + (FRACTION * (A(I+1) - A(I)))
      RETURN
      END
```

E.6

KINEMATIC WAVE ROUTING

```
      PROGRAM KINWAV
C
C.... This program will perform kinematic wave computations for
C.... overland segments and channel segments.  Physical character-
C.... istics and rainfall data are entered as prompted for.  The
C.... program is used with Problems 4.8 and 4.9.
C
      REAL LO,SO,NO,FO,DXO,LC,DXC,SC,NC,WC,DT,T,RAIN(100)
      REAL AO(100,100),QO(100,100),AC(100,100),QC(100,100),TOTAL,CBAR
      REAL YBAR,ALPHA,M,UNITS
      INTEGER NTIM,NXO,NXC,METHOD,NEW,RUN,WIDE,WIDE2
      CHARACTER*1 ANS
      CHARACTER*5 FLOW
      LOGICAL OUT1
      WRITE(*,*) '      KINEMATIC WAVE PROGRAM'
      WRITE(*,*)
C
C.... Overland only, or Overland and Channel?
    5 WRITE(*,*) '(O)verland flow only, or Overland and (C)hannel flow?'
      READ(*,'(A1)') ANS
      RUN = O
      IF ((ANS.EQ.'O').OR.(ANS.EQ.'o')) RUN = 1
      IF ((ANS.EQ.'C').OR.(ANS.EQ.'c')) RUN = 2
      IF (RUN.EQ.O) GOTO 5
C
C.... Which units for the run?
      WRITE(*,*) '(E)nglish or (M)etric Units (Enter ''E'' or ''M'')'
      READ(*,'(A1)') ANS
      UNITS = 1.486
      FLOW = '(cfs)'
      IF ((ANS.EQ.'M').OR.(ANS.EQ.'m')) THEN
         UNITS = 1.0
         FLOW = '(cms)'
      ENDIF
C
C.... Input Overland Flow Parameters.
      WRITE(*,*) 'Enter overland flow path length (ft or m)'
      READ(*,*) LO
      WRITE(*,*) 'Enter overland flow distance increment (ft or m)'
      READ(*,*) DXO
      WRITE(*,*) 'Enter overland flow slope'
      READ(*,*) SO
      WRITE(*,*) 'Enter overland flow roughness'
      READ(*,*) NO
      WRITE(*,*) 'Enter overland flow infiltration rate'
      READ(*,*) FO
```

```
C
C.... Input Channel Flow Parameters.
      IF (RUN.EQ.1) GOTO 10
      WRITE(*,*) 'Enter channel length (ft or m)'
      READ(*,*) LC
      WRITE(*,*) 'Enter channel distance increment (ft or m)'
      READ(*,*) DXC
      WRITE(*,*) 'Enter channel slope'
      READ(*,*) SC
      WRITE(*,*) 'Enter channel roughness'
      READ(*,*) NC
      WRITE(*,*) 'Enter channel width (ft or m)'
      READ(*,*) WC
C
C.... Input Timing Parameters
   10 WRITE(*,*) 'Enter total time for simulation (seconds)'
      READ(*,*) T
      WRITE(*,*) 'Enter time increment (seconds)'
      READ(*,*) DT
      NTIM = INT(T/DT)
C
C.... Input Rainfall.
      IF (UNITS.EQ.1.0) THEN
          WRITE(*,*) 'Rainfall units are cm/hr'
        ELSE
          WRITE(*,*) 'Rainfall units are in/hr'
      END IF
      DO 20 I=2,NTIM+1
          WRITE(*,*) 'Enter rainfall for period ',i-1
          READ(*,*) RAIN(I)
          IF (UNITS.EQ.1.0) THEN
              RAIN(I) = RAIN(I)/360000.0
          ELSE
              RAIN(I) = RAIN(I)/43200.0
          END IF
   20 CONTINUE
C
C.... Compute Overland Flow.
      WRITE(*,*) 'Do you want detailed overland flow output (y/n)'
      READ(*,'(A1)') ANS
      OUT1 = .TRUE.
      IF ((ANS.EQ.'N').OR.(ANS.EQ.'n')) OUT1 = .FALSE.
C
C.... Compute alpha and set m = 5/3
      ALPHA = UNITS * SQRT(SO) / NO
      M = 5.0/3.0
C
C.... Assume standard form for first time step
      METHOD = 1
      NXO = INT(LO/DXO)
```

```
C
C.... Write initial stuff
      IF (OUT1) THEN
          WRITE(*,*) 'ALPHA = ',ALPHA
          WRITE(*,*) 'DISTANCE    TIME      METHOD        AREA          FLOW'
      END IF
C
C.... Begin time loop
      DO 90 J=2,NTIM+1
C
C.... Return here if wrong method used
   30 CONTINUE
C
C.... B.C. at x = 0 is a,q = 0.0 for every time step
          AO(1,J) = 0.0
          QO(1,J) = 0.0
C
C.... Begin distance loop
          DO 60 I=2,NXO+1
              GOTO (40,45) METHOD
C
C.... Standard form
   40         AO(I,J) = (RAIN(J)-FO)*DT + AO(I,J-1) - ((ALPHA*M*DT/DXO)*
     & (((AO(I,J-1)+AO(I-1,J-1))/2.0)**(M-1.0))*(AO(I,J-1)-AO(I-1,J-1)))
              QO(I,J) = ALPHA*(AO(I,J)**M)
              GOTO 50
C
C.... Conservation form
   45         QO(I,J) = QO(I-1,J) + (RAIN(J)-FO)*DXO - (DXO/DT)*
     & (AO(I-1,J)-AO(I-1,J-1))
              AO(I,J) = (QO(I,J)/ALPHA)**(1.0/M)
C
C.... Write some output
   50         IF (OUT1) THEN
                  WRITE(*,1000) (I-1)*DXO,(J-1)*DT,METHOD,AO(I,J),QO(I,J)
              END IF
C
C.... End of distance loop
   60     CONTINUE
          IF (OUT1) WRITE(*,*)
C
C.... Compute celerity, and set form of equations to use
          TOTAL = 0.0
          DO 70 K=2,NXO+1
              TOTAL=TOTAL+AO(K,J)
   70     CONTINUE
          CBAR = ALPHA*M*((TOTAL/NXO)**(M-1.0))
          IF (CBAR.GT.DXO/DT) THEN
              NEW = 2
```

```
              ELSE
                  NEW = 1
              END IF
C
C.... If the wrong method was used, re-do the time step.
              IF (METHOD.NE.NEW) THEN
                  METHOD = NEW
              IF (OUT1) THEN
                  WRITE(*,*) 'Used wrong method ...'
                  WRITE(*,*) 'C = ',CBAR
                  WRITE(*,*) 'dx/dt = ',DXO/DT
                  WRITE(*,*) 'Re-do time step ...'
                  WRITE(*,*)
          END IF
          GOTO 30
          END IF
C
C.... End of time loop
   90 CONTINUE
C
C.... Compute channel flow.
      IF (RUN.EQ.1) GOTO 160
      WRITE(*,*) 'Do you want detailed channel flow output (y/n)'
      READ(*,'(A1)') ANS
      OUT1 = .TRUE.
      IF ((ANS.EQ.'N').OR.(ANS.EQ.'n')) OUT1 = .FALSE.
C
C.... Compute alpha and set m = 5/3 as for a wide shallow rectangle
      ALPHA = (UNITS * SQRT(SC) / NC) * (WC**(-2.0/3.0))
      M = 5.0/3.0
      WIDE = 1
C
C.... Assume standard form for first time step
      METHOD = 1
      NXC = INT(LC/DXC)
C
C.... Write initial stuff
      IF (OUT1) THEN
          WRITE(*,*) 'DISTANCE    TIME    METHOD      AREA         FLOW'
      END IF
C
C.... Begin time loop
      DO 150 J=2,NTIM+1
C
C.... Return here if wrong method used
  100 CONTINUE
C
C.... B.C. is 0.0
          QC(1,J) = 0.0
```

```
                AC(1,J) = 0.0
                QC(1,J-1) = 0.0
                AC(1,J-1) = 0.0
C
C.... Begin distance loop
          DO 130 I=2,NXC+1
                GOTO (105,110) METHOD
C
C.... Standard form
   105          AC(I,J) = QO(NXO+1,J)*DT + AC(I,J-1) - ((ALPHA*M*DT/DXC)*
      & (((AC(I,J-1)+AC(I-1,J-1))/2.0)**(M-1.0))*(AC(I,J-1)-AC(I-1,J-1)))
                QC(I,J) = ALPHA*(AC(I,J)**M)
                GOTO 120
C
C.... Conservation form
   110          QC(I,J) = QC(I-1,J) + QO(NXO+1,J)*DXC -
      &(DXC/DT)*(AC(I-1,J)-AC(I-1,J-1))
                AC(I,J) = (QC(I,J)/ALPHA)**(1.0/M)
C
C.... Write some output
   120    IF (OUT1) THEN
                WRITE(*,1000) (I-1)*DXC,(J-1)*DT,METHOD,AC(I,J),QC(I,J)
          END IF
C
C.... End of distance loop
   130    CONTINUE
          IF (OUT1) WRITE(*,*)
C
C.... Compute celerity, and set form of equations to use
          TOTAL = 0.0
          DO 140 K=2,NXC+1
             TOTAL=TOTAL+AC(K,J)
   140    CONTINUE
          CBAR = ALPHA*M*((TOTAL/NXC)**(M-1.0))
          IF (CBAR.GT.DXC/DT) THEN
             NEW = 2
          ELSE
             NEW = 1
          END IF
C
C.... If the wrong method was used, re-do the time step.
          IF (METHOD.NE.NEW) THEN
             METHOD = NEW
          IF (OUT1) THEN
             WRITE(*,*) 'Used wrong method ...'
             WRITE(*,*) 'C = ',CBAR
             WRITE(*,*) 'dx/dt = ',DXC/DT
             WRITE(*,*) 'Re-do time step ...'
             WRITE(*,*)
```

```
          END IF
          GOTO 100
          END IF
C
C.... Check width to depth ratio
          YBAR = (TOTAL/NXC)/WC
          IF (YBAR.EQ.0.0) YBAR = 0.000001
          IF (WC/YBAR.GT.1.5) THEN
              WIDE2 = 1
              ALPHA = (UNITS * SQRT(SC) / NC) * (WC**(-2.0/3.0))
              M = 5.0/3.0
          ELSE
              WIDE2 = 0
              ALPHA = (UNITS/2.08) * SQRT(SC) / NC
              M = 4.0/3.0
          END IF
C
C.... If the wrong assumption was used, re-do the time step.
          IF (WIDE.NE.WIDE2) THEN
              WIDE = WIDE2
          IF (OUT1) THEN
              WRITE(*,*) 'Used wrong alpha and m ...'
              WRITE(*,*) 'Width/Depth = ',WC/YBAR
              WRITE(*,*) 'New alpha = ',ALPHA
              WRITE(*,*) 'New m      = ',M
              WRITE(*,*) 'Re-do time step ...'
              WRITE(*,*)
          END IF
          GOTO 100
          END IF
C
C.... End of time loop
  150 CONTINUE
C
C.... Summary hydrographs
  160 WRITE(*,*)
      WRITE(*,*) 'SUMMARY OUTFLOW HYDROGRAPHS'
      WRITE(*,*)
      WRITE(*,1010)
      WRITE(*,1012) FLOW,FLOW
      WRITE(*,1020) 0.0,QO(NXO+1,1),QC(NXC+1,1)
      DO 170 I=2,NTIM+1
          WRITE(*,1020) (I-1)*DT,QO(NXO+1,I),QC(NXC+1,I)
  170 CONTINUE
 1000 FORMAT(1X,F8.2,F8.2,6X,I1,2X,F12.6,2X,F12.6)
 1010 FORMAT('          TIME      OVERLAND FLOW      CHANNEL FLOW')
 1012 FORMAT('          (sec)',9X,A5,13X,A5)
 1020 FORMAT(7X,F7.2,8X,G10.3,10X,F7.4)
      END
```

E.7

UNIFORM FLOW COMPUTATIONS

```
      PROGRAM UNIFORM
C
C.... This program assumes normal depth of flow in a channel.
C.... Rectangular, triangular, trapezoidal or circular channels
C.... can be simulated using Manning's equation.  The program
C.... will solve for either depth, flow rate, slope, or roughness,
C.... given all of the other values.  This program can be used
C.... with Problems 7.1, 7.2, 7.6, 7.7, 7.18 and 7.21.
C
      REAL C,G,Q,D,A,T,Z,RR,W,AL,TH,N,DIA,V,ALPHA
      REAL D1,D2,A1,A2,P1,P2,R1,R2,Q1,Q2,QNEW
      INTEGER ISECT,I
      CHARACTER ANS*3
      CHARACTER*4 DEPTH,WIDTH
      CHARACTER*5 FLOW
      CHARACTER*7 AREA,VELO
      WRITE(*,*)
      WRITE(*,*) '       UNIFORM FLOW PROGRAM'
      WRITE(*,*)
   10 WRITE(*,*) '(E)nglish or (M)etric units ?'
      READ(*,2000) ANS
      IF (ANS.EQ.' ') STOP
      C = 1.49
      G = 32.16
      FLOW = '(cfs)'
      DEPTH = '(ft)'
      AREA = '(sq ft)'
      WIDTH = '(ft)'
      VELO = '(ft/s)'
      IF ((ANS.EQ.'M').OR.(ANS.EQ.'m')) THEN
         C = 1.0
         G = 9.81
         FLOW = '(cms)'
         DEPTH = '(m)'
         AREA = '(sq m)'
         WIDTH = '(m)'
         VELO = '(m/s)'
      END IF
      PI = 3.14159
C
C.... Determine Channel Shape and Enter Parameters
   20 WRITE(*,*) '(R)ectangular, (T)riangular, (TR)apezoidal,',
     &' or (C)ircular ?'
      READ(*,2000) ANS
      IF ((ANS.EQ.'R').OR.(ANS.EQ.'r')) THEN
         WRITE(*,*) 'Channel width ?'
```

```
                  READ(*,*) W
                  ISECT = 1
                  GOTO 30
              END IF
          IF ((ANS.EQ.'T').OR.(ANS.EQ.'t')) THEN
                  WRITE(*,*) 'Side slope (horizontal/vertical) ?'
                  READ(*,*) Z
                  ISECT = 2
                  GOTO 30
              END IF
          IF ((ANS.EQ.'TR').OR.(ANS.EQ.'tr')) THEN
                  WRITE(*,*) 'Bottom width ?'
                  READ(*,*) W
                  WRITE(*,*) 'Side slope (horizontal/vertical) ?'
                  READ(*,*) Z
                  ISECT = 3
                  GOTO 30
              END IF
          IF ((ANS.EQ.'C') .OR. (ANS.EQ.'c')) THEN
                  WRITE(*,*) 'Diameter ?'
                  READ(*,*) DIA
                  ISECT = 4
                  GOTO 30
              END IF
          GOTO 10
C
C.... Uniform Flow Computations
   30 WRITE(*,*) 'Solve for (D)epth, (F)low rate, (S)lope, or ',
      &'(R)oughness coefficient ? (space to end)'
       READ(*,2000) ANS
       IF (ANS.EQ.' ') GOTO 20
       IF ((ANS.NE.'D').AND.(ANS.NE.'d')) GOTO 90
C
C.... Depth
   35 WRITE(*,*) 'Flow rate ?'
       READ(*,*) Q
       WRITE(*,*) 'Slope ?'
       READ(*,*) S
       WRITE(*,*) 'Mannings coefficient ?'
       READ(*,*) N
       GOTO (40,50,60,70) ISECT
   40 D = 0.0
       A = (Q/C * N/W/SQRT(S))**0.6
   45 T = A*((W+2.0*D)/W)**0.4
       IF (ABS(T-D)/T.GT.0.001) THEN
              D = T
              GOTO 45
          END IF
       D = T
```

```
      A = W*D
      V = Q/A
      T = W
      GOTO 80
 50 D = (Q/SQRT(S))**(3.0/8.0)*(N/(C*Z))**(3.0/8.0)/
    & (Z/(2.0*SQRT(Z**2+1.0)))**0.25
      A = Z*D**2
      V = Q/A
      T = 2.0*Z*D
      GOTO 80
 60 D = 0.0
      A = (Q/C*N/SQRT(S))**0.6
 65 T = A*(W+2.0*D*SQRT(Z**2+1.0))**0.4/(W+Z*D)
      IF (ABS(T-D)/T.GT.0.001) THEN
          D = T
          GOTO 65
      END IF
      D = T
      A = D*(W+Z*D)
      V = Q/A
      T = W+2.0*Z*D
      GOTO 80
 70 A1 = PI*(DIA/2.0)**2.0
      A2 = 0.5*A1
      P1 = PI*DIA
      P2 = 0.5*P1
      R1 = A1/P1
      R2 = A2/P2
      Q1 = C/N*A1*R1**(2.0/3.0)*SQRT(S)
      Q2 = C/N*A2*R2**(2.0/3.0)*SQRT(S)
      IF (Q .GT. Q1) THEN
          WRITE(*,1070)
          GOTO 35
      END IF
      IF (Q .LE. Q2) THEN
          D1 = 0.5*DIA
          D2 = 0.01*DIA
      ELSE
          D1 = DIA
          D2 = 0.5*DIA
      END IF
      D = D2
      DO 75 I = 1,200
          ALPHA = 2.0*ACOS(1.0 - (2.0*D/DIA))
          A = ((ALPHA - SIN(ALPHA))*DIA**2.0)/8.0
          P = ALPHA*DIA/2.0
          R = A/P
          QNEW = C/N*A*R**(2.0/3.0)*SQRT(S)
          IF (ABS(Q - QNEW) .LE. 0.01) GOTO 75
```

```
            IF (QNEW .LT. Q) THEN
                D2 = D
            ELSE
                D1 = D
            END IF
            D = (D1 + D2)/2.0
   75 CONTINUE
        V = Q/A
        T = DIA*SIN(ALPHA/2.0)
   80 WRITE(*,1000) D,DEPTH
        WRITE(*,1020) T,WIDTH
        WRITE(*,1040) S
        WRITE(*,1050) N
        WRITE(*,1010) A,AREA
        WRITE(*,1060) V,VELO
        WRITE(*,1030) Q,FLOW
        GOTO 30
   90 IF ((ANS.NE.'F').AND.(ANS.NE.'f')) GOTO 150
C
C.... Flow Rate
   95 WRITE(*,*) 'Depth ?'
        READ(*,*) D
        WRITE(*,*) 'Slope ?'
        READ(*,*) S
        WRITE(*,*) 'Mannings coefficient ?'
        READ(*,*) N
        GOTO (100,110,120,130) ISECT
  100 A = W*D
        R = A/(W+2.0*D)
        T = W
        Q = C/N*A*R**(2.0/3.0)*SQRT(S)
        V = Q/A
        GOTO 140
  110 A = Z*D**2
        R = A/(2.0*D*SQRT(Z**2+1.0))
        T = 2.0*Z*D
        Q = C/N*A*R**(2.0/3.0)*SQRT(S)
        V = Q/A
        GOTO 140
  120 A = D*(W+Z*D)
        R = A/(W+2.0*D*SQRT(Z**2+1.0))
        T = W+2.0*Z*D
        Q = C/N*A*R**(2.0/3.0)*SQRT(S)
        V = Q/A
        GOTO 140
  130 IF (D .GT. DIA) THEN
            WRITE(*,1070)
            GOTO 95
        END IF
```

```
      ALPHA = 2.0*ACOS(1.0 - (2.0*D/DIA))
      A = ((ALPHA - SIN(ALPHA))*DIA**2.0)/8.0
      P = ALPHA*DIA/2.0
      R = A/P
      T = DIA*SIN(ALPHA/2.0)
      Q = C/N*A*R**(2.0/3.0)*SQRT(S)
      V = Q/A
  140 WRITE(*,1000) D,DEPTH
      WRITE(*,1020) T,WIDTH
      WRITE(*,1040) S
      WRITE(*,1050) N
      WRITE(*,1010) A,AREA
      WRITE(*,1060) V,VELO
      WRITE(*,1030) Q,FLOW
      GOTO 30
  150 IF ((ANS.NE.'S').AND.(ANS.NE.'s')) GOTO 210
C
C.... Slope
  155 WRITE(*,*) 'Depth ?'
      READ(*,*) D
      WRITE(*,*) 'Flow rate ?'
      READ(*,*) Q
      WRITE(*,*) 'Mannings coefficient ?'
      READ(*,*) N
      GOTO (160,170,180,190) ISECT
  160 A = W*D
      V = Q/A
      R = A/(W+2.0*D)
      T = W
      S = (Q/C*N/A/R**(2.0/3.0))**2
      GOTO 200
  170 A = Z*D**2
      V = Q/A
      R = A/(2.0*D*SQRT(Z**2+1.0))
      T = 2.0*Z*D
      S = (Q/C*N/A/R**(2.0/3.0))**2
      GOTO 200
  180 A = D*(W+Z*D)
      V = Q/A
      R = A/(W+2.0*D*SQRT(Z**2+1.0))
      T = W+2.0*Z*D
      S = (Q/C*N/A/R**(2.0/3.0))**2
      GOTO 200
  190 IF (D .GT. DIA) THEN
          WRITE(*,1070)
          GOTO 155
      END IF
      ALPHA = 2.0*ACOS(1.0 - (2.0*D/DIA))
      A = ((ALPHA - SIN(ALPHA))*DIA**2.0)/8.0
```

```
      V = Q/A
      P = ALPHA*DIA/2.0
      R = A/P
      T = DIA*SIN(ALPHA/2.0)
      S = (Q/C*N/A/R**(2.0/3.0))**2
  200 WRITE(*,1000) D,DEPTH
      WRITE(*,1020) T,WIDTH
      WRITE(*,1040) S
      WRITE(*,1050) N
      WRITE(*,1010) A,AREA
      WRITE(*,1060) V,VELO
      WRITE(*,1030) Q,FLOW
      GOTO 30
  210 IF ((ANS.NE.'R').AND.(ANS.NE.'r')) GOTO 30
C
C.... Roughness Coefficient
  215 WRITE(*,*) 'Depth ?'
      READ(*,*) D
      WRITE(*,*) 'Flow rate ?'
      READ(*,*) Q
      WRITE(*,*) 'Slope ?'
      READ(*,*) S
      GOTO (220,230,240,250) ISECT
  220 A = W*D
      V = Q/A
      R = A/(W+2.0*D)
      T = W
      N = C/Q*A*R**(2.0/3.0)*SQRT(S)
      GOTO 260
  230 A = Z*D**2
      V = Q/A
      R = A/(2.0*D*SQRT(Z**2+1.0))
      T = 2.0*Z*D
      N = C/Q*A*R**(2.0/3.0)*SQRT(S)
      GOTO 260
  240 A = D*(W+Z*D)
      V = Q/A
      R = A/(W+2.0*D*SQRT(Z**2+1.0))
      T = W+2.0*Z*D
      N = C/Q*A*R**(2.0/3.0)*SQRT(S)
      GOTO 260
  250 IF (D .GT. DIA) THEN
          WRITE(*,1070)
          GOTO 215
      END IF
      ALPHA = 2.0*ACOS(1.0 - (2.0*D/DIA))
      A = ((ALPHA - SIN(ALPHA))*DIA**2.0)/8.0
      P = ALPHA*DIA/2.0
      R = A/P
```

```
      V = Q/A
      T = DIA*SIN(ALPHA/2.0)
      N = C/Q*A*R**(2.0/3.0)*SQRT(S)
  260 WRITE(*,1000) D,DEPTH
      WRITE(*,1020) T,WIDTH
      WRITE(*,1040) S
      WRITE(*,1050) N
      WRITE(*,1010) A,AREA
      WRITE(*,1060) V,VELO
      WRITE(*,1030) Q,FLOW
      GOTO 30
 1000 FORMAT(1X'Depth        = ',F7.2,1X,A4)
 1010 FORMAT(1X'Area         = ',F7.2,1X,A7)
 1020 FORMAT(1X'Top Width    = ',F7.2,1X,A4)
 1030 FORMAT(1X'Flow Rate    = ',F7.2,1X,A5)
 1040 FORMAT(1X'Slope        = ',F7.5,' (unitless)')
 1050 FORMAT(1X'Mannings n   = ',F7.4)
 1060 FORMAT(1X'Velocity     = ',F7.2,1X,A7)
 1070 FORMAT(1X'Pipe is too full to be considered an open channel.')
 2000 FORMAT(A3)
      END
```

E.8

CRITICAL FLOW COMPUTATIONS

```
      PROGRAM CRITICAL
C
C.... This program assumes critical flow in a channel and will
C.... solve for either depth or flow rate, given the other value
C.... and the channel characteristics.  The program may be used
C.... with Problems 7.8, 7.9, 7.18, and 7.22.
C
      REAL C,G,Q,D,A,T,Z,RR,W,TH,N,V
      INTEGER ISECT
      CHARACTER ANS*3
      CHARACTER*4 DEPTH,WIDTH
      CHARACTER*5 FLOW
      CHARACTER*7 AREA,VEL
      WRITE(*,*)
      WRITE(*,*) '      CRITICAL FLOW PROGRAM'
      WRITE(*,*)
   10 WRITE(*,*) '(E)nglish or (M)etric units ?'
      READ(*,2000) ANS
      IF (ANS.EQ.' ') STOP
      C = 1.49
      G = 32.16
      FLOW = '(cfs)'
      DEPTH = '(ft)'
```

```
        AREA = '(sq ft)'
        WIDTH = '(ft)'
        VEL = '(ft/s)'
        IF ((ANS.EQ.'M').OR.(ANS.EQ.'m')) THEN
            C = 1.0
            G = 9.81
            FLOW = '(cms)'
            DEPTH = '(m)'
            AREA = '(sq m)'
            WIDTH = '(m)'
            VEL = '(m/s)'
        END IF
C
C.... Determine Channel Shape and Enter Parameters
    20 WRITE(*,*) '(R)ectangular, (T)riangular, (TR)apezoidal',
      &' or (C)ircular?'
        READ(*,2000) ANS
        IF ((ANS.EQ.'R').OR.(ANS.EQ.'r')) THEN
            WRITE(*,*) 'Channel width ?'
            READ(*,*) W
            ISECT = 1
            GOTO 30
        END IF
        IF ((ANS.EQ.'T').OR.(ANS.EQ.'t')) THEN
            WRITE(*,*) 'Side slope (horizontal/vertical) ?'
            READ(*,*) Z
            ISECT = 2
            GOTO 30
        END IF
        IF ((ANS.EQ.'TR').OR.(ANS.EQ.'tr')) THEN
            WRITE(*,*) 'Bottom width ?'
            READ(*,*) W
            WRITE(*,*) 'Side slope (horizontal/vertical) ?'
            READ(*,*) Z
            ISECT = 3
            GOTO 30
        END IF
        IF ((ANS.EQ.'C').OR.(ANS.EQ.'c')) THEN
            WRITE(*,*) 'Diameter ?'
            READ(*,*) DIA
            ISECT = 4
            GOTO 30
        END IF
        GOTO 10
C
C.... Critical Flow Computations
    30 WRITE(*,*) 'Solve for (D)epth or (F)low rate ?'
        READ(*,2000) ANS
        IF (ANS.EQ.' ') GOTO 20
```

```
      IF ((ANS.EQ.'D').OR.(ANS.EQ.'d')) GOTO 100
      IF ((ANS.EQ.'F').OR.(ANS.EQ.'f')) GOTO 40
      GOTO 30
C
C.... Flow Rate
   40 WRITE(*,*) 'Depth ?'
      READ(*,*) D
      GOTO (50,60,70,80) ISECT
   50 Q = W * SQRT(G) * D**1.5
      A = W * D
      V = Q/A
      T = W
      GOTO 90
   60 Q = Z * SQRT(G/2.0) * D**2.5
      A = Z * D**2
      V = Q/A
      T = 2.0 * Z * D
      GOTO 90
   70 Q = SQRT(G * D**3 * (W+Z*D)**3/(W+2.0*Z*D))
      A = D * (W+Z*D)
      V = Q/A
      T = W+2.0*Z*D
      GOTO 90
   80 IF (D .GT. DIA) THEN
          WRITE(*,1050)
          GOTO 40
      END IF
      ALPHA = 2.0*ACOS(1.0 - (2.0*D/DIA))
      A = ((ALPHA - SIN(ALPHA))*DIA**2.0)/8.0
      T = DIA*SIN(ALPHA/2.0)
      Q = SQRT((A**3)/T*G)
      V = Q/A
   90 WRITE(*,1030) D,DEPTH
      WRITE(*,1000) Q,FLOW
      WRITE(*,1040) V,VEL
      WRITE(*,1010) A,AREA
      WRITE(*,1020) T,WIDTH
      GOTO 30
C
C.... Depth
  100 WRITE(*,*) 'Flow rate ?'
      READ(*,*) Q
      GOTO (110,120,130,140) ISECT
  110 D = (Q/(W * SQRT(G)))**(2.0/3.0)
      A = W * D
      V = Q/A
      T = W
      GOTO 150
  120 D = (Q/(Z * SQRT(G/2.0)))**0.4
      A = Z * D**2
```

```
      V = Q/A
      T = 2.0 * Z * D
      GOTO 150
130 D = 0.0
135 T = (Q**2 * (W+2.0*Z*D)/G)**(1.0/3.0)/(W+Z*D)
      IF (ABS(T-D)/T .GT. 0.001) THEN
          D = T
          GOTO 135
      END IF
      D = T
      A = D * (W+Z*D)
      V = Q/A
      T = W+2.0*Z*D
      GOTO 150
140 D2 = 0.01*DIA
      D1 = DIA
      D = D2
      DO 145 I = 1,200
         ALPHA = 2*ACOS(1.0 - (2.0*D/DIA))
         A = ((ALPHA - SIN(ALPHA))*DIA**2.0)/8.0
         T = DIA*SIN(ALPHA/2.0)
         QNEW = SQRT((A**3.0)/T*G)
         IF (ABS(Q - QNEW) .LE. 0.01) GOTO 145
         IF (QNEW .LT. Q) THEN
             D2 = D
         ELSE
             D1 = D
         END IF
         D = (D1 + D2)/2.0
145 CONTINUE
      IF (D .GT. DIA) THEN
         WRITE(*,1050)
         GOTO 100
      END IF
      V = Q/A
150 WRITE(*,1030) D,DEPTH
      WRITE(*,1000) Q,FLOW
      WRITE(*,1010) A,AREA
      WRITE(*,1040) V,VEL
      WRITE(*,1020) T;WIDTH
      GOTO 30
1000 FORMAT(1X,'Flow rate  = ',F7.2,A5)
1010 FORMAT(1X,'Area       = ',F7.2,A7)
1020 FORMAT(1X,'Top Width  = ',F7.2,A4)
1030 FORMAT(1X,'Depth      = ',F7.2,A4)
1040 FORMAT(1X,'Velocity   = ',F7.2,A7)
1050 FORMAT(1X,'Pipe is too full to be considered an open channel.')
2000 FORMAT(A3)
      END
```

E.9

DRAWDOWN TO A WELL IN A CONFINED AQUIFER: STEADY STATE

```
      PROGRAM DRAWDOWN
C
C.... This program determines drawdown near a well in a confined
C.... aquifer using Poisson's equation (Eqn. 8.69).  The well is
C.... considered to be located at the lower left-hand corner of
C.... the grid (the origin) and the x- and y-axes through the well
C.... are set as no-flow boundaries. The grid should be large
C.... enough that the outer nodes are outside the limit of the
C.... zone of influence of drawdown.  Due to symmetry, this grid
C.... could be considered one quadrant of a larger grid which
C.... would have the well located at its center.
C
      REAL H(25,25)
      INTEGER BOUND(25)
      REAL OMEGA,DX,T,R,RWELL,LIMIT,HO,H1,Q,TOL,MAX,ERROR
      REAL DIST, DIFF, DIFMIN
      INTEGER NI,NJ,NITER
      CHARACTER*1 ANSWER
C
      WRITE(*,*) '   DRAWDOWN NEAR A WELL IN A CONFINED AQUIFER'
      WRITE(*,*)
C
C.... Enter parameters
      WRITE(*,*) 'Compute (H)ead or (D)rawdown?'
      READ(*,'(A)') ANSWER
      WRITE(*,*) 'Enter the number of columns in the grid.'
      READ(*,*) NI
      WRITE(*,*) 'Enter the number of rows in the grid.'
      READ(*,*) NJ
      WRITE(*,*) 'Enter the distance between nodes (ft,m).'
      READ(*,*) DX
      WRITE(*,*) 'Enter the radial limit (ft,m) of the zone of ',
     &'influence.'
      READ(*,*) LIMIT
      WRITE(*,*)
      WRITE(*,*) 'Enter the transmissivity (sq ft/day,sq m/day).'
      READ(*,*) T
      WRITE(*,*) 'Enter the pumping rate (gal/day,cu m/day).'
      READ(*,*) Q
      WRITE(*,*)
      WRITE(*,*) 'Enter the relaxation factor omega (0<omega<2).'
      READ(*,*) OMEGA
      WRITE(*,*) 'Enter the error tolerance (ft, m)'
      READ(*,*) TOL
```

```
      WRITE(*,*) 'Enter the limit of iterations'
      READ(*,*) NITER
      WRITE(*,*)
      WRITE(*,*) 'Enter the initial head (ft,m) before pumping started.'
      READ(*,*) HO
C
C.... Set all nodes equal to the initial head
      DO 20 I = 1,NI+1
         DO 10 J = 1,NJ+1
            H(I,J) = HO
   10    CONTINUE
   20 CONTINUE
C
C.... Set flag at zone of influence boundary
      DO 90 I = 2,NI+1
         DIFMIN = 10000.
         DO 85 J = 1,NJ
            X = DX*(I-2)
            Y = DX*(NJ-J)
            DIST = SQRT(X**2 + Y**2)
            DIFF = ABS(DIST-LIMIT)
            IF (DIFF.GE.DIFMIN) GOTO 85
            DIFMIN = DIFF
            BOUND(I) = J+1
   85    CONTINUE
   90 CONTINUE
C
C.... Set all heads outside boundary to initial head value
      DO 100 I=2,NI+1
         DO 95 J=1,BOUND(I)-1
            H(I,J) = HO
   95    CONTINUE
  100 CONTINUE
C
C.... Start calculations and keep track of iterations
      RWELL = -Q/(DX**2)
      NUMIT = 0
   65 MAX = 0.
      NUMIT = NUMIT+1
      IF (NUMIT .GT. NITER) THEN
         WRITE(*,*) 'Max. Iterations Exceeded!'
         GOTO 115
      END IF
C
C.... Set left and bottom symmetry no-flow boundaries
      DO 70 J = 2,NJ + 1
         H(1,J) = H(3,J)
   70 CONTINUE
      DO 80 I = 2,NI + 1
```

```
              H(I,NJ+1) = H(I,NJ-1)
    80 CONTINUE
C
C.... Sweep nodes with finite difference form of Eqn. 8.69
       DO 110 I = 2,NI
          DO 105 J = BOUND(I),NJ
               R = O.
               OLDH = H(I,J)
               IF ((I.EQ.2) .AND. (J.EQ.NJ)) R = RWELL
               H(I,J) = (H(I-1,J) + H(I+1,J) + H(I,J-1) + H(I,J+1) +
     &DX*DX*R/T)/4.
               H(I,J) = OMEGA*H(I,J) + (1.0 - OMEGA)*OLDH
               ERROR = ABS(H(I,J) - OLDH)
               IF (ERROR .GT. MAX) MAX = ERROR
   105     CONTINUE
   110 CONTINUE
C
C.... Check to see if error is within tolerance.  Do another iteration if not
       IF (MAX .GT. TOL) GOTO 65
C
C.... Print results on screen
   115 WRITE(*,*)
       WRITE(*,*) 'The number of iterations made was ',NUMIT
       WRITE(*,*)
       IF ((ANSWER .EQ. 'H') .OR. (ANSWER .EQ. 'h')) THEN
           WRITE(*,*) 'The head (ft,m) array is:'
           DO 120 J = 1,NJ
               WRITE(*,1000) (H(I,J), I = 2,NI+1)
   120     CONTINUE
         ELSE
           WRITE(*,*) 'The drawdown (ft,m) array is:'
           DO 130 J = 1,NJ
               WRITE(*,1000) (HO - H(I,J), I = 2,NI+1)
   130     CONTINUE
       END IF
  1000 FORMAT(1X,13F6.2/)
       END
```

E.10

DRAWDOWN TO A WELL IN A CONFINED AQUIFER: TRANSIENT

```
       PROGRAM TRANS
C
C.... This program solves the iterative form of the governing
C.... equation for drawdown to a well under transient conditions
C.... (Eqn. 8.74).  A pumping well is assumed to be located at
```

```
C.... the bottom left-hand corner of a grid.  There is no flow
C.... across the boundaries of the grid.  The time increment is
C.... multiplied by a factor so that the initial increments are
C.... small (head is changing rapidly) and the later increments
C.... are larger (head is changing more slowly).  The time increment
C.... is limited to 5 days.  The drawdown is printed out for the
C.... last time period only.
C
      REAL HNEW(25,25),HOLD(25,25)
      REAL MAX,DX,T,S,ALPHA,TOL,HO,Q,RWELL,TIME,ERROR
      REAL H1,H2,F1,F2,TFACTOR,DT
      INTEGER NI,NJ,NITER,NT,NUMIT
      CHARACTER*1 ANSWER
      WRITE(*,*) '    NUMERICAL SOLUTION OF DRAWDOWN IN A CONFINED'
      WRITE(*,*) '         AQUIFER UNDER TRANSIENT CONDITIONS'
      WRITE(*,*)
      WRITE(*,*) 'Compute (H)ead or (D)rawdown?'
      READ(*,'(A)') ANSWER
C
C.... Enter parameters
      WRITE(*,*) 'Enter the number of columns'
      READ(*,*) NI
      WRITE(*,*) 'Enter the number of rows'
      READ(*,*) NJ
      WRITE(*,*) 'Enter the distance between nodes (ft,m).'
      READ(*,*) DX
      WRITE(*,*)
      WRITE(*,*) 'Enter the transmissivity (sq ft/day, sq m/day).'
      READ(*,*) T
      WRITE(*,*) 'Enter the storativity.'
      READ(*,*) S
      WRITE(*,*) 'Enter the pumping rate (gal/day, cu m/day).'
      READ(*,*) Q
      WRITE(*,*)
      WRITE(*,*) 'Enter alpha.'
      READ(*,*) ALPHA
      WRITE(*,*) 'Enter the error tolerance.'
      READ(*,*) TOL
      WRITE(*,*) 'Enter the limit of iterations used to solve for',
     &' drawdown.'
      READ(*,*) NITER
      WRITE(*,*)
      WRITE(*,*) 'Enter the starting time increment (day).'
      READ(*,*) DT
      WRITE(*,*) 'Enter the factor to multiply the time increment.'
      READ(*,*) TFACTOR
      WRITE(*,*) 'Enter the number of time increments to be made.'
      READ(*,*) NT
      WRITE(*,*)
```

```
        WRITE(*,*) 'Enter the initial head (ft,m) before pumping starts.'
        READ(*,*) HO
C
C....  Initialize the head arrays.  (HOLD is the head at time t while
C....  HNEW is the head at time t+1)
        DO 20 I=1,NI+2
            DO 10 J=1,NJ+2
                HNEW(I,J)=HO
                HOLD(I,J)=HO
    10      CONTINUE
    20 CONTINUE
C
C....  Define recharge to well and start calculations
        RWELL = -(Q/(DX**2))
        TIME = 0.
        DO 110 N=1,NT
C
C....  Keep track of the number of iterations needed to solve for drawdown
            NUMIT = 0
            TIME = TIME+DT
    30      MAX = 0.
            NUMIT = NUMIT+1
            IF(NUMIT .GT. NITER) GOTO 110
            DO 50 I=2,NI+1
              DO 40 J=2,NJ+1
                R = 0.0
                IF((I .EQ. 2) .AND. (J .EQ. NJ+1)) R = RWELL
                TEMP=HNEW(I,J)
                H1=(HOLD(I,J+1)+HOLD(I,J-1)+HOLD(I+1,J)+HOLD(I-1,J))/4.
                H2=(HNEW(I,J+1)+HNEW(I,J-1)+HNEW(I+1,J)+HNEW(I-1,J))/4.
                F1=(DX**2)*S/(4.*T*DT)
                F2=1./(F1+ALPHA)
                HNEW(I,J)=((F1*HOLD(I,J))+(1.-ALPHA)*(H1-HOLD(I,J))+(ALPHA*H2)
      & +  (R*(DX**2)/(4.*T)))*F2
                ERROR = ABS(HNEW(I,J) - TEMP)
                IF(ERROR .GT. MAX) MAX = ERROR
    40      CONTINUE
    50      CONTINUE
C
C....  Adjust the no-flow boundaries
            DO 60 I=2,NI+1
                HNEW(I,1)=HNEW(I,3)
                HNEW(I,NI+2)=HNEW(I,NI)
    60      CONTINUE
            DO 70 J = 2,NJ+1
                HNEW(1,J)=HNEW(3,J)
                HNEW(NJ+2,J)=HNEW(NJ,J)
    70      CONTINUE
            IF(ALPHA .LT. 0.1) GOTO 80
```

```
             IF(MAX .GT. TOL) GOTO 30
     80      CONTINUE
C
C.... Put the head values into the HOLD array in order to go to the
C.... next time period
             DO 100 I = 1,NI+2
                DO 90 J=1,NJ+2
                   HOLD(I,J)=HNEW(I,J)
     90         CONTINUE
    100     CONTINUE
C
C.... Increase the time step and go to next time period
             DT=DT*TFACTOR
             IF(DT .GT. 5.0) DT=5.0
    110 CONTINUE
C
C.... Print out head from last time period
             IF ((ANSWER .EQ. 'D') .OR. (ANSWER .EQ. 'd')) GOTO 130
             WRITE(*,*)
             WRITE(*,1000) TIME
             DO 120 J = 2,NJ+1
                WRITE(*,1010) (HNEW(I,J),I=2,NI+1)
    120 CONTINUE
             GOTO 150
C
C.... Print out drawdown from last time period
    130 WRITE(*,*)
             WRITE(*,1001) TIME
             DO 140 J = 2,NJ+1
                WRITE(*,1010) (HO-HNEW(I,J),I=2,NI+1)
    140 CONTINUE
C
   1000 FORMAT(1X,'Time = ',F5.2,/,1X,'The head array is:',/)
   1001 FORMAT(1X,'Time = ',F5.2,/,1X,'The drawdown array is:',/)
   1010 FORMAT(1X,10F8.2)
    150 END
```

Glossary

adverse slope   channel slope that slopes upward in the downstream direction

albedo   reflection coefficient

alluvium   sediment deposited by flowing rivers and consisting of sands and gravel

annual exceedances   series of n largest independent events in an n-yr period

antecedent precipitation index   measure of the soil moisture

aquiclude   relatively impermeable layer that does not allow water to flow through

aquifer   geologic formation that is saturated and that transmits large quantities of water

aquitard   low permeability layer that allows leakage to occur

area-elevation curve   curve relating surface area to a particular elevation

arithmetic mean   simple method for averaging rainfall

artesian aquifer   confined aquifer that is under pressure greater than atmospheric pressure

artesian well   well that penetrates a confined aquifer where the water level rises above the ground surface

auger hole test   field test for determining hydraulic conductivity by measuring a change in water level in a bore hole after a rapid addition or removal of a volume of water

backward difference   finite difference evaluated by looking backward in space or time from a given point

backwater curve   plot of water depth along the channel length

base flow   flow in a channel due to soil moisture or ground water

bored well   well constructed by means of a hand-operated or power-driven auger

bulk density   ratio of the oven-dried mass of a sample to its original volume

capillary rise   height to which water will rise under capillary forces

capillary water   water held above the water table due to capillary forces, caused by attraction of soil and water

card image   line of data corresponding to 80 columns of terminal input

central difference   finite difference that is centered in space or time around a point of interest

central moments   statistical moments with respect to the mean

characteristic curves   a set of curves of x or y vs. t produced by the method of characteristics

characteristic method   method of solving a set of equations by converting partial differential equations into ordinary differential equations

class interval   a range into which data may be grouped

class mark   midpoint of the class interval

coalescence process   process by which droplets of rain increase in size through collision with other droplets

collectively exhaustive   term describing events that account for all possible outcomes

collector channels   small channels that collect overland flow and carry it to larger channels

combined sewer system of conduits that carries storm runoff and domestic sewage together

computational block a set of several lines of data that describe one type of computation

condensation phase change of water vapor into liquid droplets

cone of depression plot of a surface of decline of the water table or piezometric level near a pumping well

confidence limits control curves between which a known percentage of data points is expected to fall

confined aquifer aquifer that is overlain by a confining unit consisting of a lower hydraulic conductivity

constant head permeameter laboratory device that determines hydraulic conductivity by supplying a continuous source of water at a constant head

continuous simulation model model based on long-term water balance equations

continuous variable a variable that can assume any value on the real axis

control section cross section of a channel that has a controlling structure (bridge, free outfall, or weir)

convective motion of air due to intense heating at ground level

conveyance ratio of amount of flow carried in the channel to the square root of channel slope

convolution equation equation used to derive a storm hydrograph from a unit hydrograph (add and lag method)

Courant condition limiting condition for a time step used to avoid stability problems

crest segment portion of the hydrograph that contains the peak flow

critical depth depth of water for which specific energy is minimum

critical flow flow of water through a channel for which the specific energy is minimum

critical slope channel slope for which uniform flow is critical

cumulative density function (CDF) probability of nonexceedance as a function of a continuous random variable

cumulative distribution same as cumulative density function, but for either continuous or discrete variables

cumulative mass curve graph of total accumulated rainfall or other variable vs. time

cyclonic counterclockwise motion of air due to movement of large air mass systems (in Northern Hemisphere)

deep well water well dug to a large depth from the surface

degree-day type of equation used to determine snowmelt

density mass per unit volume of a substance

detention storage reservoir that detains water for a given time and then discharges entirely downstream

deterministic model in which parameters are based on physical relations for dynamic processes of hydrologic cycle

dew point temperature temperature at which the air just becomes saturated when cooled

diffusion model flood routing model based on solving a diffusion equation

dimensionless hydrograph a general hydrograph developed from many unit hydrographs, used in the Soil Conservation Service method

direct runoff hydrograph graph of direct runoff (rainfall − losses) vs. time

discrete probability probability of a discrete event, such as rainfall occurrence

discrete variable variable that can assume only discrete values

distributed parameter model that attempts to describe physical processes and mechanisms in space and time

drainage divides boundary that separates subbasin areas according to direction of runoff

drawdown curve plot of a curve of decline of the water table or piezometric level near a pumping well

driven well well constructed by driving a series of pipe lengths into the ground

dug well well excavated by hand

Dupuit parabola parabolic shape of the water table determined by the Dupuit equation

duration length of time during which rain falls

effective flow area portion of the total cross-sectional area where flow velocity is normal to the cross section

effluent stream flow out of an aquifer into a stream during flood conditions

equilibrium discharge constant maximum outflow reached for a constant rainfall intensity that falls continuously

equipotential lines lines of equal piezometric or water table level used in the analysis of ground water flow

evaporation phase change of liquid water to water vapor

evapotranspiration the combined process of evaporation and transpiration

event type of model that simulates a single storm response

explicit method method of solving a set of equations based on previously known data

falling head permeameter laboratory device that determines hydraulic conductivity by measuring the rate of fall of water in a column

field capacity amount of water held in soil after gravitational water is drained

finite difference method used to solve differential equations by approximating them as algebraic terms over a grid

fixed format structured arrangement of data on a line with fixed column fields

flood hazard zone area that will flood with a given probability

floodway encroachment area that separates the floodway from the limits of the floodplain

flow net set of orthogonal streamlines and equipotential lines used in hydrogeologic site investigations

flowing well same as artesian well

force main a pressurized conduit

forward difference finite difference evaluated by looking forward in space or time from a given point

free format a line of data that is not in a structured arrangement but is separated by commas or spaces

frequency the number of observations of a random variable

friction slope equal to the total energy slope in open channels and calculated with Manning's equation for gradually varied flow

Froude number dimensionless parameter to characterize open channel flow

geomorphology study of parameters that describe the physical nature of a watershed

gradually varied flow open channel flow that changes gradually so that one-dimensional analysis applies in each reach

grate inlet structure used to cover an inlet to a sewer system to keep out debris

gravel pack material used near a well screen to stabilize the aquifer and provide an annular zone of high permeability

gravitational water water that will drain through the soil under the force of gravity

grout seals material used to seal a well so that water is pumped only from a predetermined layer

histogram bar graph of data

horizontal slope channel slope that is horizontal

hydraulic conductivity ratio of velocity to hydraulic gradient, indicating permeability of porous media

hydraulic jump sudden transition from supercritical flow to subcritical flow

hydraulic routing flood routing method using both the continuity equation and the momentum equation

hydraulically connected impervious area impervious area that drains directly into the drainage system

hydrograph graph of discharge vs. time

hydrologic cycle the continuous process of water movement near the earth's surface

hydrologic routing flood routing method that uses the continuity equation and a storage equation

hyetograph graph of rainfall intensity vs. time

hygroscopic water moisture that remains adsorbed to the surface of soil grains in the unsaturated zone

ice crystal process process by which water vapor condenses onto frozen nuclei to form ice crystals

implicit method method of solving simultaneous equations over a finite-difference grid system at each time step

independent term describing an event that does not influence the occurrence of another event

ineffective flow area portion of the total cross-sectional area where flow velocity is not in a downstream direction

infiltration movement of water from the surface into the soil

infiltration capacity rate at which water can enter soil with excess water on surface

influent stream flow into an aquifer from a stream during drought conditions

instantaneous unit hydrograph hydrograph produced by a unit rainfall falling for a duration D, when D approaches zero

intensity-duration-frequency curve statistical plot relating intensity, duration, and frequency of design rainfalls

interceptor conduit at downstream end of combined sewer that carries sewage to a treatment plant

interflow flow of water through the upper soil zones to a stream

intersection the occurrence of both of two events

intrinsic permeability property that indicates the ability of a unit to transmit water; a property of the porous media only

invert bottom of a channel or pipe

isohyetal method method for areally weighting rainfall using contours of equal rainfall (isohyets)

isotropic condition in which hydraulic properties of an aquifer are equal in all directions

jetted well well constructed by directing a high-velocity stream of water downward

kinematic wave wave that has a unique function relating Q and y

lag time time from the center of mass of rainfall to the peak of the hydrograph

lapse rate rate of change of temperature with elevation in the atmosphere

latent heat amount of heat that must be removed or added for a phase change to take place

leaky aquifer an aquifer confined by a low-permeability unit able to transmit water as recharge to an adjacent aquifer

left overbank floodplain area above the channel on the left-hand side of a cross section looking downstream

linear reservoir reservoir that has a linear relationship between storage and outflow

lognormal distribution for which the log of the random variable is distributed normally

low flow flow at a depth below the crown of a culvert or lowest point of the low chord of a bridge

lumped parameter model model that aggregates spatially distributed parameters

main channels large channels that carry flow from collector channels to some outlet such as a lake

major drainage system a group of major streams that carry runoff from large storms when the minor drainage system is full

median 50% value of CDF

method of characteristics see *characteristic methods*

method of moments method to fit a distribution using estimates of the mean, variance, and skewness

mild slope channel slope for which uniform flow is subcritical

minor drainage system system that carries runoff from small storms

mode value at which the probability density function is a maximum

model calibration process of parameter estimation using known data

model verification process of comparing parameter estimates against a new set of data once model has been calibrated

moisture content ratio of volume of water to total volume of a unit of porous media

monoclinal flood wave simple flood wave consisting of a step increase in discharge that moves downstream

Muskingum method method of river routing that uses the continuity equation and a linear storage relationship

mutually exclusive term describing events that cannot occur simultaneously

nonstationary varying with time

nonuniform flow flow of water through a channel that gradually changes with distance

normal bridge method bridge modeling in HEC-2 for low flow through culverts, bridges without piers, etc.

normal depth depth at which water will flow under uniform conditions

open channel flow flow of water through an open channel with a free water surface

optimum channel cross section cross section that requires a minimum flow area

orographic rainfall due to mechanical lifting of air over mountains

outliers data points that fall an "unusually large" distance from the cumulative density function

overland flow flow of water across the land surface

overland flow planes area over which runoff flows in thin sheets

pan coefficient coefficient relating evaporation from a standard pan to evaporation from a lake

parameter optimization selection of the best value for a parameter

perched water table an unconfined aquifer situated on top of a confining unit separated from a main aquifer

permeability see *hydraulic conductivity*

ϕ **index** simple infiltration method that assumes that loss is uniform across the rainfall pattern

piezometers small-diameter well that measures water table or confined aquifer pressure

piezometric surface indicator of pressure level of water in an aquifer

plotting position empirical estimate of frequency or return period used for graphing data

population complete series of random

events that belong to a given probability density function or probability mass function

porosity ratio of volume of voids to total volume of sample

porous medium geologic material that will allow water to flow through it

potential (initial) abstraction initial loss of water prior to infiltration in the SCS method of runoff analysis

potential evapotranspiration amount of water that could be lost to evapotranspiration if water supply was unlimited

potentiometric surface same as *piezometric surface*

precipitation water that falls to the earth in the form of rain, snow, hail, or sleet

pressure flow flow in a conduit with no free surface

prism storage prism-shaped volume of storage in a river reach

probability relative number of occurrences of an event after a large number of trials

probability density function (PDF) distribution of probability for a continuous random variable; integral is probability

pump device used to raise water to a higher level or to discharge water under pressure

pump test field test for determining hydraulic characteristics by pumping water from one well and observing the decline in water level in other wells

quantized-continuous continuous data presented in a discrete form due to the measurement process

random variable parameter that cannot be predicted with certainty

rapidly varied flow flow of water through a channel whose characteristics are rapidly changing

rating curve relationship between depth and amount of flow in a channel

rational method simple linear rainfall-runoff relationship that predicts peak outflow

recession curve portion of the hydrograph where runoff is from base flow

recharge water that infiltrates to an aquifer, usually from above

recurrence interval time interval in which an event will occur once on the average

regional skew coefficient skew parameter based on several different gage analyses in a region

regulator hydraulic control structure that diverts combined sewage into an interceptor

relative frequency ratio of number of observations in a class interval to total number

relative humidity ratio of water vapor pressure to the saturated vapor pressure at the same temperature

retention storage reservoir that retains some flood water without allowing it to discharge downstream

return period time interval for which an event will occur once on the average

right overbank floodplain area above the channel on the right-hand side of a cross section looking downstream

rising limb portion of the hydrograph where runoff is increasing

river routing flood routing where outflow depends on both river and floodplain storage

root zone depth to which the major vegetation draws water through a root system in soil

routing parameter coefficient that accounts for attenuation of the peak flow through a channel segment

Runge-Kutta methods numerical scheme for solving ordinary differential equations

runoff coefficient ratio of runoff to precipitation

runoff curve number parameter used in the Soil Conservation Service method that accounts for soil type and land use

S-curve method method for converting a unit hydrograph of one duration to a new duration

saturated thickness thickness of an aquifer measured from the water table or the confining layer to the bottom

saturated vapor pressure partial pressure of water vapor when air is saturated (no more evaporation can occur)

saturated zone zone of an aquifer in which the porous media is under pressure greater than atmospheric pressure

separated sewer system of conduits that carry storm runoff and domestic sewage separately

shallow well well dug less than a few meters deep

side-flow weir weir discharging to side of flow direction

simulation model model describing the reaction of a watershed to a storm using numerical equations

skew coefficient skewness divided by cube of standard deviation

skewness third central moment about the mean, a measure of asymmetry

slug test field test for determining hydraulic conductivity by injecting or removing a volume of water in a single well and observing the decline or recovery of water level

soil moisture storage volume of water held in the soil

soil water zone portion of soil from the ground surface through the root zone

special bridge method modeling in HEC-2 for bridges with pier structures or where pressure or weir flow occurs

specific energy the sum of elevation head and velocity head at a cross section referenced to the channel bed

specific yield volume of water released from unconfined aquifer per unit area per unit drop in head

St. Venant equations pair of continuity and momentum equations that govern hydraulic flood routing

standard step method numerical backwater computation that determines water surface elevations

station estimate skewness estimate using data from a single station

steep slope channel slope for which uniform flow is supercritical

stochastic model that contains a random component

storage coefficient volume of water that an aquifer releases from or takes into storage per unit surface area per unit change in piezometric head

storage delay constant parameter that accounts for lagging of the peak flow through a channel segment

storage routing flood routing in which outflow is uniquely determined by the amount of storage

storage-discharge relation values that relate storage in the system to outflow from the system

storage indication curve relationship among storage, outflow, and Δt

storage indication method reservoir routing that uses the continuity equation and a storage indication curve

storm sewer system of conduits that carries storm runoff

stream function mathematical description of the flow lines for a given set of equipotential lines

streamlines flow lines that indicate the direction of water movement in a flow net

subbasins hydrologic divisions of a watershed that are relatively homogeneous

subcritical flow of water at a velocity less than critical

subsurface stormflow flow of water through the upper soil zones to a stream

supercritical flow of water at a velocity greater than critical

surcharge condition in which the water level in a storm sewer rises above the crown of the conduit

synthetic design storm rainfall hyetograph obtained through statistical means

synthetic unit hydrograph unit hydrograph for ungaged basins based on theoretical or empirical methods

Thiessen method method for areally weighting rainfall through graphical means

tide gate structure installed at the outlet of a sewer system to prevent backflow into the conduits

time base total duration of direct runoff

time of concentration time at which outflow from a basin is equal to inflow or time of equilibrium

time of rise time from the start of rainfall excess to the peak of the hydrograph

time to peak time from the center of mass of rainfall to the peak of the hydrograph

time-area histogram bar graph of translation time vs. incremental area

time-area method simple rainfall-runoff relationship based on the concept of pure translation of runoff

total energy the sum of the pressure head, elevation head, and velocity head at a cross section

tracer test injection and observation of a chemical in an aquifer to measure rate and direction of ground water flow

transient flow flow characterized by time-variable flow

transmissivity measure of how easily water in a confined aquifer can flow through the porous media

transpiration conversion of water to water vapor through plant tissue

type curve used in Theis analysis of a pumping well in a confined aquifer

unconfined aquifer aquifer that does not have a confining unit and is defined by a water table

uniform flow equilibrium flow for which slope of total energy equals bottom slope

uniformity coefficient measure of the uniformity of grain sizes in a soil sample

union the occurrence of either of two events

unit hydrograph graph of runoff vs. time produced by a unit rainfall over a given duration

vadose zone zone of aeration that extends from the surface to the water table including the capillary fringe

vapor pressure partial pressure exerted by water vapor

variance second central moment about the mean, a measure of scale or width

velocity head energy due to the velocity of water

velocity potential product of total head and hydraulic conductivity, the gradient of which defines velocity

water divide point where the water table is at a maximum and flow does not cross

water surface profile plot of the depth of water in a channel along the length of the channel

water table   the surface where fluid pressure is exactly atmospheric or the depth to which water will rise in a well screened in an unconfined aquifer

wave celerity   speed at which a wave moves downstream

wedge storage   wedge-shaped volume of storage in a river reach that accounts for rising and falling stages

weir flow   flow at a depth greater than the top of a roadway of a bridge; flow over the top of a channel structure

well   vertical hole dug into the soil that penetrates an aquifer and is usually cased and screened

well casing   lining that prevents entry of unwanted water into a well

well screen   slotted casing that allows water to enter a well from the aquifer that it penetrates

Author Index

Subject Index

FLOW CONVERSION

UNIT	m^3/s	m^3/day	ℓ/s	ft^3/s	ft^3/day	$ac\text{-}ft/day$	gal/min	gal/day	mgd
1 m^3/s	1	8.64×10^4	10^3	35.31	3.051×10^6	70.05	1.58×10^4	2.282×10^7	22.824
1 m^3/day	1.157×10^{-5}	1	0.0116	4.09×10^{-4}	35.31	8.1×10^{-4}	0.1835	264.17	2.64×10^{-4}
1 liter/s	0.001	86.4	1	0.0353	3051.2	0.070	15.85	2.28×10^4	2.28×10^{-2}
1 ft^3/s	0.0283	2446.6	28.32	1	8.64×10^4	1.984	448.8	6.46×10^5	0.646
1 ft^3/day	3.28×10^{-7}	0.02832	3.28×10^{-4}	1.16×10^{-5}	1	2.3×10^{-5}	5.19×10^{-3}	7.48	7.48×10^{-6}
1 ac-ft/day	0.0143	1233.5	14.276	0.5042	43,560	1	226.28	3.259×10^5	0.3258
1 gal/min	6.3×10^{-5}	5.451	0.0631	2.23×10^{-3}	192.5	4.42×10^{-3}	1	1440	1.44×10^{-3}
1 gal/day	4.3×10^{-8}	3.79×10^{-3}	4.382×10^{-5}	1.55×10^{-6}	0.1337	3.07×10^{-6}	6.94×10^{-4}	1	10^{-6}
1 million gal/day (mgd)	4.38×10^{-2}	3785	43.82	1.55	1.337×10^5	3.07	694	10^6	1

VOLUME CONVERSION

UNIT	ml	liters	m^3	in^3	ft^3	gal	ac-ft	million gal
1 ml	1	0.001	10^{-6}	0.06102	3.53×10^{-5}	2.64×10^{-4}	8.1×10^{-10}	2.64×10^{-10}
1 liter	10^3	1	0.001	61.02	0.0353	0.264	8.1×10^{-7}	2.64×10^{-7}
1 m^3	10^6	1000	1	61,023	35.31	264.17	8.1×10^{-4}	2.64×10^{-4}
1 in^3	16.39	1.64×10^{-2}	1.64×10^{-5}	1	5.79×10^{-4}	4.33×10^{-3}	1.218×10^{-8}	4.329×10^{-9}
1 ft^3	28,317	28.317	0.02832	1728	1	7.48	2.296×10^{-5}	7.48×10^{-6}
1 U.S. gal	3785.4	3.785	3.78×10^{-3}	231	0.134	1	3.069×10^{-6}	10^{-6}
1 ac-ft	1.233×10^9	1.233×10^6	1233.5	75.27×10^6	43,560	3.26×10^5	1	0.3260
1 million gallons	3.785×10^9	3.785×10^6	3785	2.31×10^8	1.338×10^5	10^6	3.0684	1